PREDICTION
of Extreme Events in Nature and Society

V. I. KEILIS-BOROK
edited by A. A. Soloviev

Ori Books

Prediction of Extreme Events in Nature and Society

Author: V. I. Keilis-Borok
Editor: A. A. Soloviev

Ori Books is an imprint of Dragonwell Publishing
(www.dragonwellpublishing.com)

ISBN 978-1-940076-44-7
DOI: 10.28935/9781940076447

V. I. Keilis-Borok has worked on this book during the past few years of his life. The current text was compiled from the major draft he left behind, completed and edited by one of his closest colleagues, A. A. Soloviev.

The book's target audiences include scientists and professionals interested in learning more about V. I. Keilis-Borok's prediction methods. This book can also be used as a reference source and as a guide for experts and non-experts interested in prediction of extreme events in complex systems.

ABOUT THE AUTHOR

V. I. Keilis-Borok (1921-2013) is one of the founders of computational seismology and mathematical geophysics and a pioneer of advance predictions of extreme events in complex systems, including earthquakes, economic recessions, outcomes of elections, surges of unemployment, and crime waves. His main idea is the consideration of the Earth lithosphere as a complex hierarchical system and earthquakes – as extreme events in it. Among successful advance earthquake predictions made by his group are Irpinia earthquake in Italy (1980), Loma Prieta earthquake in California (1989), Chile earthquake of 2010, Japan earthquake of 2011, and many others.

At the end of the last century V. I. Keilis-Borok began to apply his earthquake prediction approaches on the prediction of extreme events in other complex systems: economic recessions, outcomes of elections, surges of unemployment, and crime waves. His trademark style involved exceptional organizational talent and insight that enabled him to make seemingly impossible connections between different fields of research and different groups of experts, often breaking the barriers between high theory, numerical modeling, and data analysis.

He was the first chairman (in 1964-1979) of the International Working Group on Geophysical Theory and Computers (now – the IUGG Commission on Mathematical Geophysics), and a leader of numerous scientific organizations worldwide. In the last years of his life he held a faculty position at UCLA. His outstanding achievements are recognized by memberships in many academies and international scientific organizations, including National Academy of Sciences of the USA, American Academy of Arts and Sciences, Academy of Sciences of the USSR (now – Russian Academy of Sciences), Pontifical Academy of Sciences, Academia Europea, Royal Astronomical Society, and others, as well as through awards and honors, and by the respect and devotion he inspired in his friends and colleagues.

The book contains an overview of over fifty years of work by his group on developing successful algorithms for prediction of extreme events in nature and society, including predictions of earthquakes, outcomes of elections, and socioeconomic crises.

ABOUT THE EDITOR

A. A. Soloviev belonged to V. I. Keilis-Borok's team from 1976 to 2013 and his studies included application of pattern recognition approaches to solving geophysical problems (in particular, recognition of earthquake prone-areas), magnetic dynamo problem, numerical simulation of block structure dynamics and seismicity etc. He played an important role in achievements by V. I. Keilis-Borok in the field of prediction of extreme events in socioeconomic systems.

A. A. Soloviev is a corresponding member of the Russian Academy of Sciences and participated in several national and international projects focused on prediction of extreme events and seismic hazard estimation. From 1998 to 2017 he has served as the Director of the Institute of Earthquake Prediction Theory and Mathematical Geophysics, Russian Academy of Sciences (MITPAN – IEPT RAS), founded by V. I. Keilis-Borok in 1990 in Moscow. He has collaborated with V. I. Keilis-Borok on this book until October, 2013.

TABLE OF CONTENTS

INTRODUCTION

Today's civilization is highly vulnerable to crises arising from extreme events generated by complex and poorly understood systems. Examples include natural disasters, external and civil wars, terrorist attacks, crime waves, economic downturns, and famines, to name just a few. Yet more subtle effects also threaten modern society, e.g., the inability of democratic systems to produce policies responsive to challenges, such as climate change, global poverty, and resource depletion.

Our capacity to predict the course of events in complex systems is inherently limited. However, there is a new and promising approach to predicting and understanding complex systems that has emerged through the integration of studies in the natural and social sciences and the mathematics of prediction. This book describes and analyzes this approach and its real-world applications. Originally this approach was successfully applied to natural disasters, such as earthquakes. Further developments include algorithmic prediction of electoral fortunes of incumbent parties, economic recessions, surges of unemployment, and outbursts of crimes. This leads to important inferences for averting and responding to impending crises.

Ultimately, improved prediction methods enhance our capacity for understanding the world and for protecting and sustaining our civilization.

Hierarchical complex systems persistently generate extreme events – rare, fast changes that have a strong impact on the system. Depending on connotation they are also known as critical phenomena, disasters, catastrophes, and crises. This book summarizes decades of work on the development and application of algorithmic predictions to extreme natural, socio-economic, and political events.

The studies presented here summarize over 50 years of research by teams from over 20 institutions in 12 different countries, led by V. I. Keilis-Borok and stemming from his original discoveries, open a reasonable hope to break the current stalemate in disasters' prediction, in application to multiple extreme and catastrophic events in nature and society.

BASIC SCIENCE FOR THE SURVIVAL OF HUMANITY IN THE THIRD WORLD WAR

M. Alfimov, R. Corell, V. Courtillot, V. Fortov, M. Intriligator, and V. Keilis-Borok

This article, initiated by V. Keilis-Borok, was originally published in Russian in a national newspaper "Kommersant Daily", November 29, 1997

At the time of this publication, M. Alfimov was the Chairman of the Russian Foundation for Basic Research; R. Corell was Assistant Director of the US National Science Foundation; V. Courtillot was Director of the Institute of the Physics of the Earth, Paris; V. Fortov was the Vice President of the Russian Academy of Sciences; M. Intriligator was Professor of Economics, Political Science, and Policy Studies at the University of California, Los Angeles; V. Keilis-Borok was the Director of the International Institute of Earthquake Prediction Theory and Mathematical Geophysics, Russian Academy of Sciences.

It is commonly recognized that the very survival of our civilization is threatened by dangers as great as the danger of nuclear war. They include the destabilization of the environment, the economy, and the social order due to man-made and natural disasters; the self-destruction of megacities; pandemic terrorism; the loss of control over telecommunications systems; and the depletion of natural resources. These dangers keep growing despite huge sums of money being spent on their containment. Meanwhile, modern basic science has created new revolutionary possibilities to break this stalemate, and humankind cannot afford to ignore them. Their realization requires the same spirit of responsibility, excellence, and urgency as drove the defense-oriented projects in World War II.

Hundreds of billions of US dollars are being spent worldwide in an effort to contain these dangers by the massive application of existing technologies. These efforts may prevent a part of the potential damage, but, on the whole, they are in a stalemate: the factors destabilizing our civilization prevail, the scale of possible catastrophes is rapidly growing, and the solution is generally believed to lie in spending ever more money.

Currently, a massive release of radioactivity from a nuclear waste disposal site, an earthquake in the middle of a major city, a large-scale outburst of violence, and any one of a formidable array of other quite possible disasters could cause millions of casualties, render a large part of our world uninhabitable, trigger a global economic depression, and even trigger a war in a "hot" region. In addition, each country has become vulnerable to developments in other parts of the global village that are outside its control. Meanwhile, looming on the horizon is the critical problem of sustainable development on a longer time scale.

We know from history and common sense that basic research is pivotal to breaking such a stalemate. Indeed, since ancient times basic science has time and again rescued humanity from major threats and sustained its development by creating a springboard to entirely new technologies and thereby providing "new solutions to old problems." Among recent examples are antibiotics, transistors and integrated circuits, synthetic fibers, and the green revolution, to name just a few. Thus, the hope and responsibility to break this stalemate rest not with the Treasury but rather with the Academia. Still the Treasury is more popular, as in the French proverb: "Nobody is satisfied with his wealth, everybody is satisfied with his wisdom."

Frontier research of the last years continues this tradition, discovering new possibilities to cope with many of the present dangers. To be specific, let us mention some of them.

In the area of disaster reduction, they include neutralization of the time bomb contained in radioactive wastes; the prediction of natural disasters; geo-engineering stabilization of megacities; the control of the traffic in chemical explosives; the control of telecommunication networks; and the prediction of social, economic, and political crises.

In the area of sustainable development, they include the discovery of new mineral deposits; the creation of new materials and sources of energy; the development of new forms of transportation; and the processing of large waste disposals.

Now is the time to launch a series of "superprojects" to explore these kinds of possibilities. Inevitably, they will be too unorthodox to be supported via the usual procedures. A classical example is the research that led to the development of radar in England prior to World War II. Funding of this research was at first objected to in that it detracted from the limited funds for "more immediate" goals, such as the improvement of barrage balloon systems. Funding was granted due to the interference of Winston Churchill, and radar ended up saving England in the first year of the war. Another, but this time tragic example, was the initiative of Leonid Kantorovich, later to be a Nobel Laureate in Economics. In 1941 he suggested that the Soviet authorities optimize the routing of railroad trains using the technique of linear programming that he had just invented. This proposal was rejected as an arcane irrelevancy, yet it would have increased by many times the capability of the rail system (during the war!).

Both examples illustrate the fact that research of the kind discussed here, by its very nature, cannot be established within existing programs. For example, the development of radar could not be supported by the program of barrage balloon development. On the other hand, such research would not represent much of an economic burden since it would require only a fraction of the funds already allocated to similar goals. Moreover, military R&D facilities could be engaged at the subsequent development stage, rather than undergoing, as they are now doing, costly, slow, and piecemeal conversion. As a matter of fact, one of the difficulties of starting such a project may be just the opposite of the lack of funds: the cost to start will be in the millions rather than the billions of dollars and, accordingly, the project will not be considered at the appropriate level of authority.

Research of such significance, urgency, and difficulty would require wide international collaboration, engaging the top scientists and research facilities of many nations. Such collaboration will make feasible the goals that no country can accomplish alone even if it has large resources of its own. At the very least, such collaboration will save each country considerable time and funds in R&D. The major opportunities in this respect lie in collaboration with Russia. Despite its present perils, Russian science retains immense resources, including active research groups working at the state of the art level; unique experimental facilities; and, in a broader view, leading scientific schools with strong links between abstract and experimental research.

To establish such projects would require a rare commodity: bold, innovative, and responsible initiatives. A major difficulty lies in the self-destructive dwindling of interest in basic research after each major power lost its archenemy. Another difficulty, coming from the side of the scientific community, is its fatalistic lack of persistence in its own initiatives.

Governments are aware of the stalemate, and only basic science may break it. How is it possible to make a connection between them? In peacetime there is neither a mechanism nor a tradition by which the most important and revolutionary proposals could be developed and decided upon with the speed and expertise commensurate with their urgency.

We are suggesting a simple straightforward mechanism to break the stalemate by challenging the scientific community to come up with specific initiatives. The first step would be to invite scientists to submit brief (say, five page) pre-proposals based on previous studies. Second, a panel of outstanding scientists will review these pre-proposals and award grants to work out detailed proposals. Such grants are necessary since developing detailed proposals is bound to require a number of meetings, considerable work on the text, etc. Some deviations from the usual review process will probably be necessary. For example, recent achievements of the authors of proposals should be given a larger than usual weight and the authors should be invited to discuss objections to their proposals. This will help to ensure that the most outstanding ideas are not rejected simply because they are too unusual. There are many such ideas which scientists have given up on due to the difficulty of muddling through channels.

These proposals will be the final product of this venture. The history of basic research gives us assurance that some of them will be sufficiently compelling to be funded by an appropriate source. Depending on the nature of a proposal, this source could be a government agency, an international organization, a consortium of private foundations, etc. This venture will be a major success if even one of the proposals would generate its own support; that is if the decision makers could not afford to turn it down.

Similar procedures have been successfully applied on a more limited scale. For example, the U.S. National Science Foundation had developed in this way cooperative projects with South American countries. A venture of this type is also planned by the Russian Foundation for Fundamental Research. Purdue University, in the U.S., has already established a program of grants "to support the preparation of competitive proposals", where grants are awarded on the basis of a two-page outline.

We now answer some possible objections of the skeptics.

First objection: "As Vannevar Bush put it, 'Scientific progress on a broad term results from the free play of free intellects, working on subjects of their own choice, in the manner dictated by their curiosity.' Accordingly, frontier basic research is, by definition, unpredictable, and the best way to get from it major practical results would be not to focus on them in advance." Our reply is that we speak here on such areas of research, which already show a reasonable hope of addressing critically important practical problems.

Another objection: "Governments always respect and support basic science but there are, regretfully, more urgent needs.". Our reply is that we are addressing exactly such needs.

Another: "Maybe, in these lean times, one should support only the projects, which may lead to important applications?" Our reply is that this would be as impossible as supporting not an apple tree but only the apples: basic research requires a professional environment. Moreover this would be short-sighted, since one may then miss the most important possibilities that are yet unnoticed.

Another: "Recent studies of science policy and priorities have not recommended such projects." Our reply is that what we suggest is complementary to these comprehensive studies, that involve more focused goals and a shorter time scale. Metaphorically, we are proposing a sort of commando operation.

Another: "You are exaggerating the dangers." Our reply is that the dangers we are referring to are widely recognized and are already the subject of large-scale funding. What is not recognized, however, is that they cannot be contained without basic research.

Another: "Why don't the projects of such kind appear all the time?" Our reply is that, in fact, they do appear, mainly in the private sector, such as in biotechnology, where fundamental discoveries can be picked up by industry within a few weeks.

However, in many critically important areas the responsibility rests on the agencies, operating on a *cost-plus* basis, through big business formed around them. Obviously, they have lesser built-in safeguards against reliance on costly and inadequate traditional approaches, with the resulting stalemate explained by the lack of funding and not by the necessity of frontier research and new approaches.

An objection of quite a different kind: *"What if the scientific community would not come up with such ideas?"* We cannot imagine this, and it is certainly important to try.

We are suggesting here a first step in merging the international resources of basic science in a new type of war effort: in some 30 Manhattan projects, not in a shooting war or a cold war but rather in joint defense of survival and sustainable development of our civilization.

PREDICTIVE UNDERSTANDING OF EXTREME EVENTS

By V.I. Keilis-Borok and A. Gabrielov

This lecture, given by A. Gabrielov at the symposium "Four paradigms in predicting extremes: Legacy of Vladimir I. Keilis-Borok", (part of the 30th IUGG Conference on Mathematical Geophysics, a series founded by V. Keilis-Borok) in Merida, Yucatan, Mexico, 2-4 June 2014 was based on the talks given by V. Keilis-Borok. Thus, we assign dual authorship to this presentation, transcribed from A. Gabrielov's talk.

I. Lithosphere as a complex system.

In classical geology, earthquakes have always been considered as simple and isolated events, where a crack forms in the Earth's crust and propagates across a flat surface of a tectonic fault. Foreshocks and aftershocks have been long known, but no one has ever considered the possibility of interactions between earthquakes, or attempted to view the whole structure of the lithosphere as a single complex system. These concepts, introduced by V. Keilis-Borok in the 1960s, have revolutionized our understanding of the earthquakes.

There are several reasons why the lithosphere can indeed be viewed as a complex system. Geometrically, it is a **hierarchical structure** characteristic for classical complex systems. A typical tectonic map always appears hierarchical, with major, secondary and tertiary elements – it is essentially a mess of blocks and faults. Classical geology used to study each fault separately. There used to be classifications of faults, but even in these classifications, each of their properties essentially is based on the properties of single faults.

The idea of hierarchical structure of the lithosphere comes from geomorphology (the work of E. Rantsman) and plate tectonics. In plate tectonics, the dynamics of the lithosphere is reflected in the movements of large, global-scale plates, but there are structural blocks within these plates that can be recognized on all scales. The entire lithosphere is divided into a hierarchy of blocks, from ~ 10 tectonic plates, to ~10^{25} grains of rocks. These blocks are separated by boundary zones, 10-100 times thinner. They are called *fault zones* (starting with the boundaries between the plates) that include *faults, sliding surfaces,* and, finally, *interfaces* between the grains of rocks.

Except for a few lowest ranks, these boundary zones have similar hierarchical structure with much denser fragmentation. Most densely fractured mosaic "*nodes*" are formed around intersections and junctions of these zones; this is due to collision of the blocks' corners. For brevity, a system of boundary zones and nodes is called "fault network".

Along some boundaries, movements can be slow or hidden. In these cases, geologists don't count these boundaries as faults, but they are still seen on the maps and aerial/space photos. Such boundaries with slow or hidden movements are called lineaments.

The structural blocks distinguished by geomorphology are three-dimensional. However, since their horizontal scale is very extensive compared to the thickness of the lithosphere, we consider these blocks two-dimensional, and faults as one-dimensional structures, or lines. This is a simplification, of course. The faults can be kilometers wide, and the intersections of faults are not merely geometrical points but have their own size. So, it is important to understand that even the zero-dimensional fault intersection has its own scale. This scale can be from plate boundaries to grains of rock.

In separate observations, laboratory experiments modeled earthquakes on rocks samples, and the intersections of faults in such experiments were indeed grains of rock. Thus, the global picture includes all kinds of scales.

There are some aspects of self-similarity in this structure. It is not exactly scale-invariant (nothing is, in the real world). There are ideal scale-invariant structures in mathematics, but there is a certain range of scale invariance that is observed, and thus the parameters, such as fractal dimensions, can be calculated. All this already is complex enough to be considered a complex system, because the blocks are moving, their relative movements are observed as displacements along the faults, and there are nodes where movements in different directions are observed in most complicated and least understood ways. These nodes are the nucleation points of earthquakes.

This is the geometric part. Next comes the dynamic part – **the instability of the strength and stress fields**.

It is recognized that the stress in the lithosphere is not homogeneous, there are places where stress concentrates and other places where stress is released. Stress and strength can be very variable across space. In some nodes, strength decreases, and in some nodes a lock-in effect occurs: a fault becomes locked and stress accumulates until it becomes higher than the strength of the fault, but the fault does not move. This is the most dangerous situation that may be catastrophic.

A fault network is a stockpile of instability. This instability includes both "physical" instability, originated at micro level, and "geometric" instability, controlled by the geometry of the fault network at the macro level.

Movement of the blocks is controlled by a stress - strain field. The stress, and particularly the strength, are in turn controlled by a multitude of mechanisms that include filtration of fluids, stress corrosion, fracturing, buckling, petrochemical and phase transitions, etc. Each process may cause up to 10^5 drop of strength. None of them is consistently dominant: even a grain of rock can act as viscoelastic element. Other elements can include aggregates of crystals, source/absorber of fluid, volume, heat, etc. Altogether these mechanisms turn the lithosphere into a nonlinear complex system.

A dramatic drop of strength in a fault would result in a strong earthquake.

In seismically active regions a large part of the movement is realized through earthquakes. There are few places within such regions where the movement is silent, so no earthquakes are observed, but as a rule the movement in the lithosphere is not continuous but results in bigger or smaller drop.

One striking example is the area west of Vancouver, where tectonic plates are moving in respect to each other, but the actual fault is not seen, which means that this fault is locked. It is expected that this area at some point may produce a major earthquake (with the magnitude 9 or more) that would have a catastrophic effect on the whole region.

Now we come to the problem of earthquake prediction, which is in itself contradictory and not as simple as one would expect.

II. Earthquake prediction

There are multitudes of scales in space, time, and energy, in which earthquakes happen. Because of this, the task should not be poised as an attempt to predict an earthquake of a particular magnitude in a particular place. Instead, earthquake prediction should be a process that includes several stages and scales in time and space. Thus, the overall problem of earthquake prediction is posed as the step by step reduction of the space-time window where a strong earthquake may occur.

There are **five stages of prediction**:

1) Background prediction (100 years and more, to identify the places where strong earthquakes may occur);
2) Long-term prediction (10 years or so);
3) Intermediate-term prediction (a few years);
4) Short-term prediction (less than a year);
5) Immediate prediction (days to minutes).

V. Keilis-Borok's earthquake prediction program was originally concentrated on the background predictions and identification of earthquake-prone areas. Much of this work was done in collaboration with I. M. Gelfand, F. Press, L. Knopoff, and E. Rantsman. As the next step, he moved to intermediate-term predictions. With shorter time scales, the predictions become less and less reliable. Short-term and immediate predictions are still out of reach for us, except that, when an earthquake does happen, there are now some systems that recognize the first incoming seismic waves and are able to make some adjustments to neutralize or diminish the next incoming wave. But this is more of an engineering task.

V. Keilis-Borok's main tool was pattern recognition. This tool was pretty controversial in the field. Many times he had to answer the questions: "How can you predict something when you don't have an adequate model?" or "Why don't you derive from the first principles?"

He always responded with a quotation from A. Kolmogorov:

"It became clear for me that it is unrealistic to have a hope for the creation of a pure theory [of the turbulent flows of fluids and gases] closed in itself. Due to the absence of such a theory we have to rely upon the hypotheses obtained by processing of the experimental data..."

In other words, there is no first principle in this case, no differential equations that would describe the principle.

Kolmogorov's quotation referred to the theory of turbulence, where his famous scaling law was derived from half-empirical manipulations with dimensions. His law became one of his most famous results in hydrodynamics. In fact, complete theory of turbulence still doesn't exist—and lithosphere is far more complex than fluid dynamics. But one important similarity between these two systems is that in both systems there is no complete theory that would derive behavior from first principles.

So, what do we mean when we say "prediction"?

It is surprising how many different points of view people have when they talk about earthquake predictions. V. Keilis-Borok's view was very specific, and probably unusual:

First, it is important to define the target: **what do we want to predict?** V. Keilis-Borok's goal was to predict rare extreme events. A predictor in this case would be a discrete sequence of alarms. This is opposite to another very popular point of view of probabilistic prediction. For example, seismic risk maps, regularly published by the US Geological Survey, contain probabilities with which strong earthquake may happen in certain areas. This is an entirely different approach to predictions from V. Keilis-Borok's method. Yes, it is nice to have a probability, which is a real number, but in most cases we don't have enough information to generate a real number like this. Thus, a prediction in this case cannot be a subject of classical statistics.

Pattern recognition of rare extreme events is a unique tool that gives a natural framework for integration of modeling, theory, and extensive data analysis.

The next point, introduced by V. Keilis-Borok and still not properly understood, concerns **premonitory phenomena** — something we observe before a strong earthquake happens. There are two kinds of premonitory phenomena. Some are "perpetrators", the events that start happening and eventually cause a strong earthquake to occur. However, immediate causes of a strong earthquake are often very hard to identify. The other kinds of premonitory phenomena are termed “witnesses”. Witnesses don’t cause extreme events, but they are by-products of a destabilized system that can be observed. Thus, “witnesses” might predict not an extreme event per se, but the destabilization of the system which makes it ripe for an extreme event.

One example is the proverbial “straws in the wind” that precede a hurricane.

The need for holistic approach.

It is not possible to understand a complex system by breaking it apart. Holistic approach, from the whole to details, opens a possibility to overcome the complexity itself and the chronic imperfection of the data.

Figure P1. The need for holistic approach: a fracture in a rock cannot be seen in a close-up view (left), but becomes far more evident at a distance (right).

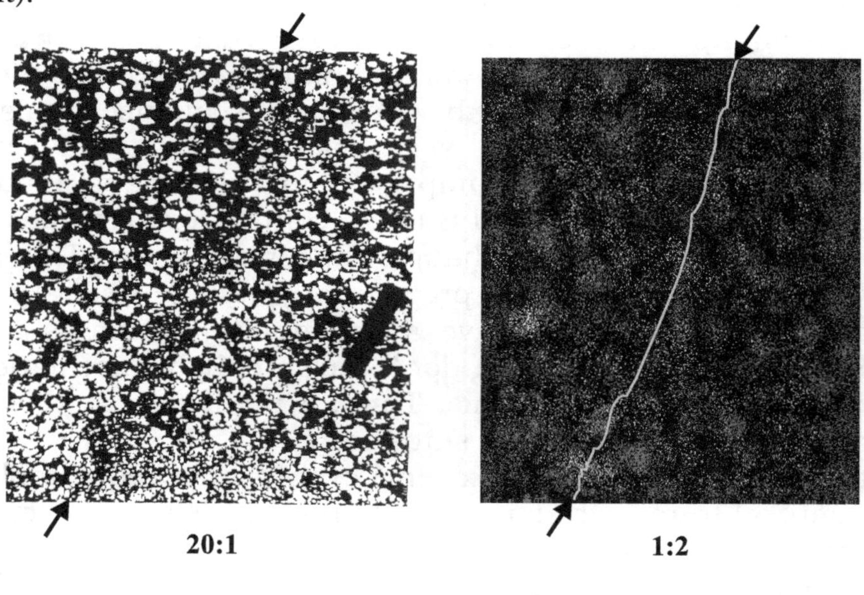

20:1 1:2

Fig. P1 is an illustration of a geological experiment done a while ago. It shows images of a rock at two different magnifications. If you look at the rock too closely (left) you don’t see anything but a chaotic mess of grains. But if you zoom out and look at the same rock on a smaller scale (right), you see a fracture, which is completely lost in the higher magnification view on the left. This was one of the main V. Keilis-Borok's themes – if you want to study the system as a whole you need to get spacial and temporal averages, smoothing things to understand large scale structures that will be lost in a too-detailed view. This smoothing is essential in prediction of strong earthquakes.

Quoting the famous American physicist M. Gell-Mann:

"If the parts of a complex system or the various aspects of a complex situation, all defined in advance, are studied carefully by experts on those parts or aspects, and the results of their work are pooled, an adequate description of the whole system or situation does not usually emerge... The reason, of course, is that these parts or aspects are typically entangled with one another... We have to supplement the partial studies with a transdisciplinary crude look at the whole."

In other words, if you study separate aspects of a complex system and then combine them, you may lose the properties of the system

Figure P2: Attacking herd of elephants as an example of a complex system behavior.

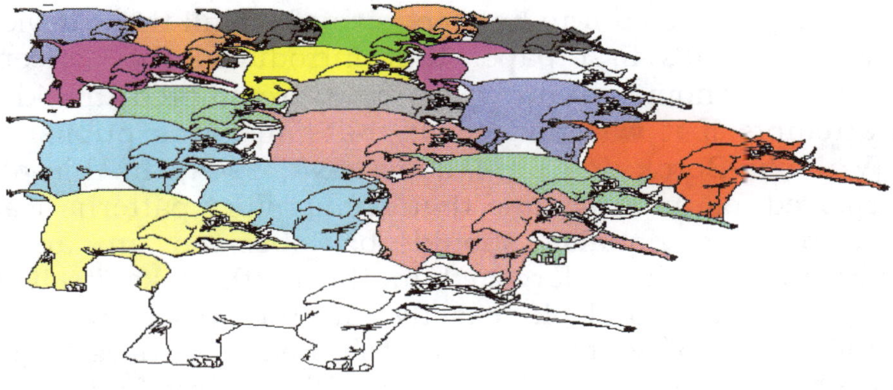

as a whole, which may only appear when elements interact with each other.

This kind of a behavior can be well illustrated by an attacking herd of elephants (Fig. P2). If you look at a single elephant, you would see him extend his trunk and raise his ears, but you may not recognize that the whole herd is about to attack. Only looking at the herd as a whole makes it obvious.

One example of why large-scale observations are essential in prediction of earthquakes is the Parkfield experiment, in the early 1980's. The experiment involved detailed observation of seismic activity in Parkfield, CA. In this area, which includes a segment of the St. Andreas fault, earthquakes have been occurring nearly every 21 years with some exceptions. The project involved installing a lot of monitoring devices to measure tectonic stress, microseismicity, fluid levels, and everything else they were able to measure, in a small area where the next Parkfield earthquake was expected to happen. Eventually the earthquake occurred 20 years later than expected, and none of the devices were able to detect anything preceding this earthquake. But during this same period of time strong earthquakes happened in several other places relatively nearby – Northridge, Loma Prieta, and others. The participants of this project ignored all these other earthquakes, waiting for another Parkfield. In reality, they should have been looking at a much larger area.

Another example concerns an attempt to predict an earthquake in Tokyo. If one occurred, it would have been very damaging for Tokyo. To predict it, a project was initiated and a lot of devices were installed to monitor different parameters in the immediate Tokyo area, but then the Kobe earthquake happened in a region where nobody expected it to happen, away from all these devices.

V. Keilis-Borok's first paper that introduced the concept of premonitory phenomena and would have pre-empted this kind of failed attempts if it was noticed at that time, was published in 1964 (V. Keilis-Borok and L. Malinovskaya, 1964). In this paper, they reported for the first time that premonitory patterns leading to a strong earthquake should be observed in an area approximately 10 times larger than the earthquake itself. It is definitely useless to look exactly at the place where the earthquake is going to happen to predict an earthquake. Historically, all efforts to do it have failed. The first stage of earthquake prediction following this discovery was originally started by V. Keilis-Borok in collaboration with the prominent Russian mathematician I. M. Gelfand, the American scientists L. Knopoff and F. Press, and a Russian geomorphologist E. Ya. Rantsman (Fig. P3). The figure shows two maps: on the right is the map of the blocks and lineaments defined by Rantsman and her group using a combination of geological, geomorphological, and tectonic data. On the left is the map of the major known tectonic faults. The points on the left-hand map, indicating the sites of the strong earthquakes that historically occurred in these areas, corresponded to the nodes identified in V. Keilis-Borok's analysis as the sites which could serve as the points of the nucleation of strong earthquakes. The goal of this work was to use historical data and a list of properties of the nodes to distinguish between the nodes where the strong earthquake may or may not nucleate. Using the two maps gives an overall similar results, but the more detailed map on the right reveals more nodes, some of them dangerous but not seen in the tectonic fault map.

Figure P4. Strong earthquakes nucleate in some "dangerous" structures (D-nodes) (from Gelfand et al, 1976).

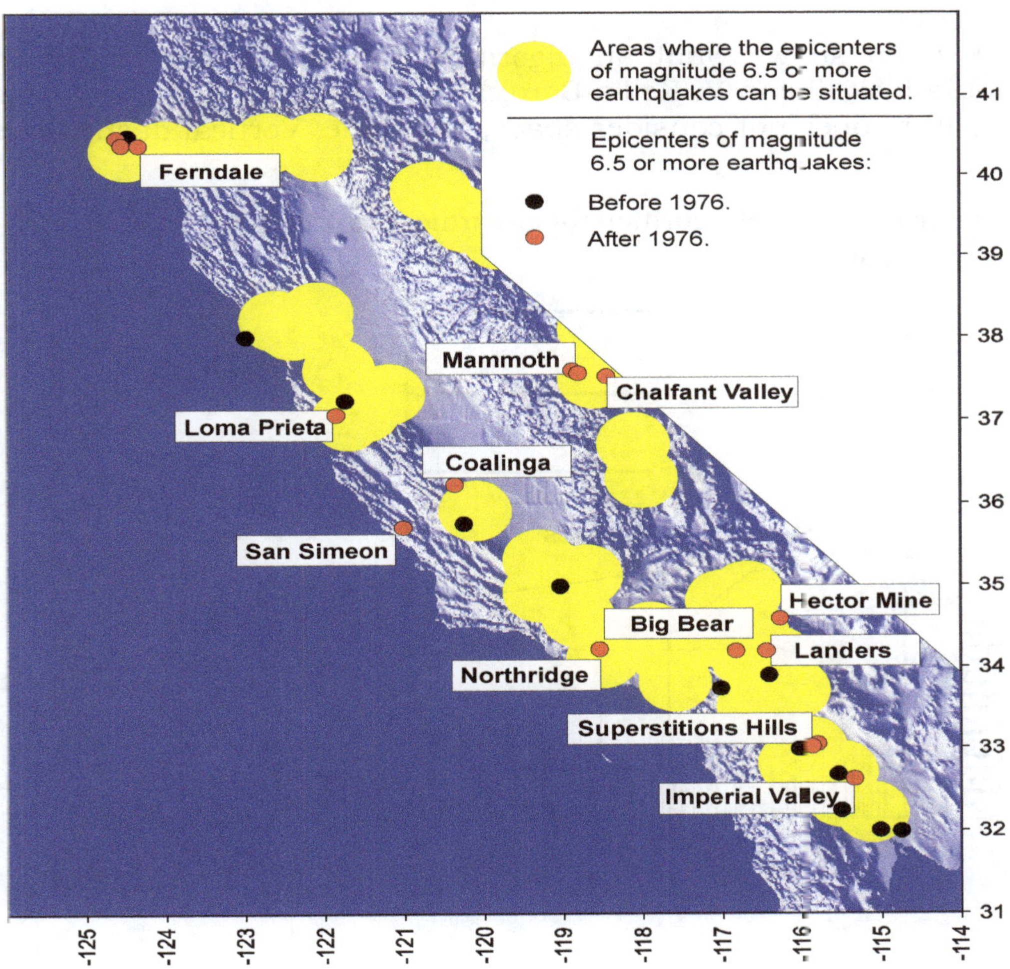

In fact, some nodes identified by this work later became the sites of strong earthquakes – for example, Northridge, which nobody expected to become the earthquake site. In California everyone was looking at another fault, and construction codes were not prepared for the Northridge earthquake. After Northridge, many other places were identified, which were not on the active faults, but were similar to the known places on the active faults. Many areas that were originally missed as potential earthquake sites were mapped – e.g., in downtown Los Angeles.

V. Keilis-Borok's group used pattern recognition to identify such "dangerous" nodes (Fig. P4). The identified nodes are shown in yellow, and black dots denote the sites of earthquakes that happened before 1976 and were used in the analysis. Red dots show the earthquakes that happened after 1976. Several of them occurred in the areas that were not previously considered dangerous.

Fig. P5 shows what an algorithm for earthquake prediction should look like in V. Keilis-Borok's approach. For simplicity, this algorithm does not consider space, only time. Vertical dashed line

Figure P5. A model algorithm for an earthquake prediction, in V. Keilis-Borok's approach.

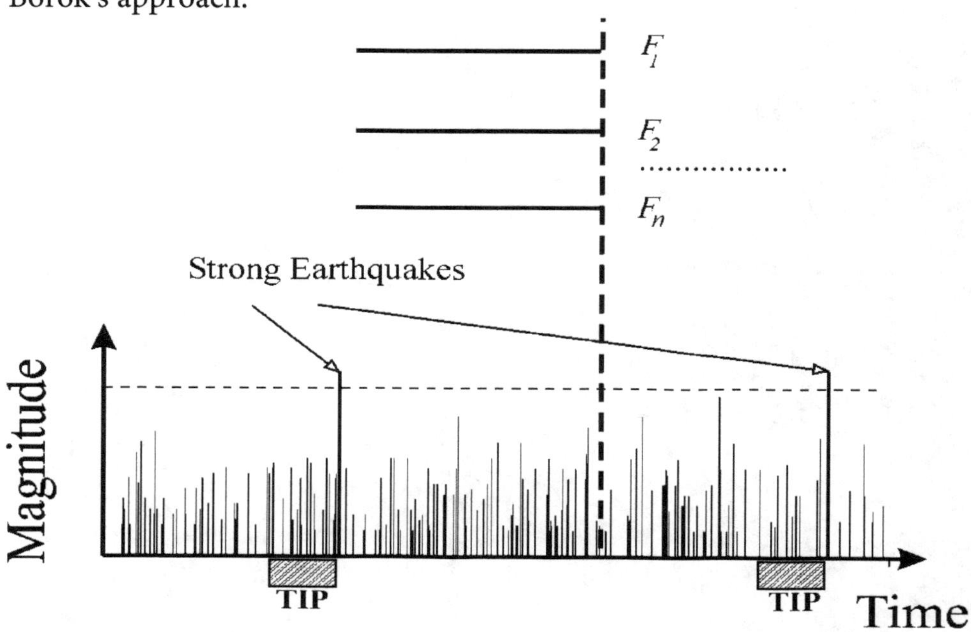

shows a given time moment. From this moment, we take a time window backwards (defined as an observation window) and compute some functionals based on observations within this window—usually some integral patterns of seismicity (e.g., total seismic activity, clustering, etc.). We next associate several numbers to our time moment and look at the historical statistics of those numbers to compare their frequency just before a strong earthquake (dangerous period), excluding the period just after (unstable period), but also considering longer periods before and after. When the actual observed values reach the extreme, alarm is declared (TIP- **t**ime of **i**ncreased **p**robability of a strong earthquake).

The main requirements for this kind of algorithm are that (i) it must be objective, (ii) enough data must be available to compute those functionals, and that (iii) the data should be homogeneous. The real predictions also consider space were the earthquake will happen.

The diagram in Fig. P5 shows two strong earthquakes and two TIPs.

Possible outcomes of predictions range in time and space, from false alarms and failures to correct alarms and successful predictions. Good and bad prediction methods could be distinguished by comparing prediction to real data.

So, how do we choose the functionals?

The choice is guided by **four paradigms**:

1. Basic types of premonitory phenomena. Premonitory phenomena in V. Keilis-Borok's studies consisted of all the weak and intermediate seismicity data in the area. Seismicity data were the most reliable and uniform data available at the time. Other data (e.g. GPS) did not exist back then, but can be used in future predictions.

2. Long - range correlations in the fault system dynamics. Premonitory phenomena are formed not only in vicinity of the incipient source but within a much larger area. Overall, one needs to look in the area at least 10 times larger than the earthquake area.

3. Partial similarity of premonitory phenomena in the diverse conditions. All the functionals are scaled, so that if you change the premonitory phenomena you can directly compare data in the areas with different levels of seismicity. The same functionals can apply to a variety of phenomena, from small rock bursts in mines, and induced seismicity in basins, as well as the largest earthquakes magnitude (M) from 4.5 to 8+ worldwide – and, possibly, to starquakes. The energy can range from erg to 10^{26} erg, and possibly to 10^{41} erg.

4. The dual nature of premonitory phenomena. Some of the premonitory phenomena are "universal", common for complex

Figure P6. An extreme event is preceded by premonitory changes of the background activity (its deviations from long-term average).

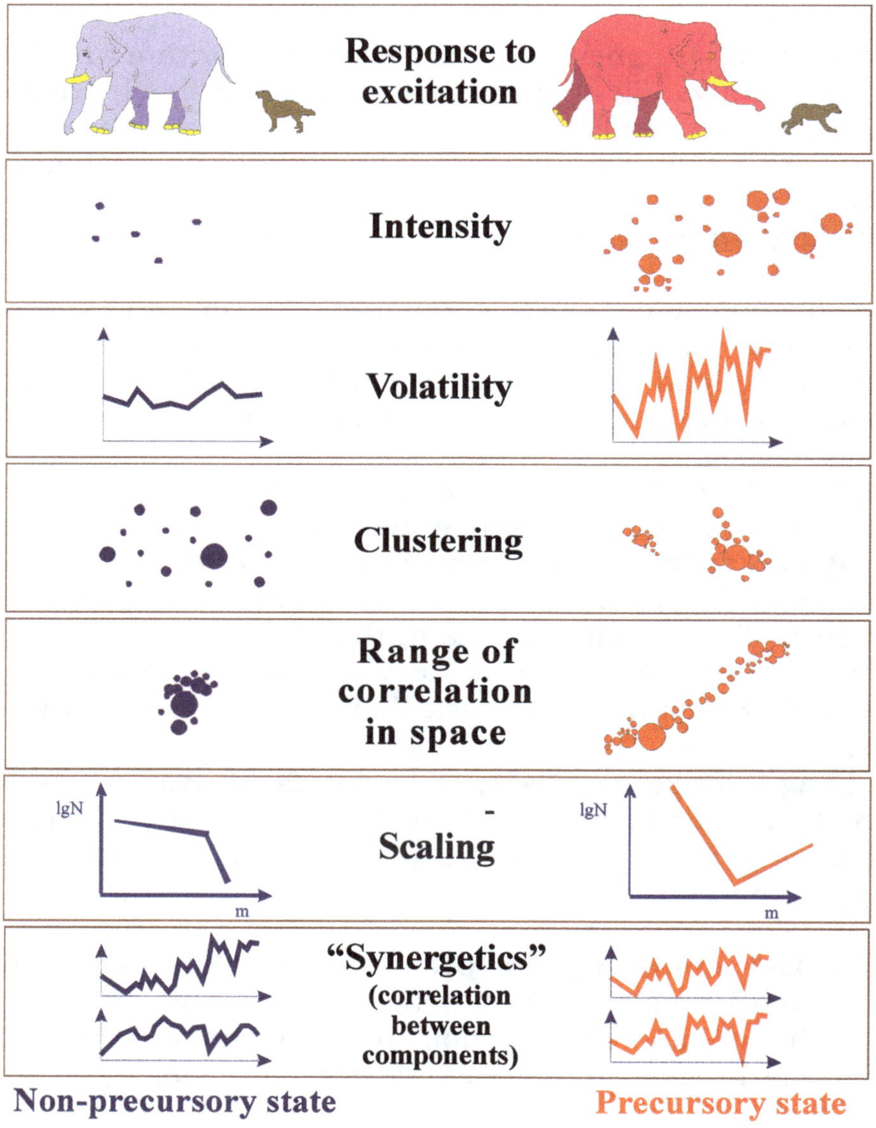

nonlinear systems of different origin; others are Earth-specific. According to the general theory of complex nonlinear systems, these systems can exhibit abrupt qualitative changes — bifurcations, phase transitions, and loss of stability. Some of these changes can serve as premonitory phenomena for critical transitions in these systems, such as earthquakes. Some of these events are tied to particular properties of the Earth and appear only in very specific settings of tectonic fault systems. The functionals should take into account these specifics.

Common premonitory patterns.

In defining the premonitory patterns, we are looking at the long-term activity in this system, and when the statistics of this activity becomes extreme, this activity becomes a premonitory pattern for an extreme event (e.g., an earthquake).

Figure. P7. Successful advance prediction of Loma Prieta earthquake in 1989, M=7.1 Light gray – area of investigation; Red - alarm reduced by MSc; Dark gray – alarm by M8; Asterisk – Loma Prieta epicenter

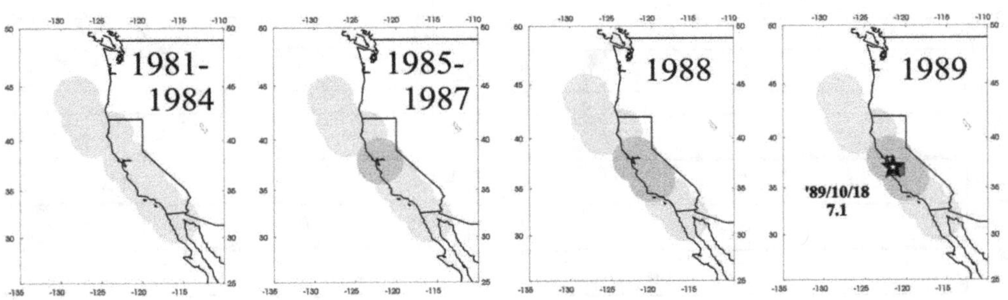

The following changes can serve as premonitory patterns (illustrated in Fig. P6):

1. Response to excitation. In normal state, a system should not show a strong disturbance in response to mild excitation. If it shows a strong reaction, this indicates that the system is unstable.

2. Intensity increase. Activity in an area becomes less uniform, concentrating in some places (not necessarily in the place where an earthquake will happen).

3. Volatility. The amplitude of activity increases.

4. Clustering. Spacial and temporal distribution of events becomes less uniform, concentrating in some places (not necessarily where the earthquake happens)

5. Range of correlation in space. When a strong earthquake approaches, moderate earthquakes tend to appear close in time at faraway places in space.

6. Scaling law. The quiet-time distribution of earthquakes is dominated by smaller ones, the dangerous times are dominated by stronger ones.

7. Correlation between components. This is not properly observed with earthquakes, but is observed in other systems (e,.g, financial markets), thus, hypothetically, this can also serve as a precursor.

Figure P8. Frontiers of similarity.

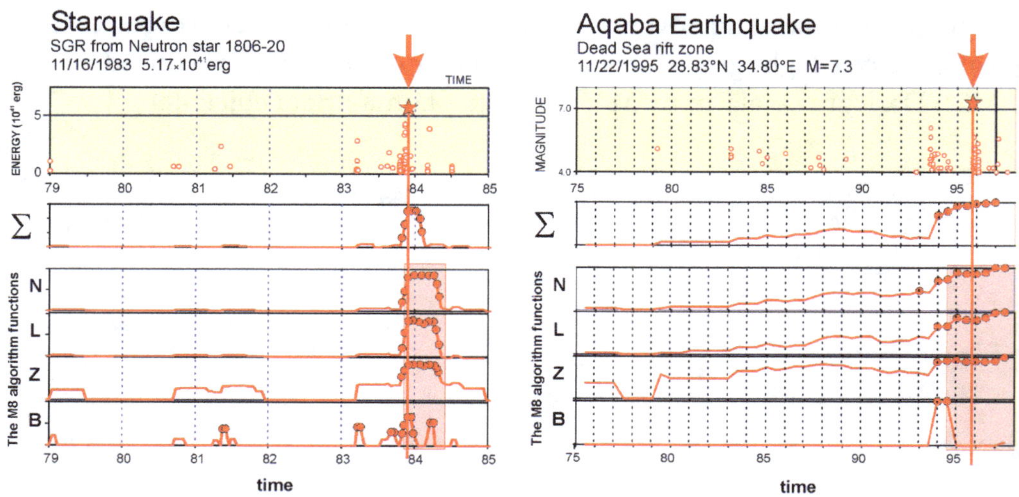

Advance predictions.

Using these approaches, V. Keilis-Borok's group has developed the algorithm M8 (for prediction of earthquakes with the magnitude of 8 and more). One of the first predictions using this algorithm was the Loma Prieta earthquake that occurred in California in 1989. This algorithm was successfully used for decades of earthquake predictions that are still ongoing right now.

The prediction was made 5 years ahead (Fig. P7). In 1981-1984 the dangerous zone was identified (gray in the leftmost panel in Fig. P7). In 1985-1987 one area in that zone became dangerous, and in 1988 two adjacent areas also showed increase seismic activity (dark gray circles in middle panels). The 1989 earthquake happened within those areas. The results of this prediction were reported in a meeting between the Russian President M. Gorbachev and the US President R. Reagan in 1988. After the earthquake happened, V. Keilis-Borok's team from Russia was urgently sent to California.

The same algorithm M8 was able to predict major busts from neutron star and an earthquake in a Dead Sea rift zone, which has fairly low seismicity (Fig. P8).

III. Socioeconomic predictions

The same approaches can be applied to predictions in socioeconomic systems. In each case, the socioeconomic system in question is considered a complex system, and the functionals are chosen by experts in the appropriate areas as the parameters that reflect the behavior of this particular system. Successful socioeconomic predictions by V. Keilis-Borok's group, outlined later in this book, include predictions of outcomes of US elections, economic recessions, and surges of unemployment and crime (see Fig. P9 and chapters 3-6 of the current book).

Predictions of the US presidential elections.

In collaboration with A. Lichtman, V. Keilis-Borok developed a successful algorithm for predictions of the US presidential elections. This prediction was based on thirteen socio-economic and political factors, later termed "thirteen keys to presidency" or "thirteen keys to the White House" (Chapter 3 section 3.3). Victory of the challenging party is predicted when 6 or more factors are in its favor. Otherwise victory of incumbent party is predicted.

Notably, this algorithm predicted the victory of Donald Trump in 2016, months in advance of the election.

IV. Using the limited accuracy predictions for disaster preparedness.

> "Of course, things are complicated....But in the end every situation can be reduced to a simple question: Do we act or not? If yes, in what way?"
>
> E. Burdick

In using predictions for disaster preparedness, it is important to remember the following:

Basic principle: escalate or de-escalate preparedness measures, according to what and where is predicted and what is the quality of prediction. Such is the standard practice in preparedness to all disasters, war included.

Diversity of damage: failure of constructions; fires; release of dangerous materials; triggering of floods, avalanches, landslides, tsunami, etc.

Socio-economic impact: disruption of vital services - supply, medical, financial, law enforcement etc.; epidemics; disruptive anxiety of population, profiteering and crime; drop of production and employment; destabilization of financial systems.

Figure P9. Socioeconomic predictions in the US.

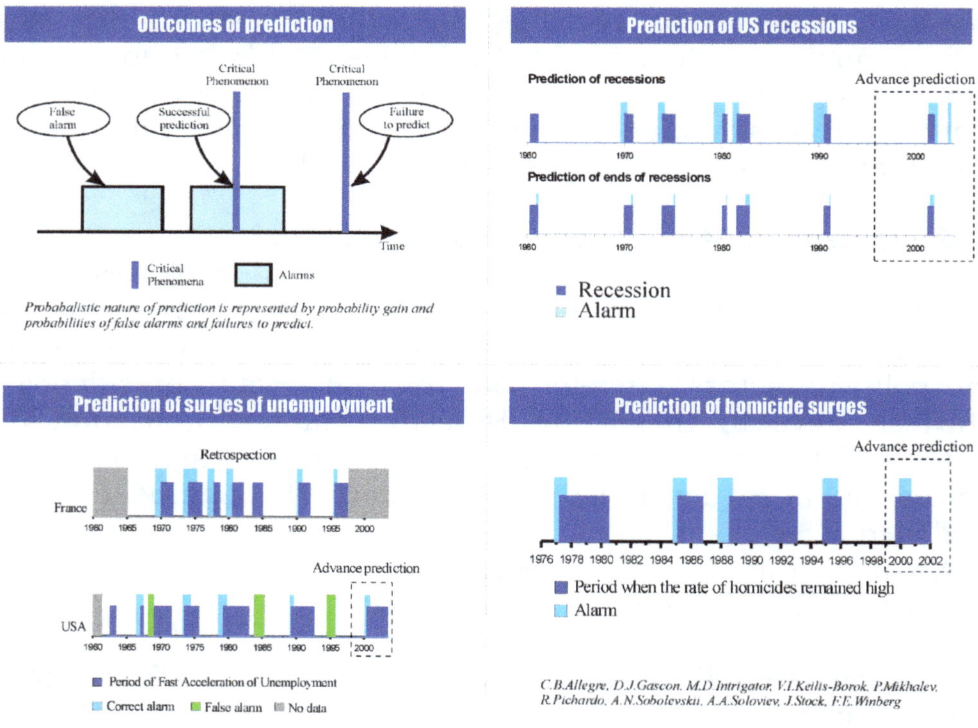

Impact by undue release of predictions: knowledge that an extreme event is about to occur may cause panic, with dramatic consequences on all levels. This was one of the reasons V. Keilis-Borok's group distributed their predictions in a closed way, to a selected group of experts worldwide.

A hierarchy of **diverse safety measures** is required by the diversity of damage from earthquakes. These include:

Background measures: restriction of land use; building codes; insurance and bonds; preparedness of civil defense type; R&D.

Temporary measures, activated in response to a time prediction: enhancement of permanent measures - safety inspections, simulation alarms etc.; partial neutralization of high - risk objects; mobilization of post - disaster services; emergency legislation, up to martial law; evacuation of population etc.

These measures are required in different forms on local, provincial, national and international levels.

Different measures require different lead time, from seconds to years, to be activated; having different cost they can be realistically maintained for different time - periods, from hours to decades; and they have to be spread over different territories - from selected points to large regions.

Different stages of earthquake prediction are useful for damage reduction.

No single measure is sufficient alone. On the other hand, many important measures are inexpensive. The accuracy of prediction should be known, but not necessarily high.

CHAPTER 1

Global Approach to Prediction of Extreme Events in Complex Systems

1.1. Holistic Approach

Natural science had, for many centuries, regarded the Universe as a completely predictable machine. As Pierre Simon de Laplace wrote in 1776, "...if we knew exactly the laws of nature and the situation of the universe at the initial moment, we could predict exactly the situation of the same universe at a succeeding moment." However, at the turn of the 20th century (1905) Jules Henry Poincare discovered, that "... this is not always so. It may happen that small differences in the initial conditions will produce very great ones in the final phenomena. Prediction becomes impossible". This concerns nonlinear (chaotic) systems that contain mutually dependent mechanisms destabilizing their behavior. The complexity is an attribute of such systems. It comprises instability and complex, but not random, behavior patterns – "order in chaos." The complex systems are not predictable with absolute precision. However, after coarse-graining (on a not- too-detailed scale), premonitory phenomena emerge, and a system becomes predictable, up to the limits (Farmer and Sidorowich, 1987; Ma et al., 1990; Kravtsov, 1993; Gell-Mann, 1994; Holland, 1995; Kadanoff, 1976; Crutchfield et al., 1986).

The holistic approach, "from the whole to details", opens a possibility to detect "universal" types of precursors overcoming the complexity itself, and the chronic imperfection of observations as well. Through robust integral description of the complex systems, it is possible to discover their regular behavioral patterns transcending their inherent complexity. This is achieved at an unavoidable price: the accuracy of prediction is limited. The holistic approach that proceeds from the whole to details is opposite to the reductionism approach that proceeds from details to the whole. It is in principle not possible "to understand a complex system by breaking it apart" (Crutchfield et al, 1986). Table 1 compares the holistic approach with the complementary (but not necessarily contradictory) reductionism approach.

Among the regular behavior patterns of complex systems are "premonitory" ones that emerge more frequently as an extreme event approaches. These premonitory patterns make complex systems predictable. The accuracy of these predictions, however, is inevitably limited, due to the systems' complexity and observational errors.

1.2. The Prediction Problem and Premonitory Patterns

If we consider extreme events that concern the complex system as a whole, then the prediction problem can be formulated as follows:

given are time series that describe dynamics of the system up to the current moment of time t and contain potential precursors of an extreme event;

to predict whether an extreme event will or will not occur during the subsequent time period $(t, t + \tau)$; if the answer is "yes", this will be the "*period of alarm.*"

As the time goes by, predictions form a discrete sequence of alarms. The possible outcomes of such a prediction are shown in Fig. 1. The actual outcome is determined unambiguously, since the extreme events are identified independently of the prediction. In some problems considered here they are routinely determined independently of our research (e.g. earthquake occurrence or an election result). In other problems they are defined by a by a separate algorithm (e.g. homicide surge).

Such "yes or no" prediction is aimed not at analyzing the whole dynamics of the system, but only at identifying the occurrence of rare extreme events. In a broad field of prediction studies this prediction is different from, and complementary to, the classical

Kolmogoroff–Wiener prediction of continuous functions, and to traditional cause-and-effect analysis.

Table 1. Two complementary approaches to prediction

"REDUCTIONISM" *(from details to the whole)*	"HOLISM" *(from the whole to details)*
Premonitory phenomena preceding an extreme event are formed:	
in the narrow time-space vicinity of the future event	in the wide time-space volume containing the future event
Premonitory phenomena are:	
specific to mechanisms controlling the behavior of the complex system	divided into: – "universal" ones common for many complex systems; – those, depending on the complex system structure; – mechanism-specific ones.
Premonitory phenomena in different complex systems and energy ranges are:	
different	to a considerable extent similar
Constitutive equations are:	
local	non-local
Triggering of extreme events is controlled:	
by the system state in the time-space vicinity of the future event	by the state of the whole system during a rather long period before the future event

The problem includes estimating the predictions' accuracy: the rates of false alarms and failures to predict, and the total duration of alarms in % to the total time, are considered. These characteristics represent the inevitable *probabilistic component* of prediction; they provide for statistical validation of a prediction algorithm and for optimizing preparedness to predicted events (e.g. recessions or crime surges).

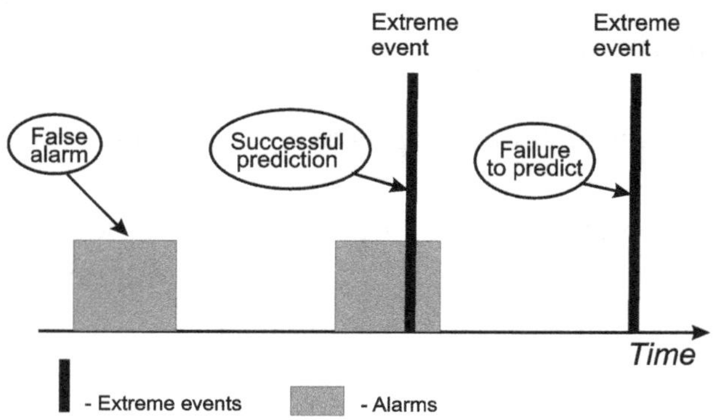

Figure 1. Possible outcomes of prediction

Twofold importance.

Prediction problem is pivotal in two fields.

— *Fundamental understanding of complex systems.* Prediction algorithms quantitatively define phenomena that anticipate extreme events. Such definition is pivotal for fundamental understanding of a complex system where these events occur, including the intertwined mechanisms of system's development and its basic features, e.g. multiple scaling, correlation range, clustering, fragmentation etc. The understanding of complex systems remains a major unsolved problem of modern science, tantamount to transforming our understanding of the natural and human world.

— *Disaster preparedness.* On the practical side, prediction is pivotal for coping with a variety of disasters, commonly recognized as major threats to the survival and sustainability of our civilization (e.g. Keilis-Borok and Sorondo, 2000; Davis et al., 2010; see also materials of G8-UNESCO World Forum on "Education, Innovation and Research: New Partnership for

Sustainable Development", http://g8forum.ictp.it). The reliable advance prediction of extreme events can save lives and contribute to social and economic stability, and to improving the governing of modern societies.

Certain systems exhibit behavior patterns that emerge more frequently as an extreme event draws near. These patterns, called premonitory patterns, signal destabilization of the system and thus an increase in the probability of an extreme event. Premonitory patterns do not necessarily contribute to causing a subsequent extreme event. Both the patterns and the event itself might be parallel manifestations of the same underlying process occurring in the complex system, formed in multiple time-, space-, and energy- scales. For this reason premonitory patterns might emerge in a broad variety of observable fields reflecting complex system dynamics, and in different scales.

Prediction algorithms are designed by analysis of the "learning material" – a sample of past critical events and the time series hypothetically containing premonitory patterns. Typically, for complexity studies we do not have a complete set of fundamental equations that govern dynamics of the system and unambiguously define prediction algorithms. In lieu of such equations "...we have to rely upon the hypotheses obtained by processing of the experimental data" (A. Kolmogorov, 1943 on transition to turbulence). A methodology for designing the prediction algorithms is the *pattern recognition of rare events* that was developed by the school of I. Gelfand for studying rare events of highly complex origin (e.g., Bongard, 1970; Gelfand et al., 1976; Keilis-Borok and Press, 1980; Keilis-Borok and Lichtman, 1993; Press and Allen, 1995; Keilis-Borok et al., 2000, 2003, 2005, 2008).

Algorithm design comprises three following steps.

1. *Detecting premonitory patterns.* Each time series considered is robustly described by the functionals $F_k(t)$, $k = 1, 2, \ldots$, capturing hypothetical patterns. Hypotheses on what these patterns may be are provided by "universal" modeling of complex systems, modeling of system-specific processes, exploratory data analysis, and practical experience, even if it may be intuitive. Pattern recognition of rare events is an efficient common framework for formulating and testing such hypotheses, their diversity notwithstanding. With a few exceptions the functionals

are defined in sliding time windows; the value of a functional is attributed to the end of the window.

2. *Discretization.* Emergence of a premonitory pattern is defined by the condition $F_k(t) \geq C_k$. The threshold C_k is chosen is such a way that premonitory pattern emerges on one side of the threshold more frequently than on another side. That threshold is usually defined as a certain percentile of the functional F_k. In such robust representation of the data, pattern recognition is akin to exploratory data analysis developed by J. Tukey (1977).

3. *Formulating an algorithm.* Prediction algorithm will trigger an alarm when certain combination of premonitory patterns emerges. This combination is determined by further application of pattern recognition procedure (Keilis-Borok and Lichtman, 1993; Press and Allen, 1995; Keilis-Borok and Soloviev, 2003).

1.3. Algorithm Reliability. Error Diagrams

Estimating reliability of an algorithm is necessary, since an algorithm inevitably includes many adjustable elements, from selecting the data used for prediction and definition of prediction targets, to the values of numerical parameters. In lieu of the closed theory a priory determining all these elements they have to be adjusted retrospectively, by "predicting" the past extreme events. The application of the methodology to known events creates the danger of self-deceptive data-fitting: As J. Von Neumann put it "*with four exponents I can fit an elephant*". The proper validation of the prediction algorithms requires three consecutive tests:

— *sensitivity analysis*: varying adjustable elements of an algorithm;

– *out of sample analysis*: applying an algorithm to past data that has not been used in the algorithm's development;

— *predicting in advance* – the only decisive test of a prediction algorithm.

Such tests take a lion share of data analysis (Gelfand et al., 1976; Gabrielov et al., 2000b; Zaliapin et al., 2003b; Keilis-Borok and Soloviev, 2003). A prediction algorithm makes sense only if its performance is (i) sufficiently better than a random guess, and (ii) not too sensitive to variation of adjustable elements. Error diagrams described below show whether these conditions are satisfied.

Definition. Error diagrams show three major characteristics of prediction's accuracy. Consider an algorithm applied during the time period *T*. During the test *N* extreme events have occurred there, and N_m of them have been missed by alarms. Altogether A alarms have been declared and A_f of them happened to be false. Total duration of alarms is D.

Performance of an algorithm is characterized by three dimensionless parameters: the relative duration of alarms, $\tau = D/T$; the rate of failures to predict, $n = N_m/N$; and the rate of false alarms, $f = A_f/A$. These three parameters are necessary in any test of prediction algorithm regardless of a particular methodology. They are juxtaposed on the error diagrams schematically illustrated in Fig. 2. Also called "Molchan diagrams" in earthquake prediction studies, they are used for validation and optimization of earthquake prediction algorithms and for joint optimization of prediction and preparedness (Molchan, 1990, 1991, 1994, 1997, 2003). In many applications parameter *f* is not yet considered. In early applications they are called ROC diagrams for "Relative Operating Characteristics" (e.g., Mason, 2003).

1.4 Prediction and Disaster Preparedness

What, if any, preparedness actions should be undertaken in response to a prediction, given its inherently limited accuracy? Methodology assisting decision-makers in choosing optimal response to prediction has been developed in Kantorovich et al. (1974) and Molchan (1991, 1997, 2003). Its numerous applications are described by Keilis-Borok et al. (2004), Davis et al. (2007, 2010), and Molchan and Keilis-Borok (2008).

Earthquakes might hurt the population, economy, and environment in many different ways, from destruction of buildings, lifelines, and other constructions, to triggering other natural disasters, economic and political crises. That diversity of damage requires a hierarchy of preparedness measures, from public safety legislation and insurance to simulation alarms, to preparedness at home, and red alert. Different measures can be implemented on different timescales, from seconds to decades. They should be implemented in areas of different size, from selected sites to large regions; can be maintained for different time periods; and belong to different levels of jurisdiction, from local to international. Such measures might complement, supersede, or mutually exclude each other. For this reason, optimizing preparedness involves comparison of a large number of combinations of possible measures (Davis et al., 2007, 2010, 2012).

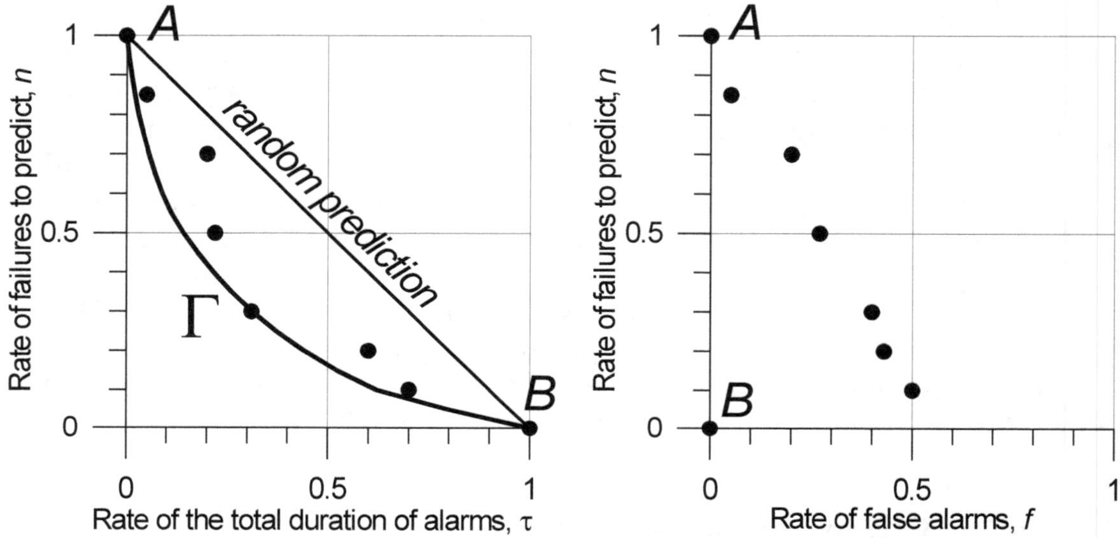

Figure 2. Error diagram. Each point shows the performance of a prediction method: the rate of failures to predict, *n*, the relative duration of alarms, τ, and the rate of false alarms, *f*. Different points correspond to different algorithms. The diagonal in the left plot corresponds to the random guess. Point *A* corresponds to the trivial "optimistic" strategy, when an alarm is never declared; point *B* marks the trivial "pessimistic" strategy, when an alarm takes place all the time; other points correspond to non-trivial predictions. Best combinations (*n*, τ) lie on the envelope of these points Γ. After Molchan (1997).

Disaster management has to take into account the cost/benefit ratio of possible preparedness measures. No single measure alone is sufficient. On the other hand, many efficient measures are inexpensive and do not require high accuracy of prediction. As is the case for all forms of disaster preparedness, including national defense, a prediction can be useful if its accuracy is known, even if it is not high.

Decision depends on specific circumstances in the area of alarm. At the same time, it also depends on the prediction quality, i.e. rate of failures-to-predict, *n*, rate of false alarms, *f*, and the fraction of time-space occupied by all alarms together, *τ*. These values are determined above. The choice of preparedness measures is never unique. Different measures may supersede or mutually exclude one another, leaving certain freedom of choice to a decision-maker (Keilis-Borok et al., 2004; Davis et al., 2007, 2010, 2012). The designer of a prediction algorithm has certain freedom to choose the trade-off between different characteristics of its accuracy (rate of failures to predict, duration of alarms, and rate of failures to predict) by varying adjustable elements of the algorithm.

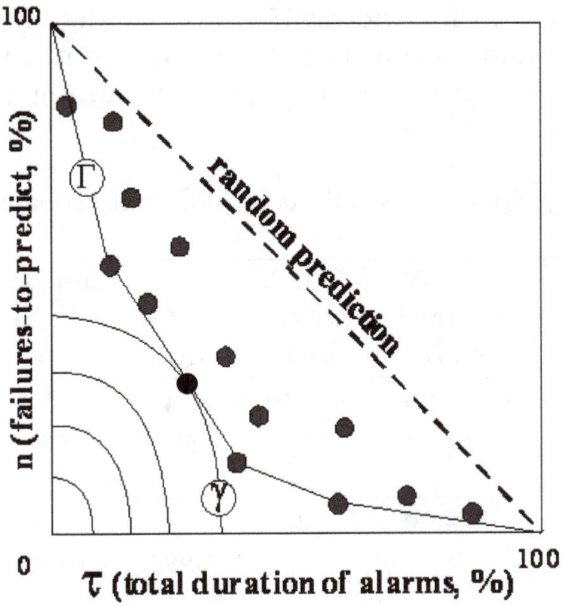

Figure 3. Joint optimization of prediction and preparedness. After Molchan (1997).

Accordingly, prediction and preparedness should be optimized jointly; *there is no "best" prediction per se* (Molchan, 1997, 2003; Molchan and Keilis-Borok, 2008). A framework for such optimization is shown in Fig. 3. Dots show points on an error diagram, Γ is their envelope. The contours show "loss curves" with constant value of prevented damage γ (see *Molchan,* 1997). Optimal strategy is the tangent point of contours Γ and γ. Thus disaster preparedness would be more flexible and efficient if prediction would be carried out in parallel with several versions of an algorithm. This has not yet been done.

Other software and know-how for assisting decision-makers in choosing the best combination is described by Kantorovich et al. (1974), Molchan (1991, 2003) and Keilis-Borok (2003). A hypothetical example is given in Fig. 4 (Davis et al., 2007). Imagine an earthquake alarm covering an area shown on the map – a part of Central California. The map shows the parts of the water supply systems in the area that are vulnerable to earthquakes. Table 2 shows the cost-efficiency of some preparedness measures (Keilis-Borok et al., 2004; Davis et al., 2007, 2010).We see that lowering the water level is justified only for a fragile reservoir and for the probability of false alarm 50% or less, but not 75%. Decision making might also require estimating distribution functions for different types of damage, casualties included (Keilis-Borok et al., 2004).

Table 2. Gain from preparedness actions for different probabilities of false alarms

	Action	DA ($1,000)	DP ($1,000)	Gain ($1,000) f = 10%	f = 50%	f = 75%
T	Lower water level in Fragile Reservoir	2,000	7,500	4,750	1,750	***-125***
T	Lower water level in Stout Reservoir	2,000	10	***-1,991***	***-1,995***	***-1,998***
T	Drain Reservoirs	16,000	7,510	***-9,240***	***-12,250***	***-14,120***

Gain G is calculated by formula $G = DP(1 - f) - DA$ where DP is damage prevented, DA – cost of action, f – probability of false alarm. T = Temporary: lasting for alarm period. Negative gain is shown in bold italic.

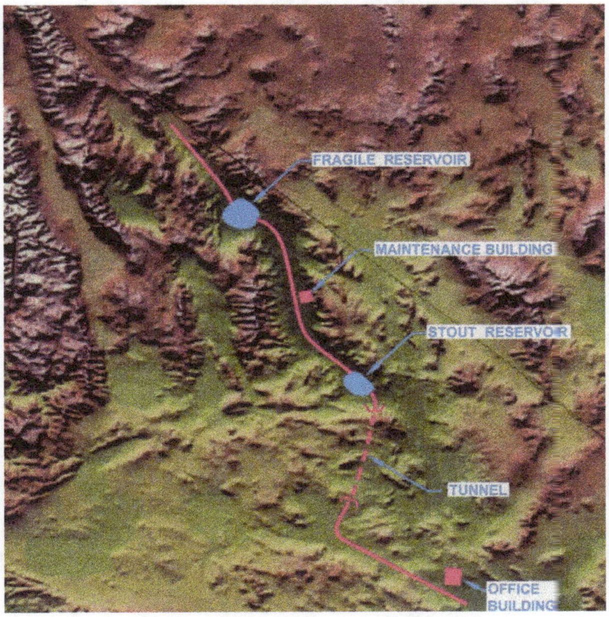

Figure 4. Schematic example: Vulnerable objects and hazards in the area of alarm.

CHAPTER 2

Earthquake Prediction

We understand earthquake prediction as prediction of time interval and geographic area where an individual future strong earthquake will occur. The strong earthquake is determined as an earthquake with magnitude $M \geq M_0$ where M_0 is a certain threshold. The magnitude range of the target earthquakes is sometimes limited by an upper boundary. Prediction is meaningful if it includes an estimated rate of false alarms. Taking into account that in the case of earthquakes the geographic area is predicted, Fig. 1 is transformed into Fig. 5.

Premonitory patterns signal destabilization of the earthquakes-prone lithosphere and thus an increase in the probability of a strong earthquake. Premonitory patterns do not necessarily contribute to causing a subsequent strong earthquake; both might be parallel manifestations of the same underlying process – the tectonic development of the Earth in multiple time-, space-, and energy- scales. For that reason premonitory patterns might emerge in a broad variety of observable fields reflecting lithosphere dynamics, and in different scales.

The process of earthquake prediction consists of consecutive, step-by-step, narrowing down of the time interval, space, and magnitude ranges where a strong earthquake has to be expected. Five stages of prediction are usually distinguished. The ***background stage*** provides maps with territorial distribution of maximum possible magnitude and recurrence time of destructive earthquakes having different magnitudes. The four subsequent stages, loosely divided, include the time-prediction; they differ in the characteristic time intervals covered by an alarm.

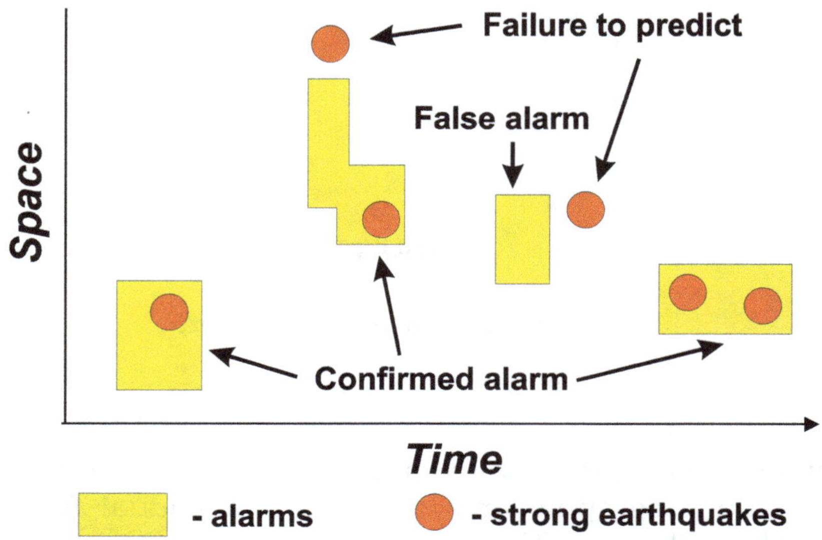

Figure 5. Possible outcomes of earthquake prediction.

These stages are as follows:

- ***long-term*** (10^1 years),
- ***intermediate-term*** (years),
- ***short-term*** (10^{-1} to 10^{-2} years), and
- ***immediate*** (10^{-3} years or less).

Such division into stages is dictated both by the characteristics of the process leading to a strong earthquake and by the needs of the earthquake preparedness in each particular region. The latter comprises an arsenal of safety measures for each stage of prediction, like in preparedness to a war.

2.1. An early example

The first premonitory seismicity pattern formally defined and featuring long-range correlations and worldwide similarity was the *"pattern Σ"*, introduced in 1964 by Keilis-Borok and Malinovskaya (1964). It comprises the premonitory increase of the total area of the ruptures in the earthquake sources in a medium magnitude range. Emergence of this pattern was captured by the function:

$$\Sigma(t/s,B) = \sum 10^{Bmi}$$

where m_i is the magnitude of ith earthquake; the sum $\sum$ is taken over all earthquakes that have occurred during the time interval (t-s, t) within the region considered; $B \approx 1$. With this value of B, the summands are coarsely proportional to the source area (with $B = 0$ and $B = 3/2$ this sum would correspond to the number of earthquakes and their total energy, respectively).

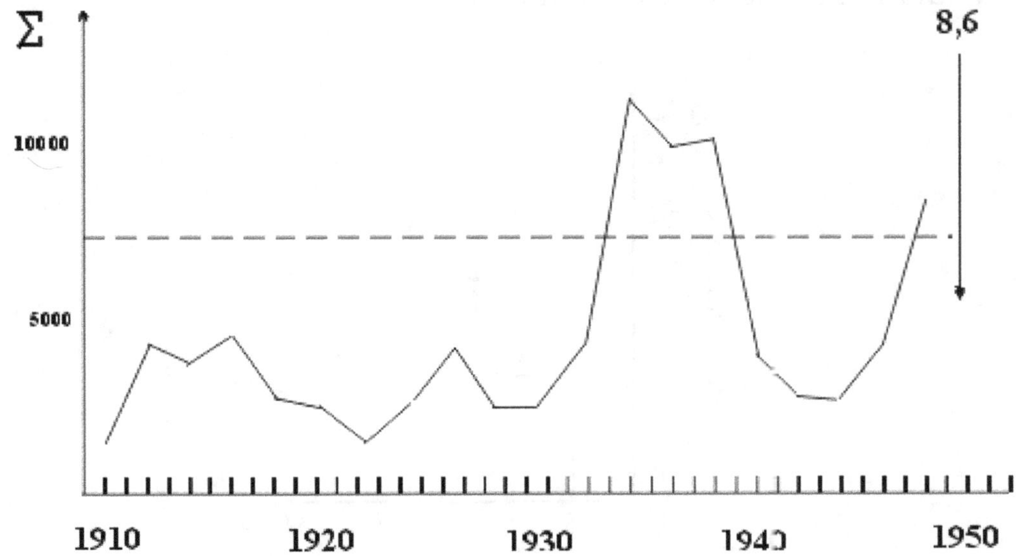

Figure 6. Illustration of the premonitory seismicity pattern Σ: rise of the function $\Sigma(t/s,B)$ before the Assam earthquake in India (1950, $M = 8.6$). The emergence of pattern Σ was captured by the condition $\Sigma(t) \geq C_\Sigma$, threshold C_Σ (shown by the horizontal line) was determined uniformly for all regions.

The change of the function $\Sigma(t)$ in periods preceding 20 strong earthquakes worldwide was investigated by Keilis-Borok and Malinovskaya (1964). It has been shown that the function $\Sigma(t/s,B)$ strongly increased one to ten years prior to each of the earthquakes considered, indicating *"a direct connection between strong earthquakes and the very large scale features of the development of the whole Earth's crust."* Fig. 6 shows an example for the catastrophic Assam earthquake in India, 1950, M=8.6. The emergence of pattern Σ was captured by the condition $\Sigma(t) > C_\Sigma$, with threshold C_Σ determined uniformly for all regions.

Pattern Σ was the first premonitory seismicity pattern that demonstrated the major features of patterns discovered later: long-range correlations and similarity. These features can be described as follows.

(i) *Long-range correlation*: The area of earthquake preparation can greatly exceed the source of the incipient earthquake. This is reflected in the large size of the areas that have to be used for calculation of the function Σ (Fig. 7).

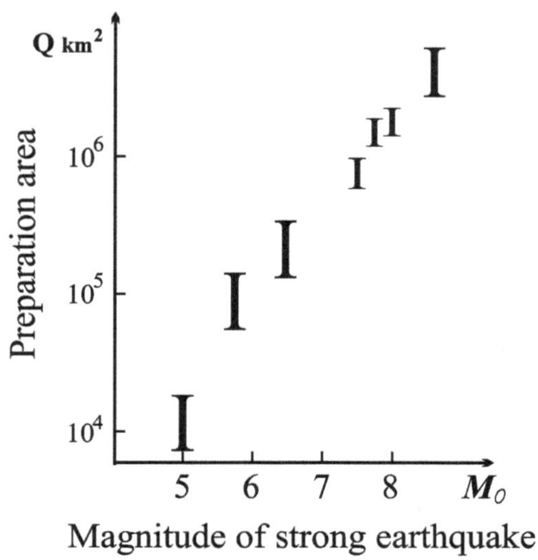

Figure 7. Long-range correlation in the formation of premonitory seismicity patterns. $Q(M_0)$ is the area, where pattern Σ was formed prior to a strong earthquake with $M \geq M_0$.

(ii) *Similarity*: Pattern Σ is self-adapting; it has a uniform definition for different magnitude thresholds M_0 of strong earthquakes targeted for prediction. Specifically,

— the area of preparation is a power-law function of M_0 (Fig. 7); and

— the threshold for identification of pattern Σ is normalized by M_0, $C_\Sigma = 0.5 \times 10^{BM0}$; this means that the area unlocked by medium magnitude earthquakes reached at least half of the area which will be unlocked by an incipient strong earthquake.

These features have been confirmed by subsequent studies and eventually evolved in the prediction paradigms.

Pattern Σ illustrates the consecutive stages in the search for prediction algorithms.

— It started with the hypothetical *premonitory phenomenon*: the rise of seismic activity, which is one of the basic characteristics of seismicity.

— That phenomenon was captured by a specific *precursor*: large area unlocked by earthquakes in a medium magnitude range. The same phenomenon is captured also by other precursors, for example by the number of earthquakes not weighted by magnitude.

— Finally, the prediction algorithm was formally defined by the function $\Sigma(t/s,B)$ and the threshold C_Σ.

2.2. Algorithms M8 and MSc ("the Mendocino Scenario")

The intermediate-term earthquake prediction algorithm M8 was designed by retroactive analysis of dynamics of seismic activity preceding the greatest, magnitude 8.0 or more, earthquakes worldwide (hence its name). Its prototype (Keilis-Borok and Kossobokov, 1984) and the original version (Keilis-Borok and Kossobokov, 1987) were tested retroactively at 143 points, of which 132 are recorded epicenters of earthquakes of magnitude 8.0 or greater from 1857-1983.

The algorithm is aimed at prediction of earthquakes with magnitude M_0 and higher from the range $M_0+ = [M_0, M_0+\Delta M]$ (where $\Delta M < 1$). Data used by the algorithm is a catalog of main shocks: $\{t_i, m_i, h_i, b_i(e)\}$, $i = 1, 2, \ldots$ Here t_i is the origin time, $t_i \leq t_{i+1}$; m_i is the magnitude, h_i is focal depth, and $b_i(e)$ is the number of aftershocks with magnitude M_{aft} or more during the first e days after a relevant main shock. Overlapping circles, with the diameter $D(M_0) = (exp(M_0-5.6)+1)^o$ in degrees of the Earth meridian (this is 384, 560, 854 and 1333 km for M_0 = 6.5, 7.0, 7.5, and 8, respectively), scan the seismic region under study. Only main shocks with $M_0 > m_i \geq M_{min}$ are considered where M_{min} is determined by the condition that the average annual number of main shocks with $m_i \geq M_{min}$ in the circle is equal to $\tilde{N}$. The main shocks are used to compute functions in the sliding time window $(t - s, t)$. The functions depict different measures of intensity in earthquake flow, its deviation from the long-term trend, and clustering of earthquakes. The functions include:

- $N(t) = N(t \mid M_{min}, s)$, the number of main shocks of magnitude M_{min} or larger in $(t - s, t)$;
- $L(t) = L(t \mid M_{min}, s, t_0)$, the deviation of $N(t)$ from the longer term trend, $L(t) = N(t) - N_{cum}(t - s)\,(t - t_0) / (t - t_0 - s)$, where $N_{cum}(t) = N(t \mid M_{min}, t - t_0)$ is the cumulative number of main shocks with $M \geq M_{min}$ from the beginning of the sequence t_0 to t;
- $Z(t) = Z(t \mid M_{min}, M_1, s, g, a, \beta)$, linear concentration of main shocks $\{i\}$ from the magnitude range (M_{min}, M_1), $M_1 = M_0 - g$ and interval $(t - s, t)$; the linear concentration is estimated as the ratio of the average source diameter l to the average distance r between sources, where source diameter of an earthquake with magnitude M is estimated by $10^{\beta(M - a)}$; and
- $B(t) = B(t \mid p, q, s_1, M_{aft}, e) = \max_{\{i\}}\{b_i(e)\}$, the maximum number of aftershocks (i.e., a measure of earthquake clustering); the sequence $\{i\}$ is considered in the trailing time window $(t - s_1, t)$ and in the magnitude range $(M_0 - p, M_0 - q)$.

Each of the functions N, L, and Z is calculated twice with M_{min} determined for $\tilde{N} = 10$ and $\tilde{N} = 20$. As a result, the earthquake sequence is given a robust averaged description by seven functions: N, L, Z (twice each), and B.

"Very large" values are identified for each function from the condition that they exceed Q percentiles (i.e., they exceed Q percent of the encountered values).

An alarm or a TIP, time of increased probability, is declared for 5 years when at least six out of seven functions, including B, become "very large"' within a narrow time window $(t - u, t)$. To stabilize prediction, this criterion is checked at two consecutive moments, t and $t + 0.5$ years. In the course of a forward application, the alarm can extend beyond, or terminate in less than 5 years when updating causes changes in the magnitude cutoffs and/or the percentiles.

The following standard values of parameters indicated above are prefixed in the algorithm M8: $s = 6$ years, $s_1 = 1$ year, $g = 0.5$, $p = 2$, $q = 0.2$, $u = 3$ years, and $Q = 75\%$ for B and 90% for the other six functions. Usually, the average diameter l of the source is estimated by $(n)^{-1} \Sigma_{\{i\}} 10^{\beta(Mi - a)}$, where n is the number of main shocks in $\{i\}$, β =0.46 to represent the linear dimension of source, and a = 0 (which does not restrict generality), and the average distance r between sources is set proportional to $n^{-1/3}$. The performance of the algorithm can be improved by estimating the linear concentration of main shocks more accurately.

Running averages are defined in a robust way, so that a reasonable variation of parameters does not affect predictions. At the same time, the discrete character of seismic data and strict usage of the prefixed thresholds result in a certain discreteness of the alarms.

It is worth mentioning that, qualitatively, algorithm M8 uses a rather traditional description of a dynamic system, adding dimensionless concentration (Z) and a characteristic measure of clustering (B) to the phase space of rate (N) and rate differential (L). The algorithm recognizes a *criterion* defined by extreme values of the phase space coordinates, as the vicinity of a system singularity. When a trajectory enters the criterion, the probability of an extreme event increases to a level sufficient to provide it effectively. The choice of the M8 criterion determines a specific intermediate-term rise, an inverse cascade (Gabrielov et al, 2000a), of seismic activity at the middle-range distance.

The second approximation prediction method "the Mendocino Scenario" (MSc) was designed by retrospective analysis of the detailed regional seismic catalog prior to the Eureka earthquake (1980, M = 7.2) near Cape Mendocino in California (Kossobokov et al., 1990). Given a TIP diagnosed for a certain territory ***U*** at time ***T***, the algorithm is designed to find a smaller area ***V*** within ***U***, where the predicted earthquake can be expected. To execute the algorithm, one needs a reasonably complete catalog of earthquakes with magnitudes $M \geq M_0 - 4$, which is lower than the minimum threshold usually used by M8. When this condition does not hold, we assume that the dynamics of earthquakes available in the database inherits behavior from lower levels of the seismic hierarchy. The detection of the MSc criteria in such a case is more difficult and might result in additional failures to predict.

The essence of MSc can be summarized as follows. Territory ***U*** is coarse-grained into small squares of size $s \times s$. Let's identify the squares by their sequence numbers j = 1, 2, ..., J. Within each square j, the number of earthquakes $n_j(k)$, aftershocks included, is calculated for consecutive, short, time windows u, months long, starting from time $t_0 = T - 6$ years onward, to include earthquakes that contributed to the TIP's diagnosis; k is the sequence number of a time window. In this way, the time-space considered is divided into small boxes (j, k) of size $(s \times s \times u)$. *"Quiet" boxes* j are defined by the condition that $n_j(k)$ is below the Q percentile of n_j. The clusters of q or more quiet boxes connected in space or in time are identified. Area $\boldsymbol{V}$ is the territorial projection of these clusters.

The standard values of parameters adjusted for the 1980 Eureka earthquake are as follows: u = 2 months, Q = 10%, q = 4, and s = $3D/16$, where D is the diameter of the circles used in the M8 algorithm.

Table 3. Scoring of M8 and M8 & MSc predictions, 1992-2010

Algorithm	Total number of target earthquakes	Number of predicted earthquakes	Space-time volume of alarm
M8	17	12	29%
M8 & MSc	17	8	15%

Qualitatively, the MSc algorithm outlines such an area of the territory of an alarm where the activity, from the beginning of the seismic inverse cascade recognized by the M8 algorithm in declaration of the alarm, is continuously high, but is infrequently interrupted for a short time. Such interruption must have a sufficient temporal and/or spatial span. The phenomenon, which is used in the MSc algorithm, might reflect the second (possibly, shorter term and, definitely, narrow-range) stage of the premonitory rise of seismic activity near the incipient source of a main shock. The anomalous quiescence used in the definition of the *precursory intermittent pattern* in the dynamics of the seismic region should not be mixed with a prolonged state of *"seismic quiescence"*, advocated by (Wyss and Habermann, 1988).

Scoring. Thus far, the algorithms have had the most success in predicting future earthquakes in the magnitude range 8-8.5 (Table 3). Statistical significance of these predictions exceeds 99%.

2.3. Other Premonitory Phenomena and Algorithm RTP

One premonitory phenomenon depicting a change in the spatial distribution of events prior to strong earthquakes was found in Southern California and then in the Lesser Antilles and on the Dead Sea Transform. Named ***"seismic reversal"*** ***(SR)***, it characterizes a period of opposite seismic activity in areas complementary to the bulk of long-term territorial distribution of low-magnitude earthquakes (Keilis-Borok et al., 1994; Shebalin et al., 1996; Shebalin and Keilis-Borok, 1999). The phenomenon happens several months prior to a strong earthquake in the area surrounding its future source. The diameter of the prediction area is of the order of 100 km.

The other two phenomena, named ROC and Accord, were designed in inspiration of evident patterns in a lattice-type "Colliding Cascades" model of interacting elements (Gabrielov et al., 2000a, 2000b). Pattern ROC describes an increase in the earthquake correlation range a few days before a strong earthquake. It was found in the Lesser Antilles (Shebalin et al., 2000), where the corresponding prediction algorithm demonstrated extreme efficiency unusual even for retrospective analysis. The pattern Accord depicts spreading of seismic activity over a fault network and is defined as a simultaneous rise of seismic activity in a sufficiently large number of neighboring fault zones. In Southern California, scaled to target magnitude 7.5, the pattern Accord emerges within a few years in the approach of each of the three largest earthquakes: Kern County, 1952; Landers, 1992; and Hector Mine, 1994, and at no other time.

The only way to answer the question whether the phenomena are really premonitory is to define prediction algorithms on their basis and to test their performance in advance prediction. All three prediction methods are fully reproducible and presented as hypotheses for testing in advance prediction. Below we describe each of them in more detail.

"Reverse Tracing of Precursors" (RTP) algorithm (Shebalin et al., 2004, 2006; Keilis-Borok et al., 2004a). The algorithm is aimed at predictions about 9 months in advance, much shorter than by the M8 algorithm. This algorithm, as its name suggests, traces precursors in the reverse order of their formation. First it identifies "candidates" for short-term precursors. These are long, quickly formed chains of earthquakes in the background seismicity. Such chains reflect an increase in the earthquake correlation range (Fig. 8). Next, each chain is examined to determine whether there had been any preceding intermediate-term precursors in its vicinity within the previous five years. If so, the chain triggers an alarm.

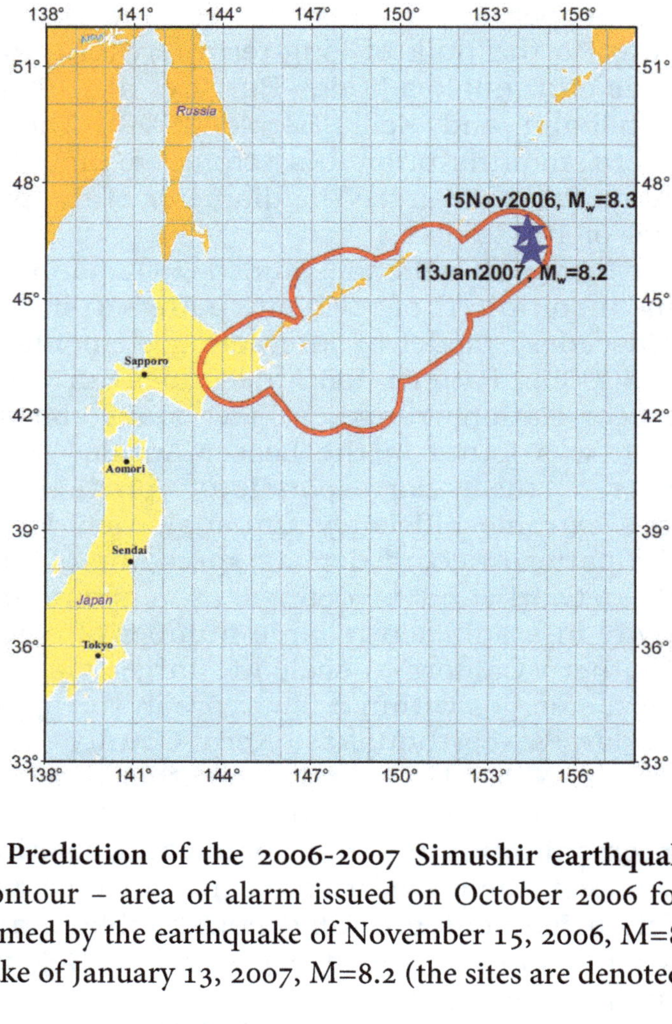

Figure 8. RTP Prediction of the 2006-2007 Simushir earthquakes in Kuril Islands. Red contour - area of alarm issued on October 2006 for 9 months. This was confirmed by the earthquake of November 15, 2006, M=8.3, followed by the earthquake of January 13, 2007, M=8.2 (the sites are denoted by stars).

The RTP algorithm has been tested by predicting future earthquakes in five regions of the world (California and adjacent regions; Central and Northern Italy with adjacent regions; Eastern Mediterranean; Northern Pacific, Japan and adjacent regions). Out of 19 alarms, five were correct and fourteen false, two of the latter being near misses occurring close to alarm areas. Five out of seven target earthquakes have been predicted (captured by alarms) and two missed. The data are still insufficient for rigorous estimation of statistical significance.

2.4 Four Paradigms in Earthquake Prediction

In the wake of the developments in the earthquake prediction, the following four paradigms have been established at the crossroad between exploratory data analysis, statistical physics, and the dynamics of fault networks (Keilis-Borok, 1994, 1996).

I. *Basic types of premonitory phenomena* comprising the variation in relevant observable fields.

II. *Long-range correlations in fault system dynamics.* Premonitory phenomena are formed not only in the vicinity of the incipient source, but also within a much wider area.

III. *Partial similarity of premonitory phenomena* in the diverse conditions, from fracturing in laboratory samples to major earthquakes worldwide, and possibly even to starquakes.

IV. *The dual nature of premonitory phenomena.* Some of them are "universal," common for complex non-linear systems of different origin; others are Earth-specific.

The paradigms discussed here have been found in the quest for premonitory seismicity patterns in the observed and modeled seismicity. There are compelling reasons to apply them also to premonitory phenomena in other relevant fields.

First Paradigm: Basic types of premonitory phenomena. The approach of a strong earthquake is indicated by the following changes in the basic characteristics of seismicity:

(i) Rise of seismic activity;

(ii) Rise of earthquake clustering in space and time;

(iii) Rise of the earthquake correlation range;

(iv) Transformation of magnitude distribution (Fig. 9);

(v) Rise of irregularity in space and time;

(vi) Reversal of territorial distribution of seismicity;

(vii) Rise of correlation between different components (decrease of dimensionality);

(viii) Rise of response to excitation.

Other relevant processes exhibit premonitory phenomena of the same types.

Patterns of the first two types, (i) and (ii), were found first in observations (Gabrielov et al., 1986; Keilis-Borok, 1990; Keilis-Borok and Shebalin, 1999) and then in models (e.g., Gabrielov et al., 2000a; Shnirman and Blanter, 2003; Soloviev and Ismail-Zadeh, 2003); patterns of the next three types, (iii) – (v), were found in the reverse order, first in models (Gabrielov et al., 2000b; Newman et al., 1995; Sornette, 2000; Shnirman and Blanter, 2003) and then in observations (Sammis et al., 1996; Kossobokov and Shebalin, 2003); reversal of territorial distribution of seismicity (vi) was found in observations and not explored yet through modeling (Kossobokov and Shebalin, 2003); the last two phenomena remain purely hypothetical so far.

Validation. Patterns of the first two types, rise of intensity and clustering, have been validated by statistically significant predictions of real earthquakes (Molchan et al., 1990; Kossobokov and Shebalin, 2003); other patterns are undergoing different stages of testing.

Reminiscence of theoretical physics. The premonitory phenomena listed above bear a resemblance to the asymptotic behavior of a nonlinear system near the point of phase transition of the second kind. However, the problem under consideration is unusual for statistical physics: the state considered is not the equilibrium state, but the *growing disequilibrium* culminated by a critical transition.

Seismicity patterns. Premonitory phenomena of each type are depicted by different seismicity patterns. Intermediate-term patterns, with characteristic duration of alarms equaling years, are the ones systematically explored. A few examples follow.

— *Measures of seismic activity*: total area of ruptures in the earthquake sources (Sect. 2.1); accumulated strain release (Bufe and Varnes, 1993; Varnes, 1989; Bowman et al., 1998); the number of earthquakes in a certain magnitude range (Wyss and Habermann, 1988; Knopoff et al., 1996; Kossobokov and Shebalin, 2003); the time period when a given number of earthquakes occurs (lower time obviously indicates higher activity) (Sykes and Jaumé, 1990), etc.

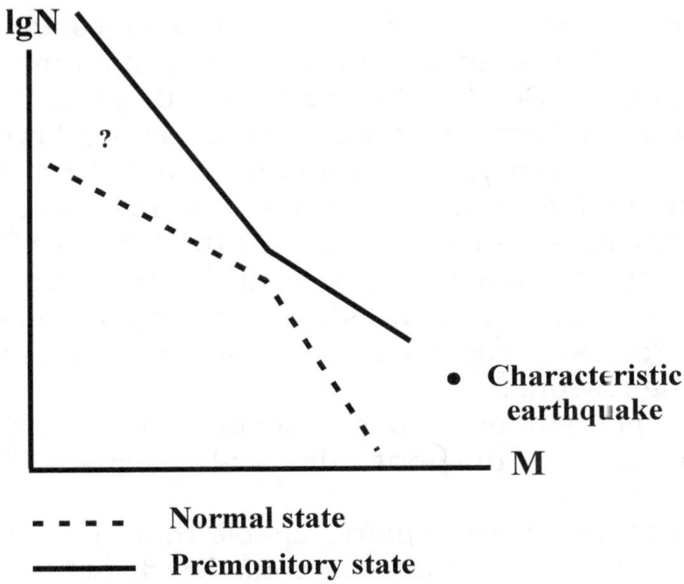

Figure 9. Scheme of premonitory transformation of the Gutenberg–Richter relation *N*(*M*). *N* is the number of earthquakes of magnitude *M* or larger. Dashed and solid lines correspond to time intervals far from a strong earthquake and close to it, respectively.

— *Measures of earthquake clustering*: the number of aftershocks closely following a medium magnitude main shock ("bursts of aftershocks," see Molchan et al., 1990; Kossobokov and Shebalin, 2003); swarms of the main shocks of medium magnitude (Caputo et al., 1977; Keilis-Borok et al., 1980; Keilis-Borok et al., 1982); swarms of relatively small earthquakes (Caputo et al., 1983) etc.

— *Measures of earthquake correlation range*: the distance between nearly simultaneous earthquakes and the number of faults with nearly simultaneous rise of activity (patterns "*ROC*" and "*Accord*", Shebalin et al., 2000; Keilis-Borok et al., 2002; Zaliapin et al., 2002); the distribution of link lengths in a single link cluster connecting earthquakes occurred in a given spatiotemporal region (Zoller et al., 2001).

— *The measures of irregularity*: the variation in magnitudes or strain release (Newman et al., 1995; Sornette and Sammis, 1995; Sornette, 2000).

— *The measures of premonitory change of magnitude distribution* (Gutenberg-Richter relation). That transformation is schematically illustrated in Fig. 9. One of its measures is the slope of the magnitude distribution in a relatively high magnitude range (pattern "Upward Bend;" Shnirman and Blanter, 2003); another measure is the difference between its slopes for lower and higher magnitudes ("pattern γ," Rotwain et al., 1997). The question mark in the figure indicates that "pattern γ," the reversal of curvature of the distribution, is yet less tested than the "Upward Bend."

Why different measures are used for the same premonitory phenomenon? These measures are certainly correlated, even by definition. Several of them are used instead of an "optimal" one for the following reasons.

— A premonitory phenomenon may have different manifestations on different timescales, spatial scales, and magnitude ranges.

— A set of measures is more reliable than a single one, due to the complexity of the processes considered, and to unavoidable noise.

— In lieu of an adequate theory, premonitory seismicity patterns have been found by heuristic analysis, and more compact definitions might just be overlooked so far.

Second Paradigm: Long-range correlations. The generation of an earthquake is not localized around its future source. A flow of earthquakes is generated by a fault network, rather than each earthquake being generated by a segment of a single fault. Accordingly, the signals of an approaching earthquake come not from a narrow vicinity of the source, but from a much wider area.

Size of areas where premonitory phenomena are formed. Let M and $L(M)$ be the earthquake magnitude and the characteristic length of its source, respectively. In the intermediate-term stage of prediction (on a timescale of years) that size may reach $10L(M)$; it might be reduced to $3\ L(M) - L(M)$ in a second approximation (Kossobokov and Shebalin, 2003). In the long-term stage, on the timescale of tens of years, that size reaches about $100L(M)$. For example, according to Press and Allen (1995) the Parkfield (California) earthquake, with *M* about 6 and *L*(*M*) ≈ 10 km "... *is not likely to occur until activity picks up in the Great Basin or the Gulf of California, about 800 km away.*"

Historical perspective. An early, and probably the first estimation of the area where premonitory patterns are formed, was obtained for pattern Σ (Fig. 7). It is noteworthy that Charles Richter, who was generally skeptical about the feasibility of earthquake prediction, made an exception for that pattern, specifically because it was based on long-range correlations. He wrote (Richter, 1964): "... *It is important that (the authors) confirm the necessity of considering a very extensive region including the center of the approaching event. It is very rarely true that the major event is preceded by increasing activity in its immediate vicinity.*" Table 4 shows similar estimations for other intermediate-term patterns.

At the same time, long-range correlations have been often regarded as counterintuitive in earthquake prediction research on the ground that redistribution of stress and strain after an earthquake in simple elastic models would be confined to the vicinity of its source ("Saint Venant principle"). Sometimes that prompted the objection: "*the earthquakes cannot trigger each other at such distances.*" The answer is that the earthquakes involved in long-range correlation do not trigger each other, but reflect the underlying large-scale dynamics of the lithosphere. More specific explanations follow.

Table 4. Estimations of the size of the area where premonitory seismicity patterns emerge

Measure	Year	$Q(L)^b$	Reference
Area of fault breaks	1964	~ 10L	Keilis-Borok and Malinovskaya (1964), Kossobokov and Shebalin (2003)
Distant aftershocks	1975	~ 10L	Prozorov (1975)
Earthquake swarms	1977	~ 5L -10L	Caputo et al. (1977)

Bursts of aftershocks, area of fault breaks, swarms	1980	~ 5*L* -10*L*	Keilis-Borok et al. (1980), Kossobokov and Shebalin (2003)
Algorithm CN[a]	1983	~ 5*L* -10*L*	Keilis-Borok and Rotwain (1990), Kossobokov and Shebalin (2003)
Algorithm M8[a]	1985	~ 5*L* -10*L*	Keilis-Borok and Kossobokov (1990), Kossobokov and Shebalin (2003)
Benioff strain release	1989	~ 5*L*	Varnes (1989), Bowman et al. (1998), Jaumé and Sykes (1999)
Algorithm SSE	1992	~ 5*L*	Levshina and Vorobieva (1992)
Number of earthquakes	1995	~ 100*L*	Press and Allen (1995)
Number of earthquakes	1996	~ 5*L*	Knopoff et al. (1996)
Near-simultaneous pairs of earthquakes	2000	~ 3*L*	Shebalin et al. (2000)
Correlation length via single link cluster	2001	~ 5*L*	Zoller et al. (2001)
Simultaneous activation of fault branches	2002	~ 10*L*	Zaliapin et al. (2002)

[a] Reference is given not to the original publication, but to the latest comprehensive reviews.

[b] *L* is the linear dimension of the approaching strong earthquake. Premonitory phenomena have been observed on a timescale of years with one exception, tens of years in the 1995 entry.

Mechanisms of long-range correlations. Correlation in earthquake occurrence at long distances greatly exceeding earthquake source dimensions is the prominent features of seismicity dynamics; it is not confined to earthquakes precursors. Among the manifestations of that correlation are the following phenomena: simultaneous changes of seismic activity within large regions (Romanowicz, 1993), migration of earthquakes along fault zones (Mogi, 1968; Vil'kovich and Shnirman, 1983; Ma et al., 1990; Kuznetsov and Keilis-Borok, 1997), and alternate rise of seismicity in distant areas (Press and Allen, 1995) and even in distant tectonic plates (Romanowicz, 1993). Global correlations have also been found between major earthquakes and other geophysical phenomena, such as Chandler wobble, variations of magnetic field, and the velocity of Earth's rotation (Press and Briggs, 1975; Keilis-Borok and Press, 1980). Several mechanisms (not mutually exclusive) have been suggested to explain the long-range correlations. They may be divided into two groups:

(i) Some explanations attribute long-range correlations to a large-scale process controlling stress and strength in the lithosphere. Such mechanisms act under different circumstances, separately or jointly. Being rather common, they make long-range correlations inevitable. Among such processes are the following:

- Microrotation of tectonic plates (Press and Allen, 1995) and crustal blocks (Soloviev and Ismail-Zadeh, 2003; Vorobieva and Soloviev, 2005); microfluctuations in the direction of mantle currents (Soloviev and Ismail-Zadeh, 2003). Each of them creates redistribution of normal and tangential stress and, consequently, redistribution of strength through a large part of the fault network.
- Migration of pore fluids in fault systems (Barenblatt et al., 1983) affects the lithosphere strength in the following ways: lubrication; stress corrosion and destabilization waves; and redistribution of hydrostatic pressure between the solid and fluid components of the fault zone.
- Hydrodynamic waves in the upper mantle (Pollitz et al., 1998) that propagate through thousands of kilometers during the decades and may trigger strong earthquakes connecting seismicity across the globe.
- Activity of creep fractures in the ductile part of the lithosphere. Deformation in the ductile part increases the

stress in the brittle part thus triggering earthquakes (Aki, 1996).

- Inelasticity and inhomogeneity of the lithosphere (Barenblatt, 1996). Due to either of the mechanisms, the redistribution of stress after fracture extends to much greater distances than in a homogeneous elastic media.

(ii) In another approach, the lithosphere is regarded as a non-linear chaotic system; then the long-range correlations are again inevitable, as a general feature of such systems in a near-critical state (Sornette and Sammis, 1995; Bowman et al., 1998; Turcotte et al., 2000).

Third Paradigm: Similarity. Premonitory phenomena are similar (identical after normalization) in extremely diverse environments and in a broad energy range. The similarity is not unlimited however and regional variations of premonitory phenomena do emerge.

Normalization. Earthquake sequences used in a prediction algorithm are normalized to ensure that the algorithm is self-adapting, i.e., it can be applied without re-adaptation in regions with different seismic regimes.

The area to which the sequence belongs to is normalized by $L(M)$ as described in Table 1.

The magnitude range is re-adapted by changing the minimum magnitude considered (e.g., Keilis-Borok and Kossobokov, 1990; Keilis-Borok and Rotwain, 1990; Kossobokov and Shebalin, 2003).

The time scale in most prediction algorithms does not depend on M; according to the Gutenberg–Richter relation, earthquakes of smaller magnitudes occur more frequently. This is not a contradiction, because the Gutenberg–Richter relation refers to a given region — the same for all magnitudes — whereas prediction is made for an area of a size proportional to $L(M)$.

This difference has a consequence which is not always recognized. Let $T_r(M)$ and $T_a(M)$ be the average return times of an earthquake of magnitude M in the whole region, and in a smaller area of linear size $L(M)$, respectively. According to the well known relations $T_r(M) \sim 10^{bM}$, and $L(M) \sim 10^{cM}$, and $T_a(M) \sim 10^{(b - cv)M}$. Here b is the slope of the Gutenberg–Richter relation, c determines the connection between the magnitude and the source dimension, and v is the fractal dimension of the cloud of epicenters. The existing estimations of parameters b (about 1), c (between 0.5 and 1), and v (between 1.2 and 2) do not contradict the hypothesis that the expression $(b - cv)$ is close to 0; accordingly, *earthquakes of different magnitudes might have about the same recurrence time in their own cells.*

Applications. Prediction algorithms, thus normalized, retain their predictive power in many cases: microfractures in laboratory samples; induced seismicity; earthquakes in subduction zones, major strike-slip fault zones, rift zones and platforms (Gorshkov et al., 1997; Keilis-Borok and Shebalin, 1999; Kossobokov and Shebalin, 2003). The corresponding seismic energy release ranges from a few ergs to 10^{25} ergs. However, performance of prediction algorithms does vary from case to case.

Frontiers of similarity, the neutron star (Kossobokov et al., 2000a). An opportunity to explore, albeit qualitatively, the frontiers of similarity is provided by registration of the 111 flashes of energy radiated from the neutron star with celestial coordinates 1806-20 in the frequency band of soft □-rays. These flashes have been probably originated by "starquakes," i.e. fractures in the crust of the neutron star. Environments of starquakes and earthquakes are summarized in Table 5.

Table 5. Environments of starquakes and earthquakes

Characteristics	Neutron star 1806-20	Earth
Composition of the crust	Lattice of heavy nuclei	Grains of rock
Radius	10 km	6371 km
Thickness of the crust	1 km	About 33 km
Density	10^{14} g/m^3	5.5□10g/m^3
Energy release	Up to 10^{46} erg	Up to 10^{25} erg
Driving forces	Magnetic field	Convection

Fig. 10 compares the emergence of premonitory patterns before a major starquake and the strong earthquake in the Aquaba gulf, 11 Dec. 1995, M = 7.3. The functions, capturing these patterns, are taken from the earthquake prediction algorithms (Kossobokov and Shebalin, 2003): Functions Σ, N, and Z capture the intensity of earthquake flow, L captures its deviation from the long-term trend, B is the measure of earthquakes clustering. The patterns work for Earth and the star, their fantastic difference notwithstanding. Only one parameter, the timescale for starquakes, had to be readjusted a posteriori.

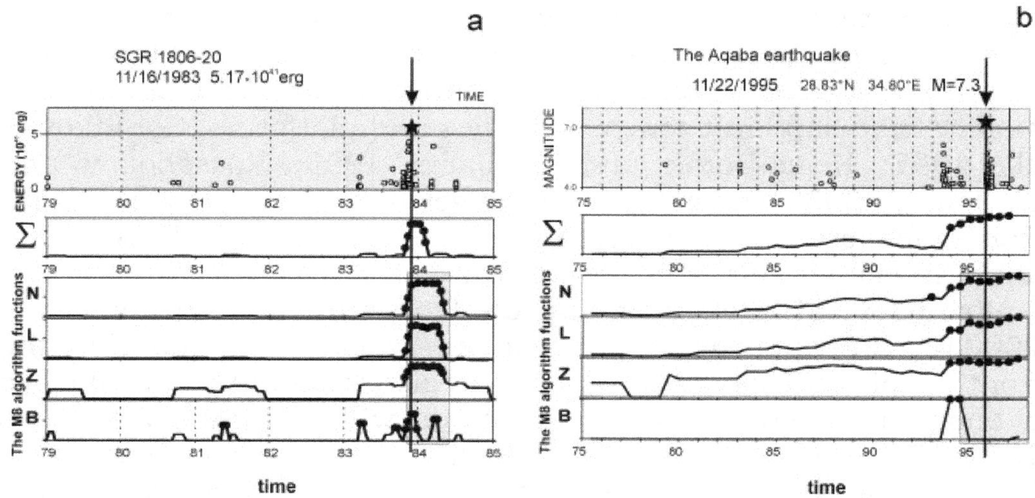

Figure 10 Frontier of similarity of premonitory seismicity patterns. (a) Starquake registered on 16 Nov. 1983, a coarse equivalent of its magnitude is about 20. (b) Aqaba earthquake, 22 Nov. 1995, $M = 7.3$. The panels show the sequence of starquakes (*left*) and earthquakes (*right*). *Stars* indicate major events that are targeted for prediction. Other panels show the functions capturing premonitory seismicity patterns defined in (Keilis-Borok and Kossobokov, 1990; Keilis-Borok and Rotwain, 1990; Kossobokov and Shebalin, 2003). Dots correspond to the functionals, depicting different characteristics of seismicity: B is the measure of clustering; other functions are different measures of seismic activity. Dots indicate the emergence of a premonitory pattern. After Kossobokov et al. (2000a).

Limitations. The performance of some prediction algorithms still does vary from region to region (see Ghill, 1994; Keilis-Borok and Shebalin, 1999; Molchan, 2003). It is not yet clear whether this is due to imperfect normalization or to Earth-specific limitations on similarity itself.

Fourth Paradigm: Dual nature of premonitory phenomena. Some premonitory phenomena are "universal," common for hierarchical complex nonlinear systems of different origins; others are specific to the geometry of the fault networks or to a certain physical mechanism controlling the strength-stress field in the lithosphere.

A. "Universal" premonitory phenomena.

The meaning of "universality." The known premonitory seismicity patterns prove to be universal in the sense that one can reproduce them in the lattice models specific not only to Earth. Obviously, seismicity and other fields relevant to earthquake prediction are Earth specific. At the same time, these fields are subject to more general laws of nature. If it is correct, then under the same clause the premonitory phenomena connected with strong earthquakes would reflect the general laws of self-organization of complex systems. The lattice models are developed in statistical physics and nonlinear dynamics (e.g., Shnirman and Blanter, 2003).

An illustration is given in Figs. 10 – 12 (after Zaliapin et al., 2003b). The model reproduces the first four types of premonitory patterns listed above in the first paradigm. So far, this is the widest set of patterns reproduced in a model. This was done with a model of "colliding cascades" introduced by Gabrielov et al. (2000) and Zaliapin et al. (2003b).

The model describes the generation of critical phenomena by colliding cascades of loading and failures. Its major features are the following:

— The model has a ternary hierarchical structure (Fig. 11a).

— The load is applied at the top of the hierarchy and is transferred downward as a direct cascade.

— Failures are initiated at the lowest level of hierarchy and propagate upward as an inverse cascade.

— The cascades are interacting (Fig. 11b). Loading triggers failures and the failures redistribute and release the load.

Eventually the failed elements "heal," thus ensuring permanent functioning of the system.

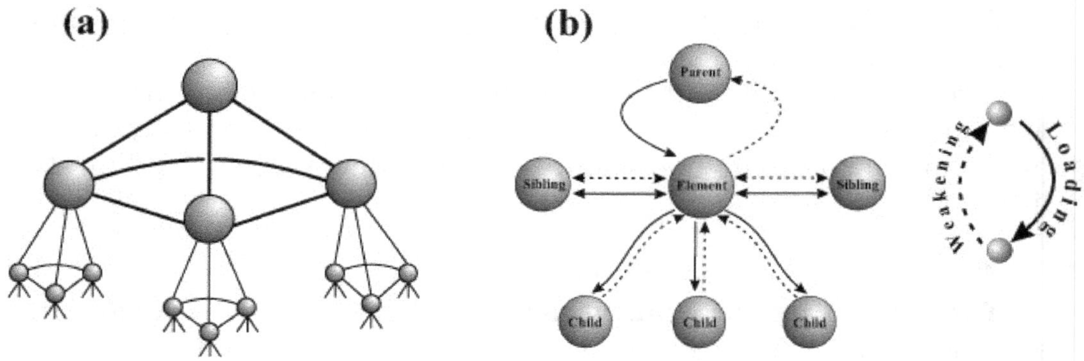

Figure 11. The structure of the colliding cascade model with branching number 3. (a) Three highest levels of the hierarchy, (b) interaction with the nearest neighbors.

In application to seismicity, the hierarchy imitates the structure of the lithosphere, where loading imitates the impact of tectonic forces, and failures imitate earthquakes; a critical phenomenon, "a major earthquake," is a failure of the top element.

Prediction of modeled seismicity. An example of an earthquake sequence generated by a colliding cascade model is shown in Fig. 12. Note that the model generates earthquakes of discrete magnitude. Prediction is targeted at the strongest synthetic earthquakes possible in the model.

Heuristic constraints. The modeled seismicity fits basic heuristic constraints, exhibiting, albeit coarsely, the major features of observed seismicity: the seismic cycle; intermittence of the seismic regime; the Gutenberg–Richter relation; aftershocks; long-range correlation; and the premonitory seismicity patterns indicated below.

Premonitory seismicity patterns. The model also reproduces the observed premonitory seismicity patterns of the first four types listed above:

(i) rise of seismic activity;
(ii) rise of clustering;
(iii) rise of correlation range;
(iv) transformation of the Gutenberg–Richter relation.

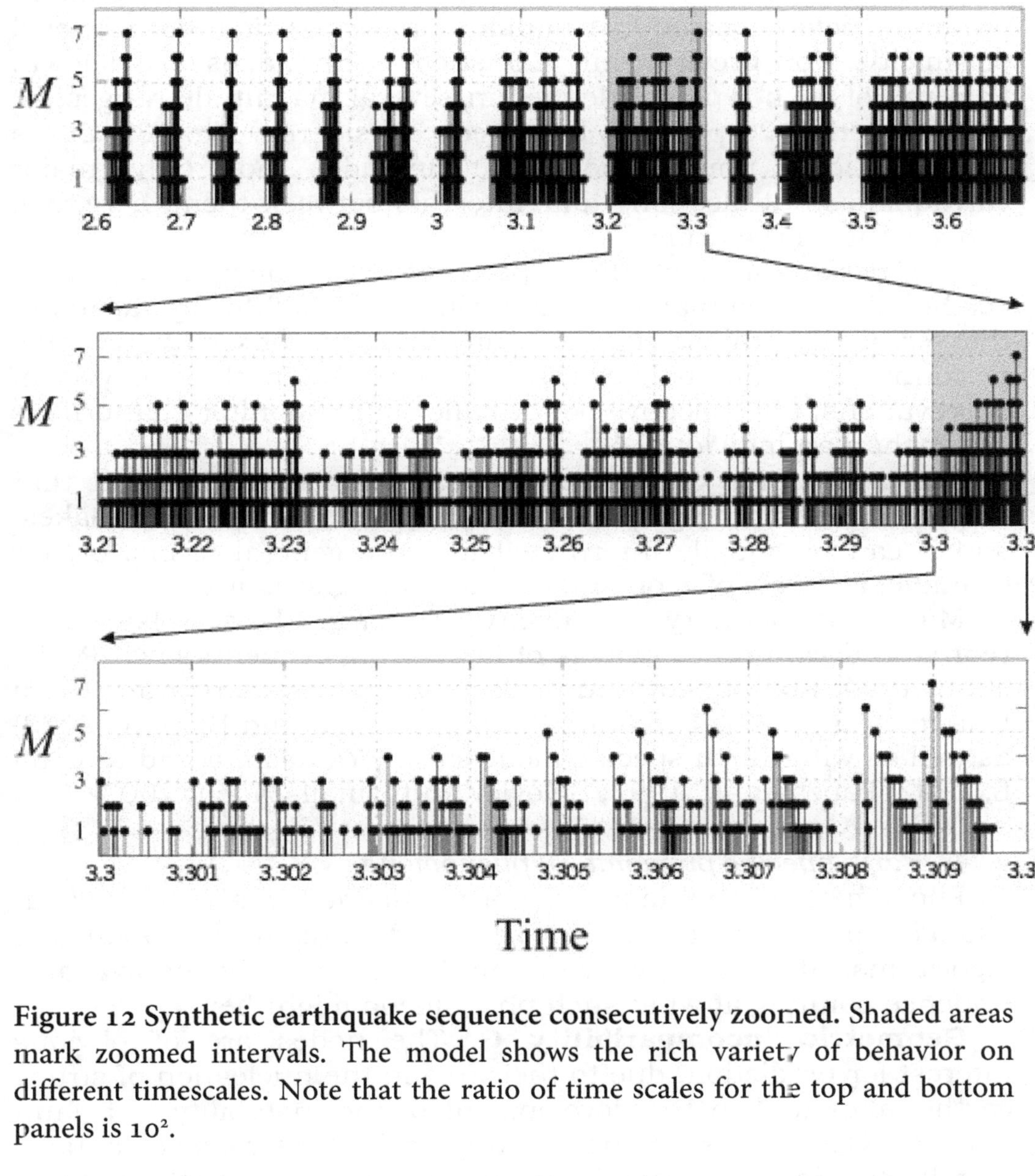

Figure 12 Synthetic earthquake sequence consecutively zoomed. Shaded areas mark zoomed intervals. The model shows the rich variety of behavior on different timescales. Note that the ratio of time scales for the top and bottom panels is 10^2.

Moreover, the premonitory rise of correlation range was found first in the colliding cascade model (Gabrielov et al., 2000a, 2000b) and then in the observed seismicity of the Lesser Antilles (Shebalin et al., 2000) and Southern California (Zaliapin et al., 2002). So far, other previously suggested patterns have not been explored with that model.

Performance. Fig. 13a shows the emergence of premonitory patterns before major earthquakes ($M = 7$) in the modeled seismicity. For brevity, only 10 such earthquakes are shown. Patterns of each type are defined in several magnitude ranges, so that altogether 23 patterns have been considered. Fig. 13b shows a continuous time period that includes not only major earthquakes but also long time intervals between them; here, one can see the false alarms.

The performance of these patterns in predictions of major earthquakes is estimated by Zaliapin et al. (2003b) by means of the error diagrams. In the analysis of observations this performance would be quite satisfactory. As in the analysis of observations, the collective performance of premonitory patterns is better than an individual one.

The universality of patterns is demonstrated by the fact that the model has nothing specific to only generation of earthquakes, and it can be equally interpreted in more general terms, or, for example, in terms of economic or engineering systems.

Many premonitory seismicity patterns have also been reproduced with other models of interacting elements (e.g., Pepke et al., 1994; Kossobokov and Carlson, 1995; Newman et al., 1995; Huang et al., 1998; Turcotte, 1999; Shnirman and Blanter, 2003) as well as with Earth specific models (e.g., Yamashita and Knopoff L, 1992; Panza et al., 1997; Soloviev and Ismail-Zadeh, 2003).

B. Earth-specific premonitory phenomena.

Phenomena of this kind are not yet defined in a way clear-cut enough for them to be incorporated directly in prediction algorithms. At present, we can only broadly outline the available evidence pointing at what such phenomena might be:

Geometric incompatibility *G*. The nodes are of obvious interest for prediction, due to their role in the nucleation of strong earthquakes and in the development of the instability of a fault network. Generally speaking, the approach of a strong earthquake is reflected in the strength-stress field, which directly controls seismicity.

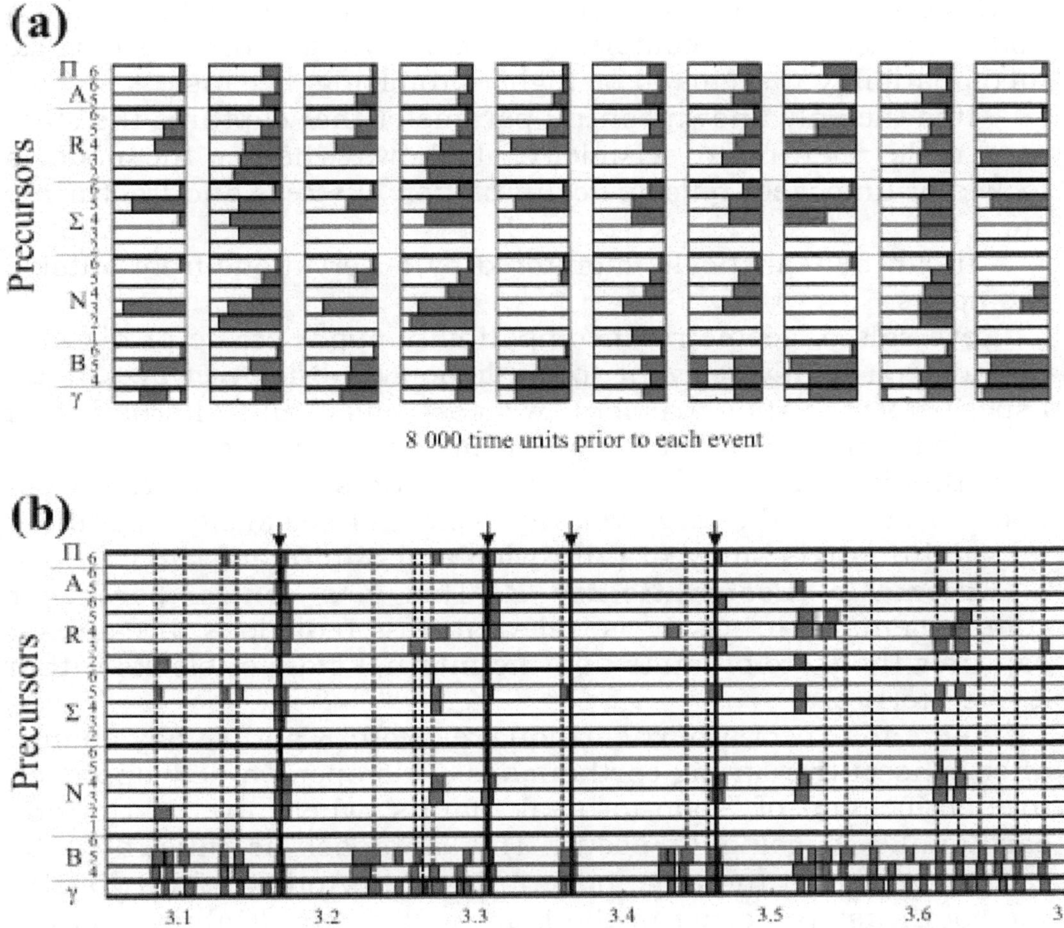

Figure 13. Collective performance of premonitory seismicity patterns (PSPs) found in observed seismicity and reproduced by colliding cascade model. After (Zaliapin et al., 2003b). The figure shows alarms generated by PSPs of the four types, discussed above: the rise of earthquake correlation range (Π_m, A_m, R_m); the rise of clustering (B_m); rise of seismic activity (Σ_m, N_m); and the transformation of the Gutenberg–Richter relation (γ). Each pattern is defined in different magnitude ranges indicated at the left side of the panel. *Shaded areas* show alarms triggered by the patterns. (a) Alarms preceding 10 major synthetic earthquakes ($M = 7$). Each box corresponds to a single major earthquake; the right edge of the box is its occurrence time. (b) Alarms during the continuous time interval. *Arrows and solid vertical lines* show the times of major earthquakes. Dashed vertical lines show the times of main shocks of magnitude $M = 6$ to illustrate their association with false alarms.

However, is hardly realistic to determine this field within a node, given its complexity. Monitoring the geometric incompatibility *G* seems to be highly promising, because:

– It is the only known control parameter that captures the state of a node relevant to seismicity; it shows whether the node is locked or unlocked; reflects accumulation of stress and fracturing, etc.

– It can be realistically monitored by observations from outside the nodes.

Reversals of geometric incompatibility might create or dissolve phenomena highly relevant to prediction (Gabrielov et al., 1996): asperities; relaxation barriers; weakest links; and alternation of seismicity and creep ("loud" and "silent" earthquakes).

If this inference is correct, these features would migrate from node to node at the large velocity typical of seismicity migration: tens to hundreds km/year (Vil'kovich and Shnirman, 1983).

Kinematic incompatibility *K*. This is a control parameter complementary to *G*. It is relevant to predictions because it captures the state of faults and the internal state of blocks within a given territory.

Where to expect a next earthquake? Both *K* and *G* may change after consecutive strong earthquakes. This opens a new approach to identification of "soon-to-break" faults, where the next strong earthquake is going to occur. This approach is illustrated by estimates of *G* in the large node in the San Andreas fault zone at the southeastern corner of the famous Big Bend (Gabrielov et al., 1996). This node lies on the junction of three groups of faults: San Andreas proper, right-lateral strike-slip faults to the north, and the thrust faults to the west. They will be called *SA*, *N*, and *W*, respectively. The impact of earthquakes in each system might be summed up as follows.

– Earthquakes in *SA* increase the kinematic incompatibility of the node, thus continuing stress accumulation there.

– Earthquakes in *N* or *W* act in the opposite way: they reduce the kinematic incompatibility, thus releasing stress.

This might explain why the last three major earthquakes occurred not at San Andreas fault, where it would seem natural to expect them, but in both of the adjacent fault groups, the Landers (1992, M = 7.6) and the Hector Mine (1999, M = 7.3) earthquakes on the *N*, and the Northridge earthquake (1994, M = 6.7) on *W*.

The very possibility of such a strong earthquake on the fault where it actually occurred was a surprise. It is noteworthy, therefore, that the epicenters of these three earthquakes lie within the nodes where, according to (Gelfand et al., 1976), earthquakes of magnitude 6.5 and above can nucleate.

Specific precursors in nodes. A much simpler but also promising possibility is to separately consider premonitory phenomena inside and outside of nodes. The difference between these two cases was observed by Rundkvist and Rotwain (1996) who applied the earthquake prediction algorithm CN in the Red Sea Rift zone. They found that during a correct alarm, the node was seismically silent, whereas during false alarms, it was active.

CHAPTER 3

Prediction of the Outcomes of the U.S. Elections

In terms of predictions, each election event, like an earthquake, can be considered as an extreme event in a socioeconomic system, with the specific that determine predictions of the election outcomes.

The elections' time is set by the law as follows.

— *National elections* are held every even-numbered year, on the first Tuesday after the first Monday in November (i.e., between November 2 and November 8, inclusively).

— *Presidential elections* are held once every 4 years, i.e. on every other election day. People in each of the 50 states and the District of Columbia are voting separately for "electors", pledged to one or another of the Presidential candidates. These electors make up the "Electoral College" which directly elects the President. Since 1860, when the present two-party system vas basically established, Electoral College reversed the decision of the popular vote only four times, in 1888, 1912, 2000, and 2016. Algorithmic prediction of such reversals is not developed so far.

— A third of *Senators* are elected for a 6-year term every election day; "mid-term' elections held in the middle of a Presidential term are considered here.

3.1. Methodology

The prediction target is an electoral defeat of an "incumbent" party, i.e. the party holding the contested seat. Accordingly, the prediction problem is formulated as whether the incumbent party will retain this seat or lose it to the challenging party (*and not whether Republican or Democrat will win*). As is shown below, this formulation is crucial for predicting the outcomes of elections considered.

Data. The pre-election situation is described by robust common sense parameters defined at the lowest (binary) level of resolution, as the *yes* or *no* answers to the questionnaires given below (Tables 6, 7). The questions are formulated in such a way that the answer *no* favors the victory of the challenging party. According to the Hamming distance analysis, the victory of the challenging party is predicted when the number *D* of answers *no* exceeds a threshold *D**.

Table 6. Questionnaire for mid-term Senatorial Elections (Lichtman and Keilis-Borok, 1989)

(Incumbency):	The incumbent-party candidate is the sitting senator.
(Stature):	The incumbent-party candidate is a major national figure.
(Contest):	There was no serious contest for the incumbent-party nomination.
(Party mandate):	The incumbent party won the seat with 60% or more of the vote in the previous election.
(Support):	The incumbent-party candidate outspends the challenger by 10% or more.
(Obscurity):	The challenging-party candidate is not a major national figure or a past or present governor or member of Congress.
(Opposition):	The incumbent party is not the party of the President.
(Contest):	There is no serious contest for the challenging-party nomination (the nominee gains a majority of the votes cast in the first primary and beats the second-place finisher at least two to one).

3.2. Mid-term Senatorial Elections

The prediction algorithm was developed by a retrospective analysis of the data on three elections, 1974, 1978, and 1982. The questionnaire is shown in Table 6. Victory of the challenger is predicted if the number of answers *no* is 5 or more (Lichtman and Keilis-Borok, 1989).

The meaning of these questions may be broader than their literal interpretation. For example, financial contributions (# 5) not only provide the resources required for an effective campaign, but may also constitute a poll in which the preferences are weighed by the money attached.

Predicting future elections. This algorithm (without any changes from year to year and from state to state) was applied in advance to the five subsequent elections, 1986 – 2002. Predictions are shown in Fig. 14. Altogether, 150 seats were put up for election. For each seat a separate prediction was made, 128 predictions were correct, and 22 – wrong.

Statistical significance of this score is 99.9%. In other words the probability to get such a score by chance is below 0.1% (Lichtman and Keilis-Borok, 1989; Lichtman, 1996, 2000). For some elections these predictions might be considered as trivial, since they coincide with prevailing expectation of experts. Such elections are identified by *Congressional Review.* Eliminating them from the score still results in 99% significance.

3.3 Presidential Elections

The prediction algorithm was developed by a retrospective analysis of the data on the past 31 elections, 1860 – 1980; that covers the period between victories of A. Lincoln and R. Reagan inclusively. The questionnaire is shown in Table 7. Victory for the challenger is predicted if the number of answers *no* is 6 or more (Lichtman, 1996, 2000).

Predicting of future elections. This algorithm (without any changes from year to year state) was applied by Prof. A. Lichtman in advance to the seven subsequent elections, 1984 – 2008 (Table 8). Predictions are shown in Table 9. All of them happened to be correct. In 2000 the decision of popular majority was reversed by the Electoral College; such reversals are not targeted by this algorithm (Lichtman, 1996, 2000).

3.4 Understanding Elections

Collective behavior. The finding that aggregate-level parameters can reliably anticipate the outcome of both presidential and senatorial elections points to an electoral behavior highly integrated not only for the nation as a whole but also within the diverse American states.

— A presidential election is determined by collective, integrated estimation of performance of incumbent administration during the previous four years.

— In case of senatorial elections the electorate has more diffused expectations of performance but puts more importance on political experience and status than in the case of presidential elections. Senate incumbents, unlike presidential ones, do not suffer from a bad economy or benefit from a good one. (This suggests that rather than punishing the party holding a Senate seat for hard times, the voters may instead regard the incumbent party as a safe port in a storm).

Similarity. For each election year in all states the outcomes of elections follow the same pattern that transcends the diversities of the situations in each of the individual elections.

The same pattern of the choice of the US President prevails since 1860, i.e. since election of A. Lincoln, despite all the overwhelming changes in the electorate, the economy, the social order and the technology of politics during these 130 years. (For example, the electorate of 1860 did not include some of the groups, which constitute 3/4 of present electorate, such as women, African Americans, or most of the citizens of the Latin American, South European, Eastern European, and Jewish descent (Lichtman, 2000).

An alternative (and more traditional) concept of American elections focuses on the division of voters into interest and attitudinal groups. By this concept the goal of the contestants is to attract maximum number of voting blocks with minimal antagonism from other blocks. Electoral choice depends strongly on the factors irrelevant to the essence of the electoral dilemma (e.g. on the campaign tactics). The drawbacks of this concept are discussed in (Lichtman, 2000; Keilis-Borok and Lichtman, 1993). In sum, the work on presidential and senatorial elections described above suggests the following new ways of understanding American politics and perhaps the politics of other societies as well.

Table 7. Questionnaire for Presidential elections (Lichtman, 1996, 2000)

KEY 1 **(Party Mandate)**: After the midterm elections, the incumbent party holds more seats in the U.S. House of Representatives than it did after the previous midterm elections.

KEY 2 **(Contest)**: There is no serious contest for the incumbent-party nomination.

KEY 3 **(Incumbency)**: The incumbent-party candidate is the sitting president.

KEY 4 **(Third party)**: There is no significant third-party or independent campaign.

KEY 5 **(Short-term economy)**: The economy is not in recession during the election campaign.

KEY 6 **(Long-term economy)**: Real per-capita economic growth during the term equals or exceeds mean growth during the previous two terms.

KEY 7 **(Policy change)**: The incumbent administration effects major changes in national policy.

KEY 8 **(Social unrest)**: There is no sustained social unrest during the term.

KEY 9 **(Scandal)**: The incumbent administration is unattained by a major scandal.

KEY 10 **(Foreign/military failure)**: The incumbent administration suffers no major failure in foreign or military affairs.

KEY 11 **(Foreign/military success}**: The incumbent administration achieves a major success in foreign or military affairs.

KEY 12 **(Incumbent charisma)**: The incumbent-party candidate is charismatic or a national hero.

KEY 13 **(Challenger charisma)**: The challenging-party candidate is not charismatic or a national hero.

0	1	2	3	4	5	6	7
			OK98				
			CO98				
			FL98				
			GA98				
			HA98	*TN02*			
			ID98	*SC02*			
			MA98	*NC02*			
			ND98	*NE02*			
			PN98	*KY02*			
			SD98	*IA02*			
			UT98	*CO02*			
			FL94	*AL02*			
			HA94	*AK98*			
			IN94	*CA98*			
			MT94	*CT98*			
			NB94	*NE98*			
			NJ94	*OR98*			
			TX94	*SC98*			
			WA94	*VT98*			
		AS98	*WV94*	*WA98*			
		KA98	*WI94*	*CT94*			
		LA98	*AK90*	*MD94*			
		MI98	*IN90*	*NV94*			
		NH98	*KN90*	*WY94*			
		MS94	*ME90*	*CO90*			
	AL98	*NM94*	*MA90*	*HA90*			
	AZ98	*ND94*	*MT90*	*KY90*			
	IO98	*RI94*	*NB90*	*MI90*			
	DL94	*VT94*	*NC90*	*AZ86*			
	MA94	*AS90*	*TX90*	*CO86*			
	NY94	*IO90*	*WY90*	*ID86*			
	AL90	*MS90*	*AR86*	*LA86*			
	DE90	*NM90*	*CA86*	*NY86*			
	IL90	*OR90*	*IL86*	*OK86*	*WI98*	*MN94*	
	LA90	*RI90*	*IN86*	*WI86*	*CA94*	*MO94*	
	OK90	*SD90*	*IA86*	**NC86**	*ID90*	*VA94*	
	SC90	*VA90*	*NH86*	**WA86**	*PA86*	*NH90*	
	TN90	*WV90*	*OR86*	**MN90**	**IL98**	**IN98**	
	HI86	*AK86*	*VT86*	**OK94**	**ME94**	**OH98**	
	OH86	*CT86*	**TN94**	**PA94**	**AL86**	**MI94**	
UT94	*SC86*	*KS86*	**TX02**	**TN294**	**FL86**	**MD86**	**KY98**
GA90	*UT86*	*KY86*	**OK02**	**NC98**	**GA86**	**NV86**	**AZ94**
NJ90	**NH02**	**ND86**	**NJ02**	**NY98**	**MO86**	**SD86**	**OH94**
0	1	2	3	4	5	6	7

OK98 – incumbent won, **KY98** – challenger won, errors are highlighted.

Figure 14 Made-in-advance predictions of the mid-term senatorial elections (1986-2002). Each election is represented by the two-letter state abbreviation with the election year shown by two last digits. Each column shows elections with certain number *D* of answers "*no*" to the questionnaire given in Table 6 (such answers are favourable to challenging party). Value of *D*, indicated at the top, is the Hamming distance from the kernel).

Table 8. Timing and source of predictions

Election	Date of prediction	Source
1984	APRIL 1982	"How to Bet in '84," *Washingtonian Magazine,* April 1982
1988	MAY 1988	"How to Bet in November," *Washingtonian Magazine,* May 1988
1992	SEPTEMBER 1992	"The Keys to the White House," *Montgomery Journal*, September 14, 1992
1996	OCTOBER 1996	"Who Will Be the Next President?" *Social Education,* October 1996
2000	NOVEMBER 1999	"The Keys to Election 2000" *Social Education*, November/December 1999.
2004	APRIL 2003	"The Keys to the White House," *Montgomery Gazette*, Apr. 25, 2003
2008	FEBRUARY 2006	"Forecast for 2008," *Foresight,* Feb. 2006,

An alternative (and more traditional) concept of American elections focuses on the division of voters into interest and attitudinal groups. By this concept the goal of the contestants is to attract maximum number of voting blocks with minimal antagonism from other blocks. Electoral choice depends strongly on the factors irrelevant to the essence of the electoral dilemma (e.g. on the campaign tactics). The drawbacks of this concept are discussed in (Lichtman, 2000; Keilis-Borok and Lichtman, 1993).
In sum, the work on presidential and senatorial elections described above suggests the following new ways of understanding the American politics and perhaps the politics of other societies as well.

1. Fundamental shifts in the composition of the electorate, the technology of campaigning, the prevailing economic and social conditions, and the key issues of campaigns do not necessarily change the pragmatic basis on which voters choose their leaders.

2. It is the governing, not the campaigning, that counts in the outcomes of presidential elections.

3. Different factors may decide the outcome of executive as compared to legislative elections.

4. Conventional campaigning will not improve the prospects for candidates faced with an unfavorable combination of fundamental historical factors. Disadvantaged candidates have an incentive to adopt innovative campaigns that break the pattern of conventional politics.

5. All candidates would benefit from using campaigns to build a foundation for governing in the future.

Table 9. Prediction of US presidential elections (up to 2008)

Years when incumbent won are shown in italic; years when challenger won – in bold.

Number of keys in favor of a challenger party									
0	1	2	3	4	5	6	7	8	9
Predictions (published months in advance)									
					*2000**				
		1984	*1988*	*2004*	*1996*	**1992**		**2008**	
Retrospective analysis									
			1964					**1980**	
			1928					**1976**	
			1916					**1968**	
			1908					**1952**	
		1944	*1900*	*1972*				**1932**	
	1956	*1940*	*1872*	*1924*	*1948*	**1912**	**1884**	**1920**	**1960**
1904	*1936*	*1868*	*1864*	*1880*	*1888**	**1892**	**1860**	**1896**	**1876***

* years when popular vote was reversed by electoral vote.

CHAPTER 4

Prediction of Economic Recessions

The five recessions in the USA from 1961 to 1996 have been considered, and it has been found that each of them was preceded by a specific pattern of 6 economic indexes, which are defined at the lowest (binary) level of resolution (Keilis-Borok et al., 2000). This pattern was present during 6 to 14 month before each recession and at no other time, suggesting a hypothetical prediction algorithm. The algorithm is exceedingly robust: the retrospectively diagnosed alarms remain about the same after variation of its adjustable numerical parameters, and of other non-unique decisions involved in its determination. It has also been found that another robust pattern of the same six macroeconomic indicators appeared within 6 months before the ends of these recessions and at no other time during the recessions (Keilis-Borok et al., 2008). The last study is a natural continuation of the previous one, aimed at predicting the start of a recession. It is found by comparing these cases that precursory trends of financial indicators are opposite during the transition to a recession and the recovery from it. To the contrary, precursory trends of economic indicators happen to have the same direction (upward or downward), but are steeper during recovery.

4.1. Prediction Targets

US National Bureau of Economic Research (NBER) has identified seven recessions that occurred in the US from 1960 to 2002 (Table 10). The starting points of a recession and of the recovery from it follow the months marked by a peak and a trough of economic activity, respectively.

Table 10. American Economic Recessions, 1960-2002 (year:month)

#	Peaks	Troughs
1	1960:04	1961:02
2	1969:12	1970:11
3	1973:11	1975:03
4	1980:01	1980:07
5	1981:07	1982:11
6	1990:07	1991:03
7	2001:03	2001:11

A peak indicates the last month before a recession, and a trough — the last month of a recession.

Prediction targets considered are the first month after the peak and the first month after the trough ("the turns to the worst and to the best", respectively). The start of the first recession, in 1960, is not among the targets, since the data do not cover a sufficient period of time preceding the recession.

4.2. Data and Their Preliminary Analysis

The data used for prediction comprise the following six monthly leading economic indicators obtained from the CITIBASE data base (abbreviations are the same, as in Stock and Watson, 1993).

IP: Industrial Production, total: index of real (constant dollars, dimensionless) output in the entire economy. This represents mainly the manufacturing industry, because of the difficulties in measuring the quantity of the output in services (such as travel agents, banking, etc.).

INVMTQ: Total inventories in manufacturing and trade, in real dollars. Includes intermediate inventories (for example held by manufacturers, ready to be sent to retailers) and final goods inventories (goods on the shelves in stores).

LHELL: Index of "help wanted" advertising. This is put together by a private publishing company that measures the amount of job advertising (column-inches) in a number of major newspapers.

LUINC: Average weekly number of people claiming unemployment insurance.

FYGM3: Interest rate on 90 day U.S. treasury bills at an annual rate (in percent).

G10FF=FYGT10–FEDFUN: Difference between the annual interest rate on 10 year U.S. Treasury bonds, and federal fund annual interest rate.

These indicators were already known (Stock and Watson, 1989, 1993), as those that correlate with a recession's approach. The first four indicators concern the economy, while the two others concern the financial market.

Premonitory behavior of an indicator may be better captured by its linear trend.

Notations:

$f(m)$ is a time series of a monthly indicator, hypothetically containing a premonitory pattern; integer m is the time in months.

$W^f(l/q,p)$ – the local linear least-squares regression of a function $f(m)$ within the sliding time window (q, p):

$$W^f(l/q,p) = K^f(q,p)l + B^f(q,p),\ q \leq l \leq p, \qquad (1)$$

integers l, q, and p stand for time in months.

Two following transformations of indicators that reflect their trends are considered:

$$K^f(m/s) = K^f(m\text{-}s,m). \qquad (2)$$

In notations of (1) this is the trend of $f(m)$ in the sliding time window s months long, $(m\text{-}s,\ m)$. It may be considered for prediction since its value is attributed to the *end* of window (m) and does not depend on the information about the future (after the month m) which would be anathema in prediction.

$$R^f(m/q) = f(m) - W^f(m/q,m\text{-}1). \qquad (3)$$

This is deviation of an index from its long-term extrapolation (since long time window (q, $m-1$) is considered in estimating W^f).

Discretization. Functions used in a prediction algorithm are considered on the lowest – binary – level of resolution, as 0 or 1, distinguishing only the values above ("large") and below ("small") a threshold T. Then the months (objects of recognition) are described by binary vectors of the same length. Discretization ensures robustness of analysis. The threshold is defined as a Q – level percentile, e.g. by the condition that the function F exceeds $T^F(Q)$ during Q% of the months considered.

4.3. Prediction of a Recession Start

Time periods. The time period from 1960 to 1996 was considered by Keilis-Borok et al. (2000) when the prediction algorithm was designed. The time of each recession and 5 months after it is eliminated from consideration, since the behavior of the indexes may have some special features during a recession and its aftermath; the interval of 5 months is chosen, because subsequent recessions are officially considered as different ones only if they are separated by at least six months. At this stage, the period before the first recession in Table 10 is also excluded (since it starts too close to the beginning of the data set), as well as the period after the sixth one. Thus, the following 5 periods are used for the search of premonitory phenomena:

W_1: 1961:08 - 1969:12 (101 months);
W_2: 1971:05 - 1973:11 (31 months);
W_3: 1975:09 - 1980:01 (53 months);
W_4: 1981:01 - 1981:07 (7 months);
W_5: 1983:05 - 1990:07 (87 months).

The combination of these periods $W = W_1 \cup W_2 \cup W_3 \cup W_4 \cup W_5$ contains 279 months. These months between recessions are considered when the discretization thresholds $T^F(Q)$ are determined.

Indicators describing the months are transformed into functions given in Table 11. By means of discretization with the values of Q given in Table 10 the functional values are reduced to "small" or "large".

The goal of the algorithm is to predict the beginning of a recession, defined, as the first month after the "peak" (table 10). Prediction algorithm, considered here, has a limited accuracy; it may indicate not a single month, but a time interval several months long, which includes the first month of a recession. Continues intervals of this kind are called "alarms". A recession is regarded as being "predicted", if it begins within an alarm and as a "failure to predict" in the opposite case. A "false alarm" is an alarm during which no recession has started.

Table 11. Functions used for the indicators and their typical values in pre-recession periods

Indicator	Function used for the indicator	Q, %	Typical value in pre-recession situation
IP	Deviation from the long-term trend $R^{IP}(m/q)^{*}$ (3)	75	Small
INVMTQ	Deviation from the long-term trend $R^{INVMTQ}(m/q)^{*}$ (3)	25	Small
LHELL	Short trend $K^{LHELL}(m/5)$ (2)	66.7	Small
LUINC	Short trend $K^{LUINC}(m/10)$ (2)	16.7	Large
FYGM3	Deviation from the long-term trend $R^{FYGM3}(m/q)^{*}$ (3)	25	Large
G10FF	Indicator itself	90	Small

*) q in the determination of $R^{f}(m/q)$ is the first month after the end (the through) of the previous recession

The whole set of months under consideration W is divided into two parts: the set W_D that includes three months before each recession, and the set $W_N = W \setminus W_D$. The value of a functional is typical for the set W_D (pre-recession situation) if a percent of months from W_D with the functional having this value is greater than this percent of months from W_N. The typical values of the functions in pre-recession situation are given in the last column of

Table 11. After this analysis all months are described by binary vectors with 6 components corresponding to the functionals listed in Table 11. Each component has been defined in such a way, that the value "1" would become typical in a pre-recession situation (more frequent when a recession is approaching). Accordingly, the description of pre-recession situations would be close to the vector (1,1,1,1,1,1), which is called the kernel. Let $\Delta(m)$ be the number of zeros in a description of the month m - its Hamming's distance from the kernel. The approach of a recession may be recognised by the small values of $\Delta(m)$. A priori this is not clear, in spite of the way the zeros are defined; for example, premonitory changes of the functions may appear not simultaneously.

Prediction algorithm. The prediction algorithm is formulated as follows (Keilis-Borok et al., 2000): *an alarm is declared for three months after each month m with $\Delta(m) \leq 2$ (regardless of whether or not this month belongs to an alarm which has already been declared).*

Alarms and recessions are juxtaposed in Fig. 15. One can see that five recessions occurring between 1962 and 2000 (#2 – 6 in Table 10) are predicted by an alarm. Duration of each alarm is between 6 and 14 months. Total duration of all alarms is 43 months, or 15.4 % of the time interval (set *W*) considered. There are no false alarms. Months between two last recessions in Fig. 15 were not included in the set *W* and were not considered in the analysis described above. The last recession in Fig. 15 (No. 7 in Table 10) is preceded by an alarm determined by the algorithm.

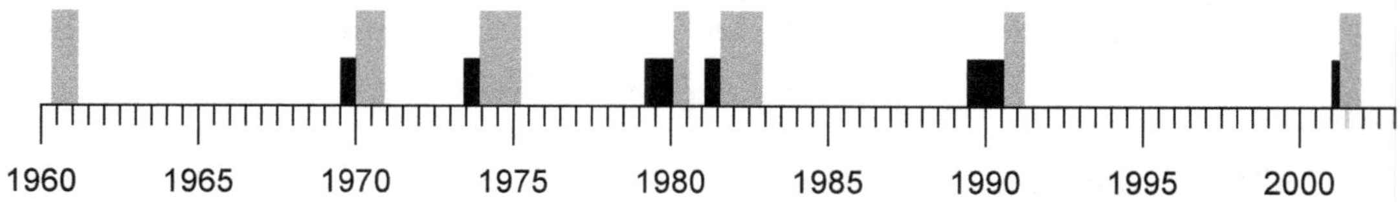

Figure 15. Alarms (black bars) and recessions (grey bars).

4.4. Prediction of Recovery from a Recession

The study on predictions of recovery from a recession (Keilis-Borok et al., 2008) is a natural continuation of the study aimed at predicting the start of a recession.

Prediction targets are defined above (see Sect. 4.1).

The data are the same six indicators that indicate the approach of a recession (see Sect. 4.2); they are analysed only within the recession period.

The time period from 1960:01 to 2002:04 is considered, when seven recessions did occur (Table 10). At the time of the study by Keilis-Borok et al. (2008) the trough for the last recession had not been yet established by the NBER and the data concerning only the first 6 recessions were used. Thus the following 6 periods are considered (for each recession a month before it (the peak) and its last month (the trough) are included).

W_1: 1960:04 - 1961:02 (11 months);
W_2: 1969:12 - 1970:11 (12 months);
W_3: 1973:11 - 1975:03 (17 months);
W_4: 1980:01 - 1980:07 (7 months);
W_5: 1981:07 - 1982:11 (17 months);
W_6: 1990:07 - 1991:03 (9 months)

The total duration of these periods $W = W_1 \cup W_2 \cup W_3 \cup W_4 \cup W_5 \cup W_6$ is 73 months. The last recession is considered separately, in Section 4.6.

As in Keilis-Borok et al. (2000), we use for prediction the values of the indicator **G10FF;** the trends of the indicators **LHELL** and **LUINC**; and deviations from the long-term trend of the indicators **IP**, **INVMTQ**, and **FYGM3** (Table 12).

Analyzing the histograms of the six functionals considered during the recession periods Keilis-Borok et al. (2008) have found that large values of **G10FF** and $K^{LUINC}(m/10)$ and small values of other functions occur more frequently as the termination of a recession approaches. To check whether this impression is correct, a robust quantitative definition of these changes is given by means of the discretization procedure described in Section 4.2. Analyzing behavior of each function, the values of Q indicated in Table 12 and the relevant values of $T^F(Q)$ have been determined.

The last column of Table 12 presents "precursory potential" for each function, i.e., the difference $P_e - P_b$, where P_e is percent of months with "precursory" value of a function in a set consisting of the last four months of each recession; P_b is percent of such months in a whole set W with the last six months of each recession eliminated. Looking at these data one can conclude that some indicators (e.g., $K^{LUINC}(m/10)$) may be used individually for determination of the precursory pattern. However the existing experience shows that using several indicators makes the result more reliable.

Table 12. Functions used for the indicators and their typical values before recovery from a recession

Indicator	Function used for the indicator	Q, %	Typical value before recovery from a recession	"Precursory potential"**, %
IP	Deviation from the long-term trend $R^{IP}(m/q)^*$ (3)	75	Small	67
INVMTQ	Deviation from the long-term trend $R^{INVMTQ}(m/q)^*$ (3)	50	Small	20
LHELL	Short trend $K^{LHELL}(m/5)$ (2)	75	Small	45
LUINC	Short trend $K^{LUINC}(m/10)$ (2)	50	Large	70
FYGM3	Deviation from the long-term trend $R^{FYGM3}(m/q)^*$ (3)	50	Small	20
G10FF	Indicator itself	33.3	Large	67

*) q in the determination of $R^f(m/q)$ is the first month after the end (the through) of the previous recession. An exception is the first recession for which q is the first month of the time period considered, 1960:01 (since the previous recession ends before this period).

**) Difference $P_e - P_b$ where P_e is percent of months with "precursory" value of a function in a set consisting of the last four months of each recession; P_b is percent of such months in a whole set W with the last six months of each recession eliminated.

After discretization of the functions, the situation during the recession is robustly described in relation to its proximity to a recession's end. This description is reduced to a monthly time series of a binary vector with 6 components. For convenience, the same code, "1", is given to the "precursory" trend of each function, regardless of whether it is above or below the threshold. Accordingly, the code "1" is given to the large values of the functions **G10FF** and $K^{LUINC}(m/10)$, and to the small values of the four other functions. Other values receive code "0".

The precursory pattern is looked for that indicates that a recession will end within a few months after this pattern appears. Continuous intervals of this kind are called "alarms." Possible outcomes of prediction based on this pattern are illustrated in Fig. 16. The probabilistic nature of the prediction is reflected in probabilities of the errors of each kind and in the percent of the time occupied by the alarms. This formulation of the prediction problem is discussed by Molchan (1997); it was applied by Keilis-Borok et al. (2000) for predicting of the start of a recession.

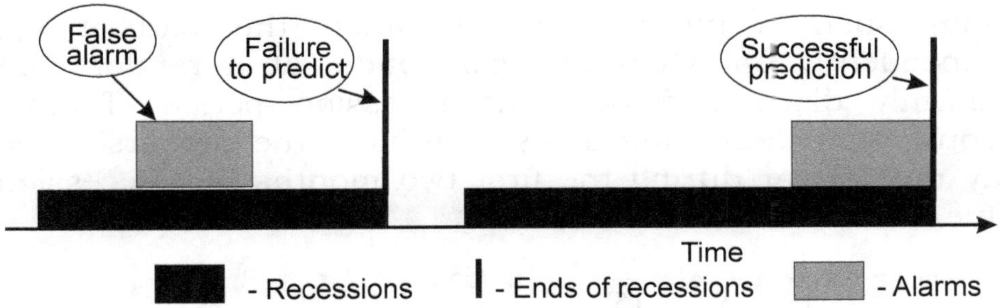

Figure 16. Outcomes of prediction of recovery from a recession.

After discretization, the monthly description of a situation during a recession is reduced to a binary vector with 6 components. Each component has been defined in such a way, that the values "1" would became more frequent when the end of a recession is approaching. Accordingly, the "ideal" situation prior to the end of a recession, when all indicators are precursory, is the vector (1,1,1,1,1,1), which is called the "kernel." Let $\Delta(m)$ be the number of zeros in a code of a month m – its Hamming distance from the kernel. A question is, whether the approach of the end of a recession may be recognized by the small values of $\Delta(m)$. A priori this is not clear, despite the way the kernel is defined; for example, precursory changes of the functions may not appear simultaneously.

Prediction algorithm. Keilis-Borok et al. (2008) have found that the end of each recession is preceded by a continuous group of months with $\Delta(m) \leq 3$. This suggests the following algorithm: *the precursory pattern appears if for three consecutive months $\Delta(m) \leq 3$ and the recession end is expected during an interval of three months long after displaying this pattern.*

The alarms (continuous intervals of this kind) obtained by this algorithm for the recessions considered are shown in Fig. 17. The end of each of the 6 recessions is preceded by an alarm and there are no false alarms. The duration of an alarm inside each recession is given in Table 13. The total duration of alarms in all 6 recessions is 16 months, which is 22% of total duration of the set W (73 months).

The duration of an alarm does not exceed 5 months; this means that the precursory pattern appears during the last 6 months of each recession. Let us estimate the probability P to obtain such result by chance, when the pattern appears independently on when recession ends, at a random moment uniformly distributed over the recession period. Taking into account that, according to its definition, the precursory pattern may not appear during the first two months of a recession, the value of P is calculated as follows

$$P = \frac{6}{9} \times \frac{6}{10} \times \frac{6}{15} \times 1 \times \frac{6}{15} \times \frac{6}{7} = 0.055 .$$

Here the denominators are the durations of the recessions (Table 10) reduced by 2. The factor corresponding to the fourth recession is 1 because it is so short that the alarm for its end may not be longer than 5 moths. The probability is encouragingly low, although additional out-of-sample tests are necessary to establish statistical significance (Molchan et al., 1990).

Table 13. Alarms before the ends of the recessions

Recessio n	Trough	The first month of an alarm and its duration (in months) before the recession end
1	1961:02	1960:12, 3
2	1970:11	1970:08, 4
3	1975:03	1975:02, 2

4	1980:07	1980:07, 1
5	1982:11	1982:07, 5
6	1991:03	1991:03, 1
7	2001:11	2001:07, 5

Recessions are numbered as in Table 10.

The approach of recession is captured (Section 4.3) by the same indicators that are used here to capture recovery from it. Behavior of the indicators during these opposite processes is compared in Table 14 and Fig. 18. One can see reverse trends of "financial" indicators; and, contrary to what might be expected, the trends of "economic" indicators have the same direction, but are steeper during recovery.

4.5 Sensitivity Analysis

The analysis given in the two previous sections, inevitably for such a heuristic one, does involve considerable data fitting, e.g. the choice of the functions, numerical parameters, etc. (Dr. W. Press, discussing an early example of such an approach, reminded the words of E. Fermi: "With four exponentials I can fit an elephant"). Therefore it is necessary to explore how stable are the results to variation of these choices. If they are sufficiently stable, one may hypothesize that the same patterns will precede the starts of the subsequent recessions and the recovery from them. The only decisive test of this hypothesis could be provided by advance predictions.

The following tests were made by Keilis-Borok at al. (2000) to check the stability of the algorithm that predicts ***a recession start***. Its application is considered below as "main variant".

"*Prediction history*". The discretization thresholds $T^F(Q)$ used by the algorithm are determined considering the whole set W (Section 4.3). In the test the thresholds $T^F(Q)$ are determined for consecutively expanding time periods ("T-sets"): W_1, $W_1 \cup W_2$, $W_1 \cup W_2 \cup W_3$, and $W_1 \cup W_2 \cup W_3 \cup W_4$. The values of $T^F(Q)$ for each period are calculated for the same values of Q that are given in Table 12. The prediction algorithm with each set of thresholds was applied to the subsequent time period up to a next recession, that is, to the sets W_2, W_3, W_4, and W_5 respectively. The alarms obtained in the experiment are shown in Fig. 19b. The alarms obtained in the main variant are shown for comparison in Fig. 19a. The agreement with the main variant is practically complete. The only difference is that the last alarm became a month longer.

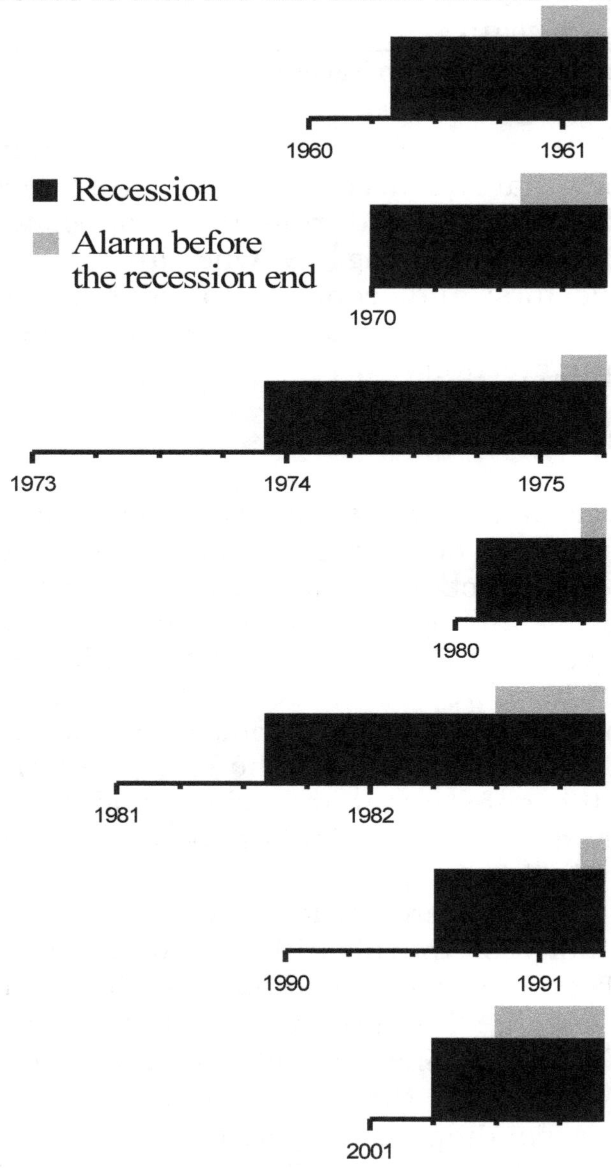

Figure 17. Results of application of the algorithm predicting recovery from a recession.

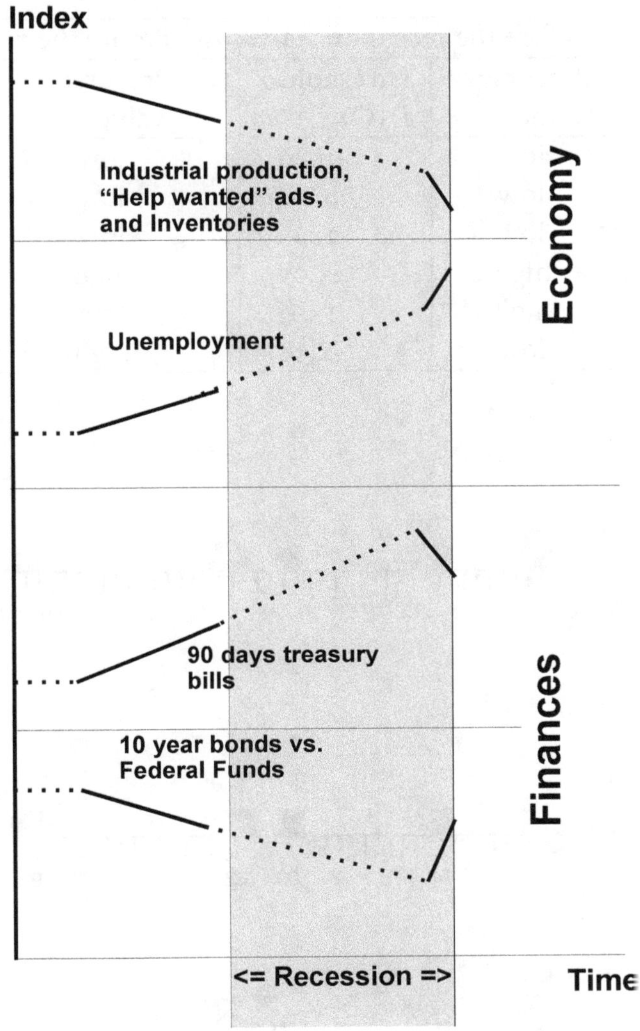

Figure 18. Premonitory changes of indicators before the start of a recession and before its end.

Table 14. Precursory behavior of indicators

Function	Before the recession start		Before the recession end	
	Precursory values	Threshold $T^F(Q)$	Precursory values	Threshold $T^F(Q)$
$R^{IP}(m/q)$	low	-0.86	low	-3.71
$R^{INVMTQ}(m/q)$	low	8.85	low	6.50
$K^{LHELL}(m/5)$	low	2.00	low	-29.71
$K^{LUINC}(m/10)$	high	25.07	high	112.15
$R^{FYGM3}(m/q)$	high	1.01	low	-0.64
G10FF	low	-1.38	high	0.82

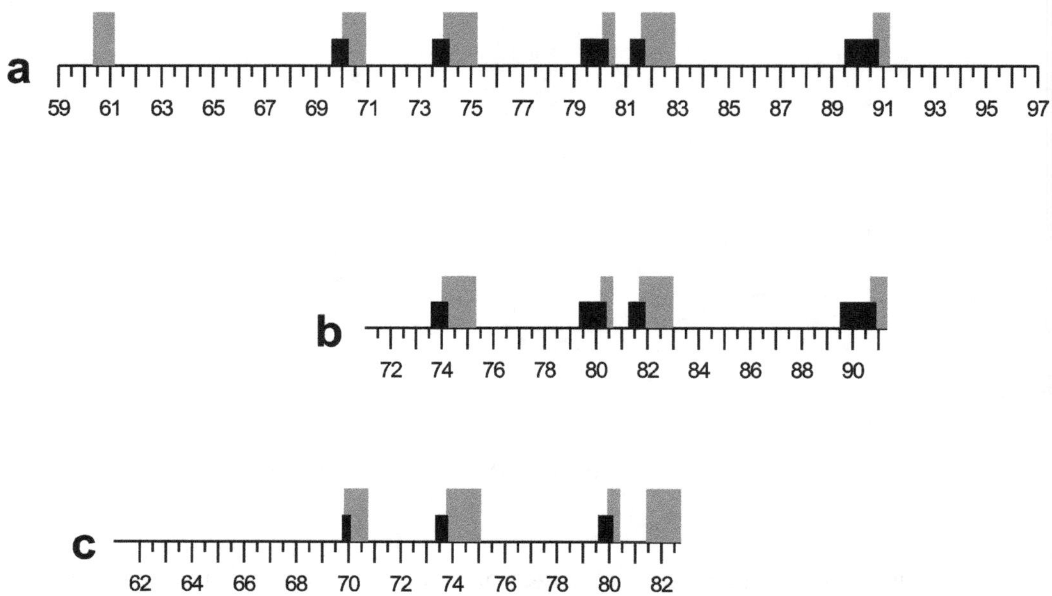

Figure 19. Alarms (shown by black bars): a - main version; b - prediction history; c - reverse prediction history.

"Reverse prediction history". In this test, the data set is expanded in the reverse order: the thresholds $T^F(Q)$ are determined consecutively for the T-sets W_5, $W_5 \cup W_4$, $W_5 \cup W_4 \cup W_3$,

and $W_5 \cup W_4 \cup W_3 \cup W_2$ and the prediction algorithm is applied to the sets W_4, W_3, W_2, and W_1 respectively. The alarms obtained in this test are shown in Fig. 19c. One can see that the recession in 1981 - 1982 is missed by the prediction; the alarms before the recessions in 1969 - 1970 and in 1980 are shorter than in the main variant.

"Change of discretization". In this test the values of Q listed in Table 11 are increased and decreased as shown in Table 15. Again, the results are similar to those in the main variant, though the changes are not insignificant: one false alarm appeared; three alarms of the main version became separated into two parts, so that in statistical tests the first parts should be counted as false alarms.

Table 15. Values of Q that were used when discretization was changed in the algorithm of recession start prediction (Keilis-Borok et al., 2000)

Function	Q in the main variant, %	Q in the test "Change of discretization", %	
$R^{IP}(m/q)$	75	66.7	80
$R^{INVMTQ}(m/q)$	25	20	33.3
$K^{LHELL}(m/5)$	66.7	50	75
$K^{LUINC}(m/10)$	16.7	14.3	20
$R^{FYGM3}(m/q)$	25	20	33.3
G10FF	90	87.5	92

"Elimination of the functions". In this test, each of the six functions used in the algorithm is eliminated from consideration, one at a time. The alarms are declared when $\Delta(m) \leq 1$ instead of 2 in the main variant. The major change with respect to the main variant is failure to predict the recession in 1981 - 1982 after either of the functions $R^{IP}(m/q)$, $R^{INVMTQ}(m/q)$, $K^{LHELL}(m/5)$, or **G10FF** is eliminated. The smallest changes are caused by elimination of $K^{LUINC}(m/10)$. When either of the functions $R^{INVMTQ}(m/q)$, $K^{LHELL}(m/5)$, $R^{FYGM3}(m/q)$, or **G10FF** is eliminated, the alarm before the recession in 1973 - 1975 starts in the same month as the recession.

"Prediction by a single function". In this test, each of the functions is used separately of others. Obviously, the alarms were declared if the value of the relevant binary component is 1. When the function $R^{IP}(m/q)$, $K^{LUINC}(m/10)$, $R^{FYGM3}(m/q)$, or **G10FF** is used, there is one failure to predict; total duration of alarms is the smallest with **G10FF**, the largest with $R^{IP}(m/q)$, and about the same with $K^{LUINC}(m/10)$ and $R^{FYGM3}(m/q)$. When $R^{IP}(m/q)$ or $K^{LHELL}(m/5)$ is used, total duration of alarms is much larger, but there are no false alarms.

Performance of different of predictions obtained in the tests is juxtaposed in the error diagram shown in Fig. 20.

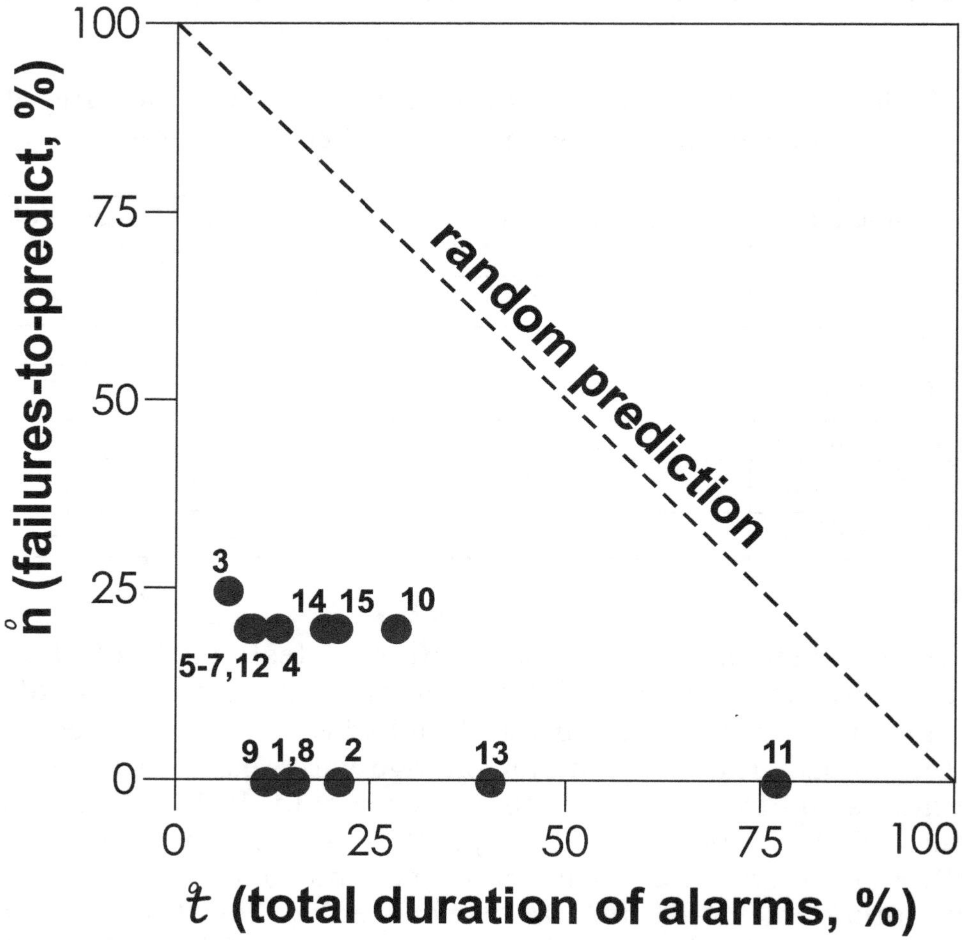

Figure 20. Error diagram: 1 - main version; 2 - prediction history; 3 - reverse prediction history; 4-9 exclusion of functions: $R^{IP}(m/q)$ (4), $R^{INVMTQ}(m/q)$ (5),

G10FF (6), $K^{LHELL}(m/5)$ (7), $K^{LUINC}(m/10)$ (8), $R^{FYGM3}(m/q)$ (9); 10-15 - prediction by a single function: $R^{IP}(m/q)$ (10), $R^{INVMTQ}(m/q)$ (11), G10FF (12), $K^{LHELL}(m/5)$ (13), $K^{LUINC}(m/10)$ (14), $R^{FYGM3}(m/q)$ (15).

The following tests were made by Keilis-Borok at al. (2008) to check the stability of the algorithm that predicts ***recovery from a recession***. Its application is considered below as "main variant".

Table 16. Changing of the alarms given in Table 13 when the functions are removed

Recession	Functions removed and the first month of alarms obtained and its duration (in months) before the recession end ("–" indicates cases where there is no difference with the result given in Table 13)					
	R^{IP} (m/q)	R^{INVMTQ} (m/q)	K^{LHELL} $(m/5)$	K^{LUINC} $(m/10)$	$R^{FYGM3}(m/q)$	G10FF
1	–	No alarm		No alarm	1961: 01, 2	No alarm
2	1970: 10, 2	1970: 10, 2		1970: 10, 2	–	–
3	–	–	–	–	–	–
4	1980: 08, 0	–	1970: 09, 3	1980: 08, 0	–	–
5	–	1982: 10, 2	–	1982: 10, 2	1982: 10, 2	–
6	–	–	1980: 08, 0	–	–	–

Recessions are numbered as in Table 10.

Elimination of the functions. As above, it is studied in this test how the prediction changes if one of the functions is removed from consideration. For that purpose, each of the 6 functions used in the prediction algorithm (Section 4.4) is removed in turn. Taking into account that in this case the length of the binary codes of months reduces to 5 the condition $\Delta(m) \leq 3$ in the algorithm formulated in Section 4.4 is changed by the condition $\Delta(m) \leq 2$. The results of this experiment are given in Table 16. The test shows stability of the prediction results to excluding of functions. Only for the first recession (1960-1961) its end is failure-to-predict when the function $R^{INVMTQ}(m/q)$, or $K^{LUINC}(m/10)$ or **G10FF** is removed. In other cases there are only some variations in durations of alarms. The greatest number of differences is in the case where the function $K^{LUINC}(m/10)$ is removed that is in accordance with the largest "precursory potential" of this function (see Table 12).

Table 17. Changing of the alarms given in Table 13 when the indicators are used individually

Recession	Functions used individually and the first month of alarms obtained and its duration (in months) before the recession end ("–" indicates cases where there is no difference with the result given in Table 13)					
	R^{IP} (m/q)	R^{INVMTQ} (m/q)	K^{LHELL} $(m/5)$	K^{LUINC} $(m/10)$	R^{FYGM3} (m/q)	G10FF
1	No alarm	1960:07, 8	No alarm	1961:01, 2	1960:07, 7 (false alarm)	–
2	1970:09, 3	–	No alarm	1970:07, 5	No alarm	1970:11, 1
3	–	No alarm	1975:01, 3	–	1975:01, 3	No alarm
4	–	No alarm	1980:06, 2	–	1980:08, 0	No alarm
5	No alarm	1981:10, 6 (false alarm) 1982:07, 5	No alarm	1982:03, 9	1981:10, 14	1982:10, 2
6	–	No alarm	1991:02, 2	–	No alarm	1991:04, 0

Recessions are numbered as in Table 10.

Prediction by a single function. The purpose of this test is to know what results would be obtained if the functions are used individually in the prediction algorithm. In this case the length of binary codes of months is 1 and the condition $\Delta(m) \leq 3$ in the algorithm formulated in Section 4.4 is changed by the condition $\Delta(m) = 0$. The results of the individual use of functions are given in Table 17. One can see that the functions used individually do not ensure good quality of prediction. Only using of $K^{LUINC}(m/10)$ does not result in false alarms or failures-to-predict, but the alarm before the end of the fifth recession (1981-1982) is too long.

Variation of the set of months used in determination of thresholds $T^F(Q)$. The thresholds of discretization $T^F(Q)$ have been determined by using a set $W = W_1 \cup W_2 \cup W_3 \cup W_4 \cup W_5 \cup W_6$ (Section

4.4) containing the months belonging to the 6 recessions under consideration. To check the sensitivity of the result to the composition of a training set the thresholds $T^F(Q)$ are computed excluding in turn the subsets W_i (i = 1, 2, ..., 6) from the training set W. The results of predictions with different values of $T^F(Q)$ are given in Table 18. These results show that the prediction is stable with respect to varying the set of months used in determination of the discretization thresholds. There is only one failure-to-predict case (the end of the second recession when the months of the third recession are excluded from the training set) and there are no false alarms.

Table 18. Changing of the alarms given in Table 13 when different training sets are used in determination of the thresholds $T^F(Q)$

Recession	Training sets and the first month of alarms obtained and its duration (in months) before the recession end ("–" indicates cases where there is no difference with the result given in Table 13)					
	$W \setminus W_1$	$W \setminus W_2$	$W \setminus W_3$	$W \setminus W_4$	$W \setminus W_5$	$W \setminus W_6$
1	1960:09, 6	1961:01, 2	1961: 01, 2	–	1960:11, 2	1961:01, 2
2	1970:11, 1	1970:07, 5	No alarm	–	1970:10, 2	–
3	–	–	–	–	–	–
4	–	–	–	–	–	–
5	1982:06, 6	1982:06, 6	–	–	1982:06, 6	–
6	–	–	–	–	–	–

Recessions are numbered as in Table 10.

Variation of parameter s in functions $K^{LHELL}(m/s)$ and $K^{LUINC}(m/s)$. The values of parameter s in functions $K^{LHELL}(m/s)$ and $K^{LUINC}(m/s)$ have been selected to be the same as those used in the algorithm for prediction of the start of a recession (Section 4.3): s = 5 and s = 10 respectively. These values are changed to see how the prediction results depend on them. When the function $K^{LHELL}(m/s)$ is calculated with s = 4 and s = 6, the alarms obtained by the algorithm are exactly the same as those given in Table 13. When the function $K^{LUINC}(m/s)$ is calculated with s = 8, there is only one difference in respect to Table 13: the first month of the alarm for the end of the first recession is 1960:11. If s = 8 for this function, then there are two differences: the first month of the alarm for the end of the first recession is 1961:03 and the first month of the alarm for the end of the third recession is 1975:01. Therefore the prediction is stable with respect to varying the value of *s*.

4.6. Experience in Advance Forecasting

The prediction algorithms described above were applied for advance forecasting after a recession occurred in 2001 (line 7 in Table 10). One recession had occurred during this period. Its first (January 2008) and last (June 2009) months were announced by NBER on 2008/12/01 and 2010/09/20, respectively. The application of the algorithm is shown in Fig. 21.

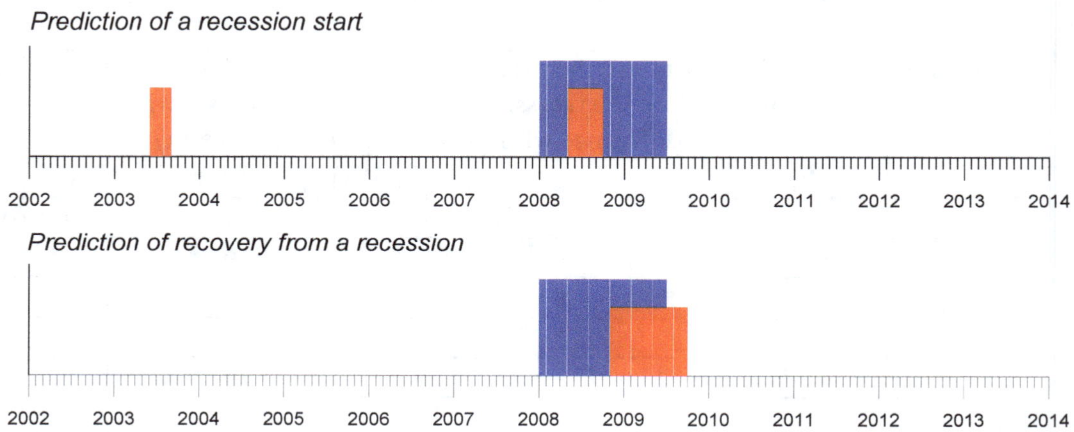

Figure 21. Advance forecasting. Blue bars show the period of the recessions, red bars – the periods of the relevant alarms.

Prediction of a recession start produced one false alarm in 2003 and the alarm on May 2008, four months later than the recession started, but six months before the announcement of its start by NBER.

An alarm for recovery from the recession started in November 2008, seven months before its last month. According to NBER the last recession ended in June 2009; this was established more than three months later. This illustrates an interesting aspect of the advance prediction. If the conclusions given above are correct, the algorithm indicates a time interval when a recession will end. However, testing the algorithm, one might not know when a recession actually ended, as that is decided later by the NBER after the comprehensive analysis. Not knowing whether the recession is still continuing, one would keep using the score Δ (applicable only within recessions) and thus might extend the alarm far into the inter-recession period. A heuristic way to overcome this difficulty could be suggested: since previous alarms lasted not more than 8 months (3.6 months on average), one can tentatively assume that an alarm will not last for more than, say, 9 months and put a 9 months limit for the duration of an alarm. An interesting question is whether the findings of this study (e.g. shown in Fig. 18) might help to recognize that a recession has ended.

CHAPTER 5

Prediction of Unemployment Surges

The study by Keilis-Borok et al. (2005) integrates the expertise in the fields of economics, labor relations, and critical phenomena in complex systems. A specific phenomenon in unemployment dynamics has been considered: episodes of a sharp increase in the unemployment rate called **"Fast Acceleration of Unemployment" or "*FAU*."** The study is a "technical" analysis that is a heuristic search of phenomena preceding *FAUs*. The methodology of *pattern recognition of infrequent events* has been used. It was developed by the artificial intelligence school of I.M. Gelfand for a study of rare phenomena of a highly complex origin, that, by their nature, limit the possibilities of using classical statistical or econometric methods. The goal of Keilis-Borok et al. (2005) was to identify by an analysis of macroeconomic indicators a robust and rigidly defined prediction algorithm of the *"yes or no"* variety indicating at any time moment, whether *FAU* should be expected or not within the subsequent *T* months. Considering unemployment in France between 1962 and 1997, a specific "premonitory" pattern of three macroeconomic indicators has been found that may be used for algorithmic prediction of *FAU*s: among seven *FAU*s identified within these years, six are preceded within 12 months by this pattern that appears at no other time. Application of that algorithm to Germany, Italy and the United States yields similar results. Such predictability reflects the fact that the economy, like many other complex systems, exhibits regular collective behavior patterns. The final test in any prediction research is advance prediction. Such predictions, for USA, have been successful.

5.1 Analysis of Data. Definition of Prediction Target

The problem: prediction of *FAU*s. Keilis-Borok et al. (2005) have considered a specific phenomenon in the dynamics of unemployment: a sharp increase in the rate of unemployment growth. Qualitatively this phenomenon is illustrated in Fig. 22. The thin line is the monthly number of unemployed $u(t)$, including seasonal variations. After smoothing $u(t)$ to eliminate these variations the function $U(t)$ is obtained. The target of the prediction is the starting time of a strong and lasting increase in the unemployment rate $U(t)$. An example is the turning point indicated by the arrow in Fig. 22. Keilis-Borok et al. (2005) termed this target *FAU*, for "Fast Acceleration of Unemployment." The goal is to design an algorithm for predicting the *FAU*s by an analysis of macroeconomic indicators.

Such an algorithm, if found and validated, may be useful in two ways: (i) As a quantitative and reproducible description of phenomena that are premonitory to *FAU*s; this would provide empirical constraints for the theoretical modeling of unemployment. (ii) As a practical tool, complementing existing methods of predicting unemployment. These are the usual twofold goals in any prediction research. Keilis-Borok et al. (2005) have developed such an algorithm for France (Section 5.2) and tested it on independent data for Germany, Italy, and USA (Section 5.3).

Methodology. This study belongs to the so-called "technical" analysis, consisting of a heuristic search for phenomena preceding *FAU*s. The alternative would be a "fundamental" analysis, focusing on "cause-and-effect" mechanisms leading to a *FAU*. The methodology that is used for technical analysis is the *pattern recognition of infrequent events*. It was developed by the artificial intelligence school of I.M. Gelfand (Gelfand et al., 1976) for the study of rare phenomena of highly complex origin. This approach differs from, but complements, classical statistical and econometric methods, such as regression analysis and ARIMA (Engle and McFadden, eds, 1994; see also Stock and Watson 1989; Klein and Niemira 1994; Mostaghimi and Rezayat 1996). For comparison of the pattern recognition with the multiple regression, see the paper by Keilis-Borok et al. (2000), on prediction of economic recessions. This methodology has been successfully applied also to predictions of the outcome of

American elections (Lichtman and Keilis-Borok, 1989; Keilis-Borok and Lichtman, 1993), as well as in seismology (Gelfand et al., 1976; Press and Briggs, 1975; Keilis-Borok and Press, 1980; Press and Allen, 1995; Keilis-Borok and Soloviev, 2003), geological prospecting (Press and Briggs, 1977), and many other fields, as given in the references in these papers. Here, the simplest version of this methodology is used, that is called the "Hamming distance" (Gvishiani and Kossobokov, 1981; Lichtman and Keilis-Borok, 1989; Keilis-Borok and Soloviev, 2003; and references therein). Its essence will be clear from the way in which the data are analyzed here.

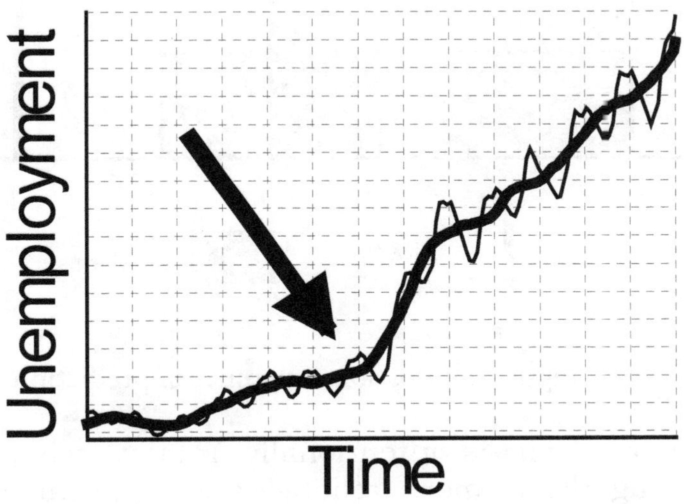

Figure 22. Fast acceleration of unemployment (*FAU*): schematic definition. Thin line – monthly unemployment; with seasonal variations. Thick line – monthly unemployment, with seasonal variations smoothed away. The arrow indicates a *FAU* – the sharp bend of the smoothed curve. The moment of a *FAU* is the target of prediction.

The specific features of this approach might be summarized as follows:

(i) It is applied to predicting not the whole dynamics of unemployment but only the relatively rare extraordinary phenomena - the *FAU*s.

(ii) A quantitative and precisely defined prediction algorithm of the *"yes or no"* variety is looked for, where, at any moment of time, the algorithm would indicate whether or not a *FAU* should be expected within the subsequent *r* months. The possible outcomes of such a prediction are illustrated in Fig. 23. The probabilistic nature of such prediction is reflected in the estimation of the probabilities of false alarms, failures to predict, and of the relative time occupied by alarms, as discussed below.

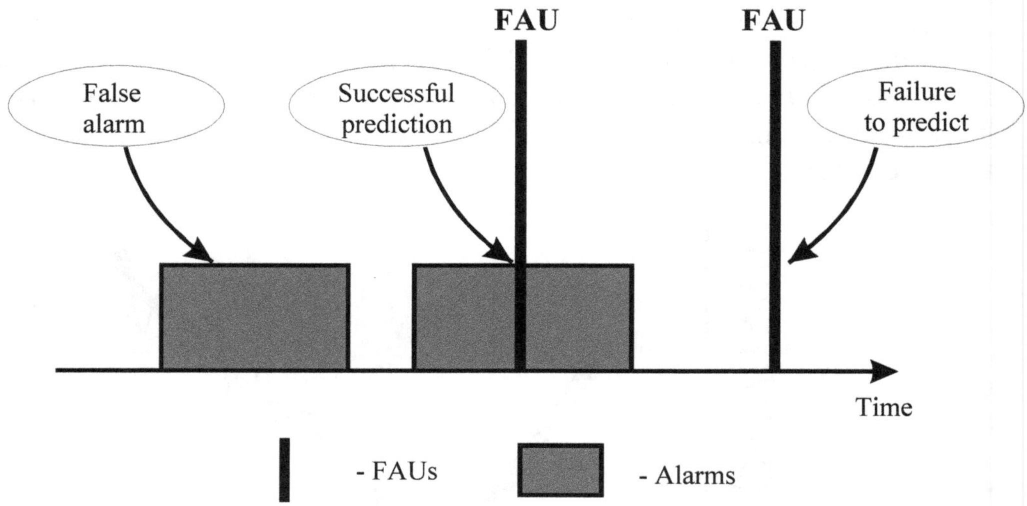

Figure 23. Possible outcomes of prediction.

(iii) The analysis is intentionally robust, which makes the prediction algorithm more reliable and applicable to different *FAU*s, despite the diversity of their specific causes. This is achieved at a price, however, in that the exactness of prediction is limited.

(iv) The analysis does not compete with, but rather complements the short-term predictions of the *"cause and effect"* kind. The cause that triggers a specific *FAU* is usually known, at least in retrospect. It may be, for example, a certain governmental decision, a change in international trade, a major event leading to changes in expectations (such as the attack on the US World Trade Center on September 11, 2001) etc. Accordingly, one might predict an imminent *FAU* (possibly not so rigidly defined) when a triggering event occurs. This does not exclude predictability of *FAU*s with a longer lead-time, as in the present study. On the contrary, it seems plausible that a *FAU* may be triggered only if and when the situation becomes "ripe" for a *FAU*; otherwise the government would not make that decision; unemployment would be less sensitive to international trade, etc. If that conjecture is correct, the approach under consideration can predict such a "ripe" situation. This may be of independent interest for the modeling and containment of unemployment, and it may help in the short-term prediction of *FAU*s.

(v) The dynamics of the economy, including changes in unemployment, may have some basic features that are common in various complex processes and found in models of statistical physics (Burridge and Knopoff, 1967; Gell-Mann, 1994; Newman et al., 1994; Allègre et al., 1995; Holland, 1995; Blanter et al., 1997; Gabrielov et al., 2000a; Keilis-Borok and Soloviev, 2003). The results presented in this section may provide for such modeling of the scaling and parameterization of the processes considered and a formal definition of the relevant critical phenomena.

The data. Databases issued by the Organization for Economic Cooperation and Development (OECD, 1997) and the International Monetary Fund (IMF, 1997) have been used. Past *FAU*s were identified by an analysis of the monthly statistics of unemployment. To explore the predictability of *FAU*s the monthly indicators listed below were analyzed. For France the data sources are sufficiently complete for the time period between January 1965 and May 1997, and it is this period that is considered here.

Composite characteristics of the French national economy:

1. *IP*: Industrial production index, composed of weighted production levels in numerous sectors of the economy, in % relative to the index for 1990.
2. *L*: Long-term interest rate on 10-year government bonds, in %.
3. *S*: Short-term interest rate on 3-month bills, in %.

The analogues of these indicators for the USA have been successfully used in research on predicting American economic recessions (Stock and Watson, 1993; Keilis-Borok et al., 2000).

Characteristics of more narrow areas of economy which are sensitive to its overall state:

4. *NC*: The number of new passenger car registrations, in thousands of units.
5. *EI*: The expected prospects for the national industrial sector.
6. *EP*: The expected prospects for selected manufacturers.
7. *EO*: The estimated current volume of orders.

The last three indicators are subjective; they distinguish the "good" from "bad" situations by polling 2,500 manufacturers, with the expectations weighted by the size of their businesses. They are available only for France.

Two indicators related to the American economy:

8. *FF/$*: Value of U.S. dollar in French francs.

9. *AR*: The state of the American economy: is it close to a recession or not? This indicator shows whether a pre-recession alarm for the USA is or is not declared by the algorithm described by Keilis-Borok et al. (2000). Note that it is not the American recessions themselves are considered here, but a certain precursor to these recessions, which includes no European indicators.

Identification of *FAUs*. Here, the definition of *FAU* that was qualitatively described above (Fig. 22) is formalized. As before $u(m)$ is the number of unemployed in a month m (m = 1, 2, ...). *FAUs* are defined as follows. First, smoothing out the seasonal variation of u function $U(m) = W^u(m/m\text{-}6, m\text{+}6)$ is obtained that is a value of the regression (1) over the time interval (m – 6, m + 6). Next, a function $F(m/s) = K^U(m+s,m) - K^U(m,m-s)$ is determined that is the difference between the linear trends in regression (1) of $U(m)$ within s subsequent months and s preceding months. This function with s = 24 months is used as a coarse measure of unemployment acceleration. Finally, the *FAUs* are defined by the local maxima of $F(m)$ exceeding a certain threshold $\boldsymbol{F}$. The time m^* and the height F^* of such a maximum are, respectively, the time and the magnitude of a *FAU*. Acceleration ends in a month m_e of the subsequent local minimum of $F(m)$.

***FAUs* in France.** Monthly unemployment in France is shown in Fig. 24. The function $F(m)$ for the time period considered, from January 1965 to May 1997, is also shown in Fig. 24. One may see in Fig. 24 that the theshold $\boldsymbol{F}$ = 4 identifies obviously outstanding peaks of $F(m)$. Seven *FAUs* identified by the condition $F^* \geq \boldsymbol{F} = 4$ are listed in Table 19. One can see that each *FAU* is the beginning of a long unemployment surge, lasting 16 – 24 months. Since it is determined after a strong smoothing of unemployment rate, the meaningful accuracy of prediction may hardly be better than about 2 months. Three "major" *FAUs*, marked in bold in Table 19, are distinctly larger than the others.

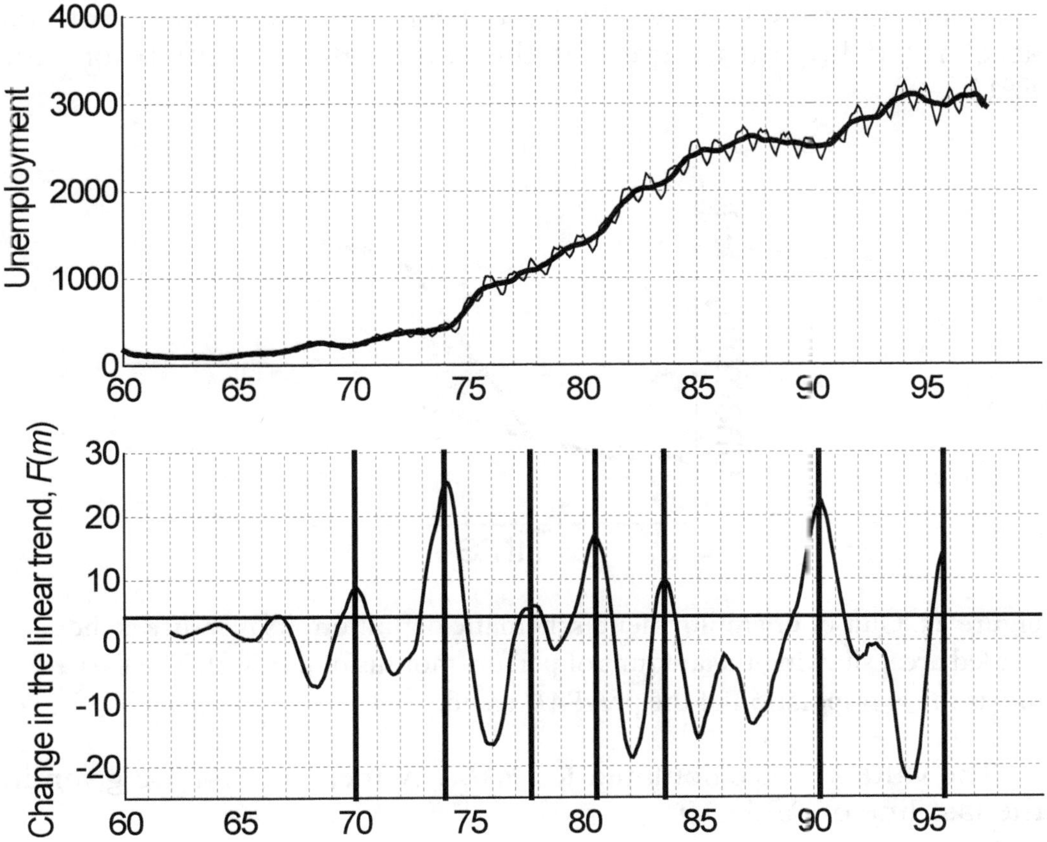

Figure 24. Unemployment in France. *Top:* Monthly unemployment, thousands of people. Thin line: $u(m)$, data from the OECD database; note the seasonal variations. Thick line: $U(m)$, data smoothed over one year. *Bottom:* Determination of *FAUs*. $F(m)$ shows the change in the linear trend of unemployment $U(m)$. *FAUs* are attributed to the local maxima of $F(m)$ exceeding threshold $F = 4.0$ shown by horizontal line. The thick vertical lines show moments of the *FAUs*.

Besides the magnitude F^*, the scale of FAU phenomenon may be characterized by the "*glut*" of unemployment G (Fig. 25)

$$G = \sum\{U(l) - W^U(l/m^* - s, m^*)\}, \qquad m^* \leq l \leq m_e. \tag{4}$$

Here m^* and m_e are as defined above the first and the last month of the *FAU*. The first term is the actual cumulative loss of employment, in person-months, during the FAU. The second term shows what this loss would have been had unemployment not accelerated but rather grew at the same rate as right before the *FAU*.

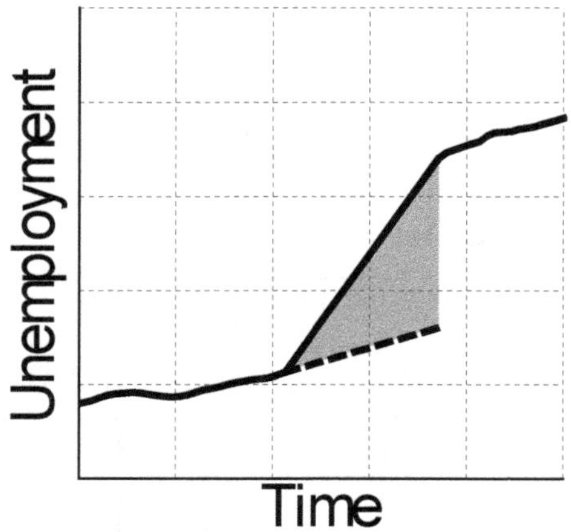

Figure 25. Glut of unemployment: schematic definition. Its measure is the shaded area, showing the amount of person-months of jobs lost. The dashed line is the extrapolation of the pre-*FAU* trend.

The values of *G* calculated for *FAU*s by means of (4) are given in the last line of Table 19.

Table 19. *FAU*s in France, 1965–1997

Time, *year: month*	1970:01	1974:01	1977:09	1980:07	1983:07	1990:05	1995:09
"Magnitude" F^*	7.6	22.6	5.3	15.7	9.2	20.3	9.4
Duration (m_e - m^*) (months)	22	24	16	20	19	21	20
G (person × months)	1993	5570	371	2760	1400	3420	2960

5.2 Hypothetical Prediction Algorithm

Premonitory trends of single indicators. Here, the "premonitory" trends of the indicators are explored, which tend to occur more frequently as a *FAU* approaches.

The trends are approximated by the regression coefficients $K^f(m/s)$ that are defined accordingly to formulas (1), (2) with f replaced by the symbol of an indicator. The value of $K^f(m/s)$ may be used for prediction, since it is attributed to the *end* of the time window where it is determined; accordingly, it does *not* depend on the information about the future. (The definition of the *FAU*s is applicable only in retrospect, two years after a *FAU* occurs in order to ensure a reliable identification of past *FAU*s).

An example of further analysis is given in Fig. 26. It shows the indicator $L(m)$ and its trend $K^L(m/s)$. The plot of $K^L(m/s)$ suggests that its peaks (i.e. the relatively steep upward trends of $L(m)$) emerge more frequently before the *FAU*s; in that sense they are "premonitory" to *FAU*s. To give a robust quantitative definition of a premonitory trend, the values of $K^L(m/s)$ are defined on the lowest level of resolution, distinguishing only the values above and below a threshold $T^L(Q)$, i.e. the discretization is made as described above in Section 4.2. The threshold is defined as a percentile of a level Q, that is, by the condition that $K^L(m/s)$ exceeds $T^L(Q)$ during Q% of the months considered.

Similar analyses of other indicators are summarized in Table 20. Premonitory behavior of these indicators has a transparent qualitative explanation.

(i) A steep rise in macroeconomic indicators (Nos. 1 – 3 in Table 20) reflects the "overheating" of the economy. For industrial production (No. 1) this may sound counterintuitive, since the rise of production is supposed to create more jobs. However, premonitory to *FAU*s is only a steep rise, above 50% percentile, which creates oversupply.

(ii) A steep decline of the next four indicators reflects the expectations of the general public (#4) and the appraisal of economic trends by the business community (#5 – 7). These indicators seem better substantiated, than, say, public opinion polls.

(iii) The last two indicators (#8 and 9) are related to the American economy. Their premonitory behavior would be more difficult to foresee – an equally plausible explanation could be probably found for the opposite behavior. Accordingly, these indicators yield particularly strong constraints on the modeling of the dynamics of unemployment.

Summing up, the description of the unemployment-relevant situation has been reduced to a monthly time series of a binary vector with 9 components, as is usual in the pattern recognition of infrequent events. For convenience, the same code, 1, is given to the "premonitory" trend of each indicator, regardless of whether it is an upward or a downward one.

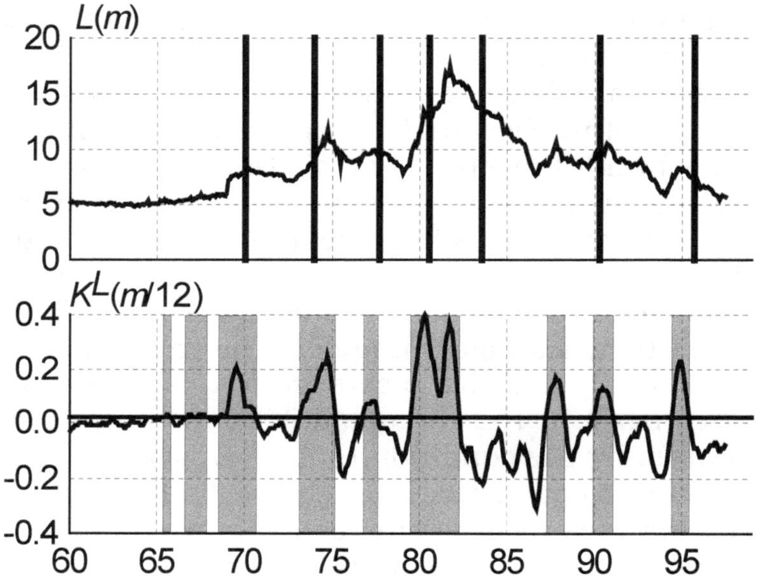

Figure 26. Robust discretization of an economic indicator. Top: $L(m)$, interest rate on long-term (10 year) governmental bonds (%). Source: OECD. Bottom: $K^L(m/12)$, linear growth of $L(m)$ in the sliding time window $(m - 12, m)$. The thin horizontal line is the upper 33% percentile for $K^L(m/12)$ so that it exceeds this threshold 33% of time. The gray bars mark time intervals when the value of $K^L(m/12)$ exceeds the threshold. The thick vertical lines show moments of *FAUs*. Note that after a quiet period before 1967 "large" values of $K^L(m/12)$ tend to appear more frequently when a *FAU* is approaching. However, with this threshold they would "predict" the *FAUs* only in combination with other indexes.

Ensemble of premonitory trends. Here, it is considered how the approach of a *FAU* is reflected in the *collective behavior* of the indicators. Its simplest description is function $\Delta(m)$ – the number of non-premonitory indicators at the month *m*. If the identification of premonitory trends is correct then the value of $\Delta(m)$ should decrease as a *FAU* approaches. By definition $\Delta(m)$ is the number of zeros in the binary code of the situation. This is the so-called "Hamming distance" between that code and the code of the "perfect" premonitory situation, when all the components are equal to 1, that is, all the trends are premonitory. This measure was used in different applications of the pattern recognition – to elections (Lichtman and Keilis-Borok, 1989), recessions (Keilis-Borok et al, 2000), and earthquakes (Keilis-Borok and Kossobokov, 1990; Vorobieva and Levshina, 1994; Keilis-Borok and Soloviev, 2003).

Table 20. Trends and thresholds

Indicator	**Premonitory trend**	***s***	***Q* (%)**
IP: industrial production index	Upward	12	50
L: interest rate, long-term bonds	Upward	12	33
S: interest rate, short-term bills	Upward	12	25
NC: new passenger cars registrations	Downward	6	33
EI: prospects for industrial sector	Downward	6	33
EP: prospects for selected manufacturers	Downward	6	33
EO: orders	Downward	6	33
FF/$: french francs per USD, exchange rate	Downward	6	33
AR: recession alarm in the U.S.	Is current		

Only the composite indicators *IP*, *L*, and *S* were considered first. The value of $\Delta(m)$ may vary in this case from 0 to 3. Change of $\Delta(m)$ through the time considered is juxtaposed with *FAU*s in Fig. 27. One can see that the minimal value $\Delta(m) = 0$ appears within 1 – 12 months before a *FAU* and at no other time. Data on Fig. 27 suggest the following hypothetical prediction algorithm: *An* *alarm is declared for 6 months after each month with* $\Delta(m) = 0$ *(regardless of whether this month belongs or not to an already determined alarm).* A waiting period of 6 months is introduced because in three cases (1977, 1980, and 1995) the premonitory pattern does not appear right before a *FAU*. Results of prediction are shown in the top row of Fig. 28. One can see, that this algorithm predicts 6 out of 7 *FAU*s, including all three major ones.

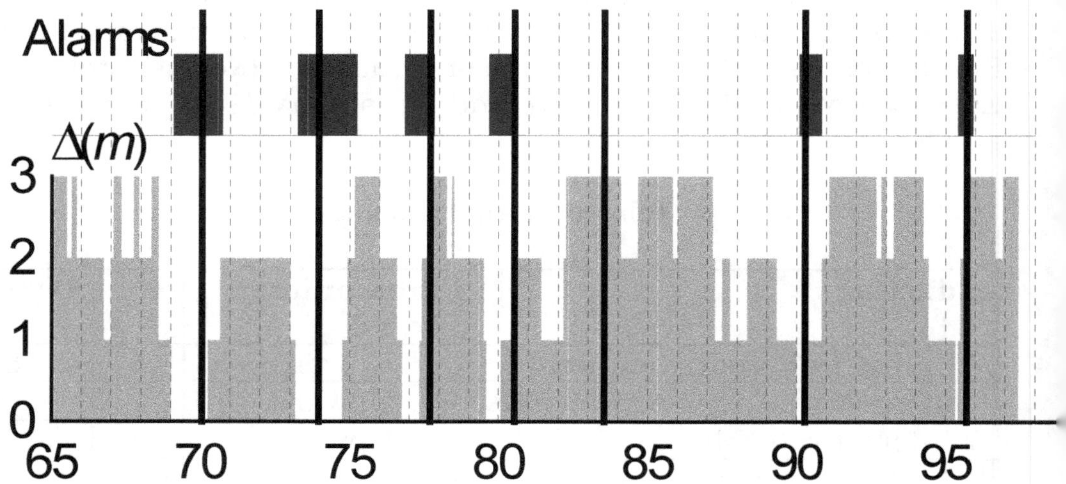

Figure 27. Collective performance of premonitory trends. Function $\Delta(m)$ is the number of *non-premonitory* trends at month *m*. Vertical lines show *FAU*s. Alarms (shown by dark gray bars) are declared for 6 months after each month when $\Delta(m) = 0$.

5.3. Stability of Prediction (Sensitivity Analysis)

The above algorithm is developed retrospectively, with a certain freedom in the ad hoc choice of indicators and numerical parameters. The results of prediction should be stable to their variations. To estimate this stability, the prediction needs to be repeated with different groups of indicators and values of parameters. In this study, the algorithm formulated above is generalized as follows: *An alarm is declared for r months after each* *month with $\Delta(m) \le D$ (regardless of whether this month belongs or not to an already determined alarm).* The hypothetical prediction algorithm formulated above corresponds to the case $D = 0$, $r = 6$.

Five other groups of indicators (Table 21) were considered. The results are compared in the error diagram in Fig. 29. Error diagrams as a tool for the evaluation of a prediction method and for the optimal choice of a response to a prediction are described by Molchan (1997, 2003). Numerical values of τ (the total duration of alarms in per cent of the total time considered), $\boldsymbol{n}$ (the number of failures to predict), and $\boldsymbol{f}$ (the number of false alarms) are given in the last three columns of Table 21.

Table 21. Variation of the groups of indicators used for prediction

Indicators	***D***	***r***	τ (%)	***n***	***f***
IP, L, S	0	6	23	1	0
IP, L, S, FF/$	0	6	13	3	1
NC, EO	0	3	31	0	5
IP, L, S, NC, EO	1	6	26	1	1
IP, L, S, NC, EO, FF/$	1	6	19	2	1
IP, L, S, NC, EO, AR	2	6	28	1	0

The last three columns show the performance of a prediction algorithm.

One can see that composite indicators (group 1) give relatively the best predictions, alone or with some other indicators. Only the group 3 gives no failures to predict; however that is achieved at the price of 5 false alarms, which are hard to accept. There were attempts to replace the indicator *EO* in groups 3 – 6 by a subjective indicator *EI* or *EP*, but in each case the number of failures to predict and false alarms increased simultaneously.

Numerical parameters that define premonitory trends (Q^I, *s*), alarms (*D, r*) and *FAU*s (***F***, *s*) were also varied.

Stability of prediction (sensitivity analysis). The stability of these predictions has been explored by repeating them with various indicators and parameters and comparing the ensuing alarms. The percentile *Q* was varied from 5 to 10% and the window *s* for determination of the regression coefficients was varied from 9 to 15 months. The outcomes of several experiments on the variation of the adjustable parameters are shown in Table 22. These experiments were made for the prediction based on the first group of indicators: *IP*, *L*, and *S* (line 1 in Table 21).

Variation of definition of FAU. Two adjustable parameters are used to define the moments of *FAU*s: the time window *s* for evaluation of the trend of unemployment and the threshold ***F*** for identification of "fast" acceleration. The analysis given above corresponds to the case *s* = 24 months and ***F*** = 4. Variation of *s* from 21 to 27 does not shift the moments of *FAUs* by more than 1-2 months – quite within the meaningful accuracy of prediction. The variation of ***F*** in the range from 3.6 to 5.0 does not lead to the emergence or disappearance of the targets of the prediction – the local maxima of *F*(*m*).

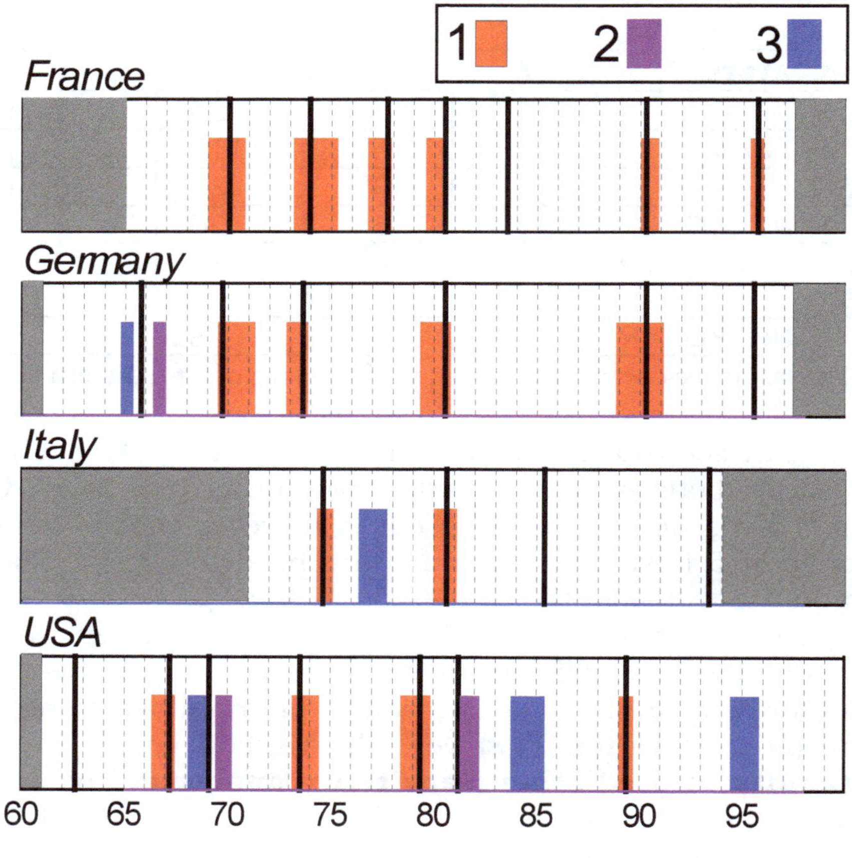

Figure 28. Retrospective prediction for four countries. *FAUs* and alarms obtained by the prediction algorithm. The thick vertical lines show the moments of *FAUs* in a country. Bars – the alarms with different outcome: 1 –

alarms that predict *FAUs*, 2 – alarms starting shortly after *FAUs* within the periods of unemployment surge, 3 – false alarms. Shaded areas on both sides indicate the times, for which data on economic indicators were unavailable. In case 2 *FAUs* are counted, for rigor sake, as not predicted. Note however that the alarms starting shortly after these *FAUs* are practically useful (see Section 5.4)

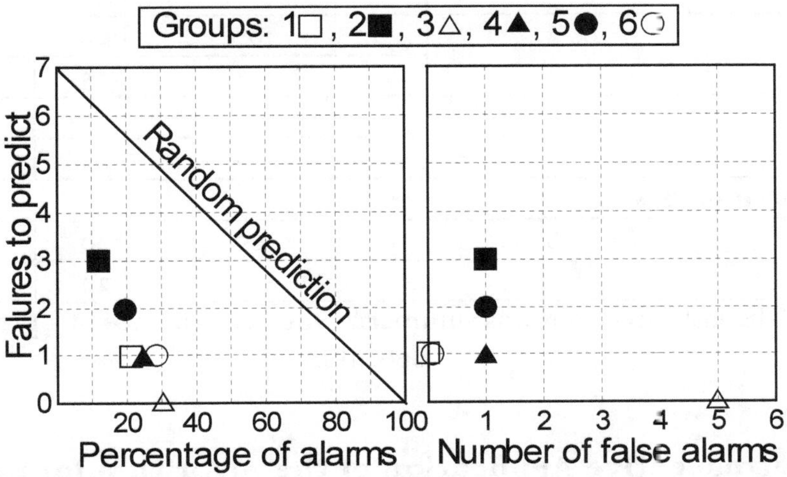

Figure 29. Error diagram for predictions with various groups of indicators. Group 1 includes *IP, L, S*; 2 - *IP, L, S, FF/$*; 3 - *NC, EO*; 4 - *IP, L, S, NC, EO*; 5 - *IP, L, S, NC, EO, FF/$*; 6 - *IP, L, S, NC, EO, AR*.

Summing up, one can conclude that the *most stable results are obtained with group 1 indicators - the three composite economic indicators*. Prediction with this group produced no false alarms; it sustains best the change of numerical parameters; and it does not gain by the addition of other indicators.

Next, the algorithm is tested on the independent data not used in its development. It is applied to Germany and Italy and to the USA.

Table 22. Variation of adjustable parameters for indicators *IP*, *L*, and *S*

#	Indicator(s), for which variation of a parameter is made	Variation of the parameter	τ (%)	*n*	*f*
Variation of *Q*					
1	*IP*	60%	23	1	0
2		40%	22	1	0
3	*L*	40%	28	1	1
4		25%	15	2	0
5	*S*	30%	30	1	2
6		20%	23	1	0
Variation of *s*					
7	*IP, L, S*	15	20	1	0
8		9	24	2	1

The last three columns summarize the performance of a prediction algorithm

5.4. Retrospective Application of the Algorithm for Several Countries

Germany and Italy. Here the algorithm formulated in Section 5.2 is applied to the data for Germany and Italy. *Exactly* the same algorithm is used, including the definition of the trends $K^f(m,m\text{-}s)$; the values of adjustable numerical parameters; and the rule for declaration of alarms. Such transfer from country to country is possible, since the algorithm is self-adaptive, where the percentiles Q^f common in each application determine the thresholds for discretization of the trends.

The results are shown in Fig. 28. Table 23 and Fig. 30 that compare the quality of prediction for different countries. Alarms preceding *FAU*s are usually continuing within the periods of fast unemployment growth (after *FAU*s), and such continuations should obviously not be regarded as false alarms. Alarms within the periods of fast unemployment growth are also not regarded as false ones. For Germany the quality of prediction is about the same as for France, while for Italy it is lower, though still better than random.

Table 23. Comparative performance of the prediction algorithm for France, Germany, Italy, and the USA

Country	*FAUs*			Alarms "not predicting" *FAUs*	
	Total	"Predicted"	"Missed"	After *FAUs*	"False"
France, 1965–97	7	6	1	none	none
Germany, 1961–97	6	4	2	1	1
Italy, 1971–93	4	2	2	none	1
USA, 1961-99	7	4	3	2	3

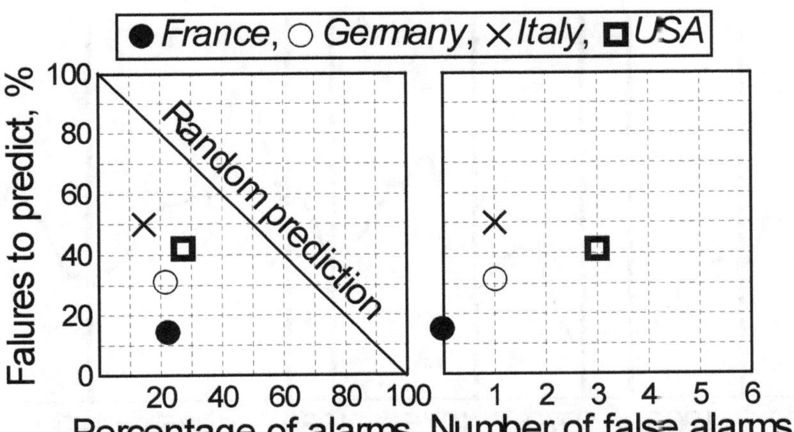

Figure 30. Error diagram for predictions for different countries.

Unemployment in the U.S. The data on monthly unemployment rates for the U.S. civilian labor force are used, as given in USDL (1999). Unlike Europe, unemployment in USA had no general trend during the years considered. One can see this in Fig. 31. The *FAUs* are the times when unemployment started to surge, that are the local minima of the unemployment rate. They are formally defined as follows. Let *R*(*m*) be the smoothed monthly unemployment rate in a month *m*. Then *R*(*m*) has the local minima in a month m^* if for j = 1, 2, 3, 4 $R(m^*-j) \geq R(m^*)$ and $R(m^*+j) > R(m^*)$. Seven such minima are identified within the period 1960-1999 in 1962:08 (9), 1967:03 (3), 1969:02 (28), 1973:07 (24), 1979:05 (19), 1981:03 (21), and 1989:05 (38). The duration of the unemployment surge is given in brackets after the corresponding months m^*, which are the targets of the prediction.

Application of the algorithm to the U.S. Indicators *IP*, *S*, and *L* have the following American equivalents. For *IP* – "industrial production, total". This is the index of the real (constant dollars) output in the entire economy (dimensionless). For *S* - interest rate on 90-day U.S. treasury bills, at an annual rate (in percent). For *L* - interest rate on 10-year U.S. treasury bonds, at an annual rate (in percent). Time series of these indicators were used for prediction of economic recessions in the U.S. (Section 4.2) and obtained from the CITIBASE, where *IP*, *FYGM3*, and *FYGT10* are their respective mnemonics.

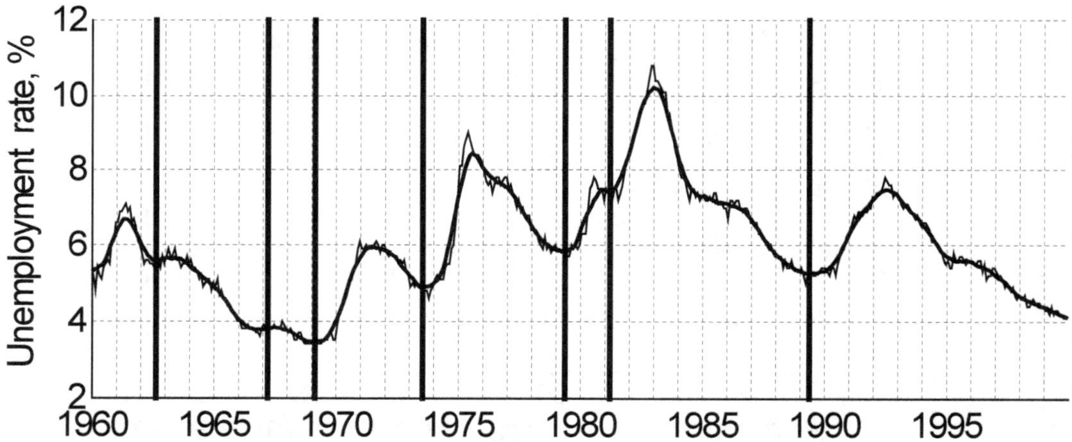

Figure 31. Unemployment rates in the U.S. Thin line: $r(m)$, original data. Thick line: $R(m)$, data after smoothing out the seasonal variations. The thick vertical lines show the moments when unemployment started to surge (local minima of smoothed unemployment rate).

Alarms and *FAU*s are juxtaposed in the bottom row of Fig. 28. One can see that 4 out of 7 *FAU*s are captured by alarms; three *FAU*s, in 1962, 1969, and 1981, are missed; and there are false alarms, in 1968, 1983, and 1994. The alarms within the periods of unemployment growth are not regarded as false ones. The total count of errors for the *USA* is given in Table 23 and Fig.24. It is worse than for European countries, though still better than random.

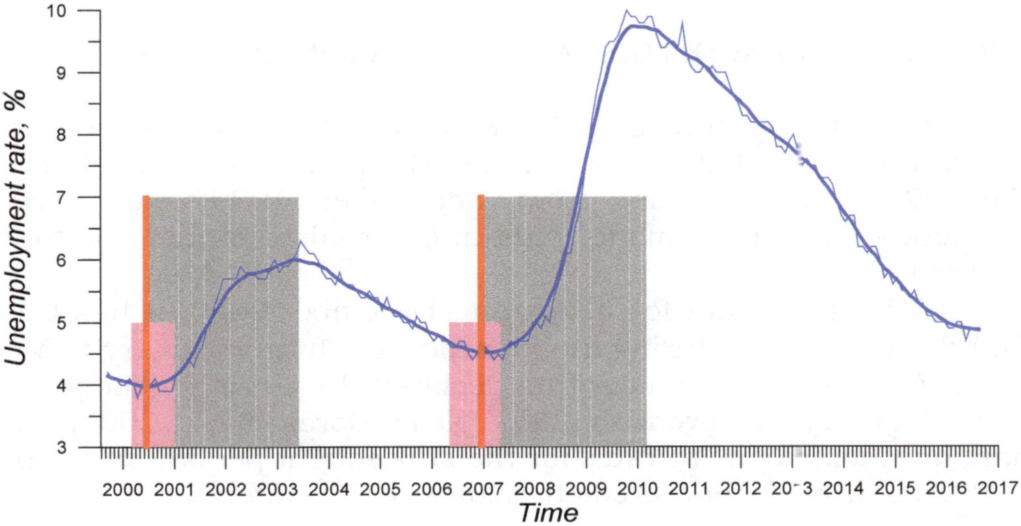

Figure 32. Prediction of future *FAU*s in the U.S. from September 1999. Thin blue curve shows monthly unemployment rate in the U.S., according to the data of the Bureau of Labor Statistics, U.S. Department of Labor. Thick curve shows this rate with seasonal variation smoothed away. Vertical red lines show prediction targets – the moments of *FAU*; grey bars – the periods of unemployment's growth; pink bars – periods of alarms.

Note, that this is a rigorous count, giving lower estimate for the algorithm's performance. Such an estimate is necessary for some purposes, e.g. evaluation of statistical significance of predictions, but for other purposes it might be misleading. Next a more practical estimate is discussed. Let us consider the count of errors from a disaster preparedness point of view. One of the alarms ended in 1968:12, a month before the *FAU*; it is counted as a false alarm and the subsequent *FAU* – as missed by prediction. Similarly, the *FAU* in 1981:03 is counted as missed, while it was followed by an alarm a month later. Since a *FAU* is a starting point of a long surge in unemployment, lasting about 20 months, a one month difference is not necessarily important for a decision-maker responding to a prediction. Moreover, this difference is within the accuracy of *FAU*s, since they are determined after considerable smoothing of the unemployment rate. Accordingly, for the end user only the three errors might be worth counting: the failure to predict in 1962 and the false alarms in 1983 and 1994.

5.5 Application to the U.S. Advance Prediction

Prediction of the future *FAU*s was launched for USA by the algorithm formulated in Section 5.2. The results are presented in Fig. 32. It shows that from September 1999 two correct predictions have been made, without ether false alarms or failures to predict.

Analyzing the data for USA up to December 2000, it has been found that $\Delta(m) = 0$ during the four months, from February to May 2000. Accordingly, the algorithm declared the alarm for the period from February to November 2000 (Keilis-Borok et al., 2005). The data on unemployment rates for the subsequent period, up to July 2002, confirmed that prediction of a *FAU* materialized in July 2000.

The next alarm was declared after April 2006 when $\Delta(m) = 0$ (Keilis-Borok et al., 2006) for the first time after 2000 during the seven months periof from April to October of 2006. In November 2006 this second prediction was filed on the web site of the Anderson School of Management, University of California, Los Angeles (http://www.uclaforecast.com). The relevant FAU occurred in December 2006.

Obviously, these confirmations are most encouraging, but much longer experimentation is necessary to validate the algorithm.

Numerous other warnings of a coming surge of unemployment in the USA did appear in 2000 and 2006, and even in the popular media. The particular feature of the prediction discussed here, however, is that it was obtained using a formal unambiguous algorithm, and that it indicates a specific time interval when the unemployment will start to surge.

5.6 Recessions and Unemployment

Table 24 shows the comparison of the periods of unemployment growth and recession in the U.S. We see that all eight American recessions during the time period, 1960-2009, did occur within the eight longest periods of unemployment growth. Three of the seven pre-recession alarms determined by the algorithm developed Keilis-Borok et al. (2000) and described in Section 4.3 begin before the relevant months m^* (Table 24). The level of unemployment, measured in a different way, was actually used for the determination of these alarms. This confirms that the prediction of unemployment is indeed not equivalent to the prediction of recessions.

Table 24. Periods of unemployment growth, recessions in the U.S., and pre-recession alarms

Local minima m^*	End of unemployment growth	Recessions	Pre-recession alarms
1959:11	1961:05	1960:05 – 1961:02	
1962:08	1963:05		
1967:03	1967:06		
1969:02	1971:06	1970:01 – 1970:11	1969:07 – 1969:12
1973:07	1975:07	1973:12 – 1975:03	1973:06 – 1973:11
1979:05	1980:12	1980:02 – 1980:07	1979:03 – 1980:01
1981:03	1982:12	1981:08 – 1982:11	1981:02 – 1981:07
1989:05	1992:07	1990:08 – 1991:03	1989:06 – 1990:07
2000:06	2003:05	2001:04 – 2001:11	2001:01 – 2001:03
2006:12	2010:02	2008:01 – 2009:06	2008:05 – 2008:09

CHAPTER 6

Prediction of Homicide Surges in Megacities

Dynamics of crimes reflects important aspects of sustainability of our society and the risk of its destabilization – a prelude to a disaster. Keilis-Borok et al. (2003) have considered a prominent feature of crime dynamics – surge of the homicides in a megacity. This study integrates the professional expertise of the police officers and of the scientists working on pattern recognition of infrequent events. The latter is a type of artificial intelligence methodology that has been successful in predicting infrequently occurring phenomena that result from highly complex processes.

Statistics of several types of crimes in the city of Los Angeles over the period 1975-2002 has been analyzed by Keilis-Borok et al. (2003). This analysis focuses on how these statistics change before a sharp and lasting rise ("a surge") of the homicide rate. The goal is to find an algorithm for predicting such a surge by monitoring the rates of different crimes.

The hope for feasibility of this goal comes from two sources. The first one is the set of available crime statistics, showing that a surge of major crimes is preceded by the rise of less severe crimes. The second one is recent research in the predictions of critical phenomena (i.e. abrupt overall changes) in various complex non-linear systems, such as those in theoretical physics, earth sciences, social sciences, etc.

Data. Out of a multitude of relevant data statistics of robberies, assaults, burglaries, and the homicides themselves is analyzed.

Results. The findings may be summarized as follows: Episodes of a rise of burglaries and assaults simultaneously occur 4 to 11 months before a homicide surge, while robberies decline. Later on, closer to the rise in homicides, robberies start to rise. These changes are given unambiguous and quantitative definitions, which are used to formulate a hypothetical algorithm for the prediction of homicide surges.

In retrospective analysis it has been found that this algorithm is applicable through all the years considered, despite substantial changes both in socio-economic conditions and in the number of crimes. Moreover, it gives satisfactory results for the prediction of homicide surges in New York City as well. Sensitivity tests show that predictions are stable to variations of the adjustable elements of the algorithm.

What was learned? The existing qualitative portrayals of crime escalation are complemented here by a quantitatively defined set of precursors to homicide surges. The same set emerges before each surge through the time period under consideration. This implies the existence of a "universal" scenario of crime escalation, independent of concrete reasons triggering each surge. These findings provide heuristic constraints for the modeling of crime dynamics and indicate promising lines of further research.

Perspective. Decisive validation of these findings requires experimentation in *advance prediction*, for which this study sets up a base. Particularly encouraging for this further research is the wealth of yet untapped possibilities: so far only a small part of the data and mathematical models that are currently available and that are relevant to crime dynamics have been used.

On the practical side, these results enhance the capability to identify a situation that is "ripe" for homicide surges and, accordingly, to escalate the crime prevention measures. In a broader scheme of things, a surge of crime is one of potential ripple effects of natural disasters. Accordingly, the risk of a natural disaster is higher in such a situation.

Prediction of a specific phenomenon in crime dynamics: a large and lasting increase in the homicide rate was considered. Qualitatively, this phenomenon is illustrated in Fig. 33; Keilis-Borok et al. (2003) called it by the acronym ***SHS*, for "Start of the Homicide Surge"**. The goal is to find a method to predict an *SHS* by monitoring the relevant observed indicators, i.e. to recognize the "premonitory" patterns formed by such indicators when an *SHS* approaches. In terms of pattern recognition, an algorithm (a "recognition rule") that solves the following problem is looked for:

given the time series of certain crime rates (or of other relevant indicators) prior to a moment of time *t*,

to predict whether an episode of *SHS* will or will not occur during the subsequent time period (t, $t+\tau$); in other words, whether the lasting surge of homicides will or will not start during that period.

If the prediction is "yes", this period will be the "period of alarm." The possible outcomes of such a prediction are illustrated in Fig. 34.

The probabilistic component of this prediction is represented by the estimated probabilities of errors – both false alarms on one side and failures to predict on the other. That probabilistic component is inevitable since we consider a highly complex non-stationary process using imprecise crime statistics. Moreover, the predictability of a chaotic system is, in principle, limited.

Such "yes or no" prediction of specific extraordinary phenomena is different from predictions in a more traditional sense - extrapolation of a process in time, which is better supported by classical theory.

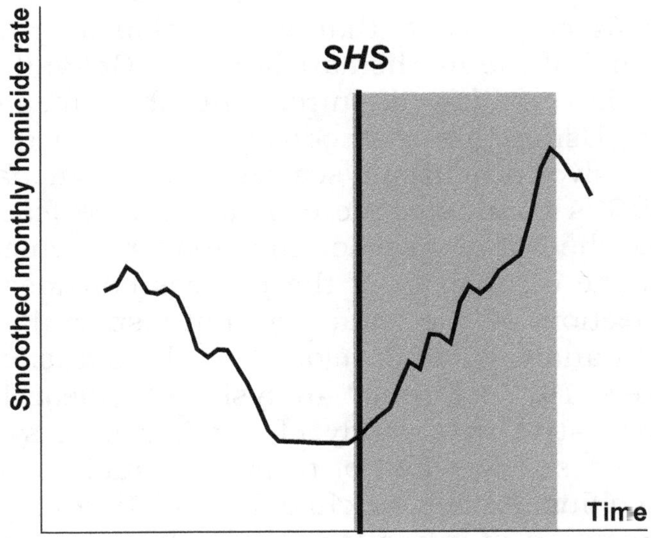

Figure 33. Target of prediction – the Start of the Homicide Surge ("*SHR*"); schematic definition. Gray bar marks the period of homicide surge.

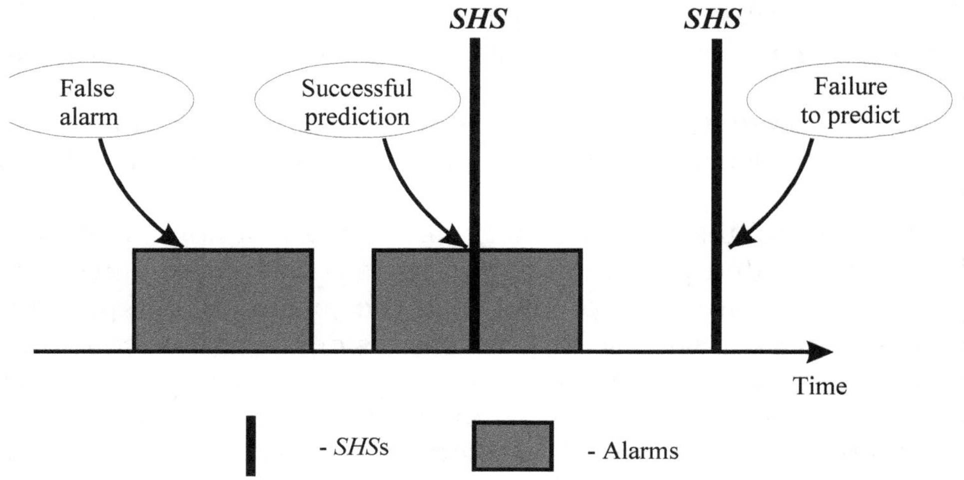

Figure 34 Possible outcomes of prediction.

The methodology used by Keilis-Borok et al. (2003) is *pattern recognition of infrequent events* that was developed by the artificial intelligence school of the mathematician I. M. Gelfand (Gelfand et al., 1976) for the analysis of infrequent phenomena of highly complex origin. Using this methodology, a so-called "technical" analysis that involves a heuristic search for phenomena preceding episodes of *SHS* is conducted here. A distinctive feature of this methodology is the robustness of the analysis, which helps to overcome both the complexity of the process considered and the chronic imperfection of the data; in that aspect it is akin to exploratory data analysis, as developed by the statistics school of J. Tukey (Tukey, 1977). Robust analysis – "a clear look at the whole" – is imperative in a study of any complex system (Gell-Mann, 1994). The surest way *not* to predict such a system is to consider it in too fine detail (Crutchfield et al., 1986).

Pattern recognition of infrequent events has been successfully used in geophysics, geological prospecting, medicine, and many other areas. Close to the present study are recent studies of the prediction of economic recessions and surges of unemployment (Keilis-Borok et al., 2000, 2005). The same pattern recognition algorithm, called "Hamming distance," is used by Keilis-Borok et al. (2003). It has been applied in these studies (Keilis-Borok et al., 2000, 2005), as well as in predictions of American elections (Keilis-Borok and Lichtman, 1993) and in seismology (e.g. Kosobokov, 1983; Vorobieva and Levshina, 1994; Vorobieva, 1999). The essence of the algorithm will be clear from the way of crime statistics analyzing.

Following is a schematic outline of our analysis.

Data comprise the monthly rates of homicides, robberies, assaults, and burglaries for Los Angeles, 1975 – 2002 (Section 6.1).

Five targets of prediction (*SHS*) are defined during the time period under consideration (Sections 6.2, 6.6). Those are the moments when a years-long trend of the homicide rate turns from a decline to a long steep rise.

Premonitory changes of crime statistics have been found as illustrated in Fig. 35. Within several months before a homicide surge, burglaries and assaults simultaneously escalate, while robberies decline (Section 6.3). On the basis of these changes a *hypothetical prediction algorithm* is suggested (Section 6.4). In retrospect, it provides a robust satisfactory prediction (Sections 6.5, 6.6). However it has to be further validated by application to independent data. As always in prediction research, *the final validation of the algorithm requires prediction in advance*, for which this study sets up a base.

Later on, closer to a homicide surge, robberies also escalate (Section 6.7). These changes will be explored elsewhere.

Common notation. The analysis focuses on trends in the crime rates. These trends are estimated by linear regression, using the following notations.

$C(m)$, $m = 1,2\ldots,$

is the time series of a monthly indicator, where m is the sequence number of a month.

$$W^C(m/q, p) = K^C(q, p)m + B^C(q, p),\ q \le m \le p, \qquad (5)$$

is the local linear least-squares regression of the function $C(m)$ within the sliding time window over the time period (q, p).

6.1. The Data

The following data sources are used:

(i) The National Archive of Criminal Justice Data (NACJD), placed on the web site:

http://www.icpsr.umich.edu/NACJD/index.html
(see Carlson (1998) for its description. This site contains data for the years 1975-1993).

(ii) Data bank of the Los Angeles Police Department (LAPD Information Technology Division); it contains similar data for the years 1990 – May 2001.

Out of numerous crime statistics given in these sources, the monthly rates of the four types of crime listed in Table 25 are analyzed: homicides, robberies, assaults, and burglaries.

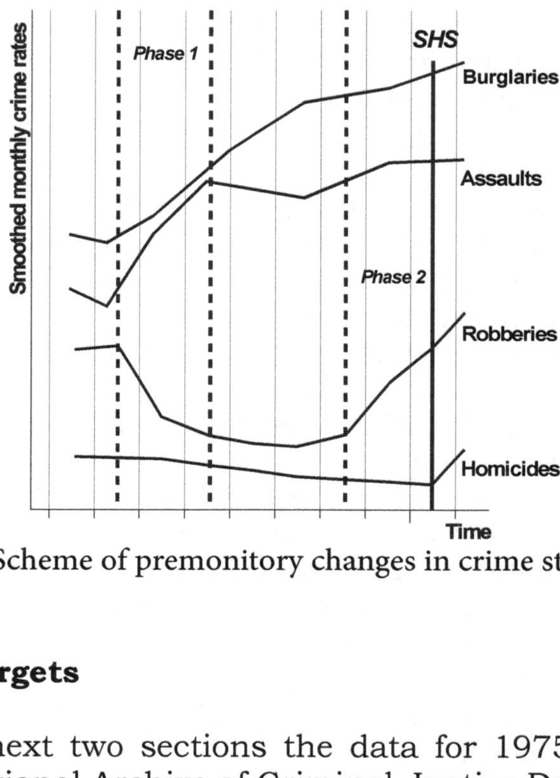

Figure 35. Scheme of premonitory changes in crime statistics.

6.2 Prediction Targets

Here and in the next two sections the data for 1975 – 1993 as taken from the National Archive of Criminal Justice Data (*Carlson, 1998*) are analyzed.

Definition. Let $H(m)$, $m = 1,2...$, be the time series of the monthly number of all homicides. Fig. 36 shows the plot of $H(m)$ in Los Angeles, per 3,000,000 inhabitants of the city. To identify the episodes of *SHS* (Fig. 33) the seasonal variations, which are clearly seen in Fig. 36, are smoothed out by replacing $H(m)$ by its linear least square regression (5): $H^*(m) = W^H(m/m\text{-}6, m\text{+}6)$. Since $H^*(m)$ is defined on the time interval $(m - 6, m + 6)$, it depends on the future. Thus, it is admissible to define prediction targets (but not precursors).

Table 25. Types of crime considered
(after Carlson (1998); abbreviations are indicated in brackets)

Homicide	Robberies	Assaults	Burglaries
• All (H)	• All (Rob) • With firearms (FRob) • With knife or cutting instrument (KCIR) • With other dangerous weapon (ODWR) • Strong-arm robberies (SAR)*	• All (A)* • With firearms (FA) • With knife or cutting instrument (KCIA) • With other dangerous weapon (ODWA)* • Aggravated injury assaults (AIA)*	• Unlawful not forcible entry (UNFE) • Attempted forcible entry (AFE)*

* Analyzed in sensitivity tests only (Section 6.5)

6.3 Premonitory Trends of Single Types of Crime

Here the monthly data on seven types of crimes out of the 13 types listed in Table 25 are analyzed. "Premonitory" trends of each crime that tend to appear more frequently as an *SHS* approaches are looked for. Prediction itself is based on the collective behavior of these trends, as analyzed in the next Section. Orientation on a *set* of precursors has been found to be rather successful in prediction research: an ensemble of "imprecise" precursors usually gives better predictions than a single "precise" precursor (Keilis-Borok and Rotwain, 1990; Zaliapin et al., 2003b).

Observation. According to police experience, the crimes considered here often rise before an *SHS*.

To smooth out seasonal variations, the plot $C(m)$ of each type of crime is replaced by its regression (1): $C^*(m) = \mathcal{W}^C(m/m\text{-}12, m)$. Regression is done over the prior 12 months and does not depend on the future, so that it can be used for prediction. These plots exhibit two consecutive patterns:

(i) First, one can see a simultaneous escalation of burglaries and assaults within several (4 to 11) months before an *SHS*; at the same time robberies are declining.

(ii) Later on, closer to an *SHS*, one can see, albeit not so clearly, a simultaneous escalation of different kinds of robberies.

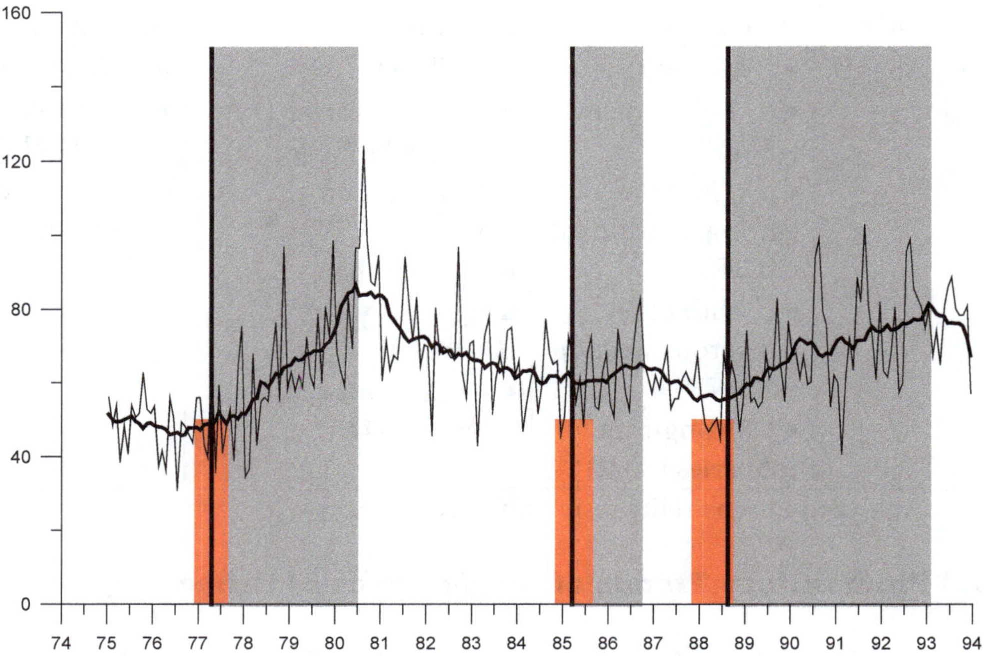

Figure 36. Total monthly number of homicides in Los Angeles city, 1975-1993. Data are taken from the National Archive of Criminal Justice Data (Carlson, 1998). Thin curve – original time series, *H(m)*, per 3,000,000 inhabitants. Thick curve - smoothed series *H*(m)*, with seasonal variations eliminated as described in Section 1. Vertical lines show the targets of prediction – episodes of *SHS* (Section 6.2). Gray bars are the periods of homicide surge. Red bars are the alarms declared by the hypothetical prediction algorithm (Section 6.4).

The first pattern is formally defined and explored in this study. The second pattern, briefly discussed in Section 6.7, will be explored elsewhere.

Discretization of crime trends. To quantify the above observation the trends of the crimes are approximated by the regression coefficients $K^C(m\text{-}s, m)$ where C identifies the type of crime. The value of K^C is attributed to the month m so that it does not depend on information on future months; therefore it can be used for prediction.

Next, following the pattern recognition approach, the trends (the values of K^C) are discretized on the lowest level of resolution: a binary one distinguishes only the trends above and below a threshold $T^C(Q^C)$. It is defined as a percentile of a level Q^C, that is, by the condition that $K^C(m\text{-}s, m)$ exceeds $T^C(Q^C)$ during Q^C percent of the months considered.

According to the above observations, it is expected that "premonitory" trends lay above the respective thresholds for assaults and burglaries, while they lay below these thresholds for robberies. One can see this in Fig. 37, showing the functions $K^C(m\text{-}12, m)$ for 7 crime types. For convenience, the same code, "1", will be given to the "premonitory" trend of each crime, regardless of whether it is above or below the threshold of discretization. The seven monthly crime statistics considered here are thus reduced to a binary vector with 7 components.

The crime statistics is discretized using the values of Q^C indicated in Table 26. The crime history, thus transformed, is given in Table 27.

Table 26. Premonitory trends for selected crime types

Crime type	Premonitory trend $K^C(m\text{-}s, m)$	s	Q^C, %	$T^C(Q^C)$
Rob	Below threshold	12	66.7	-3.69
FRob	"	12	66.7	-1.29
KCIR	"	12	50.0	1.73
ODWR	"	12	87.5	-3.87
FA	Above threshold	12	50.0	1.89
KCIA	"	12	50.0	-1.32
UNFE	"	12	50.0	1.94

See notations in the text.

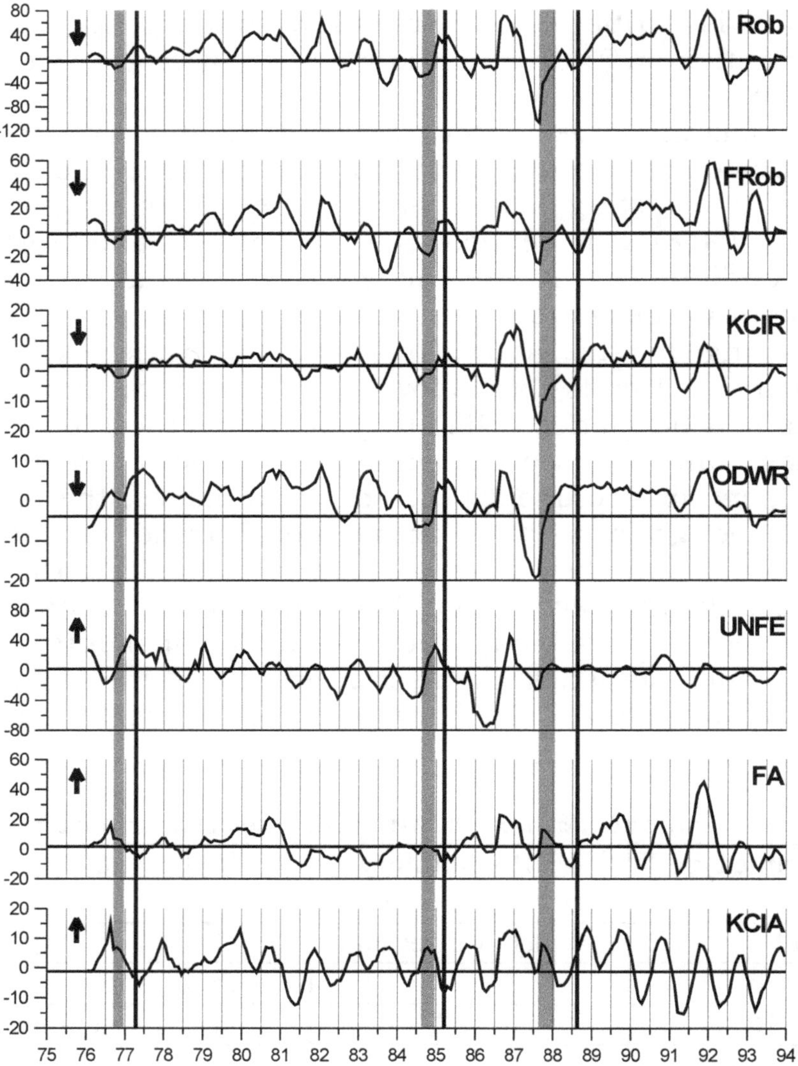

Figure 37. The regression coefficients $K^{C}(m\text{-}12, m)$ for seven crime types. See the definition in Section 6.3 and notations in Table. 25. Original data are taken from the National Archive of Criminal Justice Data (Carlson, 1998). Horizontal lines and arrows show respectively discretization thresholds and premonitory trends in accordance with Table 26. Vertical lines show episodes of *SHS*. Gray bars indicate months when $\Delta(m) \leq 1$.

Table 27. Binary codes of the trends for 7 types of crime and values of $K^H(m, m\text{-}12)$. See notations in Table 25. Discretization is defined in Table 26.

#	Year: Month	Discretized trends: Rob	FRob	KCIR	ODWR	FA	KCIA	UNFE	Δ		K^H
1	1976: 01	1	0	0	1	0	1	0	4		0.44
2	1976: 02	1	0	0	0	1	1	0	4		0.21
3	1976: 03	1	0	0	0	1	1	0	4		-0.07
4	1976: 04	0	0	0	1	1	1	1	3		-0.28
5	1976: 05	0	0	0	1	1	0	1	4		-0.35
6	1976: 06	0	1	1	1	1	0	1	2		-0.33
7	1976: 07	0	1	1	1	1	0	1	2		-1.18
8	1976: 08	0	1	1	1	1	0	1	2		-0.96
9	1976: 09	0	1	1	1	1	0	1	2		-0.87
10	1976: 10	0	1	1	1	1	1	1	1	+	-0.79
11	1976: 11	0	1	1	1	1	1	1	1	+	-0.37
12	1976: 12	0	0	1	1	1	1	1	2		0.16
13	1977: 01	0	0	0	1	0	1	0	5		0.68
14	1977: 02	0	0	0	1	0	1	0	5		0.84
15	1977: 03	0	0	0	1	0	1	0	5		0.38
16	1977: 04	0	0	0	0	0	1	0	6		0.85
17	1977: 05	0	0	0	1	0	1	0	5		0.07
18	1977: 06	0	0	0	1	0	1	0	5		0.71
19	1977: 07	0	1	0	1	0	1	0	4		0.58
20	1977: 08	0	1	0	0	0	1	0	5		0.03
21	1977: 09	0	1	0	0	0	1	0	5		0.18
22	1977: 10	0	1	1	0	1	1	1	2		-0.17
23	1977: 11	0	1	0	0	1	1	1	3		0.24
24	1977: 12	0	0	0	0	1	1	1	4		0.81
25	1978: 01	0	0	0	0	1	1	1	4		0.38
26	1978: 02	0	0	0	0	1	1	1	4		0.16
27	1978: 03	0	0	0	0	1	1	1	4		0.53
28	1978: 04	0	0	0	0	0	0	0	7		-0.09
29	1978: 05	0	0	0	0	0	0	0	7		0.55
30	1978: 06	0	0	0	0	0	0	0	7		0.20
31	1978: 07	0	0	0	0	0	0	0	7		0.57
32	1978: 08	0	0	0	1	0	0	0	6		0.66
33	1978: 09	0	0	0	1	0	1	0	5		1.19
34	1978: 10	0	0	0	0	0	1	0	6		1.07
35	1978: 11	0	0	0	0	1	1	0	5		1.72
36	1978: 12	0	0	0	0	1	1	0	5		1.81
37	1979: 01	0	0	0	0	1	1	1	4		2.52
38	1979: 02	0	0	0	0	1	1	0	5		1.66
39	1979: 03	0	0	0	0	1	1	0	5		0.88
40	1979: 04	0	0	0	0	1	1	1	4		0.98
41	1979: 05	0	0	0	0	1	0	1	5		0.66
42	1979: 06	0	0	0	0	1	0	1	5		0.27
43	1979: 07	0	0	0	0	1	0	1	5		-0.21
44	1979: 08	0	1	0	0	1	0	1	4		-0.14
45	1979: 09	0	1	0	1	1	1	1	2		-0.36
46	1979: 10	0	0	0	0	1	1	1	4		0.45
47	1979: 11	0	0	0	0	1	1	1	4		0.19
48	1979: 12	0	0	0	0	1	1	1	4		2.15
49	1980: 01	0	0	0	0	1	1	1	4		1.82
50	1980: 02	0	0	0	0	1	1	1	4		1.49
51	1980: 03	0	0	0	0	1	1	0	5		0.82
52	1980: 04	0	0	0	0	1	1	0	5		1.08
53	1980: 05	0	0	0	0	1	1	0	5		0.63
54	1980: 06	0	0	0	0	1	0	0	6		1.51
55	1980: 07	0	0	0	0	1	0	1	5		1.86
56	1980: 08	0	0	0	0	1	1	1	4		2.79
57	1980: 09	0	0	0	0	1	1	1	4		3.26
58	1980: 10	0	0	0	0	1	1	1	4		2.56
59	1980: 11	0	0	0	0	1	1	1	4		2.43
60	1980: 12	0	0	0	0	1	1	1	4		2.22
61	1981: 01	0	0	0	0	1	1	0	5		1.78
62	1981: 02	0	0	0	0	1	1	0	5		0.85
63	1981: 03	0	0	0	0	0	0	0	7		-0.57
64	1981: 04	0	0	0	0	0	0	0	7		-2.02
65	1981: 05	0	0	0	1	0	0	0	6		-2.64
66	1981: 06	0	1	0	1	0	0	0	5		-3.35
67	1981: 07	0	1	1	1	0	0	0	4		-2.51
68	1981: 08	0	1	1	1	0	0	0	4		-2.10
69	1981: 09	0	1	0	1	0	1	1	3		-0.92
70	1981: 10	0	1	0	1	0	1	1	3		-0.02
71	1981: 11	0	0	0	1	0	1	1	4		-0.10
72	1981: 12	0	0	0	1	0	1	1	4		0.09
73	1982: 01	0	0	0	1	0	1	1	4		0.70
74	1982: 02	0	0	0	1	0	0	0	6		-0.63
75	1982: 03	0	0	0	1	0	0	0	6		-0.27
76	1982: 04	0	0	0	1	0	0	0	6		-0.71
77	1982: 05	0	0	0	1	0	0	0	6		-0.97
78	1982: 06	0	0	1	1	0	0	0	5		-1.31
79	1982: 07	1	1	1	1	0	0	0	3		-1.32
80	1982: 08	1	1	1	0	0	0	0	4		-0.63
81	1982: 09	1	1	1	0	0	0	0	4		0.73
82	1982: 10	0	1	0	0	0	1	1	4		0.45
83	1982: 11	0	1	1	0	0	1	1	3		0.43
84	1982: 12	0	1	0	0	0	1	1	4		0.33
85	1983: 01	0	0	0	0	0	1	1	5		0.48
86	1983: 02	0	0	0	0	0	1	0	6		-0.35
87	1983: 03	0	0	0	1	0	0	0	6		-1.32
88	1983: 04	0	0	0	1	0	0	0	6		-0.62
89	1983: 05	0	1	0	1	0	0	1	4		-0.22
90	1983: 06	0	1	1	1	0	0	1	3		-0.73
91	1983: 07	0	1	1	1	0	0	1	3		-0.85
92	1983: 08	0	1	1	1	0	0	1	3		-0.78
93	1983: 09	0	1	1	1	0	0	1	3		-0.92
94	1983: 10	0	1	1	1	0	0	1	3		0.59
95	1983: 11	0	1	1	0	0	1	1	3		0.90
96	1983: 12	0	1	1	0	1	0	1	3		0.41

97	1984: 01	0	0	0	0	1	0	0	6		0.58
98	1984: 02	0	1	0	0	1	0	0	5		0.20
99	1984: 03	0	0	0	0	1	0	0	6		-0.93
100	1984: 04	0	0	1	0	0	0	0	6		-1.16
101	1984: 05	0	0	1	1	0	0	0	5		-0.76
102	1984: 06	1	1	1	1	0	0	0	3		-0.57
103	1984: 07	1	1	1	1	0	0	0	3		-1.04
104	1984: 08	1	1	1	1	0	0	1	2		-0.55
105	1984: 09	1	1	1	1	1	0	1	1	+	-0.21
106	1984: 10	1	1	1	1	0	1	1	1	+	-0.20
107	1984: 11	1	1	1	1	0	1	1	1	+	0.33
108	1984: 12	0	1	0	1	0	1	1	3		0.25
109	1985: 01	0	0	0	0	0	1	0	6		0.01
110	1985: 02	0	0	0	0	0	1	0	6		0.73
111	1985: 03	0	0	0	0	0	1	0	6		0.45
112	1985: 04	0	0	0	0	0	1	0	6		-0.08
113	1985: 05	0	0	0	0	0	0	0	7		-0.15
114	1985: 06	0	0	0	0	0	0	0	7		-0.51
115	1985: 07	0	1	0	0	0	0	0	6		-0.61
116	1985: 08	0	1	0	1	1	0	1	3		-0.34
117	1985: 09	0	1	1	1	1	0	1	2		-0.08
118	1985: 10	0	1	1	1	1	0	1	2		-0.11
119	1985: 11	0	1	1	1	1	0	1	2		0.19
120	1985: 12	0	1	1	1	1	0	1	2		0.70
121	1986: 01	0	1	0	1	1	0	1	3		-0.04
122	1986: 02	0	0	1	1	1	0	0	4		-0.74
123	1986: 03	0	0	1	1	0	0	0	5		0.29
124	1986: 04	0	0	1	1	0	0	0	5		0.77
125	1986: 05	0	0	1	1	0	0	0	5		0.42
126	1986: 06	0	0	1	1	0	0	0	5		0.15
127	1986: 07	0	0	1	1	1	0	0	4		-0.08
128	1986: 08	0	0	0	0	1	0	1	5		0.76
129	1986: 09	0	0	0	0	1	1	1	4		1.90
130	1986: 10	0	0	0	0	1	1	1	4		1.56
131	1986: 11	0	0	0	0	1	1	1	4		0.98
132	1986: 12	0	0	0	0	1	1	1	4		0.59
133	1987: 01	0	0	0	0	1	1	1	4		0.66
134	1987: 02	1	0	0	0	1	1	1	3		0.11
135	1987: 03	1	0	0	0	1	1	1	3		-0.43
136	1987: 04	1	0	1	1	1	0	1	2		0.14
137	1987: 05	1	1	1	1	0	0	1	2		-0.20
138	1987: 06	1	1	1	1	0	0	1	2		-0.76
139	1987: 07	1	1	1	1	0	0	0	3		-1.52
140	1987: 08	1	1	1	1	0	0	0	3		-1.48
141	1987: 09	1	1	1	1	1	0	1	1	+	-0.90
142	1987: 10	1	1	1	1	1	1	1	0	+	-0.13
143	1987: 11	0	1	1	1	1	1	1	1	+	-0.01
144	1987: 12	0	1	1	1	1	1	1	1	+	0.29
145	1988: 01	0	1	1	1	1	1	0	2		-0.01
146	1988: 02	0	0	0	1	1	1	0	4		-0.35
147	1988: 03	0	0	0	1	1	1	0	4		-0.84
148	1988: 04	0	0	0	1	0	0	0	6		-0.96
149	1988: 05	0	1	1	1	0	1	0	3		-0.69
150	1988: 06	0	1	1	1	0	1	0	3		-1.21
151	1988: 07	0	1	1	1	0	1	1	2		-0.68
152	1988: 08	0	1	1	1	0	1	1	2		-0.66
153	1988: 09	0	1	1	0	1	1	1	2		-0.13
154	1988: 10	0	1	0	0	1	1	1	3		0.27
155	1988: 11	0	1	0	0	1	1	1	3		0.04
156	1988: 12	0	0	0	0	1	1	1	4		0.15
157	1989: 01	0	0	0	0	1	1	1	4		1.38
158	1989: 02	0	0	0	0	1	1	1	4		1.41
159	1989: 03	0	0	0	0	1	0	0	6		1.13
160	1989: 04	0	0	0	0	1	0	0	6		0.62
161	1989: 05	0	0	0	0	1	0	1	5		0.21
162	1989: 06	0	0	0	0	1	0	1	5		0.34
163	1989: 07	0	0	0	0	1	0	1	5		-0.15
164	1989: 08	0	0	0	0	1	0	1	5		0.39
165	1989: 09	0	0	0	0	1	1	1	4		1.19
166	1989: 10	0	0	0	0	1	1	1	4		1.47
167	1989: 11	0	0	0	0	1	1	1	4		1.20
168	1989: 12	0	0	0	0	1	1	1	4		1.18
169	1990: 01	0	0	0	0	1	1	1	4		0.97
170	1990: 02	0	0	0	0	0	0	0	7		1.60
171	1990: 03	0	0	0	0	0	0	0	7		1.15
172	1990: 04	0	0	0	0	0	0	0	7		0.92
173	1990: 05	0	0	0	0	0	0	0	7		0.21
174	1990: 06	0	0	0	0	0	0	0	7		-0.44
175	1990: 07	0	0	0	0	1	1	0	5		0.74
176	1990: 08	0	0	0	0	1	1	1	4		1.63
177	1990: 09	0	0	0	0	1	1	1	4		1.64
178	1990: 10	0	0	0	0	1	1	1	4		2.25
179	1990: 11	0	0	0	0	1	1	1	4		1.24
180	1990: 12	0	0	0	0	1	1	1	4		0.20
181	1991: 01	0	0	0	0	0	1	0	6		0.04
182	1991: 02	0	0	0	1	0	1	0	5		-1.16
183	1991: 03	0	0	0	1	0	0	0	6		-1.38
184	1991: 04	0	0	1	1	0	0	0	5		-1.40
185	1991: 05	0	0	1	1	0	0	0	5		-1.60
186	1991: 06	0	0	1	1	0	0	0	5		-1.14
187	1991: 07	0	0	1	1	1	0	0	4		-1.10
188	1991: 08	0	0	0	1	1	0	0	5		1.12
189	1991: 09	0	0	0	1	1	0	1	4		2.63
190	1991: 10	0	0	0	0	1	1	1	4		3.04
191	1991: 11	0	0	0	0	1	1	1	4		2.86
192	1991: 12	0	0	0	0	1	1	1	4		2.60
193	1992: 01	0	0	0	0	1	1	0	5		1.84
194	1992: 02	0	0	0	0	1	1	0	5		0.95
195	1992: 03	0	0	0	0	1	0	0	6		-0.58
196	1992: 04	0	0	0	1	0	0	0	6		-0.88
197	1992: 05	0	0	1	1	0	0	0	5		-1.70
198	1992: 06	0	1	1	1	0	0	0	4		-2.75
199	1992: 07	0	1	1	1	0	0	0	4		-1.21
200	1992: 08	0	1	1	1	0	0	0	4		0.10
201	1992: 09	0	1	1	1	0	0	0	4		1.72
202	1992: 10	0	1	1	1	1	0	1	2		2.35
203	1992: 11	0	1	1	1	1	0	1	2		2.47
204	1992: 12	0	0	1	1	0	0	0	5		1.76
205	1993: 01	0	0	0	1	0	0	0	6		2.04
206	1993: 02	1	0	0	1	0	0	0	5		1.04
207	1993: 03	1	0	0	1	0	0	0	5		0.21
208	1993: 04	1	0	0	1	0	0	0	5		-0.74
209	1993: 05	1	0	1	1	0	0	0	4		-0.46
210	1993: 06	1	1	1	1	0	0	0	3		-1.05

211	1993: 07	1	1	1	1	0	0	0	3	-1.60
212	1993: 08	0	1	1	1	0	0	0	4	-1.00
213	1993: 09	0	0	0	1	0	0	1	5	-0.20
214	1993: 10	0	0	0	1	0	1	1	4	0.14
215	1993: 11	0	0	0	1	0	1	1	4	0.52
216	1993: 12	0	0	1	1	0	1	1	3	-0.12

6.4. Collective Behavior of Premonitory Trends: A Hypothetical Prediction Algorithm

Keilis-Borok et al. (2003) consider how the approach of a homicide surge is reflected in the *collective* behavior of the trends. The simplest description of this behavior is $\Delta(m)$ - the number of *non*-premonitory trends at a given month *m*. If the identification of premonitory trends is correct then $\Delta(m)$ should be low in the proximity of an *SHS*. By definition $\Delta(m)$ is the number of zeros in the binary code of the monthly situation. This is the so-called "Hamming distance" between that code and the code of the "pure" premonitory situation, {1,1,1,1,1,1,1} when all seven trends listed in Table 26 are premonitory (Gvishiani and Kosobokov, 1981; Lichtman and Keilis-Borok, 1989; Keilis-Borok et al., 2000).

The values of $\Delta(m)$ are given in Table 27. Fig. 38 shows the change of $\Delta(m)$ with time. The value of $\Delta(m)$ may vary from 0 to 7 but the minimal observed value is 1; the corresponding lines in Table 27 are marked by "+". This value appears within 4 to 11 months before an *SHS* and at no other time. An examination of the temporal change of $\Delta(m)$ in Table 27 suggests the following hypothetical prediction algorithm:

An alarm is declared for 9 months each time when $\Delta(m) \leq D$ for two consecutive months (regardless of whether these two months belong or not to an already declared alarm).

Possible outcomes of such a prediction are illustrated in Fig. 34. The condition $\Delta(m) \leq D$ means, by definition, that *D* or less trends are not premonitory at the month *m*. A count of $\Delta(m)$ in Table 27 suggests to take $D = 1$. A waiting period of 9 months is introduced because the premonitory trends do not appear right before an *SHS*. The requirement that this condition holds two months in a row makes a prediction more reliable and reduces the total duration of alarms.

The alarms obtained by this algorithm are shown in Fig. 38 by the red bars. The total duration of these alarms is 30 months, representing 14 percent of all months considered. In real prediction that score would be quite satisfactory.

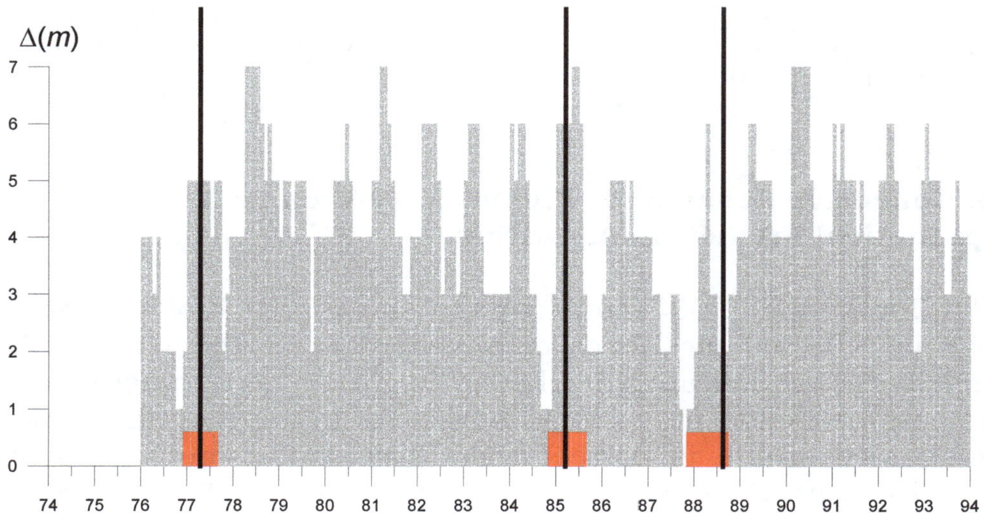

Figure 38. Homicide surges and alarms determined by the prediction algorithm. Start of a homicide surge is shown by the vertical line. Function $\Delta(m)$ is the number of crime statistics *not* showing premonitory trends at a month m. Alarms (shown by red bars) are declared for 9 months, when $\Delta(m) \leq 1$ during two consecutive month. Adjustable parameters correspond to version 10 of the algorithm (see Table 28).

6.5. Stability of Prediction (Sensitivity Analysis)

Inevitably in lieu of a set of fundamental equations for crime dynamics there is a certain freedom in the retrospective ad hoc choice of adjustable elements: the types of crimes considered, numerical parameters, such as percentiles Q^C, etc. An algorithm thus developed makes sense only if it is not too sensitive to variation of these choices; as Enrico Fermi put it, "*with four exponents I can fit an elephant*".

To explore that sensitivity Keilis-Borok et al. (2003) repeat the prediction with different sets of the kinds of crime considered and with different values for the numerical parameters. These sets are described in Table 28. The outcomes of prediction are compared on the error diagrams (Fig. 39). Molchan (1997) has introduced such diagrams as a tool for evaluating prediction methods and optimizing disaster preparedness. Their application to research in prediction of recessions and unemployment are described in (Keilis-Borok et al, 2000, 2005).

The "basic" variant (Section 6.4) is # 10 in Table 28. The variations considered are discussed below.

Variation of the percentiles Q^C, defining discretization thresholds (# 8, 9, 11, 12). Lowering them, the total duration of alarms is obviously increased, but the results of prediction do not change much and remain acceptable.

Using only two kinds of crimes (# 12) comparable results are obtained. However it would be risky to make advance prediction with only two indicators.

Table 28. Variation of the adjustable elements

<table>
<tr><td colspan="2"></td><td colspan="12">Variants</td></tr>
<tr><td colspan="2"></td><td>1</td><td>2</td><td>3</td><td>4</td><td>5</td><td>6</td><td>7</td><td>8</td><td>9</td><td>10</td><td>11</td><td>12</td></tr>
<tr><td colspan="2">Value of D*</td><td>3</td><td colspan="2">5</td><td>8</td><td colspan="2">4</td><td>3</td><td colspan="4">1</td><td>0</td></tr>
<tr><td colspan="2">Crime type</td><td colspan="12">Trend and percentile</td></tr>
<tr><td rowspan="2">Rob</td><td>PR**</td><td colspan="3">upward</td><td colspan="8">downward</td><td rowspan="4"></td></tr>
<tr><td>Q^C, %</td><td>33.3</td><td>25.0</td><td>20.0</td><td>80.0</td><td colspan="6">66.7</td><td>50.0</td></tr>
<tr><td rowspan="2">FRob</td><td>PR</td><td colspan="3" rowspan="4"></td><td colspan="8">downward</td></tr>
<tr><td>Q^C, %</td><td>80.0</td><td colspan="6">66.7</td><td>50.0</td></tr>
<tr><td rowspan="2">KCIR</td><td>PR</td><td colspan="9">downward</td></tr>
<tr><td>Q^C, %</td><td>80.0</td><td colspan="5">66.7</td><td colspan="2">50.0</td><td>66.7</td></tr>
<tr><td rowspan="2">ODWR</td><td>PR</td><td colspan="3">upward</td><td colspan="8">downward</td><td rowspan="2"></td></tr>
<tr><td>Q^C, %</td><td>33.3</td><td>25.0</td><td>20.0</td><td>80.0</td><td>66.7</td><td colspan="6">87.5</td></tr>
<tr><td rowspan="2">SAR</td><td>PR</td><td colspan="3">upward</td><td colspan="4">downward</td><td colspan="5" rowspan="4"></td></tr>
<tr><td>Q^C, %</td><td>33.3</td><td>25.0</td><td>20.0</td><td>80.0</td><td colspan="3">66.7</td></tr>
<tr><td rowspan="2">A</td><td>PR</td><td colspan="7">upward</td></tr>
<tr><td>Q^C, %</td><td>33.3</td><td>25.0</td><td>20.0</td><td>20.0</td><td colspan="3">33.3</td></tr>
<tr><td rowspan="2">FA</td><td>PR</td><td colspan="11">upward</td><td rowspan="4"></td></tr>
<tr><td>Q^C, %</td><td>33.3</td><td>25.0</td><td>20.0</td><td>20.0</td><td colspan="4">33.3</td><td colspan="3">50.0</td></tr>
<tr><td rowspan="2">KCIA</td><td>PR</td><td colspan="11">upward</td></tr>
<tr><td>Q^C, %</td><td>33.3</td><td>25.0</td><td>20.0</td><td>20.0</td><td colspan="4">33.3</td><td colspan="3">50.0</td></tr>
<tr><td rowspan="2">ODWA</td><td>PR</td><td colspan="7">upward</td><td colspan="5" rowspan="4"></td></tr>
<tr><td>Q^C, %</td><td>33.3</td><td>25.0</td><td>20.0</td><td>20.0</td><td colspan="3">33.3</td></tr>
<tr><td rowspan="2">AIA</td><td>PR</td><td colspan="7">upward</td></tr>
<tr><td>Q^C, %</td><td>33.3</td><td>25.0</td><td>20.0</td><td>20.0</td><td colspan="3">33.3</td></tr>
<tr><td rowspan="2">UNFE</td><td>PR</td><td colspan="3" rowspan="4"></td><td colspan="3">upward</td><td rowspan="2"></td><td colspan="5">upward</td></tr>
<tr><td>Q^C, %</td><td>20.0</td><td colspan="2">33.3</td><td>33.3</td><td colspan="3">50.0</td><td>33.3</td></tr>
<tr><td rowspan="2">AFE</td><td>PR</td><td colspan="3">upward</td><td colspan="6" rowspan="2"></td></tr>
<tr><td>Q^C, %</td><td>20.0</td><td colspan="2">33.3</td></tr>
</table>

*The values that give relatively best performance for that variant.
**PR is an abbreviation for the premonitory trend.

The limits of acceptable variations are reached in the other variants (# 1 – 7). Keilis-Borok et al. (2003) tried to find a premonitory *rise* of robberies, simultaneous with rise of other crimes and consider other kinds of crime; in all variants its performance remains unacceptable.

For advance prediction variants 8 – 11 might be used in parallel. Such parallel predictions might better suit the needs of a decision-maker, determining possible disaster preparedness measures (Kossobokov et al., 2000b; Zaliapin et al., 2003b).

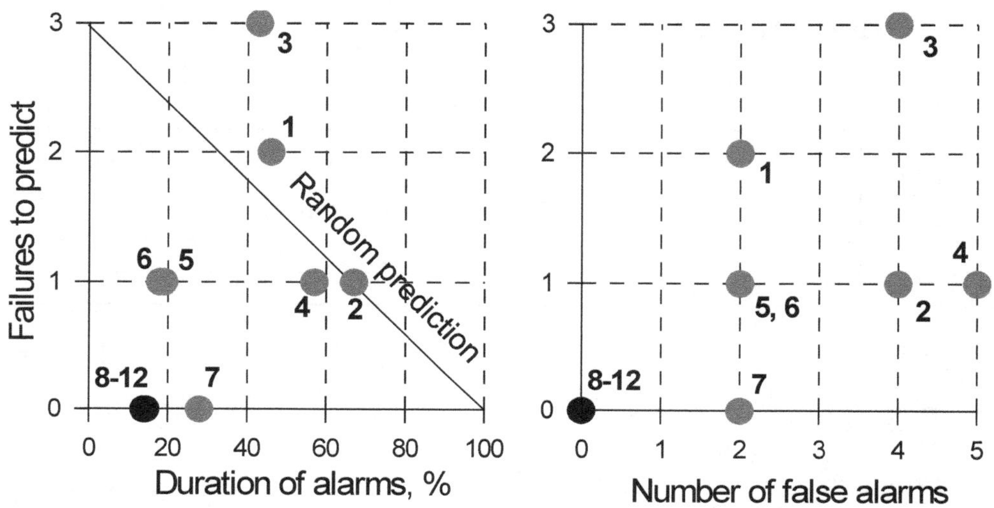

Figure 39. Error diagram. Numbers near the dots identify the variant of the algorithm in Table 28. Black dots show the variants suggested for advance prediction. See further explanations in Section 6.5.

6.6. Applications to Independent Data

Here the algorithm is tested by application to "out of sample" data not used in its development. Such tests are always necessary to validate and/or improve a prediction algorithm. Such a test is possible since the algorithm is self-adaptive: the thresholds $T^C(Q^C)$ are not fixed but are adapted to crime statistics, as the percentile of a level Q^C.

Los Angeles, 1994-2002. So far the data source (Carlson, 1998) covering the years 1975 – 1993 was used. To extend the analysis past 1993, there is the data of the LAPD Information Technology

Division, covering the time period from January 1990 to May 2002. Comparing the data for the overlapping three years it is found that they are reasonably close, particularly after smoothing.

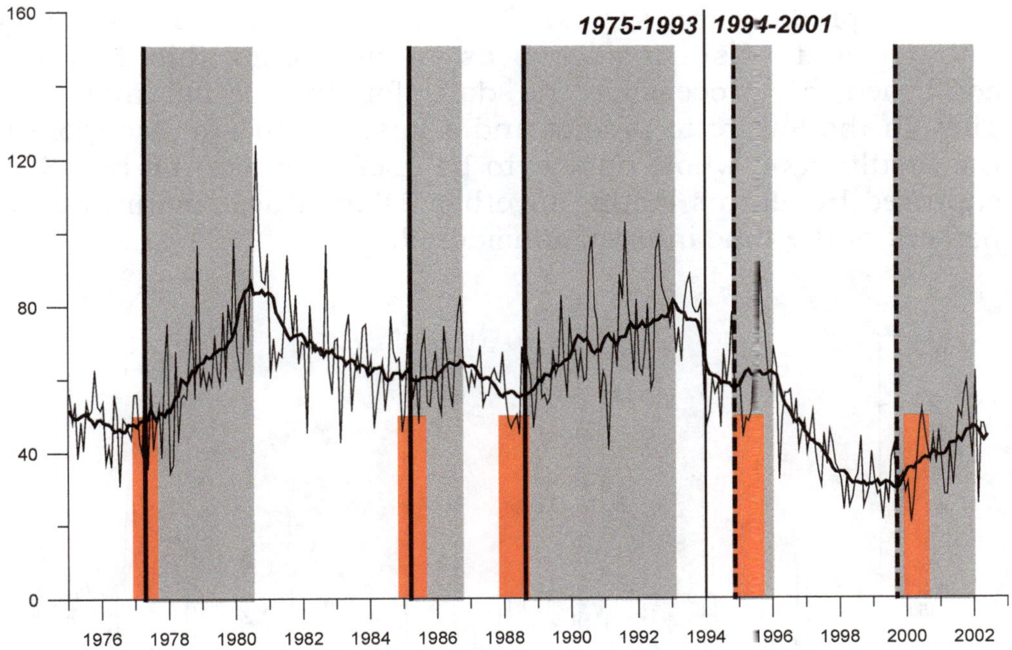

Figure 40. Performance of prediction algorithm through 1975-2002. Data from the National Archive of Criminal Justice Data (Carlson, 1998) for 1975 - 1993 have been used to develop the algorithm. It was then applied to the data from the Data Bank of the Los Angeles Police Department (LAPD Information Technology Division) for subsequent 9 years. Notations are the same as in Fig. 36. Dashed vertical lines indicate *SHS* episodes that occurred after 1993.

Fig. 40 shows the homicide rates through the whole period from 1975 to May 2002. Two *SHS* episodes are identified in the later period 1994-2001. They are indicated in Fig. 40 by dashed vertical lines. The first episode is captured by an alarm, which starts in the month of *SHS* without a lead time. The second episode is missed, since an alarm has started two months after it. This error needs to be put on the record; nevertheless the prediction remains informative: during these two months homicide rose by only a few percent, giving no indication that a lasting homicide surge had started.

New York City. Fig. 41 shows the monthly total homicide rates in New York City, per 7 million inhabitants of the city. Two *SHS* episodes (02:1978 and 02:1985) were identified. The prediction algorithm gives two alarms, as shown in Fig. 41 by red bars. One of them predicts the second *SHS*, while the first one is missed. Another alarm is considered as a false one; this has to be confirmed by processing the data for the period after 1993. Though the failure to predict and a false alarm are disappointing, the results as a whole appear to be useful: one of the two *SHS* is captured by alarms lasting together 21 months, amounting to 10 percent of the time interval considered.

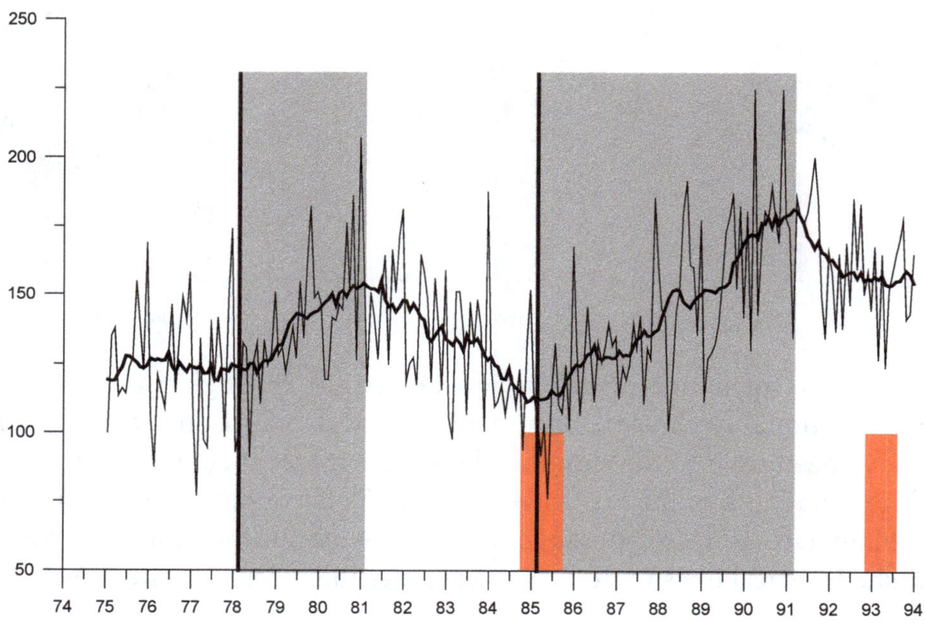

Figure 41. Application of the prediction algorithm to New York City. Notations are the same as in Fig. 36. Data are taken from the National Archive of Criminal Justice Data (Carlson, 1998). Homicide statistics is shown per 7,000,000 of inhabitants.

6.7. On a More Precise Prediction

A conjecture that could be worth exploring in the future is outlined here. Two consecutive patterns of the crimes considered have been observed. The first one precedes an *SHS* with a lead time of 4 to 11 months; it is formally defined and explored in Sections 4, 5. The second pattern emerges with a shorter lead time, promising a more accurate prediction of the time of an incipient *SHS*.

A distinctive trait of the second pattern is a steep simultaneous rise of the different types of robberies. Let us replace this pattern by a less specific one that is more broadly defined the absence of a steep decline. By definition, that pattern will be captured by the zeros in the first four columns of Table 27. Counting them, one can find that three or more emerge within 6 months before each *SHS*. This result suggests the following second approximation to the prediction algorithm described above. Consider the period of alarm declared by the algorithm; let us call it "the first phase alarm". Within that period *a "second phase alarm" is declared for 6 months after the first month when* $\Delta_1(m) \leq 1$. Here $\Delta_1(m)$ is the number of ones in the codes of the robberies (the first four columns in Table 27). In the absence of the first-phase alarm the second one is not declared.

Alarms obtained by this rule are shown in Fig. 42. The alarms become much shorter; their total duration drops to 18 months, that is, from 14 percent to 8 percent of all the months considered. Even better results could potentially be obtained by directly capturing *a rise* of robberies, but this requires weekly, if not daily, crime rates (since the lead time of the rise in robberies is relatively short).

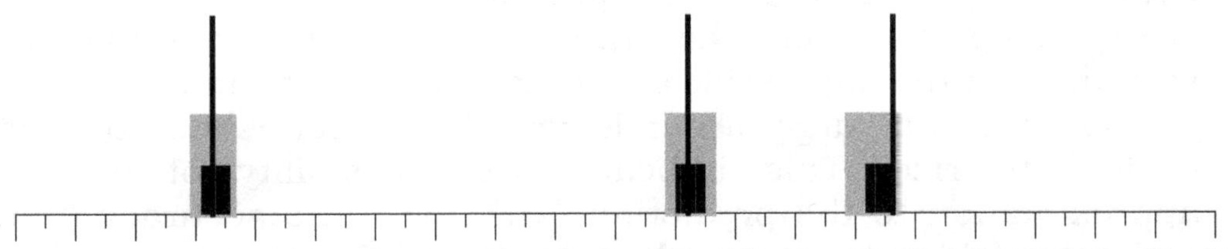

Figure 42. Possible reduction of duration of alarms. Vertical lines – starting points of a homicide surge (*SHS*). Grey bars – alarms obtained by the suggested algorithm. Black bars – alarms obtained in a hypothetical second approximation.

Using the trend of homicides themselves might provide a similar possibility. Values of the function $K^H(m\text{-}12,\ m)$, which estimates that trend (see Section 6.3) are given in Table 27 (column K^H). Within each alarm one can see the months when $K^H(m\text{-}12,\ m) > 0$. Starting alarms at these months, it would be possible to reduce further the duration of alarms without having an additional failure to predict.

6.8. Conclusions

The conclusions of Keilis-Borok et al. (2003) might be summed up as follows. Crime statistics in the city of Los Angeles for the period 1975 – 2001 was analyzed, exploring the possibility of anticipating a turn of the homicide rate from decline to a surge. It has been found that within 4 to 11 months such a turn is preceded by a specific pattern of the crime statistics: both burglaries and assaults escalate, while robberies decline, along with the homicides themselves. Both changes, escalation and decline, are not monotonic, but occur sporadically, each lasting 2-6 months.

Based on this pattern a prediction algorithm is formulated, giving it a robust and unambiguous definition. It is self-adapting to average crimes statistics, so that one could apply it to New York City as well as Los Angeles. The major limitation of this study is that, as is inevitable for an initial study, only a small number of homicide surges was available for analysis. The algorithm remains hypothetical until it is validated by advance prediction. It is encouraging, however, that those predictions are stable as to variations in the adjustable elements of the algorithm.

Closer to the surge of homicides, the robberies also turn from decline to rise. This indicates the possibility of a second approximation to the prediction, with twice the accuracy (that is with a twofold reduction in the duration of alarms).

The analysis given above captures the consecutive escalation of different crimes: first – of burglaries and assaults only, then of robberies, then of homicides. That sequence, albeit hypothetical so far, seems natural, being in good accord with previous experience in the following areas.

(*i*) The sequence reflects a more general phenomenon, commonly known in law enforcement practice: a consecutive escalation of more and more severe crimes, signaling that a surge of major crimes is approaching. A quantitative definition of a specific manifestation of this phenomenon is given by Keilis-Borok et al. (2003). Similar escalation has been found in French suburban areas (Bui Trong, 2003).

(*ii*) The sequence is also in accord with a well-known "universal" feature of many hierarchical complex systems: the rise of permanent background activity ("static") of the system culminated by a fast major change – a "critical transition". This feature happens to be common for different physical and socio-economic systems. It is reproduced by the "universal" models of hierarchical complex systems, such as those developed in theoretical physics (e.g. Yamashita and Knopoff, 1992; Gell-Mann, 1994; Holland, 1995; Newman et al., 1995; Turcotte, 1997; Allègre et al., 1998; Shnirman, and Blanter, 1998; Gabrielov et al., 2000b; Rundle et al., 2000; Sornette, 2000; Turcotte et al., 2000; Zaliapin et al., 2003b).

This feature was also observed in many very different real world systems. For example, in earthquake-prone regions the "static" includes background seismicity. Premonitory escalation of seismic activity is a well-known precursor to major earthquakes, which is used in many earthquake prediction algorithms (e.g. Keilis-Borok and Kossobokov, 1990; Kossobokov and Carlson, 1995; Newman et al., 1995; Keilis-Borok, 2002). In economy, the "static" includes various macroeconomic indicators. Their premonitory escalation has been successfully used in the prediction of recessions and surges of unemployment (Keilis-Borok et al., 2000, 2005).

The results of Keilis-Borok et al. (2003) are also in accord with a distinctive common trait of precursors established in many of these studies: premonitory evolution of background activity is not monotonic, but realised sporadically, in a sequence of relatively short intermittent changes.

The universality of premonitory phenomena is limited and cannot be taken for granted in studying any specific system. Nevertheless, it is worth exploring in crime dynamics other known types of premonitory patterns, e.g. the clustering of background activity and the rise of the correlation range (Keilis-Borok, 2002; Gabrielov et al., 2000b; Shebalin et al., 2000).

What is the place of the study of Keilis-Borok et al. (2003) in the broad field of prediction of crime dynamics? Specific features of this approach might be summed up as follows.

Keilis-Borok et al. (2003) are trying to predict not the whole dynamics of homicides, but only the relatively rare phenomena - episodes of *SHS*.

Accordingly, Keilis-Borok et al. (2003) are looking for a quantitative and precisely defined prediction algorithm of the *"yes or no"* variety: at any moment of time such an algorithm would indicate whether or not such an episode should be expected within a fixed time interval.

The analysis of Keilis-Borok et al. (2003) is intentionally robust, which makes the prediction algorithm more reliable and applicable in different circumstances. In this case the performance of the algorithm did not change through the period considered even though Los Angeles has witnessed many changes relevant to crime over this period. This stability is achieved at a price, however, in that the time of a homicide surge is predicted with limited accuracy and the duration of a surge even more so.

The approach of Keilis-Borok et al. (2003) – a heuristic "technical" analysis - is not competing with but complementary to the cause-and-effect "fundamental" analysis. The cause that triggered a specific homicide surge is usually known, at least in retrospect. This might be, for example, a rise in drug use, a rise in unemployment, a natural disaster etc. However, that does not render predictions considered in this study redundant. On the contrary, the approach might predict an unstable situation when a homicide surge might be triggered, thus enhancing the reliability of cause-and-effect predictions.

It is encouraging for further studies in this direction that only a small part of the relevant and available data that can be incorporated in the analysis were used here. Among these are other types of crimes (Bursik et al., 1990), economic and demographic indicators (Messner, 1983) and the territorial distribution of crimes. It seems worthwhile to try the same approach with other targets of prediction – e.g. surges of all violent crimes; and to other areas, e.g. separate Bureaus of the city of Los Angeles, or to other major cities. In a broader scheme of things, the analysis of Keilis-Borok et al. (2003) discriminates stable situations from unstable, where the risk of different disasters is higher.

AFTERWORD:

Lessons Learned and Outlook for the Future

While the prediction algorithms described here still need to be improved, the following findings seem promising.

1. In each case, the prediction rules are uniform, transcending the immense complexity of the processes considered, the diversity of prediction targets, and the changes of circumstances in time. For the earthquakes, premonitory seismicity patterns happen to be similar for microcracks in laboratory samples to the largest earthquakes of the world, spanning the energy range from 10^{-1} erg to 10^{26} erg. Similar patters also likely apply to the ruptures in a neutron star, 10^{41}erg. For recessions and unemployment, the similarities between them overcome the changes of economy since 1962; for FAUs, their similarities also overcome the differences between the countries.

2. Our predictions are always based on the routinely available data — global catalogs of earthquakes, major macroeconomic indicators, etc.

3. In each case predictability is achieved by extremely robust analysis. It is interesting to quote how an economist with excellent track record in recessions' prediction describes his impressions on such a robustness:

> "Prediction of recessions... requires fitting non-linear, high-dimensional models to a handful of observations generated by a possibly non-stationary economic environment....The evidence presented here

suggests that these simple binary transformations of economic indicators have significant predictive content comparable to or, in many cases, better than that of more conventional models."

J. Stock (from Keilis-Borok et. al., 2000)

The accuracy of prediction is still limited. However, only a small part of relevant models and available data has been used so far. A wealth of possibilities for better prediction remains unexplored.

REFERENCES

Aki K (1996) Scale dependence in earthquake phenomena and its relevance to earthquake prediction. Proc Natl Acad Sci USA 93: 3740-3747.

Allègre CJ, Le Mouël J-L, Ha Duyen C, Narteau C (1995) Scaling organization of fracture tectonics (SOFT) and earthquake mechanism. Phys Earth Planet Inter, 92: 215-233.

Allègre CJ, Shebalin P, Le Mouel J-L, Narteau C (1998) Energetic balance in scaling organization of fracture tectonics. Phys Earth Planet Inter 106: 139-153.

Barenblatt GM, Keilis-Borok VI, Monin AS (1983) Filtration model of earthquake sequence. Trans (Doklady) Acad Sci SSSR 269: 831-834.

Barenblatt G (1996) Scaling, Self-Similarity, and Intermediate Asymptotics, Cambridge University Press, Cambridge.

Blanter EM, Shnirman MG, Le Mouël J-L, Allègre CJ (1997) Scaling laws in blocks dynamics and dynamic self-organized criticality. Phys Earth Planet Inter, 99: 295-307.

Bongard MM (1970) Pattern Recognition. Rochelle Park, N.J.: Hayden Book Co., Spartan Books.

Bowman DD, Ouillon G, Sammis G.G, Sornette A, Sornette D (1998) An observational test of the critical earthquake concept. J Geophys Res 103: 24359-24372.

Bursik RJ, Jr, Grasmick HG, Chamlin MB (1990) The effect of longitudinal arrest patterns on the development of robbery trends at the neighborhood level. Criminology 28(3): 431-450.

Bufe CG, Varnes DJ (1993) Predictive modeling of the seismic cycle of the greater San Francisco Bay region. J Geophys Res 98: 9871–9883

Bui Trong L (2003) Risk of collective youth violence in French suburbs. A clinical scale of evaluation, an alert system. In: Beer T, Ismail-Zadeh A (eds) Risk Science and Sustainability. Kluwer Academic Publishers, Dordrecht-Boston-London (NATO Science Series. II. Mathematics, Physics and Chemistry – Vol. 112), pp 199-221.

Burridge R, Knopoff L (1967) Model and theoretical seismicity. Bull Seism Soc Am 57: 341-371.

Caputo M, Gasperini P, Keilis-Borok V, Marcelli L, Rotwain I (1977) Earthquake's swarms as forerunners of strong earthquakes in Italy. Annali di Geofisica XXX(3/4): 269-283.

Caputo M, Console R, Gabrielov AM, Keilis-Borok VI, Sidorenko TV (1983) Long-term premonitory seismicity patterns in Italy. Geophys J R Astron Soc 75: 71-75.

Carlson SM (1998) Uniform Crime Reports: Monthly Weapon-specific Crime and Arrest Time Series, 1975-1993 (National, State, and 12-City Data). ICPSR 6792, Inter-university Consortium for Political and Social Research, P.O. Box 1248, Ann Arbor, Michigan 48106.

Crutchfield JP, Farmer JD, Packard NH, Shaw RS (1986) Chaos. Sci Am 255:46–57

Davis C, Goss K, Keilis-Borok V, Molchan G, Lahr P, Plumb C (2007) Earthquake Prediction and Tsunami Preparedness. Workshop on the Physics of Tsunami, Hazard Assessment Methods and Disaster Risk Management, 14-18 May 2007, Trieste: ICTP.

Davis CA, Keilis-Borok V, Molchan G, Shebalin P, Lahr P, Plumb C (2010) Earthquake prediction and disaster preparedness: Interactive analysis. Natural Hazards Review, ASCE 11(4): 173-184.

Davis C, Keilis-Borok V, Kossobokov V, Soloviev A (2012) Advance prediction of the March 11, 2011 Great East Japan Earthquake: A missed opportunity for disaster preparedness. International Journal of Disaster Risk Reduction, 1: 17-32, doi:10.1016/j.ijdrr.2012.03.001.

Engle RF, McFadden DL (1994) (eds) Handbook of Econometrics, volume 4. North-Holland Publishing Co., Amsterdam.

Farmer JD, Sidorowich J (1987) Predicting chaotic time series. Phys Rev Lett 59: 845.

Gabrielov A, Dmitrieva OE, Keilis-Borok VI, Kossobokov VG, Kuznetsov IV, Levshina TA, Mirzoev KM, Molchan GM, Negmatullaev SKh, Pisarenko VF, Prozoroff AG, Rinehart W, Rotwain IM, Shebalin PN, Shnirman MG, Shreider SYu (1986) Algorithm of Long-Term Earthquakes' Prediction. Centro Regional de Sismologia para America del Sur, Lima, Peru.

Gabrielov AM, Keilis-Borok VI, Jackson DD (1996) Geometric incompatibility in a fault system. Proc Natl Acad Sci USA 93: 3838-3842.

Gabrielov A, Keilis-Borok V, Zaliapin I, Newman WI (2000a) Critical transitions in colliding cascads. Phys Rev E 62, 237-249.

Gabrielov AM, Zaliapin IV, Newman WI, Keilis-Borok VI (2000b) Colliding cascade model for earthquake prediction. Geophys J Int 143(2): 427-437.

Gelfand IM, Guberman ShA, Keilis-Borok VI, Knopoff L, Press F, Ranzman IYa, Rotwain IM, Sadovsky AM (1976) Pattern recognition applied to earthquake epicenters in California. Phys Earth Planet Inter 11: 227-283.

Gell-Mann M (1994) The Quark and the Jaguar: Adventures in the Simple and the Complex. Freeman and Company, New York.

Ghil M (1994) Cryothermodynamics: The chaotic dynamics of paleoclimate. Physica D 77: 130-159.

Gorshkov A, Keilis-Borok V, Rotwain I, Soloviev A, Vorobieva I (1997) On dynamics of seismicity simulated by the models of blocks-and-faults systems. Annali di Geofisica XL(5): 1217-1232.

Gvishiani AD, Kossobokov VG (1981) Proc Ac Sci USSR. Phys Earth 2: 21-36.

Holland JH (1995) Hidden Order: How Adaptation Builds Complexity. Addison-Wesley, Reading (Mass).

Huang Y, Saleur H, Sammis C, Sornette D (1998) Precursors, aftershocks, criticality and self-organized criticality. Europhys Lett 41: 43-48.

IMF (1997) International Monetary Fund, International Financial Statistics, CD-ROM.

Jaumé SC, Sykes LR (1999) Evolving to wards a critical point: A review of accelerating seismic moment/energy release prior to large and great earthquakes. Pure Appl Geophys 155: 279-306.

Kantorovich LV, Keilis-Borok VI, Molchan GM (1974) Seismic risk and principles of seismic zoning. In: Seismic design decision analysis. Department of Civil Engineering, MIT, Internal Study Report 43.

Kadanoff LP (1976) Scaling, universality and operator algebras. In: Domb C, Green MS (eds) Phase Transitions and Critical Phenomena, Vol 5a

Keilis-Borok VI, Malinovskaya LN (1964) One regularity in the occurrence of strong earthquakes. J Geophys Res 69:3019–3024.

Keilis-Borok VI, Knopoff L, Rotwain IM (1980) Bursts of aftershocks, long-term precursors of strong earthquakes. Nature 283: 258-263.

Keilis-Borok VI, Press F (1980) On seismological applications of pattern recognition. In: Allegre CJ (ed) Source Mechanism and Earthquake Prediction Applications. Editions du Centre national de la recherché scientifique, Paris, pp 51-60.

Keilis-Borok V, Lamoreau R, Johnson C, Minster B (1982) Swarms of main shocks in Southern California. In: Rikitake T (ed) Earthquake Prediction Research. Elsevier, Amsterdam.

Keilis-Borok VI, Kossobokov VG (1984) A complex of long-term precursors for the strongest earthquakes of the world. In: Proceedings 27th Geological Congress, 61, Nauka, Moscow, pp 56–66.

Keilis-Borok VI, Kossobokov VG (1987) Periods of high probability of occurrence of the world's strongest earthquakes. In: Computational Seismology, 19, Allerton Press, pp 45–53.

Keilis–Borok VI (1990) (ed) Intermediate-term Earthquake Prediction: Models, Phenomenology, Worldwide Tests. Phys Earth Planet Inter 61, special issue 1-2: 1-144.

Keilis-Borok VI, Kossobokov VG (1990) Premonitory activation of earthquake flow: algorithm M8. Phys Earth Planet Inter 61: 73-83.

Keilis-Borok VI, Rotwain IM (1990) Diagnosis of time of increased probability of strongearthquak es in different regions of the world: algorithm CN. Phys Earth Planet Inter 61: 57-72.

Keilis-Borok VI, Lichtman AJ (1993) The self-organization of American society in presidential and senatorial elections. In: Kravtsov YuA (ed) Limits of Predictability. Springer-Verlag, Berlin-Heidelberg, pp 223-237.

Keilis-Borok VI (1994) Symptoms of instability in a system of earthquake-prone faults. Physica D 77: 193-199.

Keilis-Borok VI, Rotwain IM, Shebalin PN (1994) The special redistribution of low-level seismicity in focal zones of large earthquakes. In: Keilis-Borok VI (ed) Seismicity and Related Processes in the Environment, volume 1. Research and Coordinating Centre for Seismology and Engineering, Moscow, pp 44-48.

Keilis-Borok, V (1996) Intermediate-term earthquake prediction. Proc Natl Acad Sci USA 93: 3748-3755.

Keilis-Borok VI, Shebalin PN (1999) (eds) Dynamics of Lithosphere and Earthquake Prediction. Phys Earth Planet Inter 111, special issue 3-4: 179-330.

Keilis-Borok VI, Sorondo MS (2000) (eds) Science for Survival and Sustainable Development. The Proceedings of the Study-Week of the Pontifical Academy of Sciences, 12-15 March 1999. Pontificiae Academiae Scientiarvm Scripta Varia, 98, Vatican City.

Keilis-Borok V, Stock JH, Soloviev A, Mikhalev P (2000) Pre-recession pattern of six economic indicators in the USA. J Forecast 19:65–80.

Keilis-Borok VI (2002) Earthquake prediction: State-of-the-art and emerging possibilities. Annu Rev Earth Planet Sci 30: 1-33.

Keilis-Borok V, Shebalin P, Zaliapin I (2002) Premonitory patterns of seismicity months before a large earthquake: Five case histories in Southern California. Proc Natl Acad Sci USA 99: 16562-16567.

Keilis-Borok VI (2003) Basic science for prediction and reduction of geological disasters. In Beer, T. and Ismail-Zadeh, A. (eds.), Risk Science and Sustainability, Kluwer Academic Publishers, Dordrecht-Boston-London (NATO Science Series. II. Mathematics, Physics and Chemistry – Vol. 112), pp. 29-38.

Keilis-Borok VI, Gascon DJ, Soloviev AA, Intriligator MD, Pichardo R, Winberg FE (2003) On predictability of homicide surges in megacities. In: Beer T, Ismail-Zadeh A (eds) Risk Science and Sustainability. Kluwer Academic Publishers, Dordrecht-Boston-London (NATO Science Series. II. Mathematics, Physics and Chemistry – Vol. 112), pp 91-110.

Keilis-Borok VI, Soloviev AA (2003) (eds) Nonlinear Dynamics of the Lithosphere and Earthquake Prediction, Springer-Verlag, Berlin-Heidelberg

Keilis-Borok V, Shebalin P, Gabrielov A, Turcotte D (2004) Reverse tracing of short-term earthquake precursors. Phys Earth and Planet Inter 145(1-4):75-85.

Keilis-Borok VI, Soloviev AA, Allègre CB, Sobolevskii AN, Intriligator MD (2005) Patterns of macroeconomic indicators preceding the unemployment rise in Western Europe and the USA. Pattern Recognition 38(3):423-435.

Keilis-Borok V, Soloviev A, Intriligator M, Winberg F (2006) Current prediction of the increase in the unemployment rate in the U.S. E2-C2 Meeting, Perugia, 02-05 September 2006, Golf Hotel Colle della Trinita.

Keilis-Borok VI, Soloviev AA, Intriligator MD, Winberg FE (2008) Pattern of macroeconomic indicators preceding the end of an American economic recession. J. Pattern Recognition Res., 3(1):40-53.

Klein PhA, Niemira MP (1994) Forecasting financial and economic cycles. New York, Wiley.

Knopoff L, Levshina T, Keilis-Borok VI, Mattori C (1996) Increased long-range intermediate-term earthquake activity prior to strong earthquakes in California. J Geophys Res 101(B3): 5779-5796.

Kosobokov VG (1983). Recognition of the sites of strong earthquakes in East Central Asia and Anatolia by Hamming's method. In: Keilis-Borok VI and Levshin AL (eds) Mathematical models of the structure of the Earth and the earthquake prediction, Comput. Seismol., 14. Allerton Press, New York, pp. 78-82.

Kossobokov VG, Keilis-Borok VI, Smith SW (1990) Localization of intermediate-term earthquake prediction. J Geophys Res 95(B12): 19763-19772.

Kossobokov VG, Carlson JM (1995) Active zone size vs. activity: A study of different seismicity patterns in the context of the prediction algorithm M8. J Geophys Res 100: 6431-6441.

Kossobokov VG, Keilis-Borok VI, Cheng B (2000a) Similarities of multiple fracturingon a neutron star and on the Earth. Phys Rev E 61(4): 3529-3533.

Kossobokov VG, Keilis-Borok VI, Turcotte DL, Malamud BD (2000b) Implications of a statistical physics approach for earthquake hazard assessment and forecasting. Pure Appl Geophys 157(11-12): 2323-2349.

Kossobokov V, Shebalin P (2003) Earthquake prediction. In: Keilis-Borok VI, Soloviev AA (eds) Nonlinear Dynamics of the Lithosphere and Earthquake Prediction. Springer-Verlag, Berlin-Heidelberg, pp 141-207.

Kravtsov YuA (1993) (ed) Limits of Predictability. Springer-Verlag, Berlin-Heidelberg.

Kuznetsov IV, Keilis-Borok VI (1997) The interrelation of earthquakes of the Pacific seismic belt. Trans (Doklady) Russ Acad Sci, Earth Sci Sect 355A(6): 869–873.

Levshina T, Vorobieva I (1992) Application of algorithm for prediction of a strong repeated earthquake to Joshua Tree and Landers. In: Fall Meeting AGU, Abstracts, p 382.

Lichtman AJ, Keilis-Borok VI (1989) Aggregate-level analysis and prediction of midterm senatorial elections in the United States, 1974-1986. Proc Natl Acad Sci USA 86(24): 10176-10180.

Lichtman AJ (1996) The Keys to the White House. Madison Books, Lanham

Lichtman AJ (2000) The Keys to the White House. Lexington Books Edition, Lanham

Lichtman AJ (2005) The Keys to the White House: Forecast for 2008. Foresight: The International Journal of Applied Forecasting. 3: 5-9.

Ma Z, Fu Z, Zhang Y, Wang C, Zhang G, Liu D (1990) Earthquake Prediction: Nine Major Earthquakes in China. Springer-Verlag, New York

Mason IB (2003) Binary events. In: Jolliffe IT, Stephenson DB (eds.) Forecast Verification. A Practioner's Guide in Atmospheric Science, Wiley and Sons Ltd, Chichester, pp 37-76.

Messner SF (1983) Regional differences in the economic correlates of the urban homicide rate. Criminology 21(4): 477-488.

Mogi K (1968) Migration of seismic activity. Bull Earth Res Inst Univ Tokyo 46(1): 53-74.

Molchan GM (1990) Strategies in strong earthquake prediction. Phys Earth Planet Inter 61:84-98.

Molchan GM, Dmitrieva OE, Rotwain IM, Dewey J (1990) Statistical analysis of the results of earthquake prediction, based on bursts of aftershocks. Phys. Earth Planet. Inter., 61: 128-139.

Molchan GM (1991) Structure of optimal strategies of earthquake prediction. Tectonophysics 193:267-276.

Molchan GM (1994) Models for optimization of earthquake prediction. In: Chowdhury DK (ed) Computational Seismology and Geodynamics, Vol 1. Am Geophys Un, Washington, DC, pp 1-10.

Molchan GM (1997) Earthquake prediction as a decision-making problem. Pure Appl Geophys 149: 233–237

Molchan GM (2003) Earthquake Prediction Strategies: A Theoretical Analysis. In: Keilis-Borok VI, Soloviev AA (eds) Nonlinear Dynamics of the Lithosphere and Earthquake Prediction, Springer-Verlag, Berlin-Heidelberg, pp 209–237.

Molchan G, Keilis-Borok V (2008) Earthquake prediction: Probabilistic aspect. Geophys J Int 173(3): 1012-1017.

Mostaghimi M, Rezayat F (1996) Probability forecast of a downturn in U.S. economy using classical statistical theory. Empirical Economics 21: 255-279.

Newman WI, Gabrielov AM, Turcotte DL (1994) (eds) Non-linear Dynamics and Predictability of Geophysical Phenomena, Geophysical Monograph Series, IUGG-Am Geophys. Union, Washington, DC.

Newman WI, Turcotte DL, Gabrielov AM (1995) Log-periodic behavior of a hierarchical failure model with application to precursory seismic activation. Phys Rev E. 52: 4827-4835.

OECD (1997) Main Economic Indicators: Historical Statistics 1960-1996. Paris, CD-ROM.

Panza GF, Soloviev AA, Vorobieva IA (1997) Numerical modeling of block-structure dynamics: Application to the Vrancea region. Pure Appl Geophys 149: 313-336.

Pepke GF, Carlson JR, Shaw BE (1994) Prediction of large events on a dynamical model of fault. J Geophys Res 99: 6769–6788.

Pollitz FF, Burgmann R, Romanowicz B (1998). Science 280: 1245-1249.

Press F, Briggs P (1975) Earthquakes, Chandler wobble, rotation and geomagnetic changes: a pattern recognition approach. Nature, 256: 270-273.

Press F, Briggs P (1977) Pattern recognition applied to uranium prospecting. Nature, 268: 125-127.

Press F, Allen C (1995) Patterns of seismic release in the southern California region. J Geophys Res 100(B4): 6421–6430.

Prozorov AG (1975) Changes of seismic activity connected to large earthquakes. In: Keilis-Borok VI (ed) Interpretation of Data in Seismology and Neotectonics, Comput Seismol 8. Nauka, Moscow, pp 71-82.

Richter C (1964) Comment on the paper "One Regularity in the Occurrence of StrongEarthquak es" by Keilis-Borok, V.I. and Malinovskaya, L.N. J Geophys Res 69: 3025.

Romanowicz B (1993) Spatiotemporal patterns in the energy-release of great earthquakes. Science 260: 1923–1926.

Rotwain I, Keilis-Borok V, Botvina L. (1997) Premonitory transformation of steel fracturing and seismicity. Phys Earth Planet Int 101: 61-71.

Rundkvist DV, Rotwain IM (1996) Present-day geodynamics and seismicity of Asia minor. In: Chowdhury DK (ed) Computational Seismology and Geodynamics, Vol 3. Am Geophys Un, Washington, DC, pp 130-149.

Rundle BJ, Turcotte DL, Klein W, (2000) (eds) Geocomplexity and the Physics of Earthquakes. Washington, DC: Am Geophys Union.

Sammis CG, Sornett D, Saleur H (1996) Complexity and earthquake forecasting. In: Rundle JB, Turcotte DL, Klein W (eds) SFI Studies in the Science of Complexity, vol. XXV. Addison-Wesley, Reading, Mass.

Shebalin P, Girardin N, Rotwain I, Keilis-Borok V, Dubois J (1996) Local overturn of active and non-active seismic zones as a precursor of large earthquakes in Lesser Antillean Arc. Phys Earth Planet Inter 97: 163-175.

Shebalin PN, Keilis-Borok VI (1999) Phenomenon of local "seismic reversal" before strong earthquakes. Phys Earth Planet Inter 111(3-4): 215-227.

Shebalin P, Zaliapin I, Keilis-Borok V (2000) Premonitory raise of the earthquakes' correlation range: Lesser Antilles. Phys Earth Planet Inter 122(3-4): 241-249.

Shebalin P, Keilis-Borok V, Zaliapin I, Uyeda S, Nagao T, Tsybin N (2004) Advance short-term prediction of the large Tokachi-oki earthquake, September 25, 2003, M = 8.1. A case history. Earth, Planets and Space 56(8): 715-724.

Shebalin P, Keilis-Borok V, Gabrielov A, Zaliapin I, Turcotte D (2006) Short-term earthquake prediction by reverse analysis of lithosphere dynamics. Tectonophysics 413: 63-75.

Shnirman MG, Blanter EM (1998) Self-organized criticality in a mixed hierarchical system. Phys Rev Letters 81: 5445-5448.

Shnirman M, Blanter E (2003) Hierarchical models of seismicity. In: Keilis-Borok VI, Soloviev AA (eds) Nonlinear Dynamics of the Lithosphere and Earthquake Prediction. Springer-Verlag, Berlin-Heidelberg, pp 37-69.

Soloviev A, Ismail-Zadeh A (2003) Models of dynamics of block-and-fault systems. In: Keilis-Borok VI, Soloviev AA (eds) Nonlinear Dynamics of the Lithosphere and Earthquake Prediction. Springer-Verlag, Berlin-Heidelberg, pp 71-139.

Sornette D, Sammis CG (1995) Complex critical exponents from renormalization group theory of earthquakes: Implications for earthquake predictions. J Phys I France 5: 607-619.

Sornette D (2000) Critical Phenomena in Natural Sciences: Chaos, Fractals, Self-organization, and Disorder. Concept and Tools. Springer, Berlin.

Stock JH, Watson MW (1989) New indexes of leading and coincident economic indicators. In: NBER Macroeconomics Annual, pp 351-394.

Stock JH, Watson MW (1993) A procedure for predicting recessions with leading indicators. In: Stock JH, Watson MW (eds) Business Cycles, Indicators, and Forecasting (NBER Studies in Business Cycles, Vol.28), pp 95-156.

Sykes LR, Jaumé S (1990) Seismic activity on neighboring faults as a long-term precursor to large earthquakes in the San Francisco Bay area. Nature 348: 595-599.

Tukey JW (1977) Exploratory Data Analysis. Addison-Wesley Series in Behavioral Science: Quantitative Methods. Addison-Wesley, Reading (Mass).

Turcotte DL (1997) Fractals and Chaos in Geology and Geophysics. 2nd Ed., Cambridge University Press.

Turcotte DL (1999) Seismicity and self-organized criticality. Phys Earth Planet Int 111: 275-294.

Turcotte DL, Newman WI, Gabrielov A (2000) A statistical physics approach to earthquakes. In: Geocomplexity and the Physics of Earthquakes, Am Geophys Un, Washington DC.

USDL (1999) U.S. Department of Labor, Bureau of Labor Statistics, Labor Force Statistics from the Current Population

Survey. Web site: http://stats.bls.gov/webapps/legacy/cpsatab5.htm.

Varnes DJ (1989) Predicting earthquakes by analyzing accelerating precursory seismic activity. Pure Appl Geophys 130: 661-686.

Vil'kovich EV, Shnirman MG (1983) Epicenter migration waves: Examples and models. In Keilis-Borok VI, Levshin AL (eds) Mathematical Models of the Earth's Structure and the Earthquake Prediction, Comput Seismol 14, Allerton Press, New York, pp 27-36.

Vorobieva IA, Levshina TA (1994) Prediction of a second large earthquake based on aftershock sequence. In: Chowdhury DK (ed) Computational Seismology and Geodynamics, Vol 2, Am Geophys Un, Washington, DC, pp 27-36.

Vorobieva IA (1999) Prediction of a subsequent large earthquake. Phys Earth Planet Inter, 111(3-4): 197-206.

Vorobieva IA, Soloviev AA (2005) Long-range interaction between synthetic earthquakes in the block model of lithosphere dynamics. In: Chowdhury DK (ed) Computational Seismology and Geodynamics, Vol 7, Am Geophys Un, Washington, DC, pp 170-177.

Wyss M, Habermann R (1988) Precursory seismic quiescence. Pure Appl Geophys 126: 319-332.

Yamashita T, Knopoff L (1992) Model for intermediate-term precursory clustering of earthquakes. J Geophys Res 97: 19873-19879.

Zaliapin I, Keilis-Borok V, Axen G (2002) Premonitory spreading of seismicity over the faults' network in southern California: Precursor Accord. J Geophys Res 107(B10): ESE5-1 – ESE5-15, 2221, doi:10.1029/2000JB000034.

Zaliapin I, Keilis-Borok V, Ghil M (2003a) A Boolean delay model of colliding cascades. I: Multiple seismic regimes. J Stat Phys 111(3-4): 815-837.

Zaliapin I, Keilis-Borok V, Ghil M (2003b) A Boolean delay model of colliding cascades. II: Prediction of critical transitions. J Stat Phys 111(3-4): 839–861.

Zoller G, Hainzl S, Kurths J (2001) Observation of growing correlation length as an indicator for critical point behavior prior to large earthquakes. J Geophys Res 106: 2167-2176.

APPPENDIX:

reprints of selected papers published by V.I.Keilis-Borok and his group

This appendix contains original reprints from the key publications that serve as the foundations for V. I. Keilis-Boroks prediction methods and lifetime findings. The reprints used here are listed below. Subsection I lists the reprints arranged by topics that also include some notes on their significance, with full references to the original source. Subsection II lists the same reprints in chronological order, the way they are appended in the book.

SUBSECTION I.

Reprints grouped by topic.

I. Seismic wave propagation. Inverse problems of seismology

II. New links between geodynamics, seismicity, and precursory phenomena

This paper describes the discovery of "nodes" - mosaic structures around fault intersections where instability is concentrated and strong earthquakes nucleate. Such nodes have been identified worldwide in many areas where strong earthquakes were deemed impossible; identification has been confirmed by subsequent earthquakes occurrence.

Gabrielov, A., **V.I. Keilis-Borok**, and D.D. Jackson, Geometric incompatibility in a fault system. *Proc. Natl. Acad. Sci. USA*, 1996, 93 (9): 3838-3842. (Copyright (1996) National Academy of Sciences, U.S.A.)..**p. 303**

This paper has introduced geometric incompatibility of a fault network. This is a measure of instability of fault network caused by collective behaviour of the nodes. It controls nucleation of strong earthquakes and destructive creep.

Zaliapin, I., **V. Keilis-Borok**, and G. Axen, Premonitory spreading of seismicity over the faults' network in southern California: Precursor Accord, *J. Geophys. Res.*, 2002, 107, B10: ESE5-1 – ESE5-15, 2221, doi:10.1029/2000JB000034.................**p. 337**

This paper connects precursory phenomena with collective behavior of fault zones.

III. Earthquake precursors

Keilis-Borok,V.I., and L.N.Malinovskaya, One regularity in the occurrence of strong earthquakes. *J. Geophys. Res.*, 1964, 69, 14: 3019-3024...**p. 182**

This paper has introduced "pattern Σ" that comprises the premonitory increase in the total area of the ruptures in the earthquake sources in a medium magnitude range. It was the first reported earthquake precursor formally defined that demonstrated the major features of patterns discovered later: *long-range correlations and similarity*.

Keilis-Borok,V.I., L.Knopoff, and I.M.Rotwain, Bursts of aftershocks, long-term precursors of strong earthquakes. *Nature*, 1980, 283: 259-263..**p. 250**

This paper considers a precursor capturing rise of background seismicity clustering before large earthquakes that has been validated by statistically significant predictions of real earthquakes.

IV. Nonlinear dynamics of the lithosphere and earthquake prediction

These papers introduce the concept of the *lithosphere, as a hierarchical non-linear system*, with earthquakes and other geological disasters as the critical phenomena. Keilis-Borok established the following paradigms (first regarded as strongly counterintuitive): (*i*). *Premonitory seismicity patterns*: A strong earthquake is preceded by increase of intensity, clustering, and correlation range of moderate seismicity and its deviation from the scale invariance. (*ii*). *Long-range correlations*: A strong earthquakes is prepared in the area exceeding the size of its source by the order of magnitude. (*iii*). *Similarity* of premonitory patterns in a broad range of tectonic environments and in the energy range from 1erg to 10^{26}erg. (*iv*). *Dual nature* of premonitory patterns: they are partly "universal" (common for a wide class of non-linear systems) and partly system-specific.

V. Disaster preparedness

VI. Approaches to short-term earthquake prediction

These papers describe a recently introduced prediction methodology based on tracing precursors in reverse order of their emergence. It is aimed at a considerable increase of prediction accuracy, reducing duration of alarms from years to months. Methodology currently undergoes testing by advance prediction in several regions, including California, Japan, Northern Italy, and Dead Sea rift zone.

VII. Socio-economic predictions

These papers demonstrate nonlinear dynamics approach to socio-economic predictions. In each case the experiments in advance prediction are launched. For elections the methodology is already validated by that experiment. Successful advance predictions were also obtained for recessions, unemployment, and crime waves.

VIII. Extreme events in complex systems

Gabrielov, A., **V. Keilis-Borok**, I. Zaliapin, and W.I. Newman, Critical transitions in colliding cascades. *Phys. Rev. E*, 2000, 62, 1: 1-13..**p. 324**

These papers provide statistical physics type models capturing the interaction between direct cascade of loading and inverse cascade of failures. The models reproduce major types of premonitory patterns and regime switching.

Gabrielov, A., **V. Keilis-Borok**, Ya. Sinai, and I. Zaliapin, Statistical properties of the cluster dynamics of the systems of statistical mechanics. In G.Gallavotti, W.L.Reiter, and J.Yngvason (eds.), *Boltzmann's Legacy*. European Mathematical Society, Zurich, 2008, pp. 203-215 (ESI Lectures in Mathematics and Physics)...**p. 435**

Statistical properties and critical indices related to cluster dynamics are studied in this paper. Cluster dynamics is the property of dynamics of systems of statistical mechanics when for given time the whole array of particles is decomposed onto finite clusters which move independently during a random interval of time.

Keilis-Borok,V.I., and A.A.Soloviev, Variations of trends of indicators describing complex systems: Change of scaling precursory to extreme events. *Chaos*, 2010, 20, 033104, doi:10.1063/1.3463438.............................. ....**p. 468**

This paper demonstrates a premonitory pattern common to four complex systems of different nature: relatively large magnitude of background events become more frequent before an extreme event.

SUBSECTION II.

Reprint list in chronological order, as appended below.

CHAPTER 3

SEISMOLOGY AND LOGICS

V. I. Keilis-Borok

1. Introduction

Seismology has come up against many new problems and accumulated many observations and improvements in the past 10 years — especially since the beginning of the IGY and the study of source identification. The result of this is not only an impressive body of new data: a necessity has cropped up to reexamine completely the whole of seismology — its internal structure, methods, and even its psychology. I shall try to analyze the logical structure of seismology as a whole and, on that basis, examine the revolution taking place in seismology today.

This revolution was signaled 10–15 years ago. The problems of classical seismology — an investigation of the general features of the earth's constitution, of the geography of strong earthquakes and their energetic classification — were reaching their solution. To investigate these problems it was sufficient to study a relatively small number of strong earthquakes. There were about 550 stations operating in the world; they processed approximately 100,000 records every year (of course, epicenters were not found for each record).

New problems were arising: to determine the constitution of the earth with an accuracy satisfying regional geology and theoretical physics; detailed seismic zoning and earthquake prediction; and, later, source identification. To solve these problems it was necessary to increase by 10–100 times the number of earthquakes studied, to raise the accuracy of investigation of records by 5–10 times, and to introduce new types of processing — mostly spectral analysis. Thus, for a detailed study of the crust and upper mantle, instead of one Jeffreys–Bullen's travel-time table, it is necessary to have a series of local travel-time and dispersion curves and even new types of magnitude scales.

For seismic zoning it was necessary to decrease the energetical threshold (energy minimum) of the earthquakes to be studied by approximately 10^6 times. As a result, one expedition on the Pamir–Baikal profile (40 stations) had to process 30,000 records anually; the whole seismic network of the world includes now nearly 1200 stations and must process over 1 million records per year. The accuracy of hypocenters must ensure a comparison with the geological structures: therefore, seismic zoning requires everything that is necessary to study the crust and mantle. The identification of sources requires, in addition, quantitative data on noise.

To receive all these data it is necessary to have up-to-date equipment. The material that is accumulated now is of little use at times because of the difficulty of digitalizing. The accumulated data, in turn, require new computers. Then it is necessary to apply (and frequently to work out) new theoretical methods of the mechanics of a continuous medium, of computational mathematics, and of statistics. New theories usually need additional or more exact data, and so the circle closes.

For example, loops are observed in travel-time curves. To analyze the corresponding amplitudes, it was necessary to work out the theory of waves in caustics [28]. To apply this theory, new observations were needed on the spectrum of body waves. And to define this spectrum, new equipment was needed, digital recording, and again a new theory — separation of the signal from the noise. Luckily this theory has not yet been applied in practice, and further complications are unknown.

What I have said seems trivial; evidently, the development of seismology leads to new requirements to its auxiliary means — equipment, theories, and computational technique. It is less evident, that now seismology must place new requirements on itself which are only partly a penalty for using the new methods. Here is one example.

2. Example: The Problem of the Upper Mantle and Seismology as a Whole

1. Until recently, six models of velocity distribution of the upper mantle were discussed [7, 9, 12, 22]. Naturally, a problem arose: to characterize the whole set of possible velocity distributions which could satisfy the given observations — e.g., the travel-time curve of P [29].

Mathematically the problem is to find the zone of minimum of some function in multidimensional space. This function is the difference between the observed and calculated travel times; space coordinates are the parameters of unknown velocity-depth curves — for example, the values of velocity at certain depths (between which the velocity can then be found by interpolation).

An interpretation was made of Lehmann's travel times [21] for P-waves in Europe: 115 velocity distributions were discovered (among 2000 examined) for which the standard deviation between observed and calculated travel times is less than 1 sec, and the maximum deviation is less than 1.5 sec [29]. Some of these velocity distributions are shown in Figure 1.

2. This result, nevertheless, contains certain unambiguous information: the velocity distributions discovered lie in a not too wide interval, and the average velocity is approximately the same for all distributions (the velocity increase and decrease zones mutually compensate). However, in most problems the average velocity is insufficient. Thus, amplitude-distance curves were calculated for all of the discovered velocity distributions. It turned out that, with the roughest classifications, these curves could not be divided into less than 15–20 groups at least qualitatively uniform regarding shape. For seismology this is a substantial uncertainty.

Velocity distributions, which correspond to various groups of amplitude-distance curves, differ by various sequences of sharp increase-decrease and

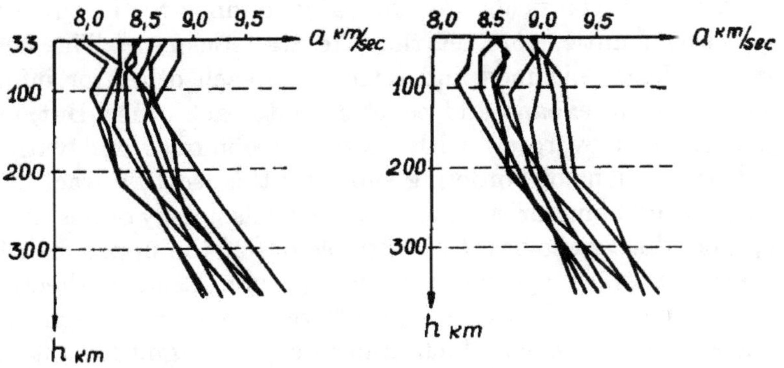

Fig. 1. Some of the 115 velocity distributions of the upper mantle, corresponding to the travel times of P-waves in Europe (after [29]). The horizontal scale is given for the first left curve. Each following curve is shifted to the right in relation to the preceding one by 0.1 km/sec.

by approximately constant velocity zones. This makes the result ambiguous for geology and physics.

To reduce the ambiguity it is necessary to refer to additional data: amplitude of body waves, especially in the shadow zone and loops of travel-time curve; dispersion of surface waves; more exact local travel times, etc. However, to obtain these data is not simple. Thus, in order to apply the amplitudes it is necessary (see Section 5): (*a*) to determine detailed magnitude scales, separately for different wave periods; (*b*) to find the local corrections for resonance in the surface layers; (*c*) to decipher travel-time loops; (*d*) to determine the absorption, simultaneously with velocity. In order to use dispersion, a study of the local variations of phase velocity must be made. To specify the travel time we must precisely determine the hypocenters, etc.

3. We see that because of the upper mantle it is necessary to dig into all branches of seismology, not for just some bits of information, but to

raise their level. This is not an unfortunate peculiarity of the upper mantle alone; it concerns any interesting problem of seismology. For example, all branches of seismology were represented in a list of required investigations [3], suggested by the United States at the nuclear test ban talks. And everything contained in that list is necessary, with slight deviations, to earthquake prediction, to studying the constitution of the earth, etc.

4. This situation has serious consequences: working in one field, it is impossible to raise simultaneously the level of all adjacent fields. In practice it is possible to make only lightning raids on them, i.e., to bypass the major difficulties and invent very narrow, often rule of thumb, methods. As a consequence, one and the same subject is studied in various fields by different methods (for example, long-period oscillations in the constitution of the earth, and in earthquake prediction).

As a result, what goes on in seismology as a whole? In the classical period its division into special fields was naturally confined to the subjects under study — crust, mantle, core, earthquake statistics, etc. The fields were more or less clearly outlined and referred to each other for information. Then they began to expand and penetrate into each other. But, instead of uniting, each field overgrew with its own techniques and with its own simplified conceptions of bordering subjects; this led to a fracturing tendency along former boundaries: thus we have a seismology of seismic activity, two long-period seismologies, two seismologies of the upper mantle (body and surface waves), magnitude seismology, etc. Each of them, in turn, divides spontaneously; for example, we have a number of magnitude scales and even a special science which transfers one magnitude into another, and then each of them into two or three earthquake energies. Exaggerating a bit, we may say, that seismologists are gradually settling on separate planets; the inhabitants of each planet consider it the center of the whole system and justify the arising discrepancies with a multitude of complications like the epicycles in Ptolemy's astronomy. The ultrabroad mind can consider all planets as centers simultaneously.

Such disunity could gradually transform seismology into a descriptive science like biology before Linnaeus. Such a science cannot exist for long in our times. A simple solution is this: as all our planets revolve around one object — a similar analysis of one and the same set of observations — let us place this analysis at the center of the whole system. In other words, to introduce a new division of investigation, which would correspond not to the objects, as they are closely connected, but to analysis of observations. In this report I shall try, first, to outline the logical scheme of interpretation of seismological data (Section 3); then to investigate the possibility of formalization of basic operation of this scheme (Sections 4–6); and last on that basis to show the possibility to distinguish a limited number of standard operations in seismology and to reduce the infinite variety of its problems to various combinations of these operations (Section 7.1). This does not mean conformism of scientific ideas. On the contrary, by liquidating ill-assorted methods it is possible to disentangle the authentic individuality of problems. Let us recall, for example, that a similar reform in the written language — a substitution of thousands of hieroglyphs by a few dozen standard letters — greatly increased the possibility of individual thought.

3. The Structure of Seismology

Studying the combined kinds of work called seismology, we can single out four consecutive stages (Fig. 2). (Further on, comments quite trivial to seismologists are given in small type.)

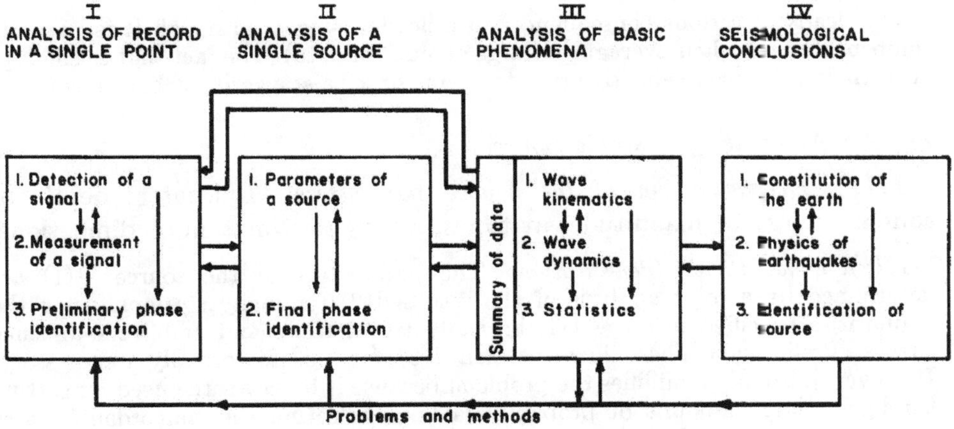

FIG. 2. The basic stages of interpretation of seismic observations.

I. An Analysis of Records at a Single Point.

I(1). Detection of a signal from a noisy background

I(2). Measurement of the parameters of separate waves or wave groups (including microseisms). Arrival times; maximal amplitudes; visual periods; polarization; duration of oscillations; and, recently, spectra and correlation functions.

I (3). Preliminary Phase Identification. At the input of this stage and consequently at the input of all seismology, there is a flow of over one million records annually (not including seismic prospecting). These are obtained by 30–40 types of instruments.

Operation I(1) is the primary one. It is intuitive for the greater part. Then a group of operations I(2) follows; these are practically never done all at once; therefore, because of various problems, we have to refer to one and the same record many times. These operations seem formalized, but only if we forget that many of them are done visually and that a statistical analysis of errors is very rarely given. Jeffreys wrote the following in this connection: "It is astonishing, that experimentors will spend months in making a series of observations, and grudge the day or so needed to present the results in an intelligible form" [13].

Phase identification I(3) is based on the parameters measured in operations I(2). The fixed ideas are used here as to what sets of parameters identify various wave groups. From time to time they are checked at later stages (III and IV).

Besides this, an intuitive "recognition" of waves is used in phase identification. The role of recognition is greatly exaggerated because, strange as it may seem, the systematic criteria for phase identification have not yet been written (except for arrival times).

In general, the data at the output of stage I are of a probability nature. Evidently their analysis in the following stages should be of the same nature: in principle we may seek not a single-valued result but a distribution of probabilities among a certain set of possible results. At the present time such an approach is difficult because the measurements in stage I are frequently made with uncoordinated methods, usually without a proper estimation of accuracy and a proper separation of intuitive operations.

This leads to serious consequences in following stages (although it is partially compensated for when averaging a large amount of data). The fact that seismology nevertheless does give effective results, seems to me a staggering achievement.

II. An Analysis of a Single Source

II(1). Determination of the source parameters. Epicenter; depth of source; energy or magnitude; at times, a rough estimation of dimensions.

II(2). Final Phase Identification. The parameters of the source II(1) are determined by a joint analysis of the results I(2) for many stations, using the regularities specified in stage III: basically travel times and amplitude-distance curves. Preliminary phase identification I(3) for this is formally unnecessary. However, it greatly simplifies the problem, because it bypasses the need for sorting through a large amount of probabilities (e.g., selecting the concordant phases *pP* and *sP* amongst the first 4 or 5 arrivals at a few dozen stations [24]. That is why, if computers are not used, operation I(3) is compulsory, even though it creates an additional threat of errors.

The parameters of the source give, in turn, additional powerful criteria to phase identification—for example, it is possible to use all travel times. That is why the final phase identification II(2) follows II(1). Strong feedback coupling arises here: if errors are found in preliminary identification I(3) we must reexamine II(2) and at times return to stage I to search for new phases or make additional measurements.

Stages I and II operate continuously as observations are received. The following two stages process the results by separate impulses as sufficient initial data are accumulated.

III. An Analysis of the Basic Phenomena

III(1). Kinematics of Seismic Waves. The travel times of body waves and dispersion curves of surface waves.

III(2). Dynamics of Seismic Waves. The change of their intensity, duration, and (recently) spectrum with the distance and depth of the source.

III(3). Statistical Relationships. Geographical distribution of the density of hypocenters with different energies (maps of "seismic activity"); the dependence of dynamics of seismic waves on the properties of the source or local geology, etc.

These operations are done more or less in parallel based on data from stages II and I(2). The general preliminary operation in stage III is to draw up a summary of these data. Unfortunately at present, summaries are only made of arrival times, hypocenters and, less frequently, the magnitude and mechanism of earthquakes.

The relationships received in stages III(1) and III(2) are also used in stages I and II. When improved specifications of these relationships appear on the output of the stage, a feedback wave of improvements in the method of operation on stages I and II arises. That is why the results received in the interval between consecutive improvements are of unequal meaning.

IV. Seismological Conclusions

IV(1). Constitution of the Earth. Distribution of the velocity of seismic waves, density, and absorption.

IV(2). Physics of Earthquakes. Seismic zoning and seismic regime.

IV(3). Source Identification.

Stage IV works out the problems for all previous operations. It intensively exchanges data with other branches of geophysics and with geology. This mutual relationship concerns the structure of geophysics as a whole and will not be discussed here.

Many conclusions at the output of stage IV are ambiguous because of the heterogeneity of the initial data. The criteria of the truth of the conclusions are taken from the internal agreement of all data used, as the possibilities for a direct check are very limited. But the amount of data gradually becomes more than can be grasped, especially with such heterogeneity. And we gradually approach a situation when, with some exaggeration, one can say that for any given, not too absurd, hypothesis one can find a self-consistent body of data.

In particular, there is a danger of creating logical cycles. For example, the conception of a two-layered crust was based on identification of P^*-phase as a head wave. But this identification was based on the conception of a two-layered earth's crust, shaken for some time only by Gutenberg's intuitive doubts, which were then confirmed by Tatel and Tuve [26]. Such cycles naturally seem especially consistent. In its turn, the development of theory expands the meaning of "consistency" to such a degree, that with ample erudition, any sample of data could be made consistent with a sufficiently complex hypothesis.

The ambiguity discussed here is solved today by accident and intuition; the methods of an objective analysis become more and more indispensable.

It turns out that the variety of seismological studies in principle can be fitted into systematic order. But in reality, as we have seen, there is no systematic order as yet: large entropy is introduced by disunity of the work.

To restore order it is not enough (although important) to understand the logical structure of seismology. This structure must be realized: first, by an introduction of unified and universal recording equipment; secondly, by a formalization of the standard elements of interpretation (which will make it possible to automate them); thirdly, by a presentation of the results of interpretation in a compact unified form. Then the seismologist could be disentangled from routine work. Magnetic-digital or digitalized records could be used as the universal equipment. A steady development of seismology without it is generally impossible. Let us now examine what elements of interpretation may be formalized and unified.

4. Formalization of Processing of Observations

1. The most intuitive operation in stage I is the separation of waves from noise. I shall show how to formalize it with an example of determination of the sign of the first arrival [23]. The problem is as follows: a signal is added to a record of noise. Determine the moment and sign (up or down) of its first arrival. This problem is solved in two consecutive operations:

(*a*) A statistical prediction of noise $\bar{f}(t)$ is calculated from the real record $f(t)$ (Fig. 3, upper trace) for a sliding time interval from t to $(t + \tau)$. This prediction is continuously subtracted from the real record, i.e., the mistake of prediction $\phi(t)$ is determined: $\phi(t) = f(t) - \bar{f}(t)$ (Fig. 3, lower trace). While the noise is being recorded, $\phi(t)$ is much smaller than $f(t)$, although it does not equal zero because of the prediction error. The

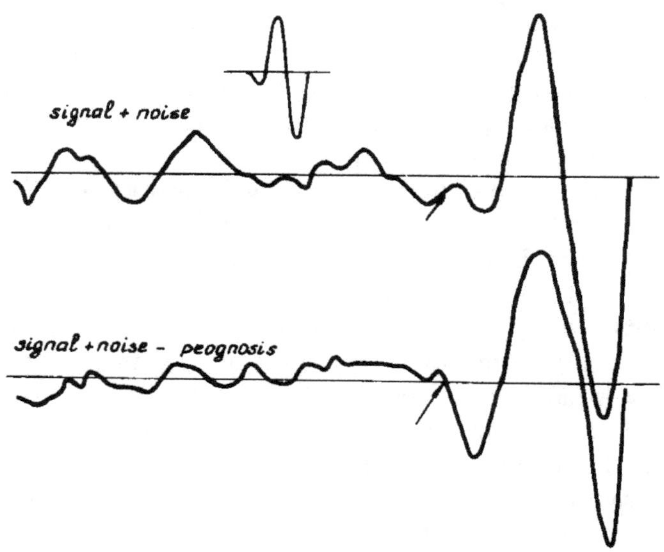

FIG. 3. Reduction of the noise by the method of statistical prediction. The arrows show the true first arrival. The shape of true arrival is shown on top.

arrival of the signal is not foreseen by the prediction; and as one can see in Figure 3, it will be clearer on background $\phi(t)$ than on the background of real noise. However, the very first peak may be camouflaged by $\phi(t)$ also if it is too weak. That is why a second operation is necessary:

(*b*) Three hypotheses are distinguished: that the peak found by operation (*a*) is the first, second, or third peak of the true signal. The statistical likelihood ratio for each hypothesis is calculated, with certain assumptions as to the shape of the true signal (it is assumed that the periods of consecutive peaks are equal, and their amplitudes grow in roughly known limits). The hypothesis with the greatest likelihood ratio is considered the true

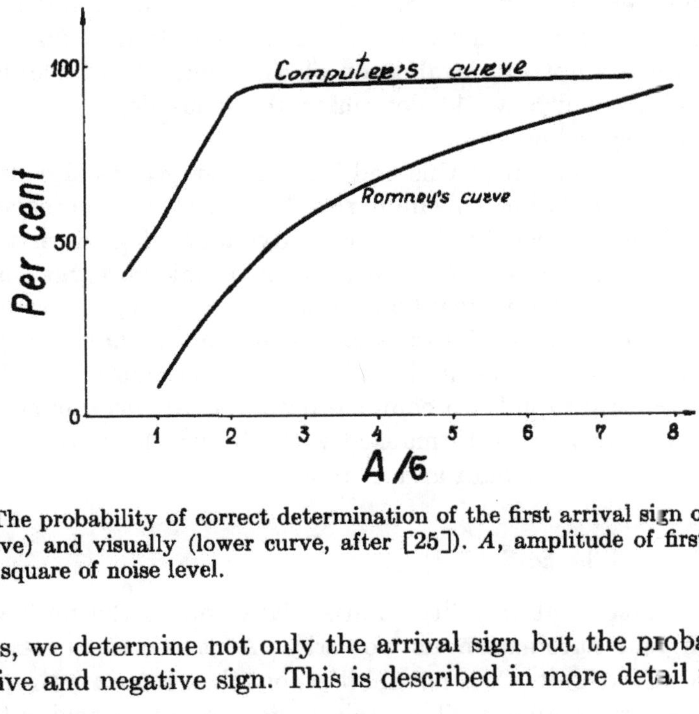

FIG. 4. The probability of correct determination of the first arrival sign on computer (upper curve) and visually (lower curve, after [25]). A, amplitude of first arrival; σ, root mean square of noise level.

one. Thus, we determine not only the arrival sign but the probabilities of the positive and negative sign. This is described in more detail in [23].

The method models the intuition of the seismologist: he also recognizes the signal by its difference from noise (operation *a*) and by the expected shape (operation *b*). A competition was held between a computer and experienced seismologists. Arrival signs were determined for the signals, artificially added to real noise, so that the correct answers were known. The percentage of correct definitions was much higher for the computer. This is confirmed by the curves of Figure 4 (results on the visual determinations are taken from Romney [25]). It is interesting to note that, in our experiments, the best results were given by the seismologist who in turn tried to repeat visually the computer's algorithm (operation *a*).

The following part of stage I, the measurement of the wave parameters, is essentially formal. Having a digital record it may be easily supplemented with a proper evaluation of accuracy. Finally, identification of the majority of phases I(3) can be easily formalized, if we refer it to stage II and to sort all possible variants on a computer [24]. Therefore, in its present volume stage I may be formalized almost completely; with a magnetic recording it can be automated as well. In more complicated situations, however, the recognition of waves is not sufficiently developed; elements of intuition here seem ineradicable.

2. An analysis of a single source (stage II) includes first of all computational and logical problems; for most of them the methods of solution have been worked out on computers. These are the determination of hypocenters [4, 24], earthquake mechanisms [11, 20, 24], and phase identification [24]. (However, the last two have been solved in a simplified manner.) Besides this, stage II includes a preliminary selection of observations or methods (for example, an evaluation of the reliability of the data or the choice of phase which would determine the focus depth). Formalization has not even begun here.

The use of the output of this and foregoing stages, as already has been said, starts with drawing up summaries. Here it is often necessary to refer to initial data (especially if we need dynamic characteristics), which destroys the advantages of the multitude of seismic observations. In order to make full use of these advantages, it is necessary to go over to electronic storage and selection of information. However, a simple expansion of information in routine (printed) bulletins and summaries would open up great possibilities: rapidly to combine data and to check hypotheses.

Stage III is also easily formalized with the previous stipulation about the criteria of choosing data and methods.

5. Constitution of the Earth (IV.1)

Here we have relatively little initial data and so the problems of formalization and automation do not arise. At times the necessary calculations are at the limit of our possibilities, but their realization on computers does not require comments. The main difficulty in this field is to find and overcome ambiguity. To do this, we must unite not only the results, but also the process of interpreting initial data (in the first place travel times, amplitudes, and dispersion curves). In order to illustrate the possibility of such an approach, let us return to the analysis of velocity distributions of the upper mantle begun in Section 2. The use of travel times was already illustrated there; let us now discuss the use of amplitudes and dispersion.

5.1. Amplitude-Distance Curves of Body Waves

As shown in [7, 8] they depend considerably on the distribution of the vertical velocity gradient along the depth. Velocity distributions determined in [29] for the upper mantle have the most varied gradients (see, for example, Fig. 1). Consequently, a comparison of observed amplitude-distance curves with calculated ones can greatly reduce ambiguity [7, 29].

However, difficulties arise here because of the unstableness of the amplitude curves: a minute change of the gradient on some depth may lead to a sharp change of amplitudes at the corresponding distances. In the observed curves, these changes must be smoothed (because of the measurement errors and the natural horizontal fluctuation of the velocity distributions), this smoothing may greatly change the shape of the curve.

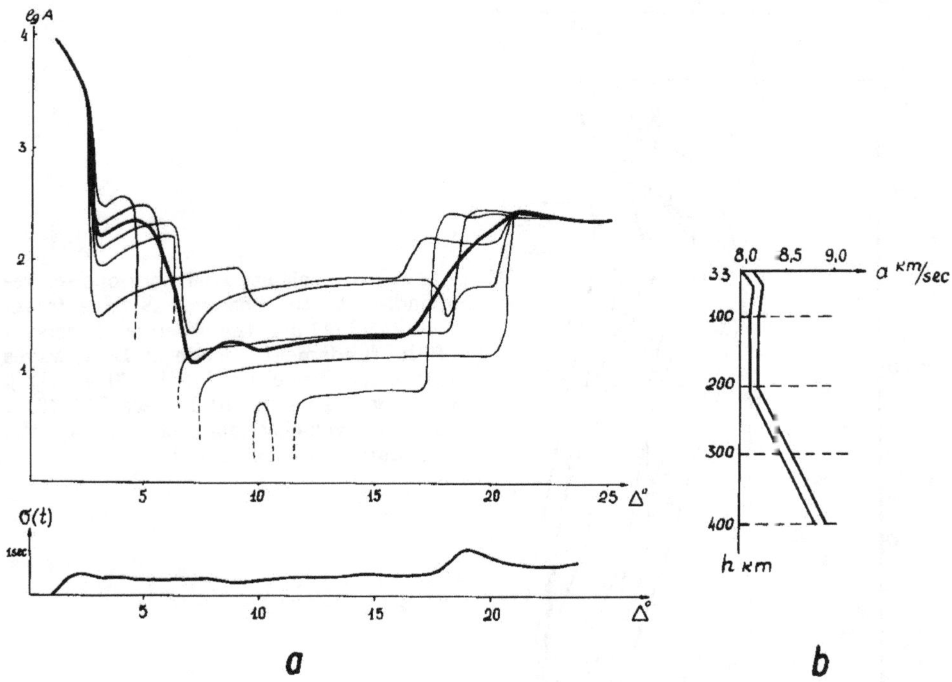

FIG. 5. (*a*) The average (thick line) and individual amplitude-distance curves; (*b*) limits of variation of *P*-wave velocity. After T. B. Yanovskaya.

A somewhat artificial example [Fig. 5*a*) illustrates this. The thin lines are amplitude-distance curves for velocity distributions which are very close to each other (they all lie inside a strip with a width of only 0.1 km/sec shown in Fig. 5*b*), and the thick line is the result of averaging a few dozen such curves. The average curve differs considerably from each separate one. The question arises: how to compare observed and calculated curves? Here, we approach a problem which resembles the problem of the computer recognizing letters written with different handwriting. Other difficulties are that the amplitudes depend on the wave period (because of the interference, nonideal elasticity, and diffraction in the shadow zone).

Therefore, instead of the ordinary amplitude curves, we must know the change of the amplitude spectrum with distance.

5.2. Dispersion of Surface Waves and Travel Times of Body Waves

Their combined interpretation can be more beneficial than the separate one. This is illustrated in Figure 6, showing the velocity of transverse waves in the upper mantle for North America. Curves 1 are the velocity distributions which correspond to the observed *S*-wave travel times [21]. Curves 2 are velocity distributions which correspond to the observed phase [5]

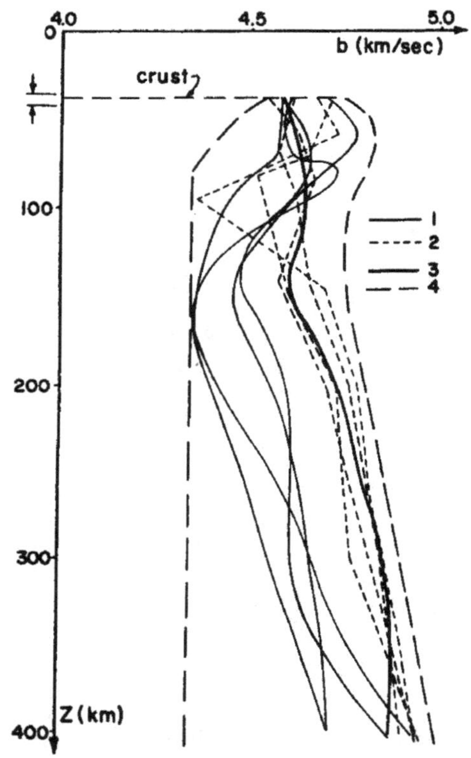

FIG. 6. Velocity distributions corresponding to the observed *S*-wave travel time [29] (1) and the observed dispersion of the fundamental mode of Love waves (2): phase velocities for 10–70 sec [5], group velocities for 10–140 sec [6]. (3) a common velocity distribution; (4) the boundary of variation of velocity.

and group [6] velocities of Love waves. (The method of determining the velocity distribution is similar to the one described in [29].) Two sets of velocity distributions nearly come into contact so that the real velocity distribution seems to be determined unambiguously. However, this conclusion would be premature: a more detailed study must be made of how the observations were obtained.

6. A Compact Characteristic of Seismic Regime (IV.2)

Here, there is no problem of formalization as in IV(1) though for another reason: there is a lot of initial data but their analysis is rather elementary.

The difficulty here is to find sufficiently compact and suggestive characteristics for the countless multitude of earthquakes. This multitude cannot be characterized either by the total number or by the total energy of earthquakes: as is well known, the number is defined only by the weakest, and the total energy only by strongest, shocks. Seismic activity — the number of earthquakes of a certain energy — is more representative, but only as the average for long time intervals. We have tried to introduce a new parameter: the weighted sum $\sum$, which is obtained by summation of earthquakes of various energies (or magnitudes) with various weights. A study was made of the changes of $\sum$ before 20 earthquakes: Central Asia (18) Assam 1950; Mediterranean 1926. (Unfortunately, a reliable determination of $\sum$ is often impossible, because of the errors in magnitude and irregularity of data on comparatively weak earthquakes.)

6.1. Definitions

By definition,

$$\sum(t) = \sum_{k=K-\delta K}^{k=K-1} N(k)P(k)$$

where k = logarithm of energy E of seismic waves; $N(k)$ = number of earthquakes with $E = 10^{k\pm1/2}$; K = value of E for the strong earthquake, before which a change of $\sum(t)$ is investigated (later this earthquake is called a "preparing" one); $P(k)$ = weight of the earthquake with $E = 10^k$; we assumed $P(k) = L^{k-k_0}$; L was assumed 4.5. (According to some very rough estimates $\sum(t)$, with this value of L, is proportional to the total area of fault planes in the sources [14, 19]. Summation was done along the sliding interval from t to $t + \Delta t$; the value of $\sum$ was attributed to the end of that interval. The values Δt and δk were selected empirically so that $\sum$ would be sufficiently stable. Earthquakes of certain areas were included in summation. It was found that before each investigated strong "preparatory" earthquake, the average seismic activity increased in a certain large area which also includes its epicenter. This area was considered as the "area of preparation" of strong earthquakes.

6.2. Some Results

Figure 7 shows the change of $\sum(t)$ before three strong earthquakes. In those and other cases we found the following:

(*a*) Before and only before strong earthquakes with a magnitude of $\bar{M}$, $\sum(t)$ has a sharp peak with a height of nearly $P_{\bar{M}} = P(K)$. (We used magnitudes instead of energies, as magnitudes are indicated in most summaries [10].)

(*b*) The time interval Δt, between the moment when the left side of the peak reaches the level of $\frac{1}{2}P_{\bar{M}}$ and the moment of the following strong

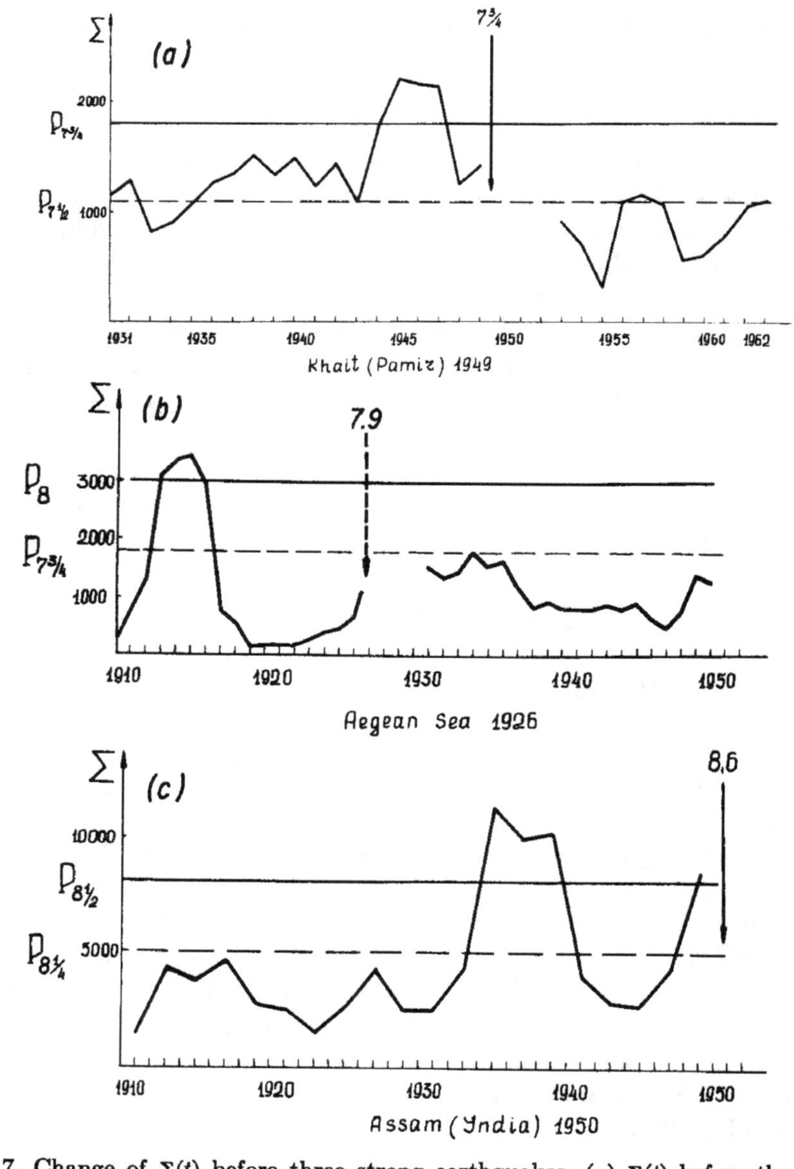

Fig. 7. Change of $\Sigma(t)$ before three strong earthquakes. (*a*) $\Sigma(t)$ before the Khait earthquake, 10 July 1949. ($\lambda = 70.1$, $\phi = 39.2$ N; h = norm, $M = 7\frac{3}{4}$). Preparation area, Pamir, Hindu-Kush, Tyan-Shyan; $\delta M = 2.5$; $\Delta t = 3$ years. (*b*) $\Sigma(t)$ before the earthquake in the Aegean Sea, 26 June 1926. ($\lambda = 27.5$ E; $\phi = 36.5$ N; $h = 100$ km, $M = 7.9$). Preparation area, East Mediterranean; $\delta M = 2$; $\Delta t = 4$ years. (*c*) $\Sigma(t)$ before the Assam earthquake, 15 August 1950. ($\lambda = 96.5$ E; $\phi = 28.5$ N; h = norm, $M = 8.6$). Preparation area, Himalayas; $\delta M = 2.5$; $\Delta t = 5$ years.

earthquake ("preparing" one), depends regularly on the magnitude $\bar{M}$ of this earthquake [Fig. 8(*a*)].

(*c*) The size of the "area of preparation" increases with $\bar{M}$ [Fig. 8(*b*)].

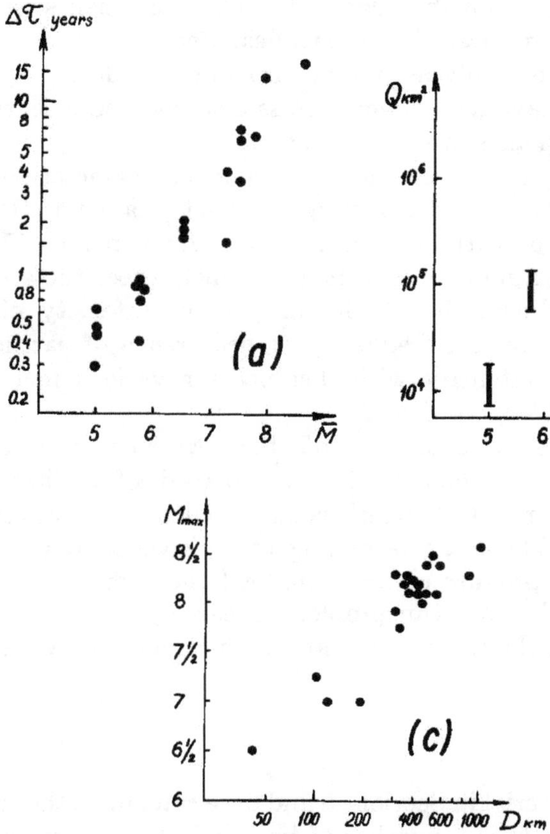

FIG. 8. (a) The values of $\Delta\bar{\tau}$; $\bar{M}$ is the magnitude of the "preparing" earthquake. (b) Dimensions of "preparation area" Q, km². $\bar{M}$ is the magnitude of the "preparing" earthquake; the lines indicate the possible variations because of ambiguous choice of the "preparation area." (c) The relation between the maximum possible magnitude M_{max} and the width of the seismic belt (D, km).

It is only natural to assume that the process which leads to its occurrence spreads over that whole area. Its size may seem unbelievably large. However, it is not more than 10–30 times larger than the linear length of the fault corresponding to the magnitude of the "preparing" earthquake [14]; this seems to be a natural relation if we do not try to explain earthquakes only by a development of local tectonic structures. (The study of earthquake mechanisms also leads to a refusal of such an explanation [16]).

(*d*) The preceding regularity leads to a hypothesis that the maximum magnitude M_{max} of earthquakes in a certain seismic belt is determined by its dimensions. Figure 8(*c*) confirms this hypothesis which is unbelievably simple in comparison to the usual attempts to explain M_{max}.

The conclusions presented here were not confirmed for some of the earthquakes (not the strongest) on the edge of the Central Asian seismic belt and for very narrow seismic areas like the Chilean one.

They are presented here only as preliminary but this does not exclude the possibility that we have really come across a simple and essential relationship between earthquakes in very large areas.

The weighted sum $\sum(t)$ is not the only compact characteristic of all earthquakes.* For example, another very compact characteristic is the probability of tremors of various intensity at a given point [17]. In a practical sense it unites a multitude of factors which affect the economics of seismoresistant buildings: the maximum possible intensity of earthquakes (seismic zoning maps); frequency of occurrence of earthquakes; the law of decrease of intensity with distance for various foci depths; local conditions.

A problem arises of a systematic search for new such characteristics. Here we are confronted with the field of formal analysis of qualitative data (geology, macroseismology, etc.), which is also needed for all of stages IV and III. Analogous problems are beginning to be investigated in mathematical linguistics, the proving of mathematical theorems on computers and the programming of music. Our problem is scarcely easier, because the logic of operations with the mentioned data has not yet been worked out.

7. Conclusion

1. Comparing the material in Sections 5 and 6 we can notice the following possibility: to bring together a number of basic problems of modern seismology into a unified, simultaneous study of the spectra and travel times of seismic waves. To do this, besides the usual processing of records, we must introduce into practice a systematic determination of spectra of various groups of seismic waves and to study them in as thorough fashion as travel times were studied in their day and with a similar aim: to find the dependence of spectra on the epicentral distance, focus depth, earthquake intensity, and local conditions. Spectra with the proper averaging [27] are very compact characteristics of wave dynamics. And if these dependences would be found in sufficient detail, then they may serve as the standard elements from which a solution of a number of major seismological problems could be built.

Thus, the dependence of the amplitude spectrum on distance is the basis for a precise and physically intelligible determination of the earthquake energy; for the application of amplitudes in studying the constitu-

* We came across it accidentally while studying the density of elastic energy near the hypocenter (it can be characterized by the relation of energy and the period of seismic waves [18]). I think, although cannot prove, that this density can become an essential characteristic of stress field in the earth's crust, as pressure or temperature is for the atmospheric investigations.

tion and the physical state of the earth; for the separation of the signal from the noise.

The change of the phase spectrum with the distance gives exact dispersion curves. The relation of the amplitude spectrum to the local conditions determines the resonance properties and the structure of the surface layers and also the corrections for energy determination. The relation of the amplitude spectrum to focus depth in principle gives the possibility to determine the focus depth within the crust, which is the key problem in the identification of sources, and also to identify the high modes of surface waves, connected with wave guides [1].

2. It is not necessary to work out this scheme any further; it is not the only one, but it shows the unlimited possibilities of unification in seismology. However, unification is only a means to control those really formal operations, which as we have seen, form the greater part of our investigations today, including even some final results. An accurate formalization of operations, even without any automation would release our thought for that activity that it is destined for.

On the other hand, the deep penetration of formal operations, together with the variety, complex mutual relations, and the partly intuitive nature of initial data, makes seismology a field where we come especially close to the general problems of modeling human activity.

3. A general discussion in this paper seems excessive; ideas on how to work would be better demonstrated by work itself. The meaning of this discussion depends on the point of view.

Evidently, equipment must be improved. Evidently, new possibilities create new ideas and new ideas, in turn, face us with new problems and force us to reestimate that which we already know. It is quite easy to wave away these meditations, for similar changes have taken place in all sciences at all times. But changes today are coming so fast that utterly new conditions are cropping up. We see torrents of information which are impossible to process, if we would think about them in the habitual way; and as a result we see the information disintegrate into an infinite number of separate parts—the random samples of what was near at hand. We see the powerful possibilities to influence reality (including the development of science) and, at the same time, the catastrophical lag of feedback coupling, which threatens the reality itself with the danger of uncontrollable development.

The only escape from this situation, and it is spreading more and more widely, is the standardization and formalization of a larger amount of the elements of reality and the penetration of probability relations into wider fields of activity.

Ideally all this in the long run should liberate us from the routine and bondage of chance. But at the same time it threatens to dissolve the very essence of our work if we do not pay enough attention to the feedback coupling, i.e., how the structure of the process of apprehending reality becomes the most active element of reality, and interacts with those changes in our surroundings which it itself creates. And on how each science—in accord with its own inherent nature—is able to answer this challenge of the times depends its future in large measure.

References

1. Z. S. Adrianova, V. I. Keilis-Borok, A. L. Levshin, and, M. G. Neygaus, Love waves and constitution of upper mantle, *Geophys. J.* (in press).
2. Z. S. Adrianova, A. L. Levshin, M. G. Neygaus, and T. M. Sabitova, The spectrum of Love waves and the depth of earthquake source in the crust, *Geophys. J.* (in press).
3. L. V. Berkner (ed.), The need for fundamental research in seismology, *Annals of the Technical Conference on the Discontinuance of Nuclear Weapons Tests* (Geneva, 1959).
4. B. A. Bolt, The revision of earthquake epicentres, focal depths and origin times using a high-speed computer, *Geophys. J.*, **3**(4) (1960).
5. I. Brune, J. E. Nafe, and J. Oliver, A simplified method for the analysis and synthesis of dispersed wave trains, *J. Geophys. Res.*, **60**(1) (1960).
6. M. Ewing, W. S. Jardetzky, and F. Press, *Elastic Waves in Layered Media* (McGraw-Hill, 1957).
7. G. V. Golikova, T. B. Yanovskaya, and B. Y. Geltchinskij, *Dok. Akad. Nauk*, **145**(2) (1962).
8. G. V. Golikova and T. B. Yanovskaya, Amplitude-distance curves of P-waves, *Report on European Seismological Commission, Jena, DDR* (1962).
9. B. Gutenberg, Wave velocities in the earth's crust, *Geol. Soc. Am. Spec. Publ. No. 62*, 19 (1955).
10. B. Gutenberg and C. F. Richter, *Seismicity of the Earth and Associated Phenomena* (Princeton, 1954).
11. J. Hodgson, A. Stevens, and A. Wickens, Computer applications to mechanism studies, Report on XIII Assembly IUGG, Berkeley, 1963.
12. H. Jeffreys, *Mon. Not. Roy. Astron. Soc., Geophys. Suppl. 4* (1939).
13. H. Jeffreys, *The Earth, its Origin, History and Physical Constitution* (Cambridge, 1959).
14. V. I. Keilis-Borok, *Annali di Geofisica, XII*(2) (1959).
15. V. I. Keilis-Borok and T. B. Yanovskaya, Rayleigh-wave spectra and depth of source in the earth's crust, *Izv. Akad. Nauk SSR, Geofiz. Ser. No. 11*, 1532 (1962).
16. V. I. Keilis-Borok (ed.), *Investigation of the Earthquake Mechanism* (N.Y., 1960).
17. V. I. Keilis-Borok, I. L. Nersesov, and A. M. Yaglom, *The Method of Estimation of Economic Effect of Seismic Resistant Buildings* (Moscow, 1962).
18. V. I. Keilis-Borok, The density of seismic energy, *Report on XIII Assembly IUGG, Berkeley* (1963).
19. V. I. Keilis-Borok and L. M. Malinovskaya, One regularity in the occurrence of strong earthquakes, *Report on XIII Assembly IUGG, Berkeley* (1963).
20. L. Knopoff, Analytical calculation of the fault plane problem, *Publ. Dominion Obs., XXIV*(10) (Canada, 1960).
21. I. Lehmann, The times of *P* and *S* in northeastern America, *Annali di Geofisica, VIII*(4) (1955).
22. I. Lehmann, Velocities of longitudinal waves in the upper part of the earth's mantle, *Ann. Geophys.*, **15**(1) (1959).
23. I. M. Moltshan, V. F. Pisarenko, and N. A. Smirnova, Some statistical methods of detection of signal in the presence of noise, *Geophys. J.*, **8**(3) (1964).
24. I. I. Pyatetskiy-Shapiro, T. S. Zhelankina, V. I. Keilis-Borok, N. V. Kondorskaya, and L. G. Pavlova, Electronic computers in seismological survey, *Report on XIII Assembly IUGG, Berkeley* (1963).

25. C. Romney, *Report on the Technical Conference on the Discontinuance of Nuclear Weapons Tests* (Geneva, 1960).
26. H. Tatel and M. Tuve, Seismic exploration of a continental crust *Geol. Soc. Amer. Spec. Paper No. 62,* 35 (1955).
27. I. W. Tukey and R. B. Blackman, *The Measurement of the Power Spectra* (Dover Publ., 1958).
28. T. B. Yanovskaya, Elastic waves in caustics, *The Problems of Dynamical Theory of Propagation of Seismic Waves*, G. I. Petrashen (ed.), *VII* (Leningrad, 1963).
29. T. B. Yanovskaya and I. I. Azbel, Determination of velocity distribution in the upper mantle as reverse mathematical problem, *Geophys. J.*, **8**(3), 313 (1964).

JOURNAL OF GEOPHYSICAL RESEARCH VOL. 69, NO. 14 JULY 15, 1964

One Regularity in the Occurrence of Strong Earthquakes

V. I. KEYLIS-BOROK AND L. N. MALINOVSKAYA

Institute of Physics of the Earth
Academy of Sciences, Moscow, USSR

Abstract. An attempt is made to characterize the process of preparation of strong earthquakes by a set of preceding weaker earthquakes in a large area. This set is described by a 'weighted sum' of earthquakes in a sliding time interval; the weight of each earthquake depends on its energy. The change of the sum before 20 strong earthquakes is investigated.

Introduction. The results described in this paper are purely preliminary but seem to be of some interest. We tried to characterize the preparation process of a strong earthquake by a whole set of preceding weak earthquakes.

This set cannot be described either by the total number or by the total energy of the earthquakes: it is well known that number represents only the weakest shocks and energy represents only the strongest ones.

'Seismic activity'—the number of earthquakes of a given energy—is more representative, but only on the average, for large time periods.

We tried to introduce a new parameter—a sum Σ—such as to include earthquakes of various energies with various weights. We shall call it 'the weighted sum.' The weights are chosen on the condition that Σ must give a good representation of the earthquakes of a given energy interval.

In this paper the behavior of Σ before some strong earthquakes is discussed.

Methodology. We define Σ as

$$\sum(t) = \sum_{k=K-\delta k}^{K-1} N(k)P(k) \qquad (1)$$

where k is the logarithm of seismic wave energy E and K is the value of k for that strong earthquake, before which $\Sigma(t)$ is investigated. We shall call this earthquake a 'prepared' one. $N(k, t)$ is the number of earthquakes with $10^{k-1/2} < E < 10^{k+1/2}$, and $P(k)$ is the weight of an earthquake with $E = 10^k$. $P(k)$ was taken in the form

$$P(k) = L^{(k-k_0)} \qquad (2)$$

The earthquakes of an area for the time interval from $t - \Delta t$ to t are included in summation in (1). Σ is attributed to the end of this interval t. The area is determined in the following way. It was found that the average seismic activity, i.e. the average density of the epicenters, increased before the investigated prepared earthquakes in a very large area, containing the epicenter of the 'prepared' earthquake.[1] The earthquakes of this area were included in the sum (1); we shall call it 'the area of preparation' of strong earthquake.

The magnitudes M are included instead of E in most seismological summaries. Thus we were obliged to use (1) in a form

$$\sum(t) = \sum_{M=M^*-\delta M}^{M=M^*-\alpha} N_M P_M \qquad (3)$$

where

N_M = the number of earthquakes of magnitude M.
P_M = $L^{\phi(M)}$, $\phi(M)$ being determined by the correlation of M and E.
$\phi(M)$ = $k - k_0$.
M^* = the magnitude of the prepared earthquake.

Other notations which will be used are

l = $\log L$.
E_0 = the total earthquake energy.
S = the fault-plane area in a source.
v = the volume of a source.
h^* = the upper limit of depth of sources included in summation.

[1] This fact is the basis of the present investigation. It was not recognized before, most likely because the areas are very large and include very different local geological structures.

t_{M^*} = the origin time of the 'prepared' earthquake.

$N^*(k)$ = the average value of $N(k)$ for many time intervals.

γ = a coefficient in a well-known relation:

$$N^*(k) = A \times 10^{-k\gamma} \tag{4}$$

Parameters. The weight factor L can be chosen by formal and physical reasons; they give concordant results. Formally the value $L = 10^{\gamma}$ is too small, $L = 10$ too large.

Indeed, from (2) and (4) we have

$$N^*(k)P(k) = A10^{\gamma(l-k)}10^{-lk_0}$$

Consequently, for $l = \gamma$, the earthquakes with various k on the average would make an equal contribution to $\Sigma(t)$. But then $\Sigma(t)$ would depend too strongly on the irregularity of the weak earthquake statistics.

The value $L = 10$ corresponds to summation of the energy. Thus, for $l = \gamma$, $\Sigma(t)$ would be unstable, and for $l = 1$ it would not be representative. In the present state of earthquake statistics we choose l near the middle of this interval. We assume $l = 2/3$ ($L \approx 4\frac{1}{2}$). It has some physical meaning.

Let us assume that E is proportional to the total energy: $E \propto E_0$. In its turn, in some conditions $E_0 \propto S^{3/2}$ [*Keylis-Borok,* 1959], and, consequently, for $l = 2/3$, $S \propto E^{2/3} \propto 10^{2/3k} \propto L^k$; i.e., Σ is proportional to a total area of fault planes.

Some authors, however, have obtained $E \propto v^{5/3}$, allowing for wave attenuation near the source [*Gurevitch et al.* 1954]. Then

$$v \propto E^{3/5} \propto 10^{3/5k}$$

That is, for $l = 3/5$ ($L \approx 4$), Σ would be proportional to the total volume of sources. (The

TABLE 1. Earthquake Data

	'Prepared' Earthquakes								
Figure No.	Date	φ	λ	h, km	M^*	'Preparation' Area, thousands km²	h^*, km	Δt, years	Magnitudes Included in Σ
1	Nov. 1937*	36.6	70.7	170	7¼	0.6–1.0 × 10^6	280	2.5	5–7
1	Nov. 1946*	41.9	71.9	Normal	7½	Pamir, Hindu-Kush,			
1	March 1949*	36 6	70.6	220	7¼–7½	West Tien Shan			
1	April 1955*	39.9	74.6	Normal	7				
1	April 1955*	40 0	74.7	Normal	(double)				
2	Aug. 1931*	38.8	70.9	Normal	6½	1.3–3 × 10^5	150	1.5	4¼–6
2	April 1941*	39.2	70.5	Normal	6½	Pamir, Hindu-Kush,			
2	Sept. 1944*	39.0	74.8	Normal	6½–6¾	South Tien Shan			
3	July 1935*	39.5	73.5	Normal	5¾	0.6–1.3 × 10^5	50	0.8	4½–5½
3	May 1939*	38.9	70.4	Normal	5¾	North Pamir, Southwest Tien Shan			
3	March 1944*	39.7	73.1	Normal	5¾				
3	April 1951*	39.1	71.6	Normal	5¾				
3	Aug. 1957*	39.3	72.9	Normal	5¾ (5½?)				
4	July 1949*†	39.2	70.8	Normal	7¾	1–1.8 × 10^6 Pamir, Hindu-Kush, Tien Shan	280	3	5–7½
5	June 1926†	36.5	27.5	100	7.9	1.3–2 × 10^6 Eastern Mediterranean	300	4	5¾–7¾
6	Aug. 1950†	28.5	96.5	Normal	8.6	2.5–5 × 10^6 Himalayas	180	5	5¾–8¼
	April 1956	39.2	71.9	Normal	4¾–5	0.8–1.8 × 10^4	40	0.5	
	July 1959	38.9	70.4	Normal	5¼	Garm-Dushanbe			
	April 1961	39.5	73.0	Normal		district			
	Jan. 1961	38.6	70.1	Normal	4¾–5				

* *Atlas of Earthquakes* [1962].

† *Gutenberg and Richter* [1954].

difference between $l = 2/3$ and $l = 3/5$ is practically insignificant.)

Of course, the connection between E, E_v, and v is known very roughly, if at all, and our consideration is not confirmation but only an illustration of a formal choice of L.

For Δt the smallest value was taken which smooths out the natural fluctuation of $\Sigma(t)$. For δM the maximum value that gives reliable values of $\Sigma(t)$ was assumed.

Results. The change of $\Sigma(t)$ before 16 strong earthquakes with $M \geq 5\frac{3}{4}$ was investigated. They are listed in Table 1. The data were taken from the *Atlas of Earthquakes* [1962] and from *Gutenberg and Richter* [1954]. Unfortunately, the latter reference contains the data only for earthquakes with $M \geq 6$; this makes it possible to determine Σ only for $M^* \geq 7\frac{1}{2}$. Besides, four earthquakes with $E = 10^{13}$ joules ($M^* \approx 5$) were investigated for the Garm-Dushanbe region, using local catalogs of comparatively weak earthquakes ($E \geq 10^{10}$ joules). The results are shown in Figures 1 to 6. Long arrows show the origin times of 'prepared' strong earthquakes. Their magnitudes M^* are indicated near the arrow.

The 'area of preparation' for a given earthquake is sometimes a part of such an area for another earthquake with larger M^*. The moments of this earthquake are shown by short arrows.

Figures 1 to 6 lead to the following conclusions:

A. The 'weighted sum' $\Sigma(t)$ has a sharp peak of an amplitude of about P_{M^*} before—and only before—an earthquake with a magnitude of M^*.

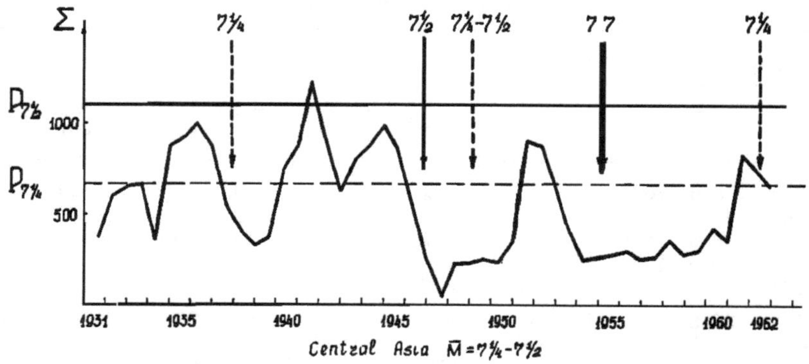

Fig. 1. $\Sigma(t)$ for central Asia; $M^* = 7\frac{1}{2}$.

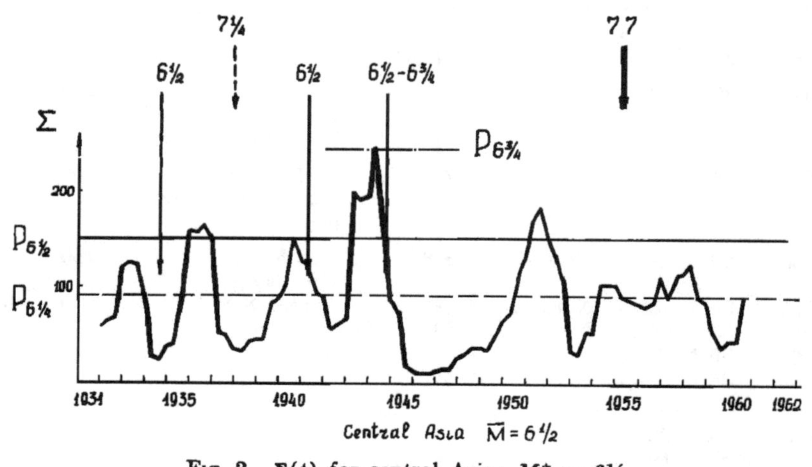

Fig. 2. $\Sigma(t)$ for central Asia; $M^* = 6\frac{1}{2}$.

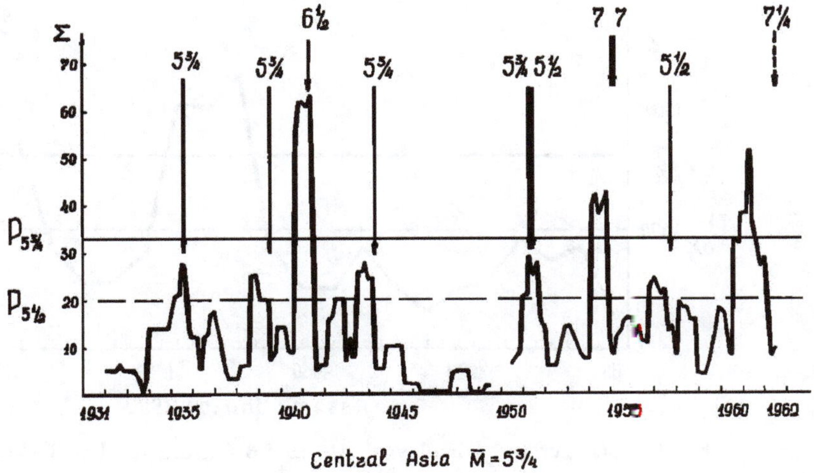

Fig. 3. $\Sigma(t)$ for Garm-Dushanbe district; $M^* = 5\frac{3}{4}$.

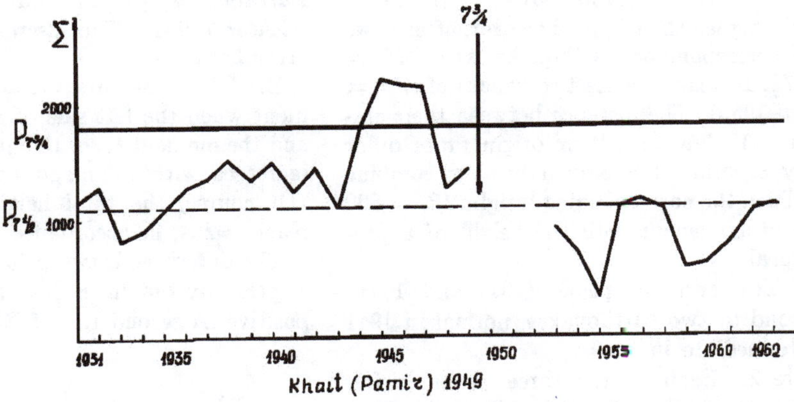

Fig. 4. $\Sigma(t)$ for central Asia; $M^* = 7\frac{3}{4}$ (Khait earthquake, 1949).

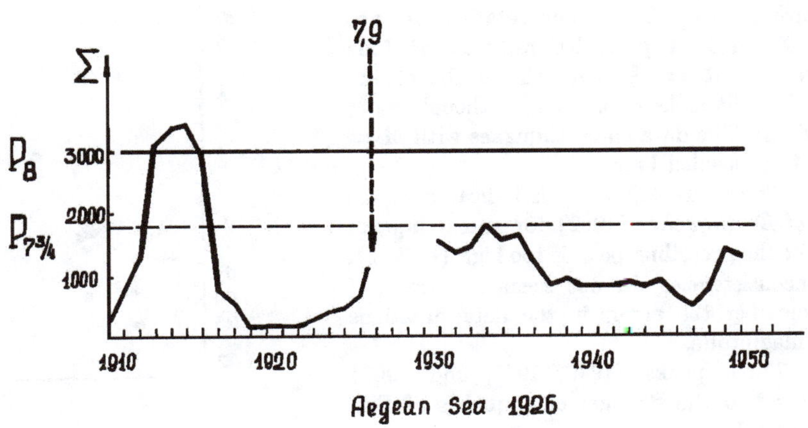

Fig. 5. $\Sigma(t)$ for the eastern Mediterranean; $M^* = 7.9$ (Aegean earthquake, 1926; $h = 100$ km).

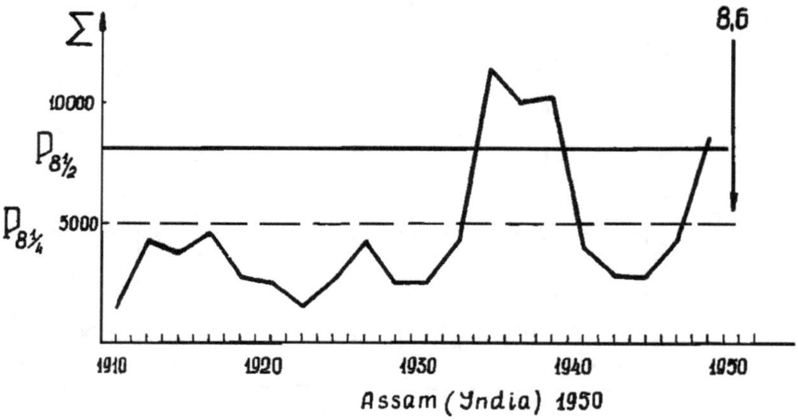

Fig. 6. $\Sigma(t)$ for the Himalayas; $M^* = 8.6$ (Assam earthquake, 1950).

This is clear for Figures 4, 5, and 6; comments are needed for the other figures:

Figure 1. (*a*) $\Sigma(t)$ in 1951 has a peak considerably higher than $P_{7\,1/4}$; however, after it we see no corresponding earthquake with $M^* = 7\frac{1}{4}$ or $7\frac{1}{2}$. Instead, we see two shocks of $M^* = 7$ (April 1955). The distance between their epicenters is 10 km, and their origin times differ by only 32 min; it is only natural to combine them. Then the corresponding weight $2P_7 \approx 800$ will be in agreement with the height of a preceding peak.

(*b*) Two adjacent peaks (1941 and 1944) correspond to two earthquakes, normal in 1941 and intermediate in 1949.

Figure 2. Each of the three peaks is followed by an earthquake with $M^* = 6\frac{1}{2}$. Two other peaks correspond to larger shocks from Figure 1.

Figure 3. (*a*) The interpretation is not clear. There is a peak before each of four shocks with $M = 5\frac{3}{4}$. But the height of the peaks is a little less than $P_{5\,3/4}$ (though more than $P_{5\,1/2}$). The data on earthquakes with $M = 3\frac{1}{2}$ to 4 are needed here.

(*b*) The value $M^* = 5\frac{1}{2}$ is indicated in the *Atlas of Earthquakes* [1962] for an earthquake in 1957; the preceding peak is too high ($\approx P_{5\,3/4}$). The inconsistencies do not seem too great, if we remember the errors in the determinations of the magnitude.

(*c*) Three peaks (1940, 1955, and 1962) correspond to the stronger earthquakes of Figures 1 and 2.

Figure 4. The peak of height $P_{7\,3/4}$ is followed by the famous Khait earthquake for which $M^* = 7.6$ is indicated in the *Atlas of Earthquakes* [1962] and in *Gutenberg and Richter* [1954]. The discrepancy is within the error limit.

B. The time interval $\Delta\tau$ between the moment when the left side of a peak reaches $\frac{1}{2}P_{M^*}$ and the moment t_{M^*} of the 'prepared' earthquake increases with its magnitude M^* (Figure 7). (Of course, the total height of a peak must reach $\approx P_{M^*}$, in accordance with A.)

The difference between t_{M^*} and the moment of a peak maximum is less regular, but always positive. A second rise of Σ immediately before

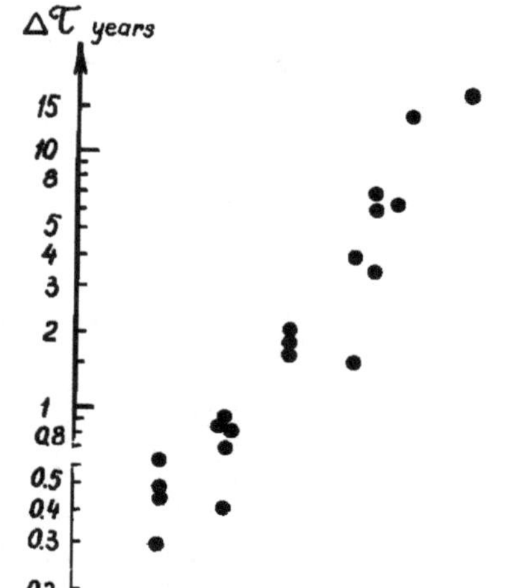

Fig. 7. Time interval between t_{M^*} and the moment when left side of peak reaches $\frac{1}{2}P_{M^*}$.

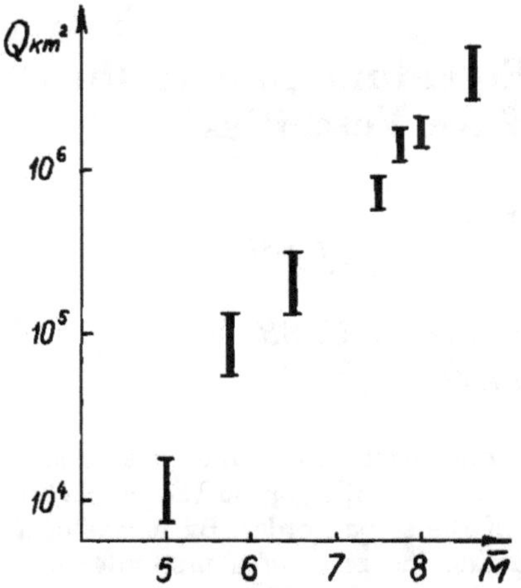

Fig. 8. The limits of preparation are Q for various M^*.

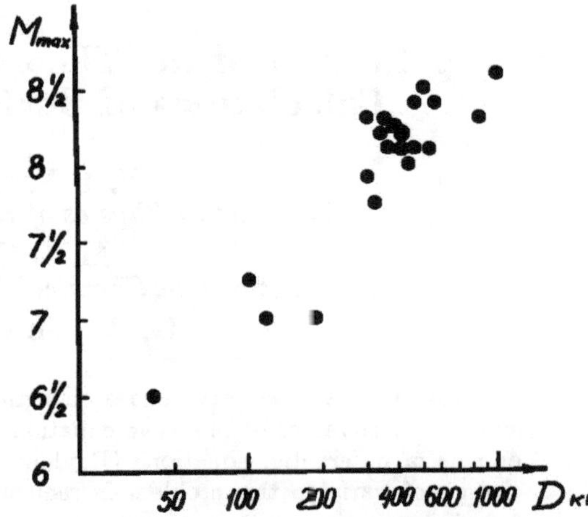

Fig. 9. The maximum magnitude M_{max} possible in various seismic belts of the world. D = the width of the belt.

t_{M^*} can be seen for two strongest earthquakes (Figures 5 and 6).

C. The 'area of preparation' Q increases with the magnitude M^* of the 'prepared' earthquake (Figure 8).

This area, of course, is limited by the dimension of the whole seismic belt. Thus the hypothesis arises:

D. The maximum magnitude, M^*_{max}, possible in a given belt is roughly determined by the width of this belt D. This is confirmed by Figure 9; M^*_{max} and D are taken from *Gutenberg and Richter* [1954].

This conclusion is surprisingly simple, compared with the usual—and natural—efforts to correlate M^*_{max} with neotectonics, modern vertical movements, etc.

E. The maximum depth h^* was determined very roughly, but seems to increase with M^* (Table 1).

Conclusion A and, consequently, all the others (except D) are not confirmed in the following cases:

(1) For earthquakes of intermediate magnitude ($M^* = 6$ to $6\frac{1}{2}$) on the edge of the seismic belt of central Asia.

(2) For very narrow seismic belts, such as the Chilean.

The results obtained are far from earthquake forecasting, but they seem to indicate a direct connection between strong earthquakes and the very large scale features of the development of the whole earth's crust and sometimes of the uppermost mantle as well. These results are purely preliminary. More data and reliable energy determinations are needed.

References

Atlas of Earthquakes in USSR, Academy of Sciences, USSR, Moscow, 1962.

Gurevitch, G. I., I. L. Nersesov, and K. A. Kusnetsov, O zakone povtoryaemosti zemletryaseniy kak sledtsvii zakonomernostey deformatsii i drobleniya, *Dokl. Akad. Nauk SSSR, 128*(7), 1163–1666, 1959.

Gutenberg, B., and C. F. Richter, *Seismicity of the Earth and Associated Phenomena,* 2nd ed., Princeton University Press, 1954.

Keylis-Borok, V. I., On estimation of the displacement in an earthquake source and of source dimensions, *Ann. Geofis., Rome, 12*(2), 205–214, 1959.

(Manuscript received August 30, 1963.)

REVIEWS OF GEOPHYSICS VOL. 3, NO. 1 FEBRUARY 1965

Application of the Theory of Eigenfunctions to the Calculations of Surface Wave Velocities*

V. I. KEILIS-BOROK
Institute of Physics of the Earth, Moscow, USSR
M. G. NEIGAUZ
Institute of Chemical Physics, Moscow, USSR
G. V. SHKADINSKAYA†

Abstract. Surface wave phase velocities for an inhomogeneous medium are usually given as eigenvalues of the wave equation. This is a differential equation (Love waves) or pair of differential equations (Rayleigh waves) of the second order. By means of a change of variable the problem is reducible to finding the zeros of a first-order differential equation or pair of differential equations of first order. The solution can be obtained in any order of surface waves, and the eigenfunctions can be obtained as well by means of this transformation. The solution is thus amenable to computation.

LOVE WAVES

1. It is known[2] that the phase velocities of Love waves are determined by finding the eigenvalues of the following equation:

$$\frac{d}{dz}\left(\mu(z)\frac{dv}{dz}\right) + (p^2\rho(z) - \xi^2\mu(z))v = 0 \tag{1}$$

subject to the boundary conditions

$$\left.\frac{dv}{dz}\right|_{z=0} = 0 \qquad v_{z=\infty} = 0 \tag{2}$$

Here p is the frequency, $\rho(z)$ is the density, and $\mu(z)$ the shear modulus; these are specified, positive functions, each having a finite number of discontinuities; $\rho(z)$, $\mu(z)$ are assumed to be constant for $z \geq Z$; ξ is the wave number.

The problem is to find values of $\xi^2 = \xi_k^2(p)$ for which a nontrivial solution for (1), (2) exists. We call this solution $v_k(z)$.

2. The following statements can be derived from the general theory of eigenvalues:

The system (1), (2) has a finite number $K_L(p)$ of eigenvalues $\xi_k^2(p)$ for any given finite p.

The system (1), (2) has one and only one eigenvalue for sufficiently small p. $K_L(p)$ increases with increase of p.

The eigenfunction $v_k(z, p^2)(k = 1, 2 \cdots)$ has $(k - 1)$ zeros in the interval $(0, Z)$ and decreases exponentially in the interval $(Z, +\infty)$. The fundamental mode will be denoted by the index $k = 1$.

* This paper was presented at the Symposium on Theory and Computers sponsored by the International Upper Mantle Committee, held in Moscow and Leningrad, May 15–21, 1964.

† In care of the Soviet Geophysical Committee, Moscow, USSR.

3. The solution of (1), (2) can be found by the method of factorization, suggested by Gelfand and Lokutsievskiy[4] and Richtmayer.[6] We use the version of this method suggested by Abramoff.[1]

We introduce a continuous function $\theta(z)$:

$$\theta(z) = \arctan\left[\frac{\mu(z)\, dv/dz}{v}\right] \tag{3}$$

Introducing this expression into (1) we obtain the following equation for $\theta(z)$:

$$d\theta/dz = -[1/\mu(z)]\sin^2\theta(z) - (p^2\rho(z) - \xi^2\mu(z))\cos^2\theta(z) \tag{4}$$

θ is continuous in the interval $(0, \infty)$.

The boundary conditions for $\theta(z)$ can be found from (2). It follows from (2), (3) that $\tan\theta(0) = 0$. We assume

$$\theta(0) = 0 \tag{5}$$

We determine the second boundary condition for some point $z = Z$. For $z \geq Z$, $\mu(z)$ is constant and $\rho(z)$ is constant,

$$v(z) = C\exp\left\{-\left(\xi^2 - \frac{p^2}{c^2(z)}\right)^{1/2} z\right\}$$

where C is a constant, $c = (\mu(z)/\rho(z))^{1/2}$.

Introducing this into (3) we have

$$\theta(Z) = -\arctan\left\{\mu(Z)\left(\xi^2 - \frac{p^2}{c^2(Z)}\right)^{1/2}\right\} - (k-1)\pi \tag{6}$$

It is easy to show that the symbol k in (6) is the number of eigenvalues of $\xi^2(p)$. Hence $\xi_k^2(p)$ is a value of ξ^2 for which the solution $\theta_k(z)$ of (4), (5), (6) exists.

Let $\theta(z, \xi^2)$ be the solution of (4) which satisfies (6). The eigenvalues ξ_k^2 will be the zeros of the function

$$F(\xi^2) = \theta(0, \xi^2)$$

We calculate $\theta(0, \xi^2)$, integrating (4) from the right to the left in the direction of decreasing z. By this method we find $\theta(z)$, corresponding to the solution of (1) which decreases as $z \to \infty$.

RAYLEIGH WAVES

The phase velocities of Rayleigh waves are given by the eigenvalues of the following vector equation:

$$\frac{d}{dz}\left(A\frac{dw}{dz} + \xi Bw\right) - \xi B^*\frac{dn}{dz} + (p^2\rho(z) - \xi^2 C)w = 0 \tag{7}$$

Here A, B, B^*, C are square matrices of second order

$$A = \begin{bmatrix}\mu(z) & 0\\ 0 & \lambda(z)+2\mu(z)\end{bmatrix} \qquad B = \begin{bmatrix}0 & \mu(z)\\ -\lambda(z) & 0\end{bmatrix}$$

$$B^* = \begin{bmatrix}0 & -\lambda(z)\\ \mu(z) & 0\end{bmatrix} \qquad C = \begin{bmatrix}\lambda(z)+2\mu(z) & 0\\ 0 & \mu(z)\end{bmatrix};$$

and w is a vector function

$$w = \begin{pmatrix} w_1 \\ w_2 \end{pmatrix}$$

$\lambda(z)$, $\mu(z)$ are the Lamé elastic parameters, and $\rho(z)$ is the density. As before, λ, μ, ρ have a finite number of discontinuities and are constant for $z \geq Z$. We consider ξ as a given parameter and p^2 as a spectral parameter.

The boundary conditions are the following:

$$A(dw/dz) + \xi Bw = 0 \quad \text{for} \quad z = 0 \tag{8}$$

$$W = 0 \quad \text{for} \quad z = \infty$$

The problem is to find values of p^2, $\rho^2 = \rho_k^2(\xi)$ for which a nontrivial solution of (7), (8) exists. It is easy to show that the boundary value problem (7), (8) is selfadjoint.

We use a version of the method of factorization suggested by Lidskiy and Neigauz[5].

Equation 7 has four linearly independent solutions.

$$W = \begin{bmatrix} w_1^{(1)} & w_1^{(2)} \\ w_2^{(1)} & w_2^{(2)} \end{bmatrix}$$

where $w^{(1)}$ and $w^{(2)}$ are two of the vectors representing the linearly independent solutions. Then we introduce a complex matrix $u(z)$ connected with $w(z)$ by the equation

$$A\frac{dw}{dz} + \xi Bw = i(E - u(z))(E + u(z))^{-1}W \tag{9}$$

It can be shown that $u(z)$ is uniquely derived from $w(z)$ and that $u(z)$ is a unitary matrix; i.e.,

$$u(z)\cdot u^*(z) = E$$

Here $u^*(z)$ is the adjoint matrix, and E is the unit matrix, $E = \begin{bmatrix} 1 & 0 \\ 0 & 1 \end{bmatrix}$

Differentiating (9) we can obtain an expression for

$$\frac{d}{dz}\left(A\frac{dw}{dz} + \xi Bw\right)$$

Substituting into (7), we obtain the following expression for $u(z)$:

$$\frac{du}{dz} = \frac{i}{2}[-(E - u)A^{-1}(E - u^*) + i\xi(E - u)A^{-1}B(E + u^*)$$
$$- i\xi(E + u)B^*A^{-1}(E - u^*) - (E + u)(p^2\rho E - \xi^2 g)(E + u^*)]u \tag{10}$$

where

$$g = \begin{bmatrix} \dfrac{4\mu(\lambda + \mu)}{\lambda + 2\mu} & 0 \\ 0 & 0 \end{bmatrix}$$

We now formulate the boundary conditions. For $z = 0$ one of the eigenvalues of $u(z)$ is equal to 1, as can be seen from (8). For $z = Z$, $u(Z)$ is determined unambiguously from $w(Z)$ if $w(Z)$ is the combination of two such linear independent solutions of (7), which decrease exponentially as $z \to \infty$. It can be shown further that the eigenvalues $p_k^2(\xi)$ of (7) and (8) are those values of p^2 for which (10) has a nontrivial solution $u(z)$, satisfying the boundary conditions at $z = 0$ and $z = Z$.

The number k of eigenvalues p_k^2 can be determined from the behavior of the eigenvalues of the matrix $u(z)$ considered as a function of z for fixed ξ, p. Let us denote the eigenvalues of $u(z)$ as $\exp(i\varphi_1(z))$, $\exp(i\varphi_2(z))$. Then $\varphi_1(z)$, $\varphi_2(z)$ are purely real functions, and, since $u(z)$ is unitary, the following statements can be proved:

$\varphi_n(z)$ ($n = 1$ or 2) increases monotonically in the vicinity of a point $z = Z_0$ such that $\exp[i\varphi_n(Z_0)] = -1$, though in general $\varphi_n(z)$ may not be monotonic.

For a given z, φ_n is a monotonic function of p^2.

For $z = Z$

$$-\pi \leq \varphi_2(z, p_{2l-1}^2) \leq 0 \qquad 0 \leq \varphi_1(Z, p_{2l}^2) \leq \pi$$

For $z = 0$

$$\exp[i\varphi_2(0, p_{2l-1}^2)] = 1 \qquad \exp[i\varphi_1(0, p_{2l}^2)] = 1 \qquad l = 1, 2 \cdots$$

These statements allow the number k of eigenvalues $p_k^2(\xi)$ ($k = 2l - 1$ or $k = 2l$) to be determined. For example, the first (fundamental) eigenvalue $p_1^2(\xi)$ is a value of p^2 for which $-\pi < \varphi_2(z, p^2) \leq 0$ and $\varphi_2(0, p^2) = 0$. For the second eigenvalue $p_2^2(\xi)$ we have $0 \leq \varphi_1(z, p^2) < \pi$ and $\varphi_1(0, p^2) = 2\pi$. In general, we find the $(2l)$th eigenvalue from the equation

$$\varphi(p^2) = \varphi_1(0, p^2) - 2l\pi = 0$$

and the $(2l - 1)$th eigenvalue from the equation

$$\varphi(p^2) = \varphi_2(0, p^2) - 2(l - 1)\pi = 0$$

To find $\varphi_1(0, p^2)$, $\varphi_2(0, p^2)$ for given p^2, (10) is integrated from the right ($z = Z$) to the left.

CONCLUSION

The method described seems preferable to the usual one, where the medium is approximated by a system of homogeneous layers and the phase velocities are determined as the zeros of determinant of boundary conditions. The advantages of the method described are: the order of a mode can be simply determined; the parameters of the medium can be approximated by any adequate functions; and the calculation of the eigenfunctions is now quite simply performed.

A program for a high speed computer is now in operation. Seismological applications have been published[2, 3] thus far only for Love waves.

REFERENCES

1. Abramoff, A. A., One version of the method of factorization (in Russian), *Zh. Vychislitelnoi Mat. Mat. Fiz.*, *1*(2), 1961.
2. Adrianova, Z. S., V. I. Keilis-Borok, A. L. Levshin, and M. G. Neigauz. Love waves in the vertically unhomogeneous medium (in Russian).
3. Adrianova, Z. S., V. I. Keilis-Borok, A. L. Levshin, and M. G. Neigauz. Love waves and structure of upper mantle, *Geophys. J.* 1964.
4. Gelfand, I. M., and O. V. Lokutsievskiy, Method of 'factorization' for solution of finite-differences equations, Supplement to the book by S. K. Godunov and V. S. Ryabenkiy. *Introduction to the Theory of the Finite-Differences Schemes* (in Russian), Moscow, 1963.
5. Lidskiy, V. B., and M. G. Neigauz, Factorization method for self-adjoined systems of second order, *Zh. Vychislitelnoi Mat. Mat. Fiz.*, *2*(1), 1962.
6. Richtmayer, R. D., Finite-differences method for boundary problems.

Physics of the Earth and Planetary Interiors, 11 (1976) 227–283

PATTERN RECOGNITION APPLIED TO EARTHQUAKE EPICENTERS IN CALIFORNIA

I.M. GELFAND [1], Sh.A. GUBERMAN [1], V.I. KEILIS-BOROK [2], L. KNOPOFF [3], F. PRESS [4], E.Ya. RANZMAN [5], I.M. ROTWAIN [6] and A.M. SADOVSKY [2]

[1] *Institute of Applied Mathematics, Academy of Sciences, Moscow (U.S.S.R.)*
[2] *Institute of Physics of the Earth, Academy of Sciences, Moscow (U.S.S.R.)*
[3] *Institute of Geophysics, University of California, Los Angeles, Calif. (U.S.A.)*
[4] *Department of Earth and Planetary Sciences, Massachusetts Institute of Technology, Cambridge, Mass. (U.S.A.)*
[5] *Institute of Geography, Academy of Sciences, Moscow (U.S.S.R.)*
[6] *Laboratory of Mathematical Methods in Biology, Moscow University, Moscow (U.S.S.R.)*

(Received October 3, 1975; accepted for publication November 21, 1975)

"This is the only way it clicks ...
It doesn't make sense otherwise ...
But this is just a theory, isn't it? ...
Call it any name you like. It's good enough for me."

Dashiel Hammett

Gelfand, I.M., Guberman, Sh.A., Keilis-Borok, V.I., Knopoff, L., Press, F., Ranzman, E.Ya., Rotwain, I.M. and Sadovsky, A.M., 1976. Pattern recognition applied to earthquake epicenters in California. Phys. Earth Planet. Inter., 11: 227–283.

A pattern recognition procedure is explained which uses geological data and the earthquake history of a region, in this case California, and learns how to separate earthquake epicenters from other places. Sites of future earthquake epicenters are predicted as well as places where epicenters will not occur. The problem is formulated in several ways and control experiments are devised and applied in order to test the stability of the procedures and engender confidence in the results. Some of the combinations of geological features which the computer recognized as significant discriminants are discussed.

1. Introduction

Pattern recognition is not a new subject but its application to geological and geophysical problems has been minimal. This is surprising because the methods of pattern recognition, though paradoxically simple, can extract more results from a body of data, by a more intensive analysis, than many alternative procedures.

Geology and geophysics are essentially experimental disciplines in which large amounts of data are accumulated and are in need of winnowing, codification, correlation, and interpretation. More so than in many other fields, hypotheses flow from data analysis, and prediction is important in a practical sense as well as a means of hypothesis testing. Pattern recognition offers a powerful tool for achieving these results. Yet care must be taken in its application, for pattern recognition itself is an unexplored field lying somewhere between logic and statistics. Its procedures are not yet formalized and the user must be on guard against self-deception.

In this paper we examine the possibility that geological patterns can be recognized which distinguish places where epicenters of large earthquakes have occurred in the past and can occur in the future, from other places which have not been and will not be the sites of epicenters. The primary data for recognition come from California, though insights and procedures stem from earlier applications of these methods to central Asia and Anatolia. If we are successful we will have predicted the sites of future epicenters (but not the times) and we will have learned something about the geological environment of epicenters.

2. Method

2.1. The problem

We place the problem of recognition in the following form: a set of objects is given, with each object described by the answers to a questionnaire. Each object belongs to one and only one of several (usually two) classes. Our goal is to find which class each object belongs to. To solve this problem we first need to go through a "learning phase" using examples of objects of each class as "learning material".

Let us make this specific for our particular problem. Objects are points on a map, representative of a local geographic region. The problem is to find those objects where the epicenters of strong earthquakes may occur in the future. Note that we mean the epicenter or place where the earthquake is initiated and not the entire length of faulting. We shall call these objects D, "dangerous". The rest of the objects we shall call N, "non-dangerous". By strong we mean earthquakes with magnitude M not lower than some threshold M_0; in the text below, "earthquake" and "epicenter" refer to strong earthquakes unless indicated otherwise.

A specific difficulty arises because the learning material consisting of known epicenters is mixed. Because of errors in locating epicenters, objects called D in the learning phase may really be N and vice versa. Also some objects which are classified N in the learning phase are really D because they will be the sites of future epicenters. It is these objects we wish to identify by pattern recognition in order to predict future epicenters. This leads us to the following problem: the objects are divided a priori into classes I and II – correspondingly close to and far from known epicenters; the goal is to recognize objects D in each class.

2.2. The algorithm

The recognition algorithm consists of three stages: learning, voting and control experiments.

Learning: certain critical questions or combinations of questions, called the distinctive features of objects D and N, are determined by analyzing the learning material, as will be described in the next two sections.

Voting: the numbers n_D and n_N of distinctive features characteristic of D and N are determined for each object. Recognition is based on their difference $\Delta = n_D - n_N$. We recognize an object as D if $\Delta \geqslant \bar{\Delta}$, where $\bar{\Delta}$ is some threshold. Making $\bar{\Delta}$ smaller will reduce the number of D objects which will be misidentified as N. At the same time more "false alarms" will occur, i.e. more N will be misidentified as D.

Control experiment: the stability and reliability of recognition are tested by numerical (logical) experiments, in order to convince ourselves that the results are significant since recognition cannot be proven either mathematically or statistically. This is a decisive stage which takes up most of the effort, after the questionnaire is formulated and answered.

Learning is based on the algorithm CORA-3. This algorithm in slightly different form has been used in several other problems (Bongard et al., 1966). For our particular problem it can be formulated as follows. The objects, used in learning, are divided into classes I and II, correspondingly close to and far from known epicenters. As stated earlier, each class may contain objects D and N:

class I is $\{D_I, N_I\}$

class II is $\{D_{II}, N_{II}\}$

Here D_i and N_i are objects D and N, respectively, in the ith class; i = I and II. In other words, D_I and N_I are objects D and N, which lie close to at least one known epicenter. D_{II} and N_{II} are the objects D and N, which lie far from all known epicenters. Each object is described in binary code, which contains the answers to the questionnaire. All binary digits and the combinations of two and three digits are considered as traits of an object.

We repeat this definition more formally.
The object A is described in binary code:-

$$A = A_1 A_2 \ldots A_L \qquad (1)$$

where the component A_i is the answer to the ith question in the questionnaire, expressed as 1 for yes and 0 for no. *The trait* τ is an array of six integers:

$$\tau = p, q, r, P, Q, R \quad (2)$$

Here $p = 1, 2, ..., L; q = p, p + 1, ..., L; r = q, q + 1, ..., L$; L is the same as in object (1); $P = 0$ or 1; $Q = 0$ or 1; $R = 0$ or 1. If $q = r$, then $Q = R$; if $p = q = r$, then $P = Q = R$; if $q = p$, then $Q = P$. An object (1) has the trait (2), if $A_p = P; A_q = Q; A_r = R$. Here A_p is the pth component in (1) etc. In the cases $p = q = r$ and $p \neq q$, $q = r$ the trait corresponds to one and two components of (1), respectively. The cases $p = q$, $q \neq r$ are redundant.

As an example, the object $A = 0110$ has traits as represented in Table I.

Thus the value of *PQR* in the trait with *pqr* equal to 222, for example, is the response to question $i = 2$ in A and has been identified in the *pqr* array by the value of P; the values of q and r are redundant and have been set equal to 2 to provide a complete syntactical statement. Similarly, the value of *PQR* in the trait with *pqr* equal to 233, for example, is the response to both questions $i = 2$ and 3 in A and is identified in *pqr* by the value of p and q; the value of r is redundant. The value of *PQR* in trait with *pqr* equal to 134 is the response to questions $i = 1, 3$ and 4 in A.

Thus, an object (1) has $L + C_{L2} + C_{L3}$ traits, where C_{LM} is the number of combinations of L things taken M at a time ($C_{L1} = L$). Eight different traits are possible for fixed p,q,r, if $p \neq q \neq r$: $(P,Q,R) = (0,0,1)$ or (0,1,0) etc. up to (1,1,1). If $p \neq q$, $q = r$, only four different traits are possible for fixed p,q: $(P,Q) = (0,0)$, (0,1), (1,0) or (1,1). If $p = q = r$, only two different traits are possible for fixed p: $P = 0$ or 1.

Therefore, the total number of different traits which are possible in objects of length L is:

$$2L + 4C_L^2 + 8C_L^3$$

The features of D and N are selected from among the traits. A *feature* of D is a trait which is present relatively frequently in class I and relatively infrequently in class II. A feature of N is defined analogously. The formal rule defining features is given in Table II.

Next we eliminate equivalent and weaker features, whose definitions follow. Consider two features of a class and the objects of this class only. Some set of these objects has the first feature, another has the second. If these sets coincide, the features are *equivalent*. If one set is part of another, the corresponding feature is *weaker* than the other. If $M_n/M_u \geq 1 - \epsilon$, these features are called "ϵ-equivalent". Here M_n is the number of objects which belong to both sets; M_u is the total number of objects which belong to both or to any one of the sets. Evidently, for equivalent features, $M_n = M_u$, i.e. $\epsilon = 0$. The definition of ϵ-equivalence is used later.

We eliminate all but one feature from each group of equivalent features, and all features which are weaker than some other one. The features remaining after this elimination are called distinctive features and are used in voting. As mentioned above voting consists of tallying the number of D and N distinctive features at each object of class I, II and III.

Class III consists of objects not used in learning.

TABLE I

Traits for the object $A = 0110$

p	q	r	P,Q,R
1	1	1	0
2	2	2	1
3	3	3	1
4	4	4	0
1	2	2	0,1
1	3	3	0,1
1	4	4	0,0
2	3	3	1,1
2	4	4	1,0
3	4	4	1,0
1	2	3	0,1,1
1	2	4	0,1,0
1	3	4	0,1,0
2	3	4	1,1,0

TABLE II

Definition of features* in learning by algorithm "CORA-3"

A feature of objects	Should be found in the class I	II
D	in $> k_1$ objects	in $< \bar{k}_1$ objects
N	in $< k_2$ objects	in $> k_2$ objects

* Each feature is a unit or a combination of two or three units in the binary code of the object.

We could divide all objects into classes I and II only, according to whether or not the point lies near an epicenter. However, we separate some objects into class III if we are uncertain about their assignment or if we use them for control experiments. Objects in class III are considered in voting, however.

The algorithm assumes that the number of epicenters is sufficient for the following two hypotheses to hold:

Hypothesis 1: most objects, which are far from known epicenters, are N.

Hypothesis 2: most objects near known epicenters are D.

The second hypothesis will be replaced by a weaker one, when we use another algorithm for learning called "CLUSTERS", which we describe next.

Algorithm "CLUSTERS" is fully described for the first time in this paper; a brief description was given in Gelfand et al. (1974a). The objects of class I are subdivided into clusters. Each cluster corresponds to a certain epicenter and includes those objects which are close to this epicenter. An object may be included in several clusters. Class II is specified as before. Thus:

Class I is $\{C[D_I^C, N_I^C]\}$

Class II is $\{D_{II}, N_{II}\}$

Here C is a cluster of objects near some epicenter; D_I^C and N_I^C are the objects D and N in the cluster C. The problem is to recognize D_I and D_{II}.

As in CORA-3 all traits of all objects in the learning materials are considered in the learning phase. We assume that a cluster has some trait if at least one object in the cluster has this trait. (In another version this condition is more flexible in that at least N/K objects in the cluster must have this trait. Here N is the number of the objects in the cluster, and K is some constant common for all clusters; naturally, if $N/K < 1$ it is replaced by 1.) With these definitions the features are selected by the rules specified in Table III.

As before equivalent and weaker features are then eliminated. Although they can be determined exactly as in CORA-3, we preferred to use the clusters rather than objects in testing for equivalent or weaker D features. After this elimination we are left with the distinctive features of D and N, which are used in voting. The voting is carried out separately for each object, whether or not it has been assigned to a cluster.

TABLE III

Definition of features* in learning by algorithm "CLUSTERS"

A feature of objects	Should be found in the		
	clusters C **	single objects in clusters (class I)	single objects outside the clusters (class II)
D	in $\geqslant k_1$ clusters	in $\geqslant k_3$ objects	$\leqslant \bar{k}_1$ objects
N	–	in $\leqslant \tilde{k}_2$ objects	$\geqslant k_2$ objects

* Each feature is a unit or a combination of two or three units in the binary code of the object.

** A cluster has some trait if at least one object in the cluster has this trait.

This voting enables us to recognize the objects D near each known epicenter (D_I) and far from all of them (D_{II}). In particular we can divide each cluster into D_I and N_I. The essential advantage, compared to CORA-3 is that objects D_I need not dominate the clusters. They may even be in the minority in many clusters, or in every cluster if $k_3 = 0$.

Formally this algorithm differs from CORA-3 only by the introduction of an additional threshold (compare Tables II and III). It may seem therefore, that the algorithm constrains the learning materials more severely. In fact, the opposite occurs, since k_3 in Table III can be much smaller than k_1 in Table II. Frequently we even take $k_3 = 0$. In effect we have replaced hypothesis 2 by a much broader one:

Hypothesis 3: most of the known epicenters have at least one D object nearby.

Hypotheses 1 and 3 seem to be close to minimally necessary ones. Without them our problem is hopeless.

2.3. Control experiments

The main control experiment will now be described. Let us neglect the data on a specified part of the list of earthquakes. These earthquakes can be the most recent ones or can be selected otherwise. Objects which are near the epicenters of these and only these earthquakes will be reassigned in the learning stage to class II. We apply the same algorithms of learning and voting and check whether these objects will be recognized as D_{II}. If the earthquakes we have neglected are the most recent ones we refer to this experiment as experiment EH ("earthquake history"), since it is equivalent to learning on the basis of an early part of an earthquake catalog and testing our ability to predict the subsequent seismic history, i.e. recognize the earthquakes in the later part of the catalog.

We must also take care that the number of objects identified as dangerous does not grow unduly large "in the future". To check this possible instability, we continue the EH experiment into the future, reassigning into group I all objects recognized as D_{II}. This is referred to as "experiment EF" ("earthquake future"). This procedure is evidently iterative and can be carried out on both experiments EH and EF.

As a further check we have compared our results against those obtained using variations of the data, the questionnaire and its answers, the criteria for the selection and classification of objects, the catalog of epicenters, and the values of numerical parameters, including k_i and $\bar{k}_i$. We have also used tests involving random variation and even random generation of data, the use of different algorithms of recognition, and the application of the characteristic features identified in one region to recognition in another, but seismically similar region.

3. Strike-slip earthquakes of California

3.1. The problem

Here we consider strong ($M \geqslant 6.5$) strike-slip earthquakes in the region shown in Fig. 1a. The catalog of earthquakes is given in Table IV. We assume that the epicenters are associated with major strike-slip faults (referred to as faults in what follows). The system of faults is depicted in Fig. 1a. The *Tectonic Map of the United States* (USGS, 1962) is the prin-

TABLE IV

Epicenters of strike-slip earthquakes with $M \geqslant 6$

N	Year	Day, month	Latitude (°N)	Longitude (°W)	*M*	Closest point (see Fig. 1)
1	1836		37.5	121.9	>7	*3*
2	1836		37.7	122.1	>7	*4*
3	1836		37.8	122.3	>7	*5*
4	1857		34.7	118.8	>7	*11*
5	1906	18, IV	38.25	122.95	8.25	*2*
6	1911	1, VII	37.25	121.75	6.6	*6*
7	1918	21, IV	33.75	117	6.8	*13*
8	1922	10, III	35.75	120.25	6.5	*10*
9	1923	22, I	40.5	124.5	7.2	*23*
10	1923	23, VII	34	117.25	6.25	*12*
11	1926	22, X	36.75	122	6.1	*7*
12	1933	11, III	33.6	118	6.25	*22*
13	1934	8, VI	36	120.5	6.0	*9*
14	1934	30, XII	32.25	115.5	6.5	*20*
15	1934	31, XII	32	114.75	7.0	*21*
16	1937	25, III	33.5	116.5	6.0	*15*
17	1940	19, V	32.7	115.5	6.7	*19*
18	1942	21, X	33	116	6.5	*18*
19	1948	4, XII	33.9	116.4	6.5	*14*
20	1951	8, X	40.25	124.5	6.0	*1*
21	1952	22, XI	35.8	121.2	6.0	*8*
22	1954	19, III	33.28	116.18	6.2	*17*
23	1968	9, IV	33.4	116.2	6.9	*16*

cipal source of fault information, with a few additions on our part. The problem is to find where on these faults earthquake epicenters may fall.

3.2. The objects

Objects of recognition are points on faults. We consider points of the following three kinds:

(1) Projection of each epicenter onto the closest fault. Epicenters of all strike-slip earthquakes with $M \geq 6.0$ (Table IV) are used. We call these points centers. They have ordinal numbers from *1* to *23* in Fig. 1b.

(2) Points on faults at distances of 25 km from centers. They have ordinal numbers from *24* to *61* in Fig. 1b. To locate these points we drew circles of 25-km radius around each center. The intersections of the circles with the faults are these points of the second kind. We call these points associates of corresponding centers. For example, point *60* is an associate of point *22* in Fig. 1b. If it occurs that two centers lie on the same fault with distance in the range $25 < r \leq 50$ km from each other, then the two associates between them lie closer than 25 km. In this case the two associates are replaced by one point in the middle. Points *38* and *48* are examples of such points.

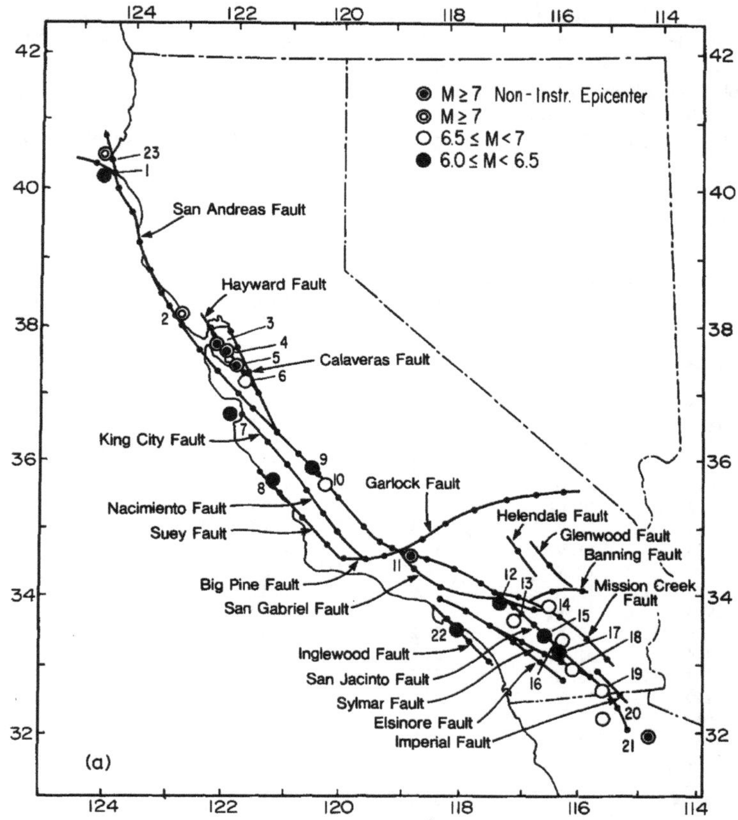

Fig. 1. (a) Major strike-slip faults of California.

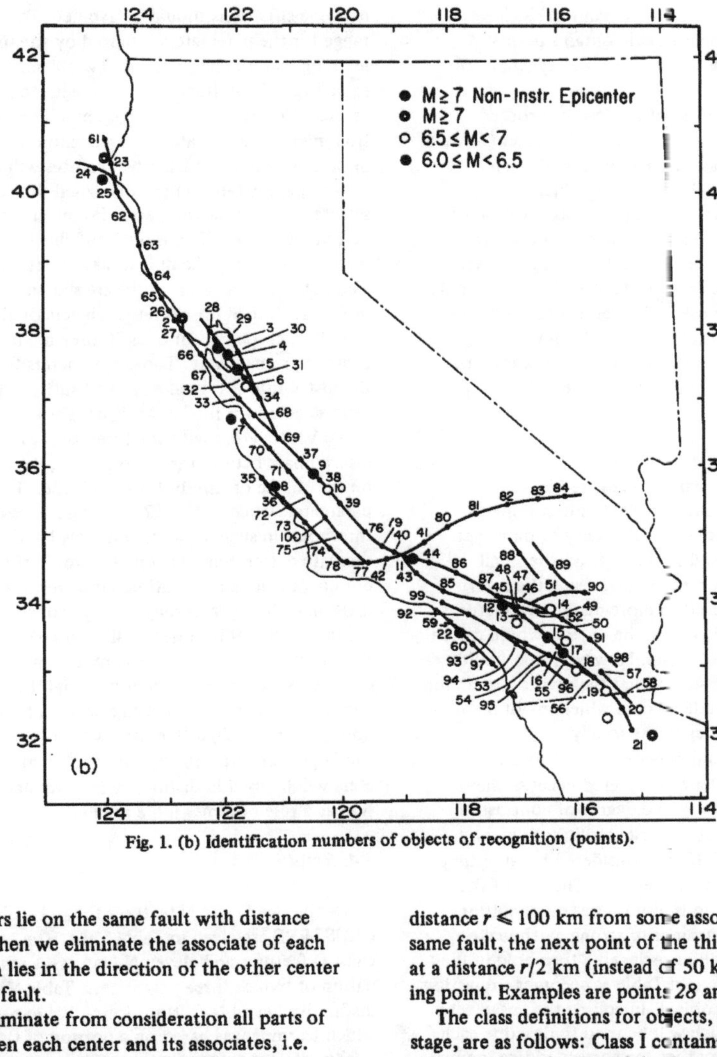

Fig. 1. (b) Identification numbers of objects of recognition (points).

If two centers lie on the same fault with distance $r \leqslant 25$ km, then we eliminate the associate of each center which lies in the direction of the other center on the same fault.

(3) We eliminate from consideration all parts of faults between each center and its associates, i.e. parts of the faults within the above mentioned circles of radius 25 km. On the remaining parts of each fault we place the points at intervals of 50 km, beginning with the northern end of the fault. These are points of the third kind.

If it occurs that a point of the third kind is at the distance $r \leqslant 100$ km from some associate point on the same fault, the next point of the third kind is placed at a distance $r/2$ km (instead of 50 km) from the preceding point. Examples are points *28* and *59*.

The class definitions for objects, in the learning stage, are as follows: Class I contains the centers and their associates (points of the first and second kind), corresponding to earthquakes with $M \geqslant 6.5$. Each of these centers together with its associates forms a cluster. Two clusters were eliminated in the learning stage because the corresponding epicenters occur too far from the nearest fault. Points *16* and *20* are the centers of these clusters. The remaining 13 clusters were used for learning. Class II contains points of the third kind (which are at a distance $r > 25$ km from all centers).

With this classification of objects we proceed through a learning stage in which each cluster (formed of a center and two associate points on the same fault) is in class I and the remaining objects are in class II. We are uncertain about the classification of objects in clusters associated with earthquakes for which $6.0 \leqslant M < 6.5$, so we place them in class III. Some associates were recognized as D. These and only these associates were used with centers to form clusters in the final variant of learning, leading to the results described in later sections. Voting was carried out for all 100 points shown in Fig. 1b.

3.3. The data

We considered the parameters listed in Table V for use in characterizing the objects. Some comments on their choice are in order. The diversity of data that might be used is limited in this case by the small number of known epicenters. Increasing the assortment of data would lead to more distinctive features and in the subsequent voting the objects would divide more readily into two classes. However, this would also increase the probability that the division is random. Thus, we cannot include all data which might be relevant. Our choice depends strongly on physical and geological intuition and experience. Parameters 1–6 and 19–24 (Table V) are considered because they characterize the intensity and degree of contrast of neotectonic movements represented in topography. Parameters 8–11 and 18 are considered because they characterize fracturing of the crust. The rest of the parameters reflect various hypotheses on conditions favorable for the occurrence of strong earthquakes.

Each parameter will be assigned a threshold in the form of an inequality, and Table V becomes a questionnaire applied to each object. In coding the answers to the questionnaire which will be used to describe each object, "1" will signify that the corresponding inequality is satisfied, "0" indicates that it is not satisfied at the object.

Two questions now arise. Which of the parameters in Table V should be used for recognition and what discrete thresholds should we assign to them? Two extreme situations should be avoided. Too narrow a range for the parameters is barred by the limited learning material because too few objects will fall in each interval. On the other hand, with too few parameters and/or overly wide ranges, much needed information might be eliminated and recognition will fail because few if any distinctive features will be found.

The parameters and their assigned thresholds were selected in the following way: For each parameter we determined a one-dimensional distribution of its occurrence among the objects as a function of the value of the parameter. These are shown as histograms in Fig. 2. Dashed lines refer to objects of class I and solid lines to objects of class II; they are more likely D and N, respectively. Those parameters for which the distributions turned out to be sufficiently different were selected for further analysis. They are listed in Table VI, together with the thresholds assigned to discriminate between the two groups. Some comments on the choice of thresholds are in order. For some parameters, such as No. 12, we could have chosen the thresholds in such a way that classes I and II would separate better. Such parameters would then play a dominant role in recognition. However, the sharp division of histograms may be random, since the learning material is rather small. Experience tells us that in such cases many errors in recognition of D_{II} objects can occur. We have found that it is safer to choose the thresholds in such a way that objects of both groups are slightly mixed with no group forming too large a majority in any interval. Then more parameters will be used in distinctive features and recognition as a rule becomes more reliable.

3.4. Results

Distinctive D and N features found by algorithm CLUSTERS are given in Table VIIA. The table lists eight D features and eleven N features, each a combination of two or three parameters. Table VIIB lists distinctive D and N features for an EH experiment in which learning was based on a portion of the earthquake catalogs extending through the year 1922.

Results of voting are given in Table VIII and in Table IX under the column for 1948. We note that recognition is formally successful in two ways: many distinctive features of D and N are found and voting divides the objects into two rather distinctive groups,

TABLE V

Characterization of objects: List of parameters considered

N	Parameter name
1	maximal elevation, h_{max} (m)
2*	minimal elevation, h_{min} (m)
3*	elevation difference, $\Delta h = h_{max} - h_{min}$ (m)
4*	distance between points with minimum and maximum elevation, l_1
5*	"gradient", $\Delta h/l_1$
6*	relative area of soft sediments, q (%)
7*	type of rocks (igneous I, metamorphic M, sediments S)
8	distance to closest fault, r_1 (km)
9	distance to closest intersection of faults, R_1 (km)
10	distance to closest end of fault, R_2 (km)
11	distance to closest end or intersection of faults, r_2 (km)
12	distance to geothermal zone, r_3 (km)
13	distance to a region of large precipitation, R_3 (km)
14	distance to closest large water reservoir, r_6 (km)
15	distance to closest occurrence of Franciscan rocks along the fault, R_4 (km)
16	distance to the closest spreading center, r_4 (km) (40.5°N, 126.6°W or 33.2°N, 115.6°W)
17	distance to the reference point (intersection of San Andreas and Big Pine faults), r_5 (km)
18*	the number of unnamed faults on the *Tectonic Map of U.S.* (USGS, 1962), N_1
19*	the types of relief (surface morphology) featured across the fault
20*	the number of changes of the types of relief along the fault, n_2
21*	the types of relief featured along the fault
22**	maximal elevation, H_{max} (m)
23**	minimal elevation, H_{min} (m)
24**	elevation difference, $\Delta H = H_{max} - H_{min}$ (m)
25**	distance between points with minimum and maximum elevation, l_2
26**	"gradient", $\Delta H/l_2$
27**	the number of unnamed faults on the *Tectonic Map of U.S.* (USGS, 1962), n_1
28**	the types of relief (surface morphology), featured across the fault
29**	the types of relief featured along the fault
30**	the number of contacts between rocks of different age on the *Geological Map of North America*, (USGS, 1965), n_3
31***	the number of parallel faults, n_4
32**	the number of faults, n_5
33***	the number of ends and intersections of faults, n_6
34***	the number of intersections, N_2
35	the angle between the fault and the dominant structural trend in the region

* Measured inside the circle r = 12.5 km.
** Measured inside the circle r = 25 km.
*** Measured inside the circle r = 50 km.

with only a small number of "neutral" results. The latter may be seen from the fact that out of 100 objects, 84 have the distinctive features of only D or only N, and only two points in all clusters have $\Delta < 0$. Experience with recognition based on real as well as random data (see below) indicates to us that these results are encouraging, though nothing more.

The threshold of recognition was chosen as follows:

$$\bar{\Delta} = \Delta^* - \tau, \qquad \Delta^* = \min_c [\max \Delta_c]$$

Here max Δ_c is the maximal value of Δ for points in the cth cluster; $\min_c$ signifies the smallest value among all clusters. We subtract τ to decrease the probability of missing a D point; although this increases the probability of false alarms, this is preferable in our problem.

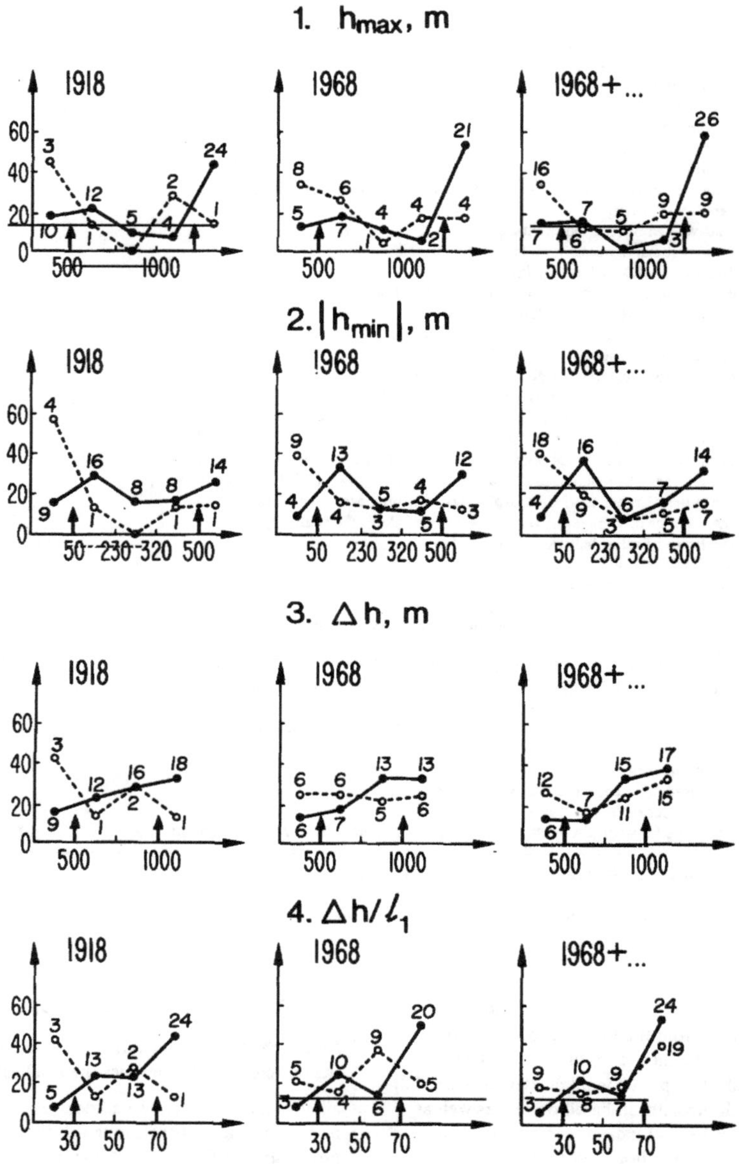

Fig. 2. For caption see p. 245.

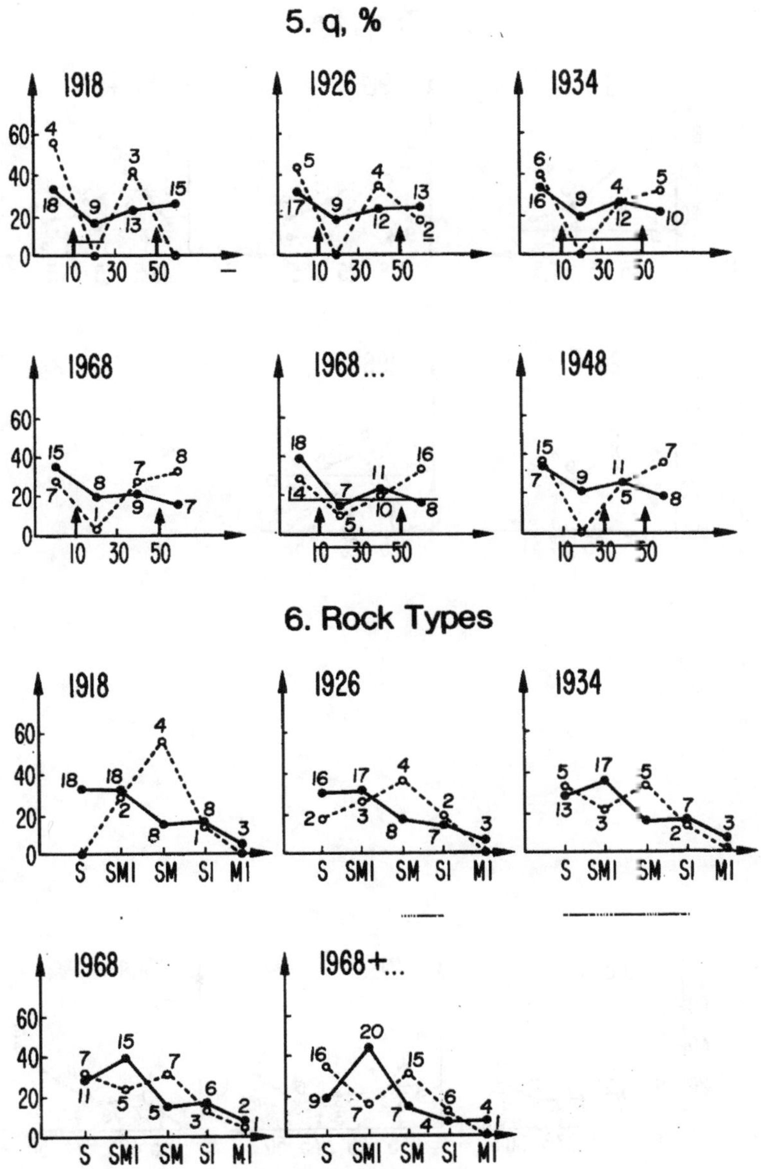

Fig. 2. For caption see p. 245.

7. r_1, km

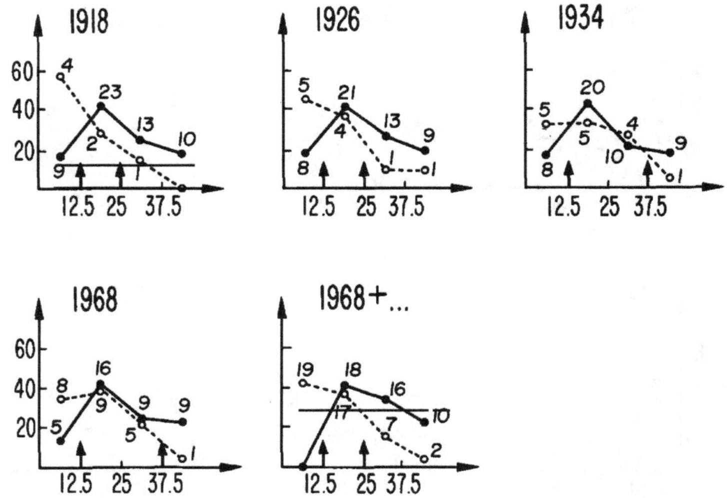

8. r_2, km

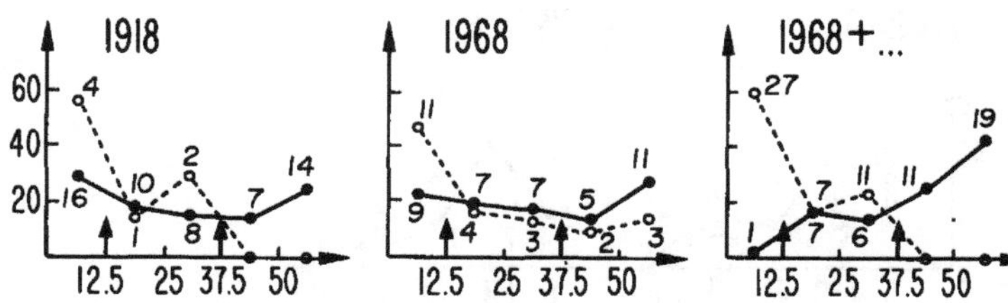

Fig. 2. For caption see p. 245.

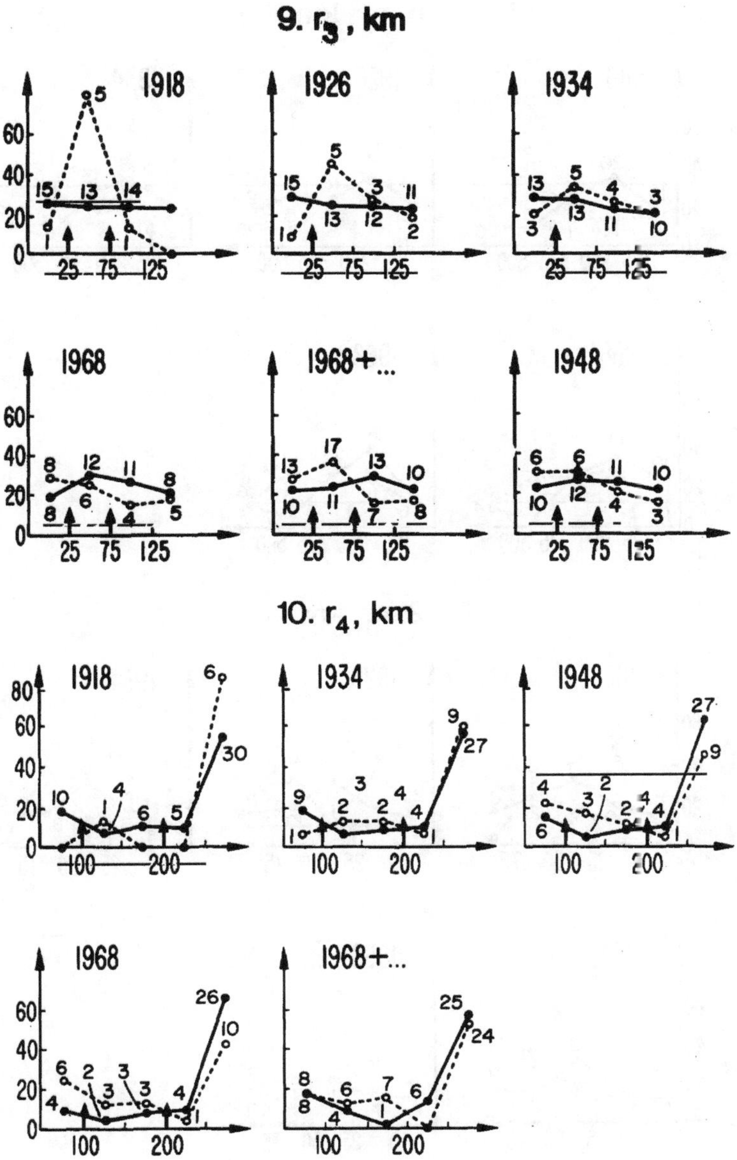

Fig. 2. For caption see p. 245.

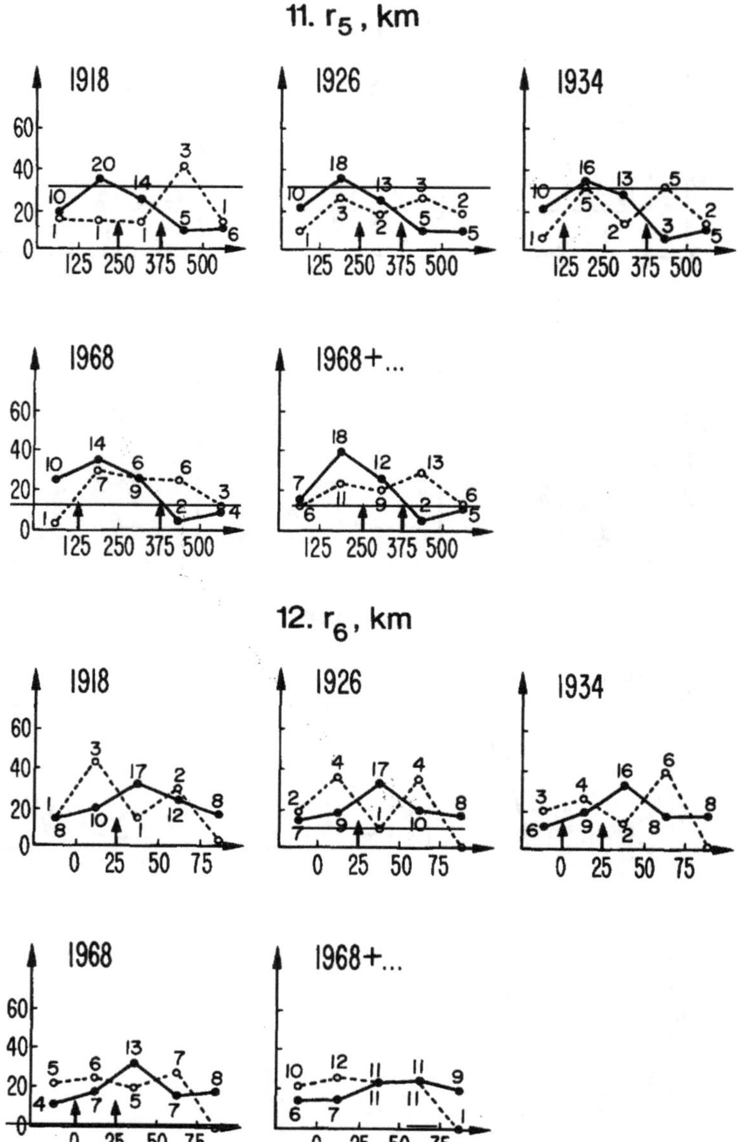

Fig. 2. For caption see p. 245.

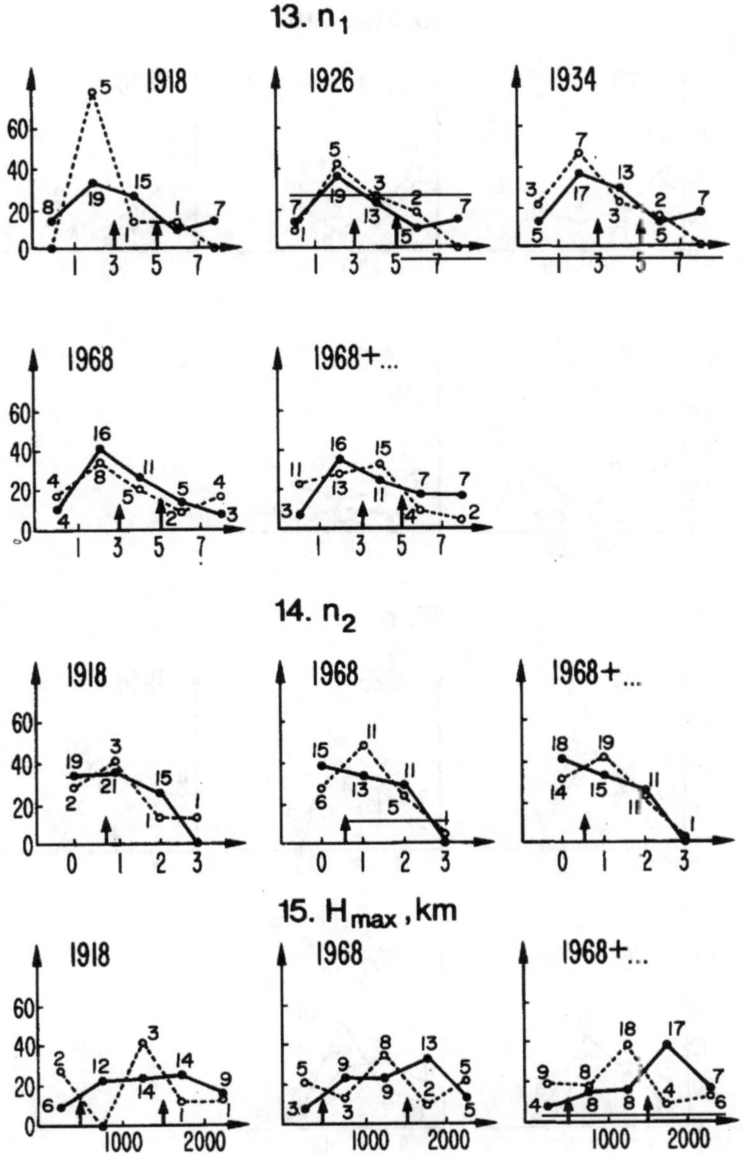

Fig. 2. For caption see p. 245.

16. H_{min}, m

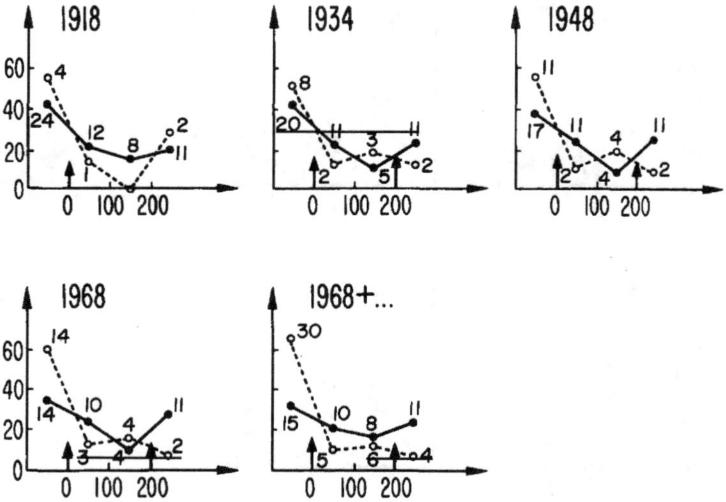

17. n_3

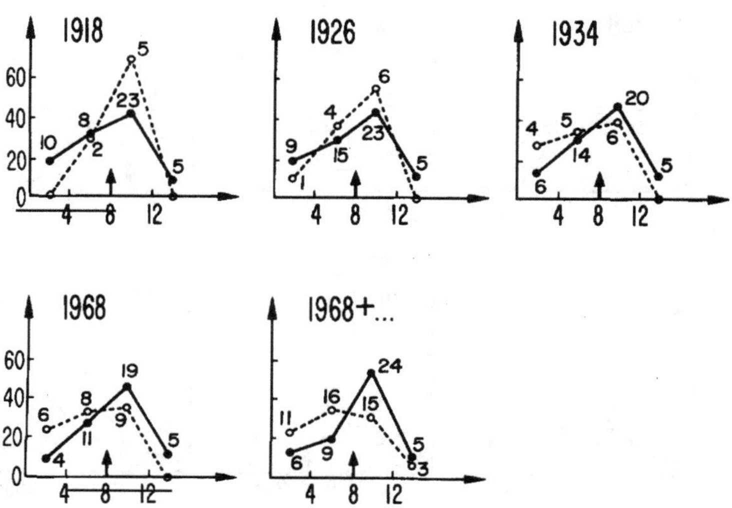

Fig. 2. For caption see p. 245.

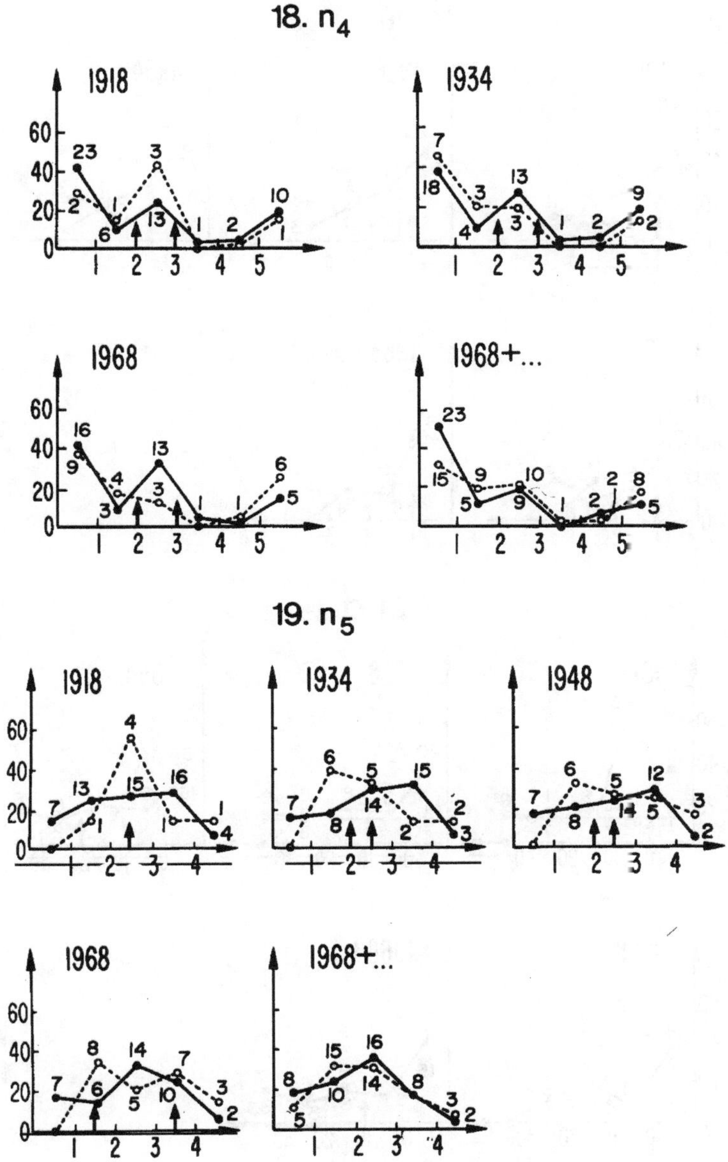

Fig. 2. For caption see p. 245.

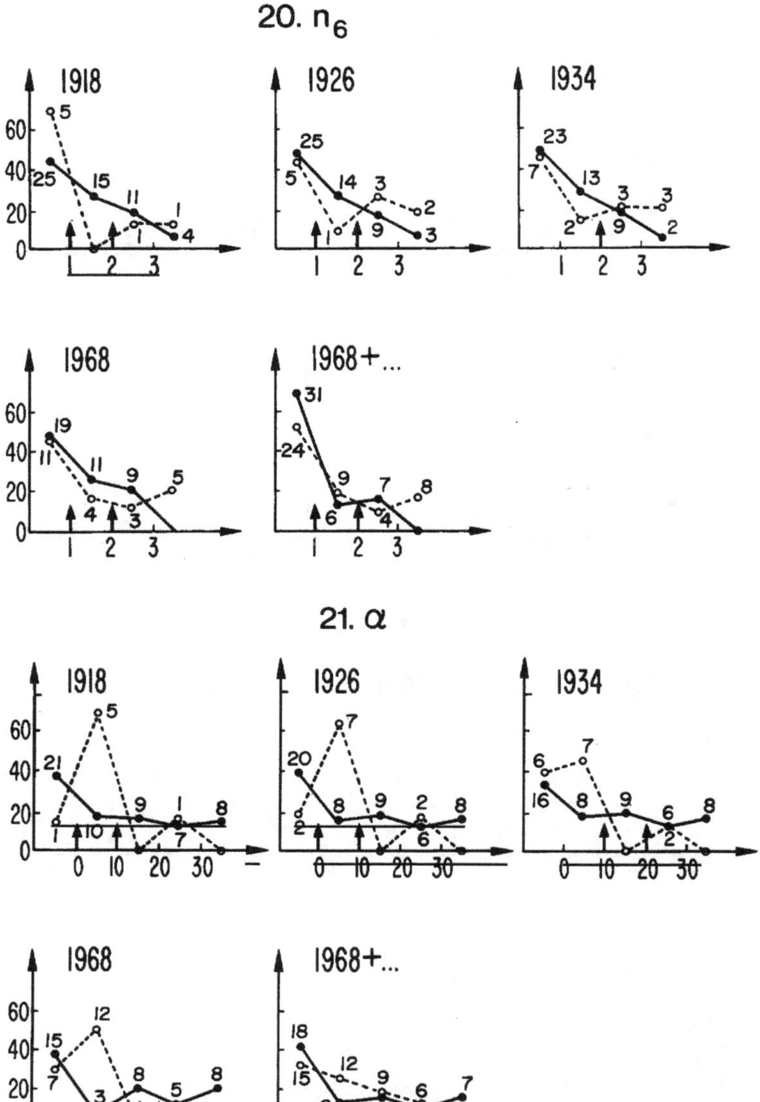

Fig. 2. For caption see p. 245.

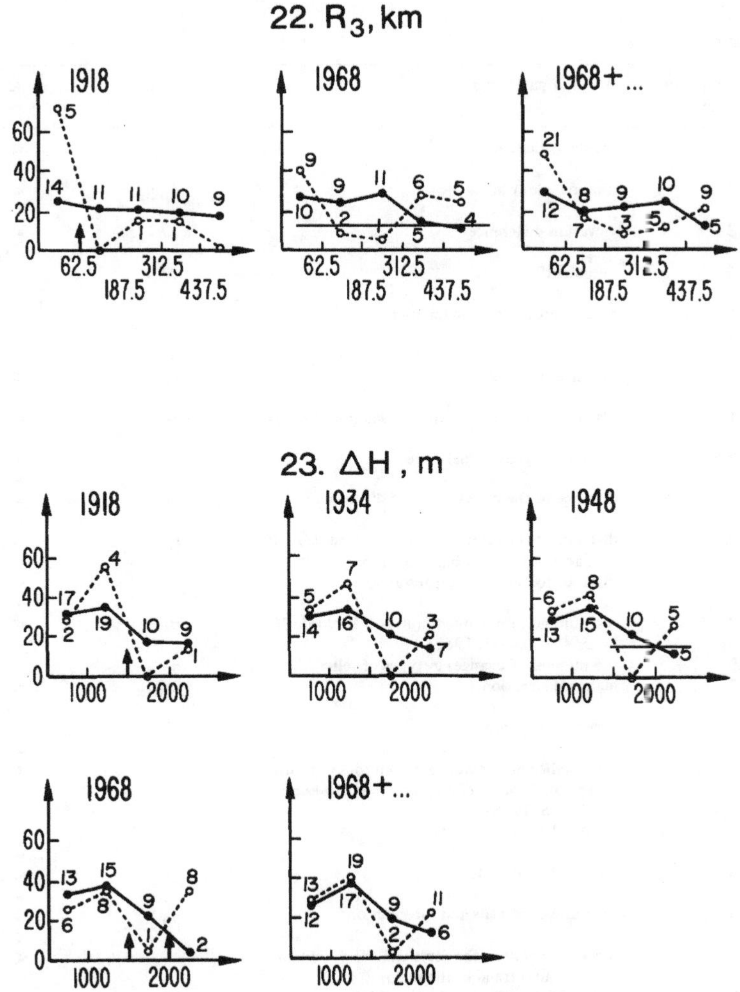

Fig. 2. One-dimensional distributions of parameters for points on major strike-slip faults of California shown in Fig. 1. 1968 = Last year in earthquake catalog; earthquakes up to this year inclusive are used in the separation of objects into classes; other dates are used to denote terminal years of EH experiment; 1968+... refers to the control experiment EF with objects recognized as D, after learning on full catalog, assigned to class I; h_{min}, etc. are the ordinal number and symbol for parameter as used in Table VI. The values of parameters are indicated on the horizontal axis; the dashed line is the histogram for points of the first kind (near known epicenters); the solid line is the histogram for points of the third kind (far from known epicenters); the assigned thresholds are shown by arrows; the number next to each point refers to the number of objects for which the values of parameters fall into the given interval shown on the abscissa; the ordinate is the same number expressed as a percentage of the total number of objects in the corresponding class.

TABLE VI

Parameters used for recognition

N	Number in Table V	Name of parameter	Symbol	Assigned threshold
1	1	maximal elevation	h_{max}	≤ 500 ≤1,250
2	2	minimal elevation	h_{min}	≤ 50 ≤ 500
3	3	elevation difference	Δh	≤ 500 ≤1,000
4	5	"gradient"	$\Delta h/l_1$	≤ 30 ≤ 70
5	6	relative area of soft sediments	q	≤ 10 ≤ 50
6	7	type of rocks	–	presence of I
7	8	distance to closest fault	r_1	≤ 12.5 ≤ 37.5
8	11	distance to closest end or intersection of faults	r_2	≤ 12.5 ≤ 37.5
9	12	distance to geothermal zone	r_3	≤ 25 ≤ 75
10	15	distance to the closest zone of divergence	r_4	≤ 100 ≤ 200
11	17	distance to the reference point (intersection of San Andreas and Big Pine faults)	r_5	≤ 125 ≤ 375
12	14	distance to closest water reservoir	r_6	≤ 0 ≤ 25
13	27	the number of unnamed faults on *Tectonic Map of U.S.* (USGS, 1962)	n_1	≤ 3 ≤ 5
14	20	the number of changes in types of relief	n_2	= 0
15	22	maximal elevation	H_{max}	≤ 500 ≤1,500
16	23	minimal elevation	H_{min}	≤ 0 ≤ 200
17	30	the number of contacts between rocks of different age on *Geological Map of North America* (USGS, 1965)	n_3	≤ 8
18	31	the number of parallel faults	n_4	≤ 2 ≤ 3
19	32	the number of faults	n_5	≤ 1 ≤ 3
20	33	the number of ends and intersections	n_6	≤ 1 ≤ 2
21	35	the angle between the fault and the dominant structural trend in the region (°)		≤ 10 ≤ 20
22*	13	distance to a region of large precipitation	R_3	≤ 62.5
23*	24	elevation differences	ΔH	≤1,500

The rule for coding the description of a point is the following: "1" means that the inequality in the last column is fulfilled.
* These parameters were used in control experiments only with learning based on earthquakes prior to 1934.

TABLE VII

Distinctive D and N features

(A) Based on entire earthquake catalog for points located on faults:

Feature number	Maximum elevation, h_{max} (meters)	Gradient $\Delta h/l_1$	Relative area of soft sediments, q %	Type of rocks	Distance to closest end or intersection of faults, r_2, km	Distance to nearest geothermal zone, r_3, km	Distance to nearest body of water, r_6, km	Minimum elevation, H_{min} (meters)	Number of contacts between rocks of different age on geologic map	Number of faults n_5	Angle between fault and dominant regional strike (deg.)
	1	4	5	6	8	9	12	16	17	19	21
D											
1					≤37.5	≤75					≤10
2				no I	≤37.5						≤10
3				no I	≤37.5					≤3	
4			>10		≤12.5					>1	
5					≤37.5			≤200	≤8		
6			>10		≤37.5			≤0			
7				no I	≤37.5	>25					
8			≤50		≤12.5	>25					
N											
1							>20		>8	≤3	
2					>37.5			≤0		≤3	
3				I		>25				≤3	
4	>500			I						≤3	
5					>12.5		>20		>8		
6					>12.5	>75			>8		
7				I		>75			>8		
8					>37.5	>25		≤200			
9			>10		>37.5			≤200			
10	>500				>37.5			≤200			
11	>500		>10		>12.5						

(B) Based on earthquake catalog extending through 1922:

Feature number	Maximum elevation, h_{max} (meters)	Minimum elevation, $\lvert h_{min} \rvert$ (meters)	Elevation difference, Δh, (meters)	Gradient, $\Delta h/l_1$	Type of rocks	Distance to nearest fault, r_1, km	Distance to closest end or intersection of faults, r_2, km	Number of contacts between rocks of different age on geologic map	Distance to the zone of large precitation, r_3, km	Elevation difference ΔH, (meters)
	1	2	3	4	6	7	8	17	22	23
D										
1	≤1250				SM					≤1500
2			≤1000				>37.5	>8		
3			≤1000			>12.5		>8		
4	≤1250					≤12.5		>8		
5	≤1250			>30				>8		
N										
1	>500							≤8	≤62.5	
2			>500					≤8		
3	>500				no SM			≤8		
4	>500				no SM		>37.5			
5	>1250					>12.5				

Parameter number corresponds to Table VI

The points recognized as D with this threshold and $\tau = 1$ are shown in Fig. 3a. It is encouraging that these points do not occur erratically, but form a limited number of compact segments. This might be expected however, because of the spatial continuity of some of the parameters. We see from Fig. 3a that significant parts of major faults are recognized as N. These should be segments where epicenters will not occur, though faulting could extend into these regions from epicenters on adjacent segments of the fault. Among the N segments the major branch of the San Andreas fault north of San Francisco, between the points *62* and *26,* seems especially surprising as does the Garlock fault. D segments concentrate at five areas from north to south between points *61–25* (lat. 41–40°N), *27–7* (38.5–37°N), *35–43* (36–34.5°N), *45–97* (34.2–33°N), and *18–21* (33–32.2°N). We shall discuss these areas later.

3.5. Control experiments

3.5.1. Experiment EH

In conducting this experiment the histograms of parameters are redetermined (see Fig. 2) because cataloged earthquakes after a specific year are not used in learning; their corresponding clusters are disbanded and their points are placed in class II. This may lead to the selection of new thresholds and even of other parameters. In practice, however, such changes will not occur at each step of the EH experiment, because the histograms are stable by definition and because the thresholds are chosen rather roughly. The histograms, corresponding to several steps, from 1918 to 1966 are compared in Fig. 2. The only changes that occur from histogram to histogram are the following: learning on the basis of earthquakes through 1918 we would assume different thresholds

TABLE VIII

Results of voting for points on major California faults – main variant

n_N \ n_D	0	1	2	3	4	5	≥6
≥7	$\tilde{70}$, $\tilde{74}$, $\tilde{80}$, $\tilde{83}$, $\tilde{89}$, $\tilde{90}$ 82, 84, 86, 91						
6	$\tilde{69}$, $\tilde{88}$						
5	36, 37, 81						
4	$\tilde{65}$, $\tilde{66}$, $\tilde{73}$, $\tilde{94}$, $\tilde{95}$ 16, 44, 52, 54, 76						
3	$\tilde{62}$, $\tilde{63}$, $\tilde{64}$, $\tilde{67}$, $\tilde{87}$ 17, 22, 50, 53, 60						
2	$\textcircled{9}$ $\textcircled{26}$ $\tilde{71}$ $\tilde{92}$, $\tilde{96}$ 15, 38	$\tilde{85}$	$\tilde{72}$				
1	41	$\textcircled{51}$ $\tilde{68}$	$\tilde{75}$, 7, 33, 45, 46	$\textcircled{55}$	$\textcircled{6}$	$\textcircled{20}$ 8, 25	
0	$\tilde{77}$, $\tilde{79}$	$\textcircled{11}$ $\textcircled{13}$ $\textcircled{47}$ $\tilde{93}$ 42	*2* *27* 30 34 *40* *43* $\textcircled{48}$ $\textcircled{49}$ $\tilde{100}$	*12* $\tilde{78}$	$\textcircled{18}$ $\textcircled{3}$ $\textcircled{10}$ $\textcircled{23}$ $\textcircled{24}$ $\textcircled{28}$ $\textcircled{58}$ $\textcircled{61}$	*1* *19* $\tilde{98}$ *21* $\textcircled{31}$ $\tilde{99}$ *39* *56* 59 *57* 29	*4* $\textcircled{5}$ $\textcircled{14}$ *32* $\tilde{97}$ 35

Circled number = number of a point in a cluster; italicized number = number of a point in a cluster which has largest value of $n_D - n_N$ in its cluster; ordinary number with tilde = number of a point not in a cluster used in learning stage; ordinary number without tilde = number of a point not in a cluster, not used in learning stage.
$k_1 = 7, k_2 = 9, \tilde{k}_1 = 4, \tilde{k}_2 = 1$ and $k_3 = 0$.

for seven parameters (No.'s 6, 7, 11, 12, 16, 19 and 21) and select for recognition two more parameters (22 and 23). The changes in the values of the thresholds are indicated in Table X.

The results of recognition in the EH control experiment are shown in Table IX and in Figs. 3b and c. They are successful since by and large the results based on a partial catalog agree with those based on the full catalog (Fig. 3a). At least one point in each cluster was recognized as D before the corresponding earthquake occurred. Moreover, at least one point in each cluster was recognized as D at all steps since 1911 with three exceptions: clusters corresponding to the earthquakes of 1923, 1940 and 1942 were not recognized with learning through 1918 and 1923.
However, at least one object in these clusters was recognized as D before the actual earthquake for all other steps in the EH experiment. The recognition of the objects in classes II and III also appears to be stable.

3.5.2. Experiment EF

The results shown in Table IX also illustrate the stability of our procedures. The identification of only seven points changed. Four D_{II} points are reclassified as N_{II} and three N_{II} points are reclassified as D_{II}. Thus, our D objects will neither wander nor proliferate over all of California, but strong earthquakes are predicted for the limited number of D_{II} points identified.

3.5.3. Inverse EH experiment

In this experiment we eliminated earthquakes from learning in the order of their occurrence, the first to occur, the first eliminated, beginning with 1836. In this and the following experiments we did not redetermine the histograms of parameters although we verified the stability of recognition by varying the learning materials.

The results of this experiment are shown in Table XI. Recognition remains successful until the 1906 earthquake. The points near its epicenter are not

TABLE IX

Recognition results for control experiments EH and EF

Number of point on Fig. 1	Year of earthquake which defines cluster	Learning based on earthquakes from 1836 through the year								
		1911	1918	1922	1923	1934	1940	1942	1948	1948+...
					Clusters $\{D_1 + N_1\}$:					
14	1948	+	+	+	+	+	+	+	+	+
49		+	+	+	+	o	o	o	+	+
51		o	+	+	o	+	o	o	+	+
18	1942	+	+	+	o	+	+	+	+	+
55		+	o	o	o	o	o	+	+	+
56		+	+	+	o	+	+	+	+	+
19	1940	+	+	+	o	+	+	+	+	+
56		+	+	+	o	+	+	+	+	+
57		+	+	+	o	+	+	+	+	+
58		+	+	+	o	+	+	+	+	+
21	1934	+	o	+	o	+	+	+	+	+
20		o	+	+	+	+	+	+	+	+
23	1923	o	o	+	+	+	+	+	+	+
1		+	o	+	+	+	+	+	+	+
24		o	o	+	+	+	+	+	+	+
61		o	o	+	+	+	+	+	+	+
10	1922	o	o	o	+	+	+	+	+	+
9		+	o	+	o	o	o	o	o	o
39		o	+	+	+	+	+	+	+	+
13	1918	o	+	+	+	+	+	+	+	+
12		+	+	+	+	+	+	+	+	+
47		+	+	+	+	+	+	+	+	+
6	1911	+	+	+	+	+	+	+	+	+
31		+	+	+	+	+	+	+	+	+
32		+	+	+	+	+	+	+	+	+
34		+	+	+	+	+	o	+	+	+
2	1906	+	+	+	+	+	+	+	+	+
26		+	+	+	o	+	o	o	o	o
27		+	o	+	+	+	+	+	+	+
11	1857	+	+	+	o	+	o	o	+	+
40		+	+	+	+	+	+	+	+	+
43		o	+	+	+	+	+	+	+	+
3	1836	+	+	+	+	+	+	+	+	+
28		+	+	+	+	+	+	+	+	+
4		+	+	+	+	+	+	+	+	+

TABLE IX (*continued*)

Number of point on Fig. 1	Year of earthquake which defines cluster	Learning based on earthquakes from 1836 through the year								
		1911	1918	1922	1923	1934	1940	1942	1948	1948+...
4	1836	+	+	+	+	+	+	+	+	+
3		+	+	+	+	+	+	+	+	+
5		+	+	+	+	+	+	+	+	+
30		+	+	+	+	+	+	+	+	+
5	1836	+	+	+	+	+	+	+	+	+
4		+	+	+	+	+	+	+	+	+
30		+	+	+	+	+	+	+	+	+
32		+	+	+	+	+	+	+	+	+
		Class II, $\{D_2 + N_2\}$:								
62		+	o	+	o	o	o	o	o	o
63		o	o	+	o	o	o	o	o	o
64		o	o	o	o	o	o	o	o	o
65		o	o	o	o	o	o	o	o	o
66		+	o	+	o	o	o	o	o	o
67		+	o	+	o	o	o	o	o	o
68		+	+	+	o	o	o	+	+	+
69		o	o	o	o	o	o	o	o	o
70		o	o	o	o	o	o	o	o	o
71		o	o	o	o	o	o	o	o	o
72		o	o	o	+	o	o	o	+	o
73		o	o	o	o	o	o	+	o	o
74		+	o	+	o	o	o	o	o	o
75		+	o	o	o	o	+	+	+	+
77		+	o	o	o	+	o	+	+	+
78		+	o	o	o	o	o	+	+	+
79		o	+	+	o	+	+	+	+	o
80		o	o	o	o	o	o	o	o	o
83		o	o	o	o	o	o	o	o	o
85		o	+	o	o	o	o	o	o	+
87		o	+	o	o	o	o	o	o	o
88		o	+	o	o	o	o	o	o	o
89		o	+	+	o	o	o	o	o	o
90		o	o	o	o	o	o	o	o	o
91		o	o	o	o	o	o	o	o	o
92		+	+	+	o	o	o	o	o	o
93		+	+	+	+	+	o	o	+	+
94		o	o	o	o	o	o	o	o	o
95		o	o	o	o	o	o	o	o	o
96		o	o	o	o	o	o	o	o	o
97		o	o	+	+	+	+	+	+	+
98		o	+	+	+	+	+	+	+	+
99		+	o	+	+	+	+	+	+	+
100		+	o	o	o	o	o	o	+	+

TABLE IX (continued)

Number of point on Fig. 1.	Year of earthquake which defines cluster	Learning based on earthquakes from 1836 through the year								
		1911	1918	1922	1923	1934	1940	1942	1948	1948+...
		Class III (objects for voting but not for learning):								
7		o	+	+	o	+	+	+	+	+
8		+	o	o	+	+	+	+	+	+
15		o	+	o	o	o	o	o	o	o
16		o	+	o	o	o	o	o	o	o
17		o	+	+	o	o	o	o	o	o
22		o	o	+	o	o	o	o	o	o
25		o	o	o	+	+	+	+	+	+
29		+	+	+	+	+	+	+	+	+
33		+	+	+	+	o	o	+	+	+
35		+	o	o	+	+	+	+	+	o
36		+	o	+	o	o	o	o	o	o
37		+	+	o	o	o	o	o	o	o
38		+	o	o	o	o	o	o	o	o
41		o	+	o	o	o	o	o	+	o
42		+	+	+	+	+	+	+	+	o
44		o	+	+	o	o	o	o	o	o
45		+	+	+	o	o	o	o	+	+
46		+	+	+	o	o	o	o	+	+
48		+	+	+	+	+	+	+	+	+
50		o	+	+	o	o	o	o	o	o
52		o	o	o	o	o	o	o	o	o
53		o	o	o	o	o	o	o	o	o
54		o	o	o	o	o	o	o	o	o
59		+	+	+	+	+	+	+	+	+
60		o	o	+	o	o	o	o	o	o
76		o	o	o	o	o	o	o	o	o
81		o	o	o	o	o	o	o	o	o
82		o	o	o	o	o	o	o	o	o
84		o	o	o	o	o	o	o	o	o
86		o	o	o	o	o	o	o	o	o

+ = object recognized as D; o = object recognized as N; year refers to the data through which earthquakes in the catalog were used for learning; last earthquake used was in 1948; last column (1948+...) refers to experiment EF; blank lines separate the clusters, each identified by the year of the earthquake associated with the cluster; vertical lines show the time when the cluster was included in learning; to the left of this line the points of this cluster were included in class II during the learning stage.

recognized as D_{II}, after we transfer them in the learning stage to class II. However, after we also eliminate the 1911 epicenter, a D point is again recognized in each cluster. Thus, on the basis of learning on seven earthquakes (1918–1948) we recognized as dangerous the sites of the earthquakes for the period 1836–1911, except for 1906.

3.5.4. Uniformity of the region

Several areas were selected in Table IX, each corresponding either to a group of epicenters or to a part of a group. The epicenters in these areas were assumed to be unknown and the corresponding clusters were eliminated from learning. The results, shown in Table XII, indicate that recognition is reasonably

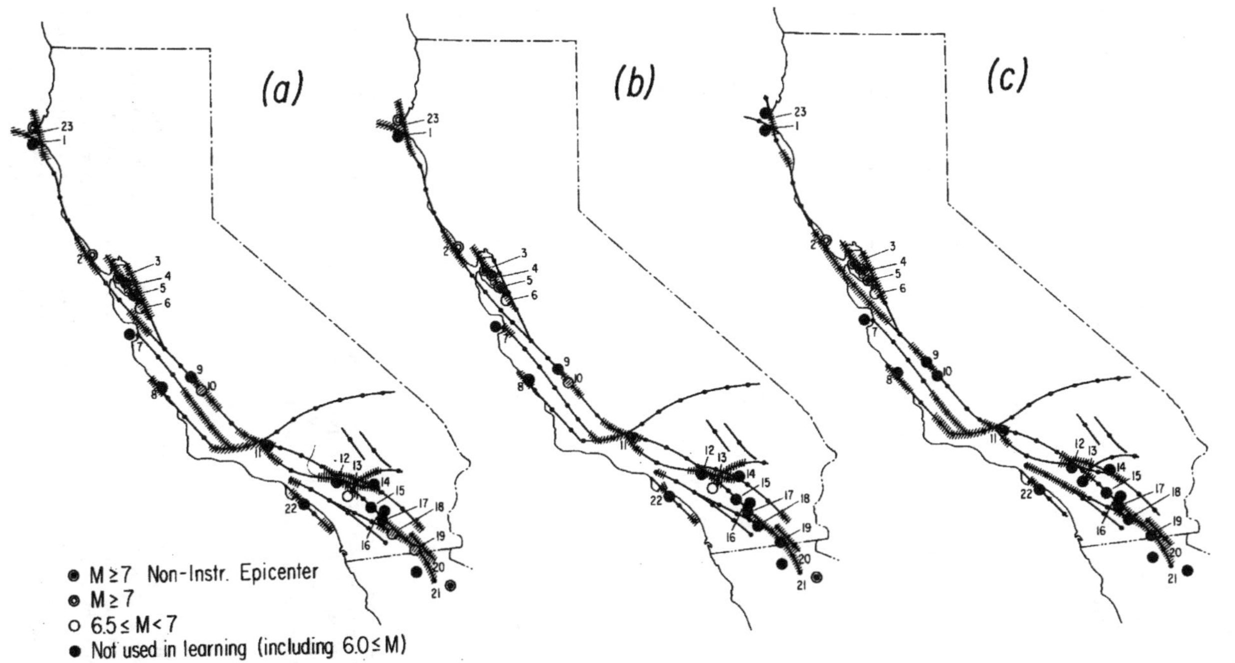

Fig. 3. Results of recognition for strike-slip earthquakes in California. Hachured regions of faults are recognized as dangerous. All earthquakes except those shown by solid circles were used in learning.
(a) All earthquakes $M > 6.5$ used in learning except those shown as solid.
(b) All earthquakes $M \geq 6.5$, prior to 1934 are used in learning.
(c) All earthquakes $M > 6.5$, prior to 1911 are used in learning.

TABLE X

Change in thresholds of discretization in experiment EH

Number in Table VI	The name of parameter	Thresholds for learning through the year	
		1968	1918
6	type of rocks	I	not SM
7	distance to closest fault, r_1 (km)	≤ 12.5	≤ 12.5
		≤ 37.5	≤ 25
11	distance to the reference point (intersection of San Andreas and Big Pine faults), r_5 (km)	≤125	≤ 250
		≤375	≤ 375
12	distance to closest water reservoir, r_6 (km)	= 0	
		≤ 25	≤ 25
16	minimal elevation, H_{min} (m)	≤ 0	≤ 0
		≤200	
19	the number of faults	≤ 1	≤ 2
		≤ 3	
21	the angle between fault and dominant strike in the region, α (°)	≤ 10	= 0
		≤ 20	≤ 10
22*	distance to a region of large precipitation, R_3	–	≤ 62.5
23*	elevation difference, ΔH (m)	–	≤1,500

* Parameter used only in learning with earthquakes prior to 1934.

stable to variations in learning material. They also indicate how homogeneously known earthquakes should cover a D area for successful recognition. At least one point near each eliminated epicenter was recognized as D in variants 1, 5, 7 and 8. These variants correspond to elimination of earthquakes in the southern and northern parts of the regions under study, namely, the Mendocino fault (variant 1), the intersection of the Garlock and San Andreas faults (variant 5), the San Jacinto Mountains (variant 7), and the Salton Sea depression (variant 8). With the other variants, 2, 3, 4 and 6, where earthquakes in the central area between San Francisco and Parkfield were eliminated, the corresponding D objects were not recognized.

3.5.5. Variations of the set of parameters

In this control experiment we eliminated each parameter (one at a time) from the data set and repeated the learning and voting. The results are shown in Table XIII. The order, in which the parameters are listed, corresponds to their increasing role in recognition. We see that r_2, the distance to a fault end or an intersection of faults, is most important. The importance of r_2 is confirmed in other ways. For example, to get a decent number of distinctive features with r_2 eliminated, we had to relax k_2 to 7 and $\bar{k}_2$ to 2. The objects separate poorly in the voting stage, and the results of recognition are unstable with respect to changes in the threshold $\bar{\Delta}$. The reason for the importance of this parameter is discussed in a later paragraph.

Recognition is stable with respect to the elimination of the other parameters, as Table XIII shows.

3.5.6. Variation of the algorithm

In this control experiment we use the algorithm CORA-3 instead of CLUSTERS. In this case class I is composed of the centers of the clusters; class II is unchanged. The results are very similar to those discussed before. This experiment confirms that a somewhat different algorithm for our recognition procedures leads to similar results. At the same time it confirms the advantage of the algorithm CLUSTERS for our particular case because the results obtained by CORA are much less stable to variations in the learning material, and especially to changes in the threshold $\bar{\Delta}$.

TABLE XI

Recognition results for inverse EH experiment (see Table IX for notation)

Number of point on Fig. 1	Year of earthquake which defines cluster	Year of first earthquake allowed for in learning						
		1836	1836	1836	1857	1906	1911	1918
14		+	+	+	+	+	+	+
49	1948	+	+	+	+	o	o	+
51		+	+	o	+	o	+	+
18		+	+	+	+	+	+	+
55	1942	+	+	+	o	+	+	+
56		+	+	+	+	+	+	+
19		+	+	+	+	+	+	+
56	1940	+	+	+	+	+	+	+
57		+	+	+	+	+	+	+
58		+	+	+	+	+	+	+
21	1934	+	+	+	+	+	+	+
20		+	+	+	+	+	+	+
23		+	+	+	+	+	+	+
1	1923	+	+	+	+	+	+	+
24		+	+	+	+	+	+	+
61		+	+	+	+	+	+	+
10		+	+	+	+	+	+	+
9	1922	o	o	o	o	+	+	+
39		+	+	+	+	+	+	+
13		+	+	+	+	+	o	+
12	1918	+	+	+	+	+	+	+
47		+	+	+	+	+	+	+
6		+	+	+	+	+	+	+
31	1911	+	+	+	+	+	+	+
32		+	+	+	+	+	+	+
34		+	+	+	+	+	+	+
2		+	+	+	+	+	o	o
26	1906	o	o	o	o	+	o	+
27		+	+	+	+	+	o	o
11		+	o	+	+	o	o	+
40	1857	+	+	+	+	+	+	+
43		+	+	+	+	+	+	+
3	1836	+	+	+	o	+	o	+
28		+	+	+	+	+	o	+
4		+	+	+	+	+	+	+
3		+	+	+	c	+	o	+
5	1836	+	+	+	+	+	+	+
30		+	+	o	o	o	o	+

TABLE XI (*continued*)

Number of point on Fig. 1	Year of earthquake which defines cluster	Year of first earthquake allowed for in learning						
		1836	1836	1836	1857	1906	1911	1918
5		+	+	+	+	+	+	+
4	1836	+	+	+	+	+	+	+
30		+	+	o	o	o	o	+
32		+	+	+	+	+	+	+
62		o	o	o	o	o	+	+
63		o	o	o	o	o	+	+
64		o	o	o	o	o	o	+
65		o	o	o	o	o	o	+
67		o	o	o	o	o	o	+
68		+	+	+	+	o	+	+
69		o	o	o	o	o	o	+
71		o	o	o	o	+	+	+
72		+	+	o	o	o	+	+
74		o	o	o	o	o	o	+
75		+	+	+	+	+	o	+
77		+	+	+	+	+	+	+
78		+	+	+	+	+	+	+
79		+	+	+	+	+	o	+
85		o	o	o	o	o	o	+
87		o	o	o	o	o	o	+
88		o	o	o	o	o	o	+
90		o	o	o	o	o	o	+
91		o	o	o	o	o	o	+
92		o	o	o	o	+	o	+
93		+	+	+	+	o	o	+
94		o	o	o	o	o	o	+
95		o	o	o	o	o	o	+
96		o	o	o	o	o	o	+
97		+	+	+	+	+	+	+
98		+	+	+	+	+	+	+
99		+	+	+	+	+	+	+
100		+	+	+	+	o	o	+
7		+	+	+	+	+	+	+
8		+	+	+	+	+	+	+
15		o	o	o	o	o	o	+
16		o	o	o	o	o	o	+
17		o	o	o	o	o	o	+
22		o	o	o	o	+	+	+
25		+	+	+	+	+	+	+
29		+	+	+	+	+	+	+
33		+	+	+	o	o	o	+
35		+	+	+	+	+	+	+
36		o	o	o	o	o	o	+
37		o	o	o	o	o	+	+
38		o	o	o	o	+	+	+
41		o	o	o	o	o	o	+
42		+	+	+	+	+	+	+

TABLE XI (*continued*)

Number of point on Fig. 1	Year of earthquake which defines cluster	Year of first earthquake allowed for in learning						
		1836	1836	1836	1857	1906	1911	1918
44		o	o	o	o	o	o	+
45		+	+	o	+	+	+	+
46		+	+	o	+	+	+	+
48		+	+	+	+	+	+	+
50		o	o	o	o	o	o	+
52		o	o	o	o	o	o	+
53		o	o	o	o	o	o	+
54		o	o	o	o	o	o	+
59		+	+	+	+	+	+	+
60		o	o	o	o	+	+	+
76		o	o	o	o	o	+	+
84		o	o	o	o	o	o	+

The earthquakes are eliminated cumulatively from learning in the order of their occurrence: first one, then two, etc. Year in top row refers to the first earthquake allowed for in learning. The rest of the notation is analogous to Table IX.

TABLE XII

Recognition in the control experiment "uniformity of the region" (elimination of groups of clusters from learning). Parentheses indicate the eliminated clusters. Otherwise the notation is the same as in Table IX

No. of point in Fig. 1	Number of variant							
	1	2	3	4	5	6	7	8
Clusters:								
14	+	+	+	+	+	+	(+)	(+)
49	+	+	+	o	+	+	(+)	(+)
51	+	o	o	o	o	+	(+)	(+)
18	+	+	+	+	+	+	+	(+)
55	o	o	o	+	+	+	+	(+)
56	+	+	+	+	+	+	+	(+)
19	+	+	+	+	+	+	+	(+)
56	+	+	+	+	+	+	+	(+)
57	+	+	+	+	+	+	+	(+)
58	+	+	+	+	+	+	+	(+)
21	+	+	+	+	+	+	+	(+)
20	+	+	+	+	+	+	+	(+)
23	(o)	+	+	+	+	+	+	+
1	(+)	+	+	+	+	+	+	+
24	(o)	+	+	+	+	+	+	+
61	(o)	+	+	+	+	+	+	+

No. of point in Fig. 1	Number of variant							
	1	2	3	4	5	6	7	8
10	+	+	+	+	+	(o)	+	+
9	o	o	o	+	o	(o)	o	+
39	+	+	+	+	+	(o)	+	+
13	+	o	o	+	o	+	(+)	+
12	+	+	+	+	+	+	(+)	+
47	+	+	+	+	+	+	(+)	+
6	+	+	+	(o)	+	o	+	+
31	+	+	+	(o)	+	+	+	+
32	+	+	+	(o)	+	+	+	+
34	+	+	+	(o)	+	+	+	+
2	(+)	(o)	(o)	(o)	+	+	+	+
26	(o)	(o)	(o)	(o)	o	o	+	+
27	(+)	(o)	(o)	(o)	+	+	+	+
11	+	+	+	+	(o)	+	+	+
40	+	+	+	+	(+)	+	+	+

TABLE XII (*continued*)

No. of point in Fig. 1	Number of variant							
	1	2	3	4	5	6	7	8
43	+	o	o	+	(+)	+	+	+
3	+	+	(o)	(o)	+	+	+	+
28	+	o	(o)	(o)	+	+	+	+
4	+	+	(+)	(+)	+	+	+	+
4	+	+	(+)	(+)	+	+	+	+
3	+	+	(o)	(o)	+	+	+	+
5	+	+	(+)	(o)	+	+	+	+
30	+	+	(o)	(o)	+	+	+	+
5	+	+	(+)	(o)	+	+	+	+
4	+	+	(+)	(+)	+	+	+	+
30	+	+	(o)	(o)	+	+	+	+
32	+	+	(+)	(o)	+	+	+	+
Group II:								
62	o	o	o	+	o	o	o	o
63	o	o	o	+	o	o	o	o
64	o	o	o	+	o	o	o	o
65	o	o	o	o	o	o	o	o
66	o	o	o	o	o	o	o	o
67	o	o	o	+	o	o	o	o
68	+	o	o	o	o	+	+	+
69	o	o	o	o	o	o	o	o
70	o	o	o	o	o	o	o	o
71	o	o	o	+	o	o	o	+
72	o	o	o	o	+	o	o	o
73	o	o	o	o	o	o	o	o
74	o	o	o	o	o	o	o	o
75	+	o	o	o	+	+	+	o
77	+	o	o	+	+	+	+	o
78	+	+	+	+	+	+	+	o
79	+	o	o	+	o	+	o	o
80	o	o	o	o	o	o	o	o
83	o	o	o	o	o	o	o	o
85	o	o	o	o	o	o	o	o
87	o	o	o	o	o	o	o	o
88	o	o	o	o	o	o	o	o
89	o	o	o	o	o	o	o	o
90	o	o	o	o	o	o	o	o
91	o	o	o	o	o	o	o	o
92	o	o	o	+	o	o	o	o
93	+	+	o	o	o	+	+	+
94	o	o	o	o	o	o	o	o
95	o	o	o	o	o	o	o	o
96	o	o	o	o	o	o	o	o
97	+	+	+	+	+	+	+	+
98	+	+	+	+	+	+	+	+
99	+	+	+	o	+	+	+	+
100	+	o	o	o	+	+	+	+
Group III:								
7	+	o	o	o	+	+	+	+
8	+	+	o	+	+	+	+	+
15	o	o	o	o	o	o	o	o
16	o	o	o	o	o	o	o	o
17	o	o	o	o	o	o	o	o
22	o	o	o	+	o	o	o	o
25	+	+	+	+	+	+	+	+
29	+	+	+	o	+	+	+	+
33	+	o	o	o	+	o	+	+
35	+	+	+	+	+	+	+	+
36	o	o	o	+	o	o	o	o
37	o	o	o	o	o	o	o	o
38	o	o	o	+	o	o	o	+
41	o	o	o	o	o	o	o	o
42	+	+	+	+	+	+	+	+
44	o	o	o	o	o	o	o	o
45	+	+	+	o	+	+	+	+
46	+	+	+	o	+	+	+	+
48	+	+	o	+	+	+	+	+
50	o	o	o	o	o	o	o	o
52	o	o	o	+	o	o	o	o
53	o	o	o	o	o	o	o	o
54	o	o	o	o	o	o	o	o
59	+	+	+	+	+	+	+	+
60	o	o	o	+	o	o	o	o
76	o	o	o	o	o	o	o	o
81	o	o	o	o	o	o	o	o
82	o	o	o	o	o	o	o	o
84	o	o	o	o	o	o	o	o
86	o	o	o	o	o	o	o	o

3.5.7. Recognition of randomly described objects

The binary codes, which describe points, are replaced by random binary numbers in this control experiment. Learning proceeds using the algorithm CLUSTERS. In a first step we use the same k_l and $\bar{k}_l$, as with the real data described earlier. In this case 32 distinctive D features and no N features are found, an unsatisfactory result. This is not sufficient to

TABLE XIII

Control experiment: change of recognition due to the elimination of parameters

After elimination of the parameter	Recognition of the following points changed:		Class in learning
	D changed to N	N changed to D	
h_{max}, r_6, n_3, or n_5 (only one at a time)		*85* or *96*	II
α	*72*		II
		41	III
r_3		*87* and *96*	II
		41	III
type of rocks	*72*	*96*	II
	25 and *35*		III
q	*51*		I
	72	*85* and *92*	II
	45 and *46*		III
H_{min}	*51*	*26*	I
	72		II
		22 and *60*	III
r_2	*49* and *6*	*9* and *26*	I
	68, 77, 79 and *100*	*66, 67, 71, 87, 62, 63, 64, 65* and *92*	II
	45 and *46*	*17, 22, 36, 38, 41, 52, 60* and *76*	III

reject the possibility of recognition with random data, because the values of k_i and $\bar{k}_i$ in the processing of real data are not predetermined; they are varied until sufficient and comparable numbers of distinctive D and N features are obtained. The only limitation is that k_i should be sufficiently large and $\bar{k}_i$ not too large. Therefore, k_i and $\bar{k}_i$ should be varied in this control experiment. With $k_1 = 12$, $k_2 = 9$, $\bar{k}_1 = 5$, $\bar{k}_2 = 2$ we find twelve distinctive features of D and fifteen of N. However, the separation of points by voting is rather vague, as is seen by comparing Tables VIII and XIV. Each feature occurs in fewer points; for example, each distinctive D feature is found at an average of fourteen points, whereas for real data the corresponding number is 21.

Also, the results are unstable, in that experiment EF results in twelve new D points in contrast to three points for real data. Furthermore the D segments of major faults are more scattered.

In this way the experiment with random data supports the results found with real data. We also learn that reliable results show up in a high degree of separation of the objects by voting.

3.6. Geological implications

To understand the geological implications of this work we pose the question: what makes a place D and not N? To answer this question, we review the parameters which make up the D and N features. We must review not only distinctive features, but also their equivalent and ϵ-equivalent features which were identified but not used in voting. Let us recall the definition of these features. Consider two features of some class F_1 and F_2. The sets of objects of the same class, which have F_1 or F_2, are called S_1 or S_2, respectively. F_1 and F_2 are equivalent, if S_1 and S_2 coincide. They are ϵ-equivalent, if only the ϵth fractions of the

TABLE XIV

Results of voting after learning on random data

n_N \ n_D	0	1	2	3	4	5	6	≥7
≥7	$\widetilde{64}$, $\widetilde{75}$, $\widetilde{79}$ $\widetilde{87}$, $\widetilde{95}$, $\widetilde{98}$ 42							
6	15							
5	36	$\widetilde{62}$, 76	$\widetilde{88}$	7				
4	$\widetilde{73}$, $\widetilde{91}$	$\widetilde{69}$	$\widetilde{67}$, 33, 35		(47)			
3	$\widetilde{78}$, $\widetilde{85}$ $\widetilde{92}$, 60	(39) $\widetilde{72}$ 16, 29, 44	(26) 52	$\widetilde{77}$, $\widetilde{90}$	84			
2	$\widetilde{63}$, $\widetilde{66}$ $\widetilde{68}$	$\widetilde{71}$, 46 54	(43) 59 81	(57)	(9) $\widetilde{97}$	$\widetilde{70}$, 8	(6)	
1	(40)	(19) $\widetilde{74}$ 22, 38	(1) $\widetilde{83}$ $\widetilde{100}$, 25, 37, 50	(55) (12) (27) $\widetilde{93}$, $\widetilde{94}$	(18) (61) (31) 45		(51) $\widetilde{99}$	
0		(58) (13) (3) (5) 48, 86 $\widetilde{96}$, $\widetilde{89}$	(4) $\widetilde{80}$ 82	(23) 53	(14) (21) *10* 17, 41	*32* *34*	*56* *20* *24*	*49* *2* *11* *28* *30* 65

$k_1 = 12$, $k_2 = 7$, $\bar{k}_1 = 5$ and $\bar{k}_2 = 2$ (k_i and $\bar{k}_i$ are relaxed compared to main variant).

objects is S_1 and S_2 do not coincide (the fraction is relative to the total number of objects in S_1 and S_2; each object is counted once). Equivalent and ϵ-equivalent features, like distinctive features, carry significant information of geologic interest. They were not used in voting in order not to bias the vote by repeated contributions by some subsets of objects.

Although 21 parameters indicated in Table VI were considered in the learning stage, only eleven were selected for incorporation in distinctive features. Since we are examining features for their "geological content" we might as well broaden the number, using more parameters in the process. We do this by rerunning the algorithm, relaxing the k_i slightly. We also used objects recognized as D or N rather than points assigned to class I or II in this new learning stage aimed at finding more features. We will next discuss 177 distinctive, equivalent and ϵ-equivalent features, involving all 21 parameters of Table VI, found this way. Table XV summarizes these features.

First consider the left-hand side part. Parameters are indicated in the uppermost row, with notation the same as in Tables V and VI. The second row indicates the thresholds for the parameters. All parameters used in the original learning stage (Table VI) are considered; parameters, which were selected for recognition in this earlier stage (Table VIIA) are marked by an asterisk. The features are divided into groups, separated by thin horizontal lines. Each group consists of one distinctive feature (from Table VIIA) and its equivalent and ϵ-equivalent features. A distinctive feature is indicated by a heavy number at the top of the group. The digit "1" means that the parameter must satisfy the threshold condition in row 2; "0" means that it must not. The numbers to the left of the table indicate the ordinal number of the distinctive feature in Table

TABLE XV

Distinctive features, examined for geological content for points located on faults. Left-hand side displayed in tabular form, right-hand side as flow diagram. (See paragraph 3.6 for explanation.)

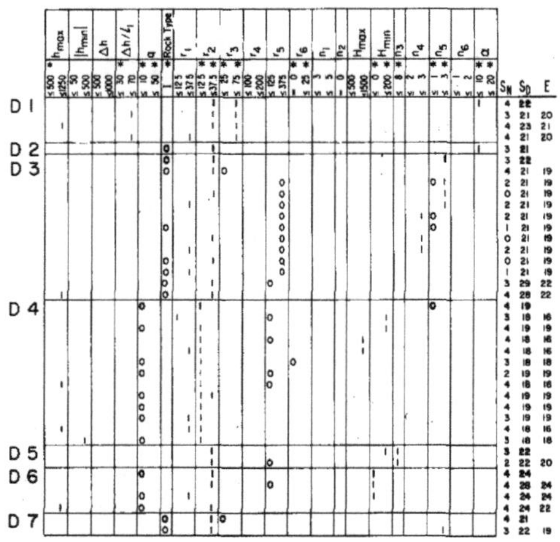

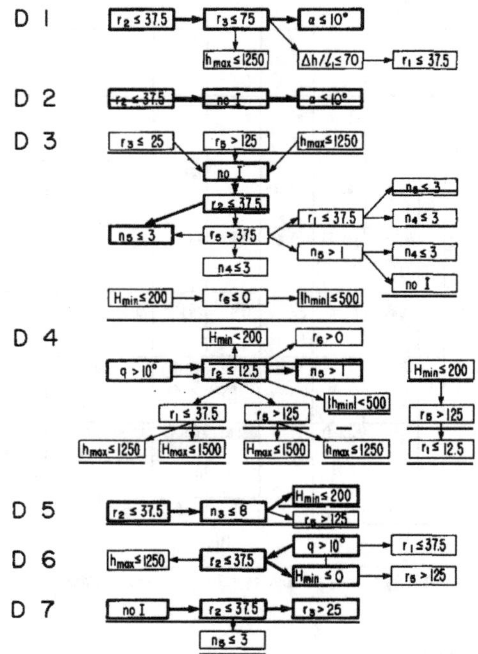

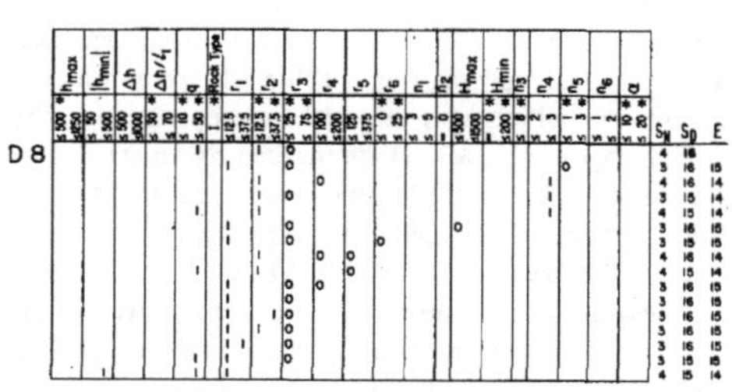

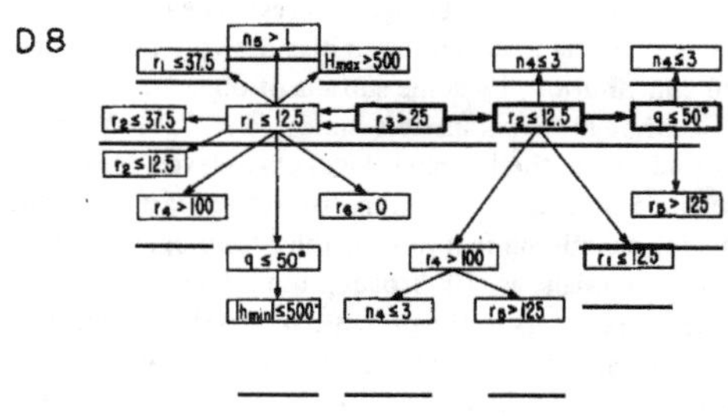

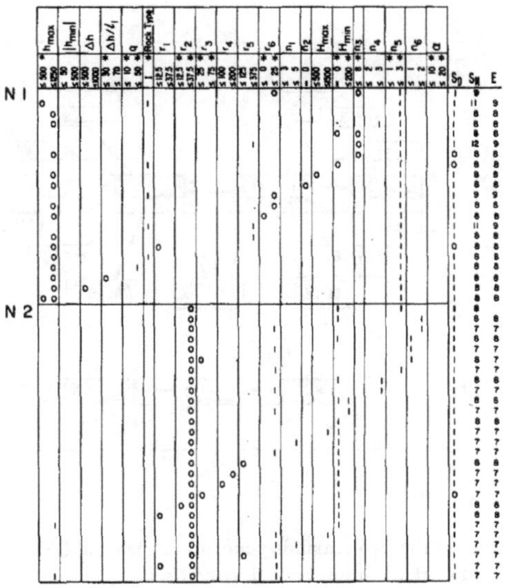

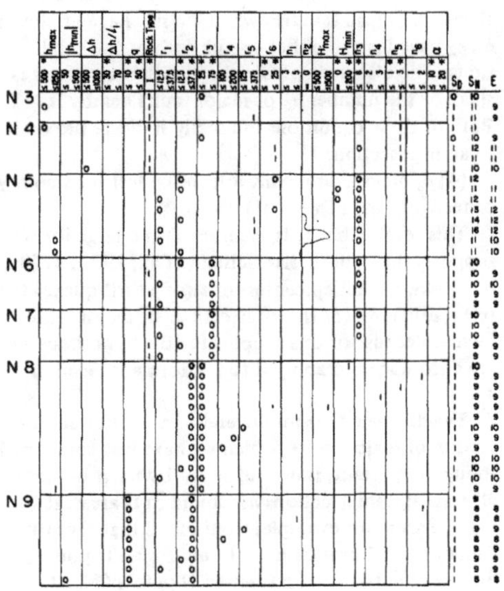

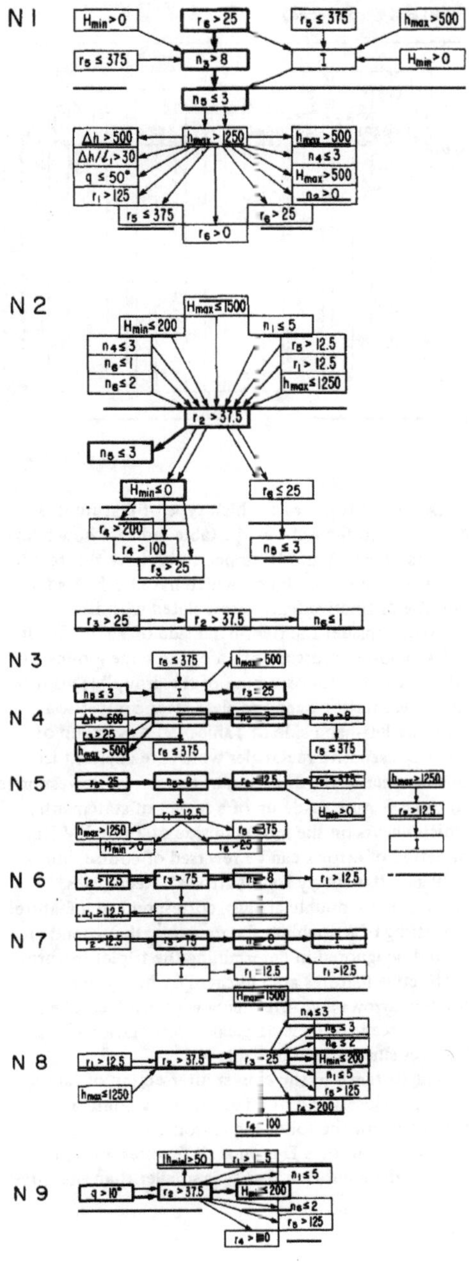
N 1
$H_{min}>0$
$r_6>25$
$r_5\leq 375$
$h_{max}>500$
$r_5\leq 375$
$n_5>8$
I
$H_{min}>0$
$n_5\leq 3$
$\Delta h>500$
$h_{max}>1250$
$h_{max}>500$
$\Delta h/l_1>30$
$n_4\leq 3$
$q\leq 50°$
$H_{max}>500$
$r_1>125$
$n_2>0$
$r_5\leq 375$
$r_6>25$
$r_6>0$
N 2
$H_{max}\leq 1500$
$H_{min}\leq 200$
$n_1\leq 5$
$n_4\leq 3$
$r_5>12.5$
$n_6\leq 1$
$r_1>12.5$
$n_6\leq 2$
$h_{max}\leq 1250$
$r_2>37.5$
$n_5\leq 3$
$H_{min}\leq 0$
$r_6\leq 25$
$r_4>200$
$r_4>100$
$n_5\leq 3$
$r_3>25$
$r_3>25$
$r_2>37.5$
$n_6\leq 1$
N 3
N 4
N 5
N 6
N 7
N 8
N 9

TABLE XV (*continued*)

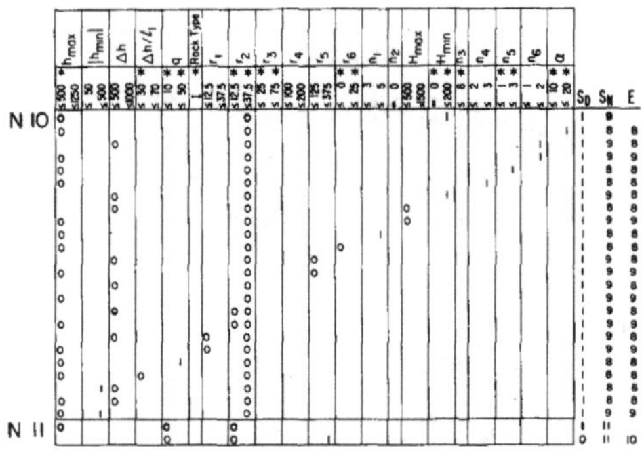

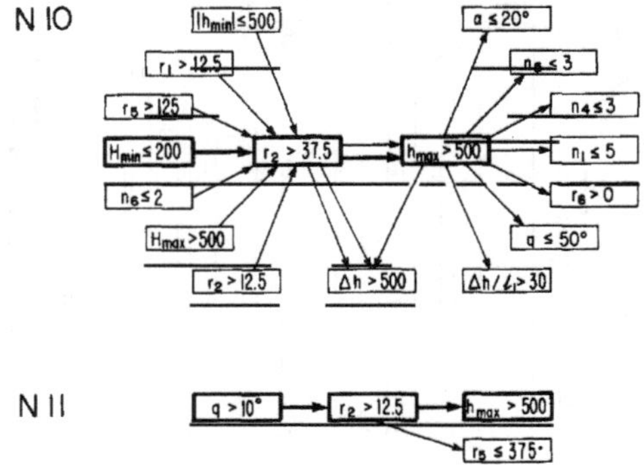

VIIA. D and N indicate which type of feature it is. S_D and S_N to the right of the table indicate how many objects of class I and II, respectively, have this feature. E is the number of objects which have both this feature and the distinctive feature associated with it.

Now consider the right-hand side of Table XV. It shows another representation of the same groups which we call "the structure of the group". This form of representation becomes clear after comparison with the left-hand side of Table XV. Any group of three consecutive rectangles which are adjacent to one another in the direction of the arrows corresponds to some feature made up of a triplet of statements, which occurs on the left-hand side of Table XV (the direction of arrows can be reversed of course, but only simultaneously for all arrows). Rectangles, connected by double arrows, correspond to a feature consisting of a doublet of statements; the second arrow should be ignored in determining the triplet features. Distinctive features are indicated by heavy lines for the two arrows and three rectangles involved. We can now proceed to the geological interpretation of the features summarized in Table XV.

The distance to the closest intersection of faults, or to the closest end of a fault (r_2) is a dominant parameter for the following reasons:

(1) In 6 out of 8 D-groups all features, except one, contain the condition that r_2 is smaller than the threshold; in the two remaining exceptions, D3 and D8, half of the features has this condition.

(2) In 8 out of 11 N-groups at least two features have the condition that r_2 is larger than the threshold.

(3) In the groups D3 and D8, the exceptions mentioned in (1), all features which do not have limitations on r_2 have instead either one or both of the conditions: The distance r_1 from the closest major fault is small or the number n_5 of major faults nearby is > 1. Both of these conditions indirectly indicate proximity to an intersection.

(4) r_2 plays a dominant role in the test for stability of recognition (Table XIII).

This result (that r_2 is small for D and large for N) supports the hypothesis (Gelfand et al., 1972, 1973a, b, 1974a, b) that epicenters of strong earthquakes tend to be situated near the intersections of major lineaments. Because of this support for the hypothesis we are encouraged to apply it to California in Section 4.

The distance r_2 from the end of a fault or an intersection of major faults, which we have just considered, though important, is not sufficient by itself in that not all ends and intersections are found in D areas. It may be seen, for example, from Fig. 2 that the situation $r_2 < 37.5$ km occurs not only for all D points, but also for 14 out of a total 44 N points (Fig. 2, histogram No. 8, 1968+). This is why r_2 must be used only in combination with other parameters in forming features.

The statistics of some of the parameters of D and N features are summarized in Table XVI. The table does not consider r_2 and r_1, which have already been discussed in order to highlight other features. In Table XVI it is interesting to compare features of three types:

D: points which are always close to the end of or an intersection of faults
NC: N points which may be close to the end of or an intersection of faults (r_2 not specified or $\geqslant$12.5 km only)
NF: N points, which are far from the end of and an intersection of faults ($r_2 \geqslant 37.5$ km)

The difference between D and NC is the difference between points D which are certainly near intersections and such points N which may be close to intersections. The difference between D and NF is the difference between D points and points far from ends and intersections, which are always N. We note first of all that some measure of elevation is small for most D points and large for NC points. For NF points the features show both large and small elevations, signifying that no clear dependence on height is established far from intersections.

An examination of geological maps of the region suggests that most of the other characteristics of D features in Table XVI may also be interpreted as an indirect indication of low elevation. These include the absence of igneous rocks, proximity to a large area of recent sediments, and the small number of contacts (n_3) on a geological map. At first sight the last parameter behaves contrary to expectations in that large n should be associated with intensive fracturing of the crust, which is a characteristic for D

TABLE XVI

Summary of the role of different parameters (excluding r_2 and r_1) in D and N features

Parameter		In how many D		In how many NC				In how many NF	
				(with $r_2 > 12.5$ km)		no condition on r_2		(with $r_2 > 37.5$ km)	
		groups	features	groups	features	groups	features	groups	features
Elevation is (h_{max}, or h_{min}, or	small	6	17	–	–	–	–	5	30
H_{max}, or H_{min})	large	1	1	2	5	1	18	2	19
Igneous rocks	present	–	–	3	7	1	5	2	7
(I) are	absent	3	10	–	–	–	–	–	–
q is	small	1	5	–	–	1	1	1	1
	large	2	11	1	2	–	–	1	9
Large water reservoir	close	2	2	–	–	–	–	1	11
(r) is	far	–	–	1	2	1	4	1	1
α is	small	2	2	–	–	–	–	1	1
	large	–	–	–	–	–	–	–	–
Number of contacts (n_3) on	small	1	2	–	–	–	–	–	–
geological map is	large	–	–	3	13	1	4	1	1
Reference point	close	–	–	2	3	1	3	1	2
(r_5) is	far	5	18	–	–	–	–	4	6

(Gelfand et al., 1972, 1973a, b, 1974b). However, only neotectonic fracturing is characteristic of D according to Gelfand et al. (1972, 1973a, b, 1974a, b), whereas n_3 represents the integrated tectonic history. A survey of places with large and small n_3 indicates that small n_3 tend to go with subsidence and low elevations.

The small distance r_6 to a large water reservoir (parameter 12, Table VI) is another characteristic of D. This indirectly implies relatively low elevation, and recalls some modern ideas on the role of water in triggering earthquakes. However, for many points small r_6 indicates proximity to the ocean, so that these interpretations are by no means firm.

Small values of the angle α between the strike of the fault and the dominant strike of the San Andreas system is characteristic of D. This suggests the reasonable notion that the strike-slip earthquakes considered in this paper tend to occur away from bends in the San Andreas fault system. Earthquakes with dip-slip components might be more characteristic of bends in the fault.

Large distance from the reference point (defined by the intersection of the San Andreas and the Big Pine faults) is characteristic of D, perhaps for the reason that earthquakes with dip-slip tend to occur in the Transverse Ranges which are adjacent to a bend in the San Andreas fault.

Other parameters which play a role in D or N features occur. We defer offering possible explanations for them. Among these parameters are the distance r_3 to a geothermal zone, the number n_4 of parallel faults. Parameters which are present in the same way in both D and N features (for example $n_4 \leqslant 3$) are puzzling. Perhaps they were selected as features by chance, due to a random interplay of events, independent of their geological meaning. Perhaps they take on meaning only in combination with the other parameters they link up with, which are different for D and N features.

The qualitative conclusion which emerges is that D areas are characterized by proximity to the end or to an intersection of major faults in association with low relief and often with some kind of downward neotectonic movement, expressed in topography and geology. Apparently, we did not formulate an adequate single parameter for such movements, so that the indication shows indirectly through several different parameters.

Also the fact that minimal and maximal elevations are simultaneously small for many D points suggests that D points for these strike-slip earthquakes are often characterized by relative low relief or subsidence on a background of weak uplift. N points near the intersections or ends of major faults are characterized by higher elevations or uplift with less contrast in relief.

4. Intersections of major lineaments

4.1. The problem

The results of Section 3 (see paragraph 3.6) suggest the hypothesis that the epicenters of strong earthquakes in California are associated with intersections of major lineaments (Gelfand et al., 1972, 1973a, b, 1974a, b; Briggs and Press, in preparation). More specific formulation of this hypothesis as well as the definition of lineaments is given in the Appendix. The major lineaments of California and adjacent regions, based on a synthesis of published data (Atwood, 1940; Richter, 1958; USGS, no date, 1962, 1965; Wright and Frey, 1965; Cook, 1966; Hamilton and Myers, 1966; Hill, 1966; Thompson, 1966; Dickinson and Grantz, 1968; King, 1969; Hain, 1971), are shown in Fig. 4. Explanation and justification is provided in Ranzman (in preparation). The transverse lineaments in Fig. 4 may raise some doubts, since they are to a large extent based on geomorphic evidence. We believe that ERTS photographs of the region support these interpretations (Ranzman, in preparation).

The objects of recognition, according to the hypothesis, are now the intersections of the lineaments. The problem, formulated exactly as in paragraph 2.1, is to find such intersections (D_I and D_{II}) near which the epicenters of strong earthquakes may occur. However, new objects are defined. A larger area is considered, and dip-slip and strike-slip earthquakes are not distinguished. The last difference is not significant, since the area contains only two strong dip-slip earthquakes (1952, near intersection *165*; and 1971, near intersection *119*; the first epicenter is near that of a strong strike-slip earthquake). In the numerical experiments (paragraph 4.5) we eliminated these two earthquakes from learning. The distinctions between recognition of San Andreas and Basin—Range earthquake sources are considered in Briggs and Press (in preparation).

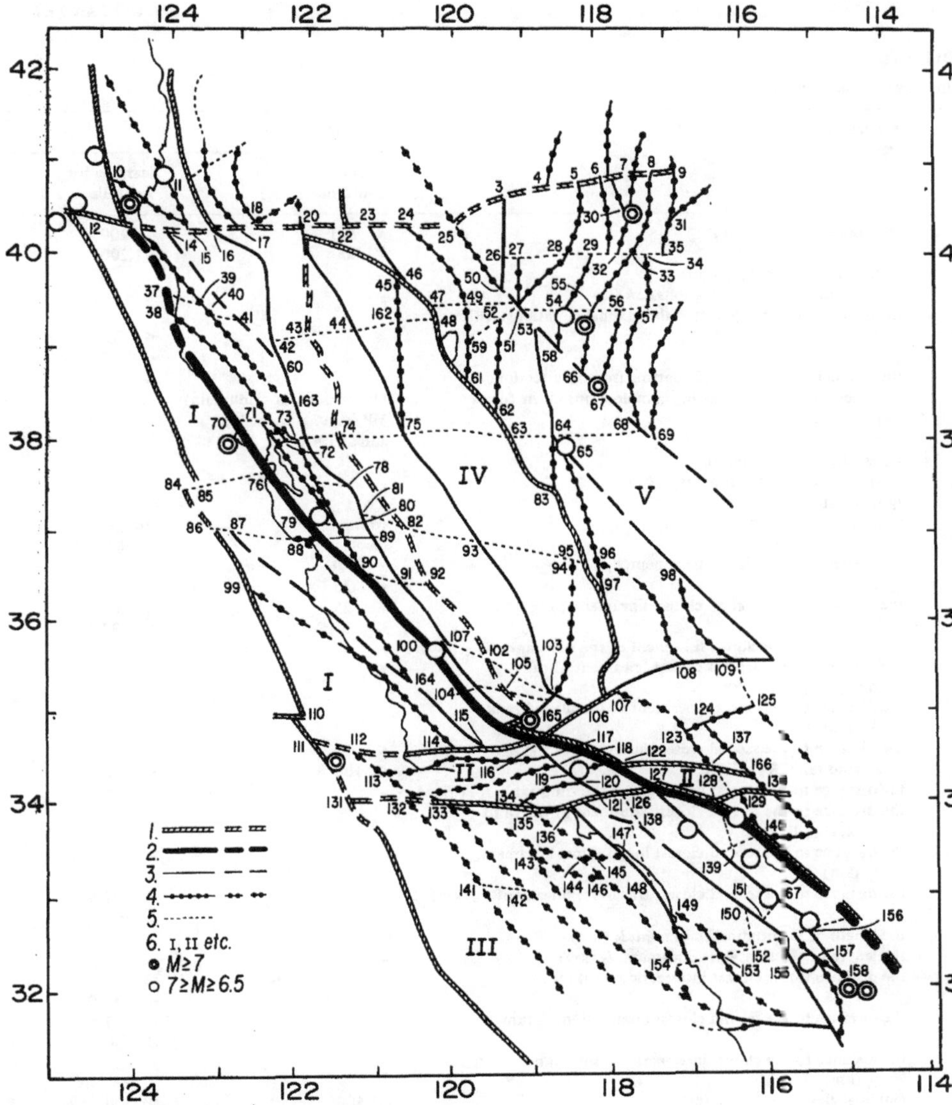

Fig. 4. Major lineaments of California (for definitions, see Appendix).
(1) Lineaments of first rank (boundaries of morphostructural countries).
(2) Major branch of San Andreas system; solid segments are reasonably certain, dashed segments are uncertain, underwater or buried under soft sediments.
(3) Lineaments of second rank (boundaries of morphostructural countries) megablocks.
(4) Lineaments of third rank (boundaries of morphostructural countries) blocks.
(5) Transverse lineament.
(6) The number of morphostructural country in the description of the scheme (second paragraph of Appendix); epicenters are shown as in legend.

TABLE XVII

Parameters of intersections

N	Parameter	Parameterization: learning for California	Parameterization: learning for Anatolia
1	absolute elevation, h_1 (m)	$\leqslant$ 200 $\leqslant$1,200	$\leqslant$ 200 $\leqslant$1,200
2*	maximal elevation, h_{max} (m)	–	
3*	maximal difference of elevations, Δh (m)	–	
4*	the distance between points with maximal elevations, l_1 (km)	–	
5*	$\Delta h/l_1$	–	$\leqslant$ 75 $\leqslant$ 125
6*	the number of lineaments, departing from intersection, n_1	–	$\leqslant$ 3
7*	morphology (most contrasting combinations in the forms of relief)	MM (mountain–mountain) MP (mountain–plain) and all the rest	
8*	types of tectonic structures	–	
9*	relative area of soft sediments, q (%)	–	
10	type of intersection	x or $\perp$	
11	how pronounced is the lineament	strong feeble	strong feeble
12	the distance to the closest lineament, r_1 (km)	= 0 $\leqslant$ 25	= 0
13	the distance to the second closest lineament, r_2 (km)	$\leqslant$ 25 $\leqslant$ 37.5	$\leqslant$ 25 $\leqslant$ 37.5
14	the distance to the closest lineament of the first rank, $R_{1.1}$ (km)	–	
15	the distance to the second closest lineament of the first rank, $R_{1.2}$ (km)	–	
16	the distance to the closest lineament of the second rank, $R_{2.1}$ (km)	–	
17	the distance to the second closest lineament of the second rank, $R_{2.2}$ (km)	$\leqslant$ 50 $\leqslant$ 100	$\leqslant$ 50 $\leqslant$ 100
18	the distance to the closest lineament of the third rank, $R_{3.1}$ (km)	–	
19	the distance to the second closest lineament of the third rank, $R_{3.2}$ (km)	–	
20	the distance to the second closest longitudinal lineament, r_3 (km)	–	= 0 < 25
21	the distance to the second closest transverse lineament, r_4 (km)	–	< 62.5 < 100
22	the length of longitudinal lineament, l_2 (km)	–	
23	the length of the transverse lineament, l_3 (km)	–	
24	the distance to the closest intersection, r_5 (km)	= 0 $\leqslant$ 37.5	= 0 $\leqslant$ 37.5
25	the distance to the second closest intersection, r_6 (km)	–	$\leqslant$ 37.5 $\leqslant$ 50
26	the distance to the closest intersection inside some cluster, r_7 (km)	–	
27	maximal elevation, H_{max} (m)	$\leqslant$1,500	$\leqslant$1,500 $\leqslant$3,000
28	minimal elevation, H_{min} (m)	$\leqslant$ 100	
29**	maximal difference of elevations, ΔH (m)	–	$\leqslant$2,000 $\leqslant$3,000
30**	the distance between points with maximal and minimal elevations, l_4 (km)	–	

TABLE XVII (continued)

N	Parameter	Parameterization	
		learning for California	learning for Anatolia
31**	$\Delta H/\ell_4$	≤ 40	
32**	relative area of soft sediments, Q_2 (%)	–	
33**	morphology	–	
34**	the number of intersections, n_2	≤ 4 ≤ 7	≤ 4 ≤ 7
35***	the weighted rank of lineament, M	≤ 15 ≤ 17	≤ 15 ≤ 17
36	the distance from San Andreas fault, l_5 (km)	–	–

Thresholds are indicated only for parameters used in learning.

* Parameter is determined in a circle of 12.5-km radius around intersection.

** Parameter is determined in a circle of 62.5-km radius around intersection.

*** $M = \sum_{j=1}^{n} (m_i/62.5)(62.5 - R_{ij})$ where i is the rank of the lineament; m_i is the weight of the rank ($m_1 = 5, m_2 = 4, m_3 = 3$); R_{ij} is the distance to the jth lineament of the rank i; n is the number of lineaments in the circle of 62.5-km radius around the intersection.

We apply recognition to the land part of California – the territory between the latitude of the Mendocino fault on the north, the Mexican border on the south and the axis of Great Valley on the east. In Fig. 4 this territory is bounded by lineaments *76–13–20–106–107–123–125* and the oceanic shore to the south of intersection *76*.

4.2. The data

4.2.1. Definitions

As in Section 3, we use for recognition only certain parameters which roughly characterize tectonic fracturing of the crust or the intensity and degree of contrast of neotectonic movements. This selection does not imply lack of significance to other factors, such as seismicity, geophysical anomalies, tectonic history, the structure of the crust, etc. The parameters used for recognition are listed in Table XVII. Parameters 6, 11, 21, 24–26, 34 and 35 characterize tectonic fracturing. Parameters 1–5, 7, 9 and 27–33 show the intensity and degree of contrast of the vertical neotectonic movements. Parameter 10 characterizes the presence of strike-slip faults. Some additional comments follow.

4.2.2. Conditions on lineaments

All lineaments are continuous and may change only at intersections; if the rank of a lineament changes at an intersection we consider it as two lineaments; if the strike of a lineament changes by more than ν degrees we consider it as two lineaments (in practice we took $\nu = 15°$).

4.2.3. Separate intersections

If the distance between intersections was less than δ we replaced them by one intersection at the midpoint (in practice we took δ = 12.5 km or 5 mm on a map with scale 1:2,500,000; not more than three intersections were merged by this rule).

4.2.4. Vicinity of intersections

Parameter 1 is determined directly at the point of intersection. Parameters 2–11 are determined within the circle of radius ϵ_1 = 12.5 km with a center at the intersection. Parameters 27–35 are determined in a larger circle of radius ϵ_2 = 62.5 km.

4.2.5. Continuity of parameters

Since the lineaments are drawn approximately, all

parameters are defined in such a way that they are continuous with respect to displacement of lineaments. They should also be continuous in the arbitrary parameters ϵ_1 and ϵ_2. Parameters 12–19 and 24–34 may become discontinuous if an intersection is formed by three or more lineaments, since after the displacement of any lineament several new intersections will appear and the considered characteristics will change sharply. To preserve the continuity we specify the definition of these parameters in the following way. Let us imagine that we displace the axes of lineaments slightly in such a way that they will not be parallel, will intersect only in pairs and that a T-intersection will not become a crossing. Each parameter converges to some limit when this displacement decreases. This limit is taken as the value of the parameter. In practice it is equivalent to the following rules (Table XVII):

Parameter numbers $N = 12$, 13 and 24: $r_1 = r_2 = r_5 = 0$ if more than two lineaments depart from intersection.
$N = 34$: $n_2 = 1, 3$ and 6 if the number of intersecting lineaments is 2, 3 and 4, respectively.
$N = 25$: $r_6 = 0$ if more than four lineaments intersect.
$N = 14$, 16 and 18: $R_{i,1} = 0$ if a lineament of rank i goes through the intersection.
$N = 15$, 17 and 19: $R_{i,2} = 0$ if more than one lineament of rank i goes through the intersection.

For example intersection *116* has $r_1 = r_2 = 0, n_3 = 6$; intersection *143* has $r_1 = 0, r_2 \neq 0, n_3 = 2$.

4.2.6. The length of lineaments (parameters 22 and 23).

Suppose that all the longitudinal lineaments are placed in order of rank, with highest rank first. Lineaments of the same rank are placed in order according to their length, with greatest length first. After that ordering the first length will be l_2; l_3 is specified similarly.

4.2.7. One-dimensional distributions

The parameters specified in the preceding section can be determined from the scheme of lineaments (Fig. 4) and topographic and tectonic maps. Their one-dimensional distributions are shown in Fig. 5. Dashed lines correspond to the intersections closest to epicenters; solid lines correspond to all the remaining intersections. We see that the histograms for the

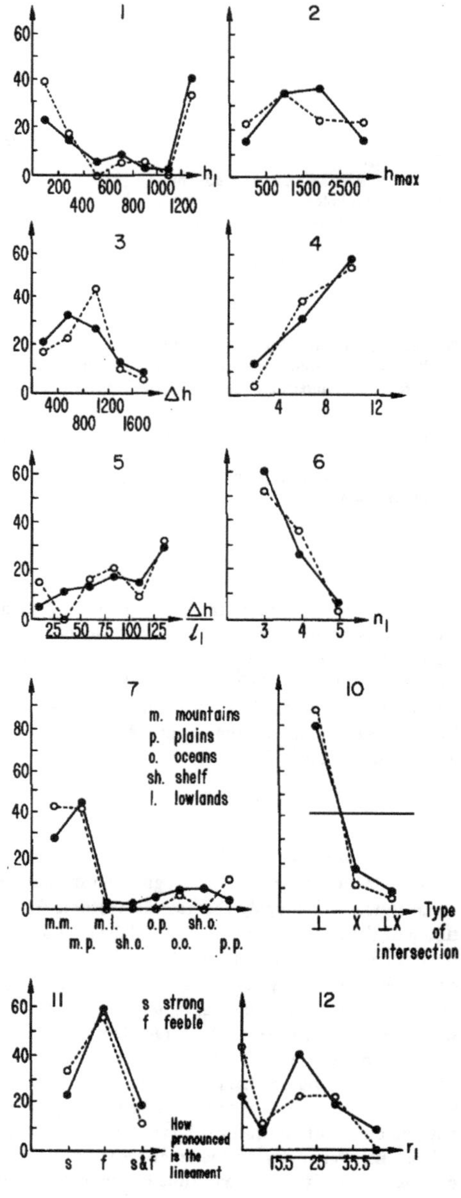

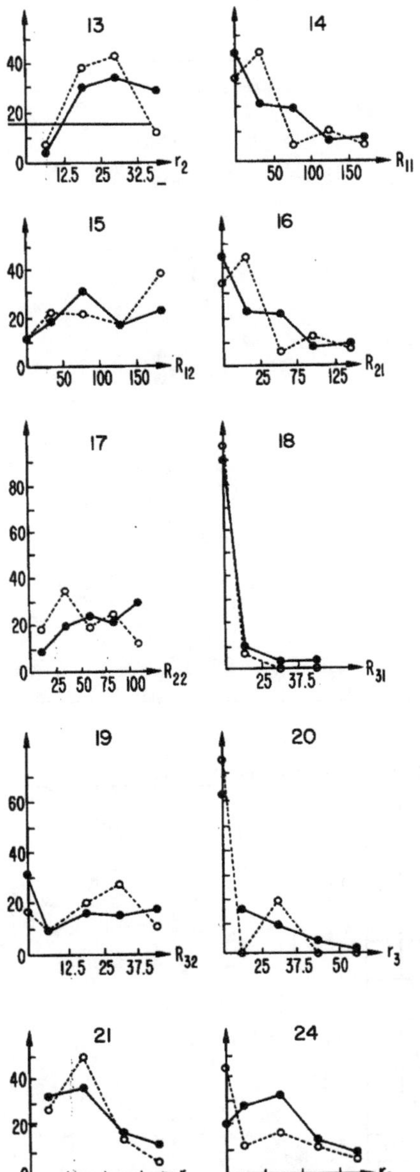

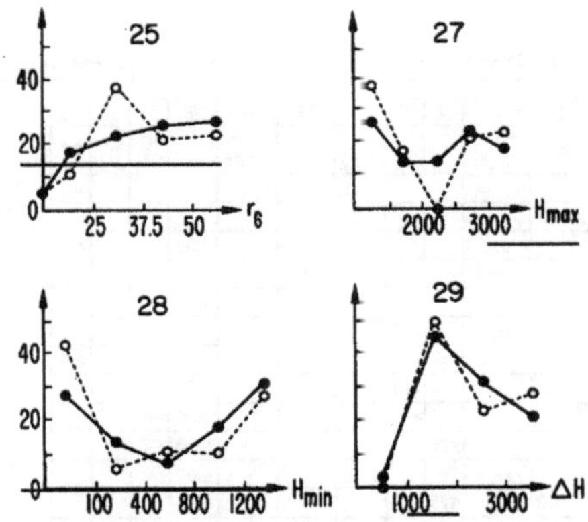

Fig. 5. Examples of one-dimensional distributions of parameters for intersections of lineaments shown in Fig. 4. The number and notation of parameters correspond to Table XVII. Otherwise, the notation is the same as in Fig. 2.

two groups are very similar for parameters 2–4, 7–9, 14–16, 18, 19, 22, 23, 28 and 30–33. These parameters were not used. The parameter l_5, distance from the San Andreas fault, was not selected for the opposite reason that it would dominate the characteristic features; the histograms separate too well. Since most known epicenters are close to this fault, the results of recognition would be trivial. Discrete values were assigned to the remaining parameters as shown in the third column of Table XVII.

Fig. 5 contains the histograms for the whole area shown in Fig. 4. Histograms based only on California data separate better, allowing thresholds such that the two groups have few or no common values of some parameters. This is undesirable for the reason indicated in paragraph 3.3; this is why we assumed thresholds corresponding to the data for the entire area in Fig. 4 although recognition was carried out for only a portion of the area.

4.3. Results

Algorithm CLUSTERS was used. Intersections within 50 km of an epicenter formed a cluster. The division of intersections between the clusters and classes II and III in learning is indicated in Table XIX.

TABLE XVIII

Distinctive D and N features for points located at intersections of lineaments

		1	7	12	13	17	24	27	28	31	34	35
		h_1	Morphology	r_1	r_2	$R_{2,2}$	r_5	H_{max}	H_{min}	$\Delta H/l_4$	n_2	M
D	1					>50						>17
	2	≤200										>17
	3		m,m						≤100			>15
	4		no m,m				≤12.5					>15
	5	>200					≤12.5					>15
	6		m,p		>25							>15
	7		m,m	≠0								>15
	8				>25				≤100		>4	
	9				>25				≤100	>40		
	10			≤25					≤100	>40		
	11		no m,m			≤50			≤100			
	12	≤200		≤25					≤100			
	13			≤25	>25			≤1500				
N	1							>1500				≤15
	2					>50					≤4	≤17
	3							>1500	>100			≤17
	4					>50		>1500				≤17
	5		no m,p				≤37.5					≤17
	6		no m,p			≤100						≤17
	7			≤25			≤37.5					≤15
	8	≤1200	no m,p									≤15
	9							>1500			≤4	
	10						≤37.5				≤4	
	11					≤100				≤40	≤4	
	12	≤1200	no m,m								≤4	

m - mountains   p - plains

TABLE XIX

Results of voting for points at intersections of lineaments in California (see Table VIII for definitions of symbols)

n_N \ n_D	0	1	2	3	4	5	6	≥7
≥10	$\widetilde{40}$, $\widetilde{123}$, $\widetilde{124}$, $\widetilde{134}$, 145, $\widetilde{148}$, $\widetilde{149}$	$\widetilde{92}$						
9								
8								
7		$\widetilde{37}$, $\widetilde{125}$, $\widetilde{154}$						
6	$\widetilde{91}$	$\widetilde{38}$, $\widetilde{114}$						
5	(10) $\widetilde{41}$, $\widetilde{42}$, $\widetilde{43}$, $\widetilde{82}$							
4	$\widetilde{39}$, 60, 135, 144, $\widetilde{153}$	143, 146	$\widetilde{107}$					
3	141, $\widetilde{147}$ (166)	$\widetilde{132}$	$\widetilde{102}$, $\widetilde{104}$, $\widetilde{115}$	(100)		*158*		
2	$\widetilde{74}$ (150)	$\widetilde{17}$, $\widetilde{152}$ (164)						
1	142	(88) 133	$\widetilde{90}$, $\widetilde{130}$	137				
0	(70) $\widetilde{76}$ (78) (79) $\widetilde{106}$, $\widetilde{121}$	$\widetilde{15}$ (118) (119) (120)	(14) $\widetilde{18}$, $\widetilde{36}$, 138 $\widetilde{127}$ (136) (89)	*13* $\widetilde{72}$, $\widetilde{126}$ (151) 163 (167)	$\widetilde{16}$ 168 *80* (81) *101* (116) *117* *122* *77*	$\widetilde{20}$ *71* $\widetilde{73}$ (103) (105) (128) (129) (139)	*155* *156* *157* *165*	*140*

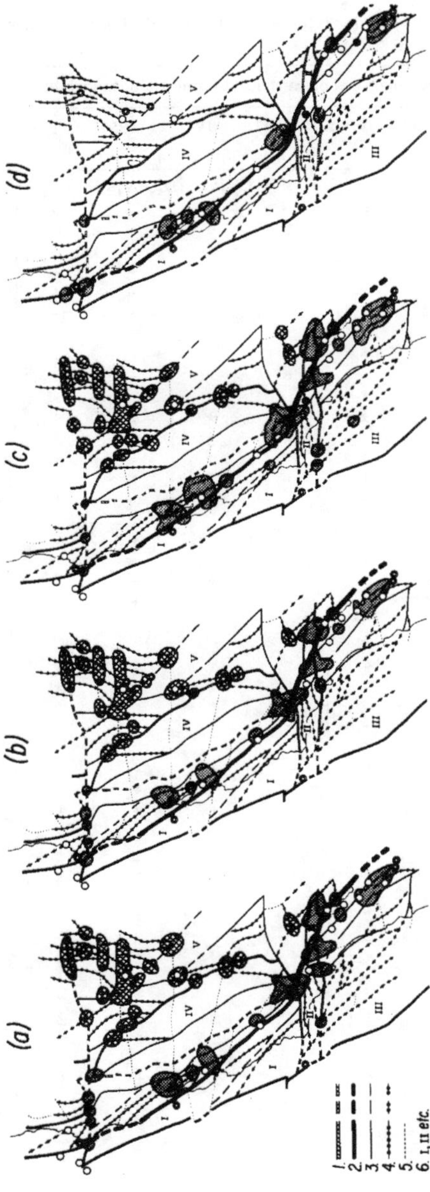

Fig. 6. Results of pattern recognition for intersections of major lineaments of California.
Dotted shading surrounds areas recognized as D for California. Crossed shading surrounds areas recognized as D for Sierra Nevada and adjacent parts of Basin and Range.
(a) Experiment EF.
(b) Main variant; all known earthquakes up to 1974 are used in learning.
(c) Experiment EH: all known earthquakes up to 1948 are used in learning.
(d) Experiment EH: all known earthquakes up to 1940 are used in learning.

TABLE XX

Recognition for control experiments EH and EF. The notation is the same as in Table IX. Uncertain ("neutral") voting: ± corresponds to $\Delta = \bar{\Delta}$; $\underline{o}$ corresponds to $\Delta = \bar{\Delta} - 1$

Number of intersection on Fig. 4	Year of earthquake which defines cluster	Learning based on earthquakes from 1836 through the year							
		1934	1940	1942	1948	1952	1968	1971	1973+...
					Clusters $\{D_1 + N_1\}$:				
119		o	o	o	o	o	o	o	o
120		o	o	o	o	o	o	o	+
118	1971	o	o	o	o	o	o	o	o
136		+	+	o	+	+	+	+	+
117		o	o	o	+	+	+	+	+
122		o	o	o	+	+	+	+	+
140	1968	o	+	+	+	+	+	+	+
139		o	o	o	+	+	+	+	+
165		o	+	+	+	+	+	+	+
116		o	o	o	±	+	+	+	+
103	1952	o	o	o	±	+	+	+	+
105		o	+	+	+	+	+	+	+
117		o	o	o	+	+	+	+	+
128		o	o	o	±	+	+	+	+
129		o	o	o	±	+	+	+	+
139	1948	o	o	o	+	+	+	+	+
140		o	+	+	+	+	+	+	+
166		o	o	o	o	o	o	o	o
167		o	o	+	+	+	+	+	+
151	1942	o	o	+	+	+	+	+	+
150		o	o	o	±	o	$\underline{o}$	o	$\underline{o}$
155		o	+	+	+	+	+	+	+
156		o	+	+	+	+	+	+	+
157	1940	o	+	+	+	+	+	+	+
155		o	+	+	+	+	+	+	+
158	1934	+	+	+	±	±	±	±	o
157		o	+	+	+	+	+	+	+
13		±	±	+	+	+	+	±	±
10	1923	±	±	o	o	o	o	o	o
14		$\underline{o}$	o	o	$\underline{o}$	±	±	±	o
101		$\underline{o}$	o	±	+	+	+	+	+
164	1922	+	±	$\underline{o}$	±	$\underline{o}$	±	o	o
100		o	o	o	$\underline{o}$	o	$\underline{o}$	o	o
158	1915	+	+	+	±	±	±	±	o

TABLE XX (*continued*)

Number of intersection on Fig. 4	Year of earthquake which defines cluster	Learning based on earthquakes from 1836 through the year							
		1934	1940	1942	1948	1952	1968	1971	1973+...
80		±	±	+	+	+	+	+	+
79		o̲	o̲	o	±	o̲	±	o	o
89		±	±	+	+	+	+	±	+
88	1911	±	±	±	+	±	+	o	o
81		o̲	o	+	+	+	+	+	+
77		o̲	±	+	+	+	+	+	±
78		o̲	o̲	o̲	±	o̲	±	o	o
70	1906	o̲	o̲	o̲	±	o̲	±	o	o
71		+	+	+	+	+	+	+	+

Number of intersection on Fig. 4	Year of earthquake which defines cluster	Learning based on earthquakes from 1836 through the year						
		1940	1947	1948	1954	1954	1959	1973+...
				Class II $\{D_2 + N_2\}$:				
3		o	+	+	o	o	o	o
5		±	+	+	+	+	±	+
6		+	+	+	+	+	+	+
8		+	+	+	+	+	±	+
9		o	±	±	+	±	o	o
25		o	o̲	±	+	±	o	+
26		o	+	+	+	+	+	+
27		o	+	+	+	+	+	+
46		+	+	+	+	±	±	±
50		+	+	+	+	+	±	+
57		+	+	+	+	+	±	+
59		+	±	+	+	+	±	+
61		o	+	+	+	o	o	o
68		+	+	+	+	+	±	+
69		o	+	+	+	+	±	+
94		o	±	o	o	o	o	o
95		+	+	+	+	+	±	+

The remaining intersections were recognized only as N.

The characteristic features are shown in Table XVIII.

The results of voting are shown in Tables XIX and XX, and in Fig. 6b.

The threshold for recognition was chosen as follows: $\bar{\Delta} = \Delta^*$ with Δ^* having the same definition as in paragraph 3.4.

We note that the recognition is successful in that most of the intersections are divided by voting into two rather distinctive groups, with only a small number of intersections given a "neutral" vote. Each cluster has at least one D intersection (Table XX). The majority of D and N intersections belong to class I (clusters) and to class II, respectively. The results are also encouraging for the following reasons.

(1) Non-instrumental data for the 19th century

show four earthquakes in the area of interest with estimated magnitude 7 or more. Epicenters of all these earthquakes are close to intersections recognized as D.

(2) Among D are the intersections closest to epicenters, with only two exceptions. One of the exceptions is the San Francisco earthquake of 1906. Intersection *70* which is closest to the 1906 epicenter was recognized as N; however, intersection *71*, which is slightly farther away, was recognized as D. The second exception is the epicenter of the 1971 earthquake, which is of the dip-slip variety. The closest intersection *119* is recognized as N; intersection *136* is recognized as D in the same cluster.

4.4. Interchangeability of distinctive features from different regions

Let us apply to California the distinctive features obtained for Anatolia, and vice versa. In this way we check our ability to recognize intersections near which epicenters of strong earthquakes actually fall, using criteria developed elsewhere. This is an important control experiment, since the features in the two

TABLE XXI

Distinctive features for recognition in Anatolia (Gelfand et al., 1974a)

		1	5	6	10	11	12	13	17	20	21	24	25	27	29	34	35
		h_1	$\Delta h / l_1$	n_1	inter-section	lineament strength	r_1	r_2	$R_{2,2}$	r_3	r_4	r_5	r_6	H_{max}	ΔH	n_2	M
D	1											12.5-37.5					>9
	2			=3								>12.5					>9
	3					s or f										>2	
	4				$\perp$											>2	
	5								≤100				≤100			>2	
	6				$\perp$				≤100				≤100		≤3000		
	7									≤50					≤3000		
	8								≤100				≤100		>2000		
	9									>12.5			≤100	≤3000			
	10								≤100				≤100	>2000			
	11								≤100			>12.5	≤100				
	12		>150									>12.5	≤100				
	13					s or f				≤50		>12.5					
N	1			>3		s and f										≤4	
	2	>200								>50						≤2	
	3											>37.5			>3000		
	4							>25						>3000	>2000		
	5					f							>62.5	>3000			
	6						>25	>62.5						>2000			
	7					s and f							>62.5				
	8	>200								>50			>62.5				
	9	>200						>62.5					>62.5				
	10	>200											>62.5				
	11	>200								>50		>37.5					
	12	>1000	≤300			f						>37.5					
	13					f	>25		≤100								
		1000 ↓ 1200	150 300 ↓ ↓ 75 125				25 ↓ 12.5	62.5 ↓ 37.5					62.5 100 ↓ ↓ 37.5 50	2000 ↓ 1500		2 4 ↓ ↓ 4 7	9 ↓ 15

s = strong f = feeble

regions were obtained completely independently. Success would have important geological implications.

The characteristic features, obtained in Gelfand et al. (1974a, b) for Anatolia, are shown in Table XXI. They were applied to California with one natural exception: the parameter thresholds were reset to values appropriate for California.

The results of voting, shown in Fig. 7c, are quite good. We first note that the intersections in California receive many votes; the average ($n_D + n_N$) is 8, despite the fact that the distinctive features were derived from Anatolian data. The intersections are distinctly divided by voting into two groups, so that recognition is stable with respect to the choice of $\bar{\Delta}$.

Second, with $\bar{\Delta} = 0$ (the same threshold used in Gelfand et al. (1974a, b), for Anatolia) we recognize as D all intersections closest to each epicenter, even number *70* near the 1906 epicenter.

Third, we recognize as D all the same intersections, as in paragraph 4.3 as well as a few additional ones.

We can reverse the procedure and apply to Anatolia the characteristic features found for California (Table XVIII). Again, we changed the thresholds assigned to the parameters. The results are analogous. The intersections have many votes and are distinctly divided by voting. With $\bar{\Delta} = 0$ we recognize the intersections near each known epicenter (Fig. 8) except one (near intersection *33*). D areas are in good agreement with those found in Gelfand et al. (1974a, b), where learning was based on Anatolian data.

The ability to transfer the criteria of high seismicity from Anatolia to California seems natural, since most of the Anatolian earthquakes are connected with strike-slip movements, as is the case in California. However, the complete recognition of intersections near all known epicenters and the good agreement with results based on Californian data is astonishing and implies some generality to the distinctive features which were found. This particular control experiment argues against the criticisms that the pattern recognition results were fortuitous, or simply a restatement of the idea that future epicenters occur near previous ones.

For the Sierra Nevada and Basin and Range (regions *IV* and *V* in Fig. 4) the Anatolian criteria give negative results in that no intersections are recognized as D in the vicinity of 2 out of 5 epicenters. This failure is expected since the eastern California–Nevada regions are dominated by dip-slip focal mechanisms. This result underlines the success for California.

4.5. Control experiments

Experiment EH (Figs. 6c, d and Table XX) shows quite satisfactory results. Learning from earthquakes through the year 1940, we would have recognized at least one dangerous intersection in the cluster around the epicenter of each subsequent strong earthquake. After the 1942 earthquake, we would miss only one cluster, namely that corresponding to the 1971 earthquake. This cluster would be recognized by learning on all shocks up to and including the 1948 earthquake. To save time we did not redetermine the histograms of parameters at each step as we did in paragraph 3.5. The parameters and their thresholds were the same for each step. Therefore the results of this experiment only illustrate the stability of recognition with respect to the learning material.

Experiment EF (Fig. 6a) also shows the stability of recognition in that only four more intersections were recognized as D (*120*, *1*[illegible], *130* and *133*); and two intersections were transferred to N (*14* and *158*).

To test the sensitivity of results to the particular algorithm used, CORA-3 was applied. Class I contained the intersections closest to an epicenter, class II included all the rest. The threshold $\bar{\Delta} = 0$ was chosen. The results of recognition are shown in Fig. 9b. They basically confirm the results of paragraph 4.3. Experiment EH was successful up to 1942.

In another test we assumed other thresholds for parameters, obtained from histograms based on California data only. The results of recognition by algorithm CLUSTERS are in satisfactory agreement with paragraph 4.3. The control experiment results were good, though somewhat degraded. Thus our results are stable to changes in the thresholds, but the thresholds assumed in paragraph 4.3 are preferable.

4.6. Clarifying experiments

4.6.1. Epicenters or total fault breaks?

Does the epicenter have some special meaning, or is it just a random point on a fault break, say the place where by chance the stress first reaches the critical fracture strength? To check this we replaced the epicenters by the intersections on the entire fault

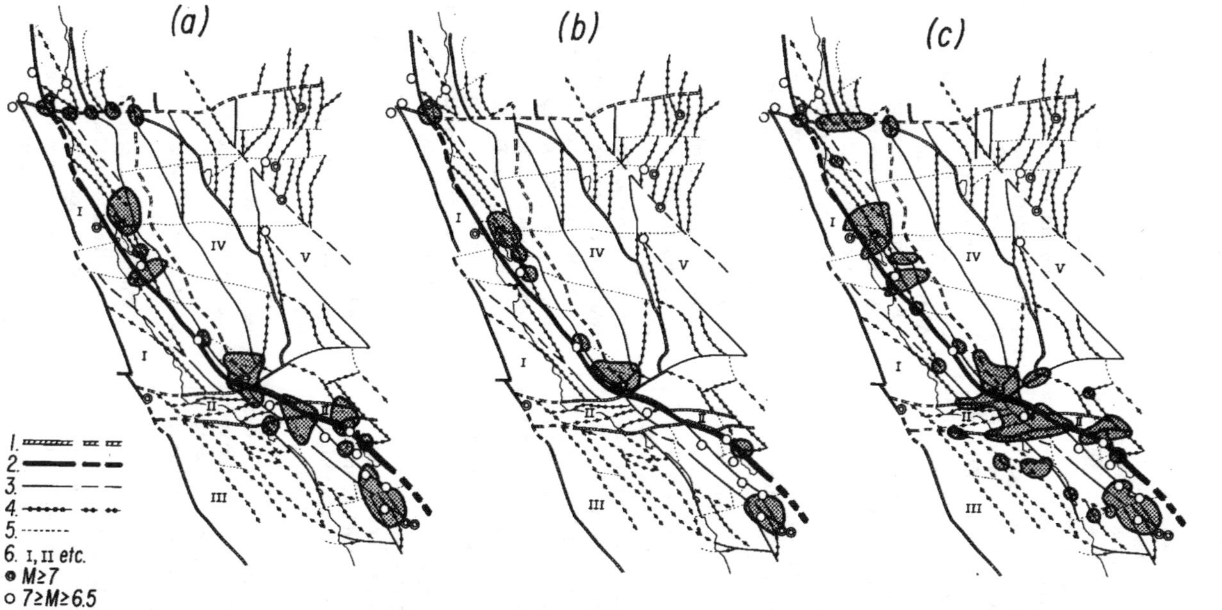

Fig. 7. Comparison of D areas for intersections of major lineaments for California, using different learning materials. Notation is the same as in Fig. 4.
(a) Main variant; same as Fig. 6b (see paragraph 4.3).
(b) Two dip-slip earthquakes eliminated (see paragraph 4.6).
(c) Learning for Anatolia (Table XXI).

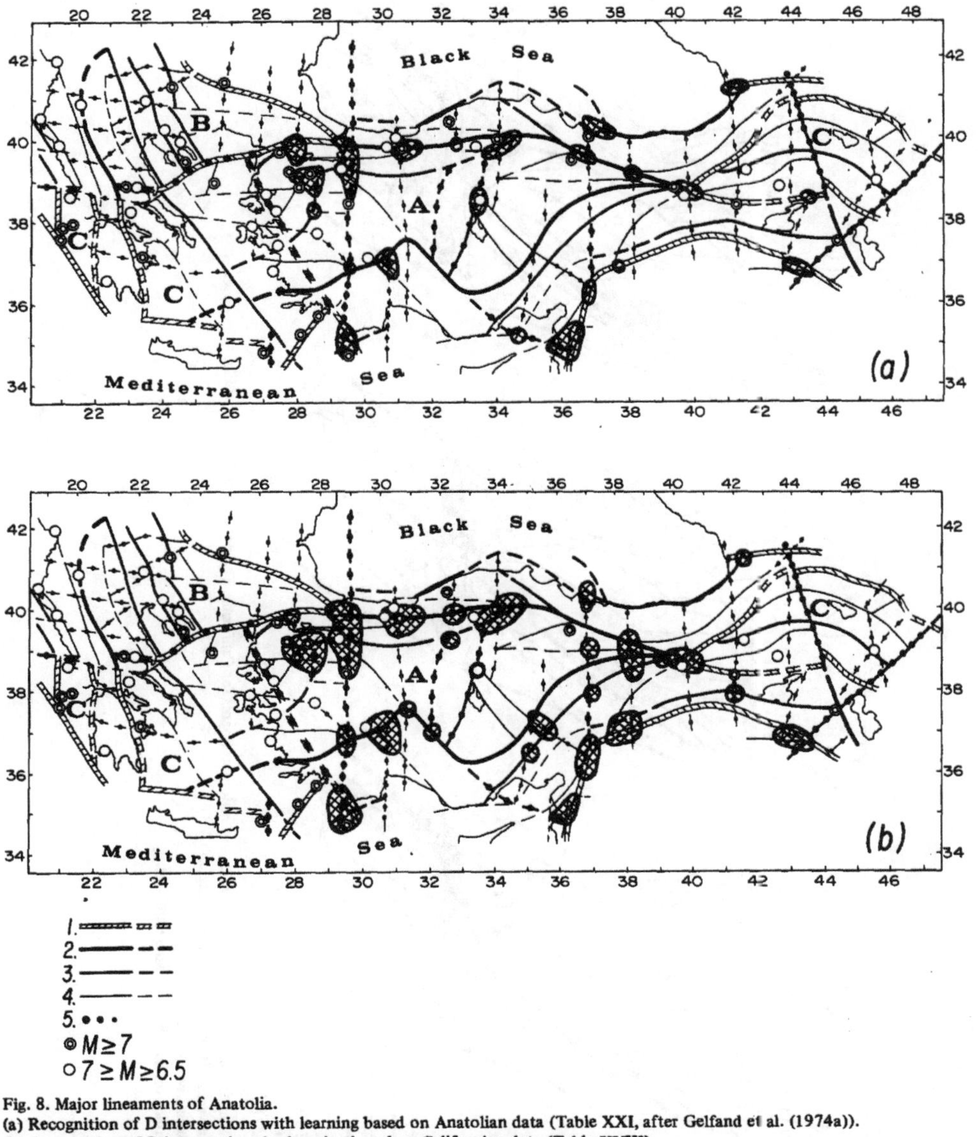

Fig. 8. Major lineaments of Anatolia.
(a) Recognition of D intersections with learning based on Anatolian data (Table XXI, after Gelfand et al. (1974a)).
(b) Recognition of D intersections by learning based on Californian data (Table XVIII).
Legend: lineaments (after Gelfand et al., 1974a): *1* = bounding the entire region (rank 1); *2* = bounding morphostructural countries (rank 1); *3* = bounding megablocks (rank 2); *4* = bounding blocks (rank 3); *5* = transverse.

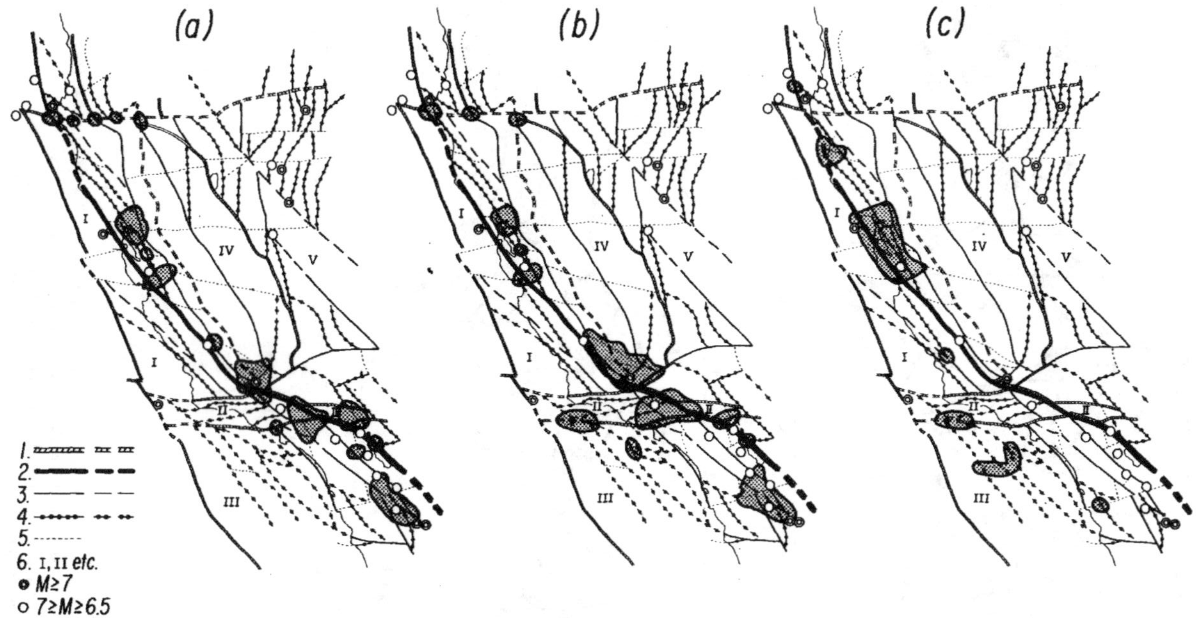

Fig. 9. Comparison of D areas for intersections of major lineaments for California. The notation is the same as Fig. 4.
(a) Main variant (same as Fig. 6b, see paragraph 4.3).
(b) Learning by CORA-3.
(c) Learning on fault breaks (paragraph 4.6).

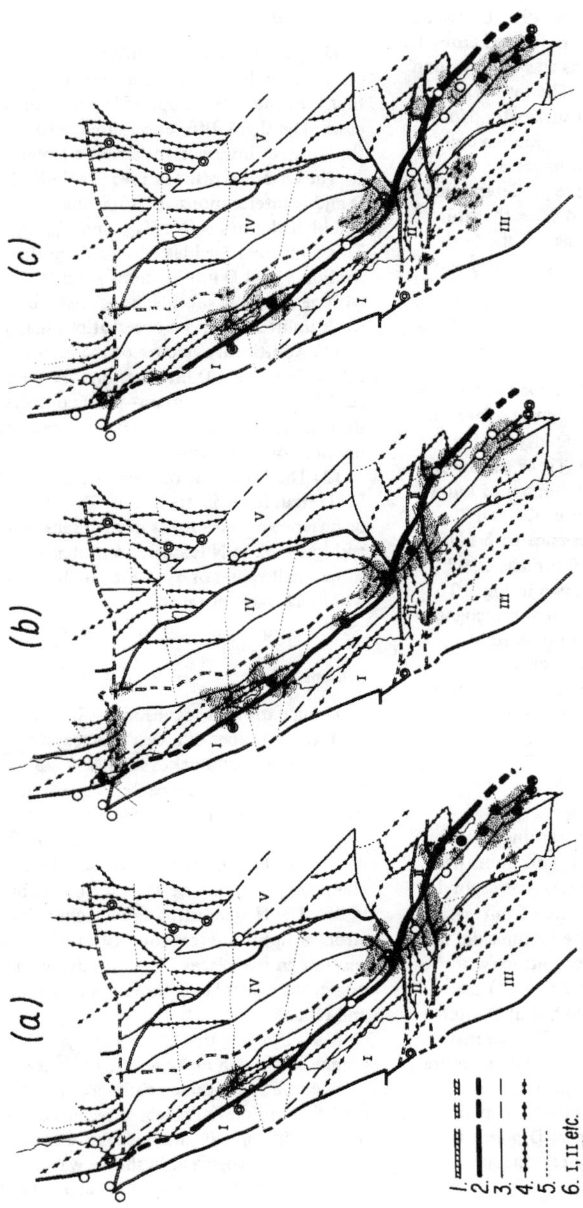

Fig. 10. Comparison of D areas for intersections of lineaments. Control experiment consists of eliminating groups of earthquakes. Epicenters eliminated from learning are filled in solid colors, the corresponding D areas are shaded. Lineaments defined in same way as Fig. 4. The experiment is aimed at recognition of intersections around the eliminated epicenters.

break of the San Francisco earthquake, 1906 (between intersections *13* and *89*). Clusters were formed around each of these intersections, exactly as was done around epicenters in the preceding sections. The results of recognition, shown in Fig. 9c, are unsuccessful. No intersections are recognized as D near 9 known epicenters out of 12. The three exceptions are the epicenters near the fault break, which are predetermined by learning. The next experiment indicates that this failure is due not to an uneven coverage of the territory by the objects of class I, but to the replacement of epicenters by the fault break.

4.6.2. Uniformity of the region

The analogous experiment was described in Section 3. The purpose is to see if we can recognize dangerous areas even if the epicenters in their vicinity are eliminated during the learning stage. Three cases were considered as shown in Fig. 10. The number of known strong earthquakes in the three cases is 6, 6 and 7, respectively. According to our experience with EH experiments, this number is marginal for successful learning. Nevertheless, the results shown in Fig. 10 are quite good in that at least one point was recognized as dangerous in almost all clusters not used in learning. The only exceptions are: two clusters near intersections *101* and *13* in Fig. 10b; one cluster near *158* in Fig. 10a. This confirms the conclusion reached in the analogous experiment in Section 3.

4.6.3. Effect of dip-slip earthquakes

Two of the earthquakes, used for learning, are of dip-slip variety (years 1952 and 1971; epicenters near intersections *165* and *119*). The clusters, corresponding to these earthquakes, were eliminated from learning in this experiment. The distinctive features did not change noticeably. The results of recognition (with the same threshold as in main variant, $\bar{\Delta} = \Delta^*$) are shown in Fig. 7b. We see that the number of D intersections decreased significantly, from 25 in the main variant to 17. Only three points outside of clusters are recognized as D (*72, 73* and *104*), slightly expanding already known clusters. Fig. 6 shows that the main effect was the elimination of a group of D points in the Transverse Ranges, a geologically reasonable result.

4.7. Comparison with Section 3

D areas recognized for intersections and for points on major faults (Figs. 3 and 6) basically coincide in that the same five groups of D-points show up in both cases. The slight differences are due to the fact that the study of intersections covers a larger territory.

The characteristic features (Table XVII) show that D and N intersections are often characterized by low and high elevations (h_1, H_{min} and H_{max}), respectively. This reinforces the idea derived from the results in Section 3 that D intersections are often associated with neotectonic subsidence against a background of weak uplift. Two more observations may be made.

(1) Additional evidence of the importance of vertical movements at D intersections are the large gradients ($\Delta H/\ell_4$) and the mountain–plain contacts which show up in many D features, with opposite conditions holding for N features.

(2) The fracturing of the crust is more pronounced for D than for N intersections. This is shown by the occurrence of lineaments of high rank in D features and low rank in N features. Also many N features show small values of n_2 – the number of intersections within 62.5-km distance.

5. Conclusions

Using pattern recognition, we have divided California into two types of areas: D areas where epicenters of strike-slip earthquakes with $M \geqslant 6.5$ can occur and N areas where these epicenters cannot occur. Distinctive features for each of these areas were identified. The D areas are characterized by proximity to intersections or ends of major faults and by relatively low elevation or subsidence against a background of weaker uplift. The N areas often show higher elevation or stronger uplift but less contrast in relief. D areas concentrate in five places. These results were tested against numerous control experiments and seem to be stable and reliable.

It is surprising that the relatively simple and crude parameters which make up the distinctive features suffice to achieve these results. Perhaps large earthquakes occur on faults which penetrate significantly into the lithosphere and fault movements on such a scale can be recognized in simple ways. Perhaps parameters based on systems of faults and other types of lineaments summarize more geophysical and geological information than their simplicity implies. Perhaps the result that epicenter locations occur at identifiable places along faults and not at arbitrary locations on faults implies the existence of a special property or agent in the epicentral region which acts to initiate rupture. This special property might be one which concentrates stress to weaken rock, or to "lubricate" a fault.

Our results are not useful as they stand for earthquake hazard mitigation since they identify epicenters and not the entire length of fault rupture, which may extend well beyond a D area. However a search for short-term precursors may be promising in D areas and future efforts in earthquake control by fluid injection and withdrawal could make use of D and N information.

Over the years many efforts have been made to characterize earthquake-prone regions by the occurrence of certain relationships in the field. For example, "earthquake country" has been characterized by topographic and geomorphic forms, by intersecting faults and by other ways in the geological literature. Our procedure seems to be a more precise formulation of many ideas which have been vaguely recognized in the past. It allows one to get more definite results, providing a method for recognizing differences between geologically different areas rather than describing each area by itself. It employs past experience, intuition and/or hypotheses; but it uses a logical framework to formulate explicitly the pertinent and reliable parts of these elements. The procedure consists of five steps.

(1) Preliminary separation of objects into groups.

(2) Separation of promising parameters by histogram analysis.

(3) The combination of parameters into features, singlets to triplets, and the analysis of all possible features to identify distinctive ones.

(4) The use of distinctive features in recognition.

(5) The analysis of distinctive features for their geological or geophysical meaning.

(6) The design of control experiments to prevent self-deception.

Statistical procedures are alternative methods to pattern recognition but in this case they are difficult to apply because the number of strong earthquakes is inadequately small.

In the future we intend to apply pattern recognition to time prediction rather than space prediction and to other geophysical applications.

Acknowledgements

This study is a product of the U.S.A./U.S.S.R. Working Group on Earthquake Prediction. The participation of Drs. L. Knopoff and F. Press was supported by grants with the U.S. Geological Survey, Department of the Interior.

Appendix – Major morphostructures and lineaments of California and adjacent regions

1. Basic definitions

Morphostructures are structures in the earth's crust (or its upper part) expressed in present day topographic relief. Tectonic and geomorphic maps often describe structures, based only on either geological or geomorphic evidence. For the study of earthquakes, however, actively developing structures are important. These structures are expressed in topographic relief; geomorphic evidence may therefore be important for their reconstruction, together with geological data.

Rank of morphostructure. The following subdivision seems adequate for tectonically and seismically active regions. Each region is divided into morphostructural countries (I), countries into megablocks (II), and megablocks into blocks (III). The rank assigned to each type of morphostructure is indicated by brackets.

Countries differ by the type of orogenesis and/or by the large-scale tectonic features.

Megablocks inside a country differ either by the dominant type of relief, or by the average parameters of the relief (such as heights of peaks, dominant strikes, relative area occupied by basins and ranges, etc.); or by the pattern of the main elements of relief.

Blocks inside megablocks differ in neotectonic history; for example, large separate ranges or depressions in contrast to a complex of smaller ranges, depressions and valleys can be outlined as separate blocks.

Morphostructural provinces. In some large countries, such as the Pamir and Tien-Shan regions, where the history of relief is long and complicated, an intermediate morphostructure – a province – was introduced (Gelfand et al., 1973a). A province is a group of megablocks which have similar pre-orogenic histories, and accordingly, similar general features of relief. The same first rank, as for countries, was assigned to provinces. (For example, an internal depression inside a platform may be transformed after activation into a large intermountain basin and regarded as a province.)

Lineaments. The boundary zones between morphostructures are called lineaments. They have the same rank as the morphostructures which they separate. Lineaments of the first rank have widths from about 10 to 50 km, perhaps more; they may be thousands of kilometers in length. A lineament is the surface trace of an active fault (or flexure), penetrating deeply into the earth's crust and perhaps through the entire lithosphere. Different parts of a lineament may be expressed in different ways: geological, geomorphic or both; the evidence for a lineament may not necessarily be as direct as faults shown on a tectonic map. Let us describe some of the evidence for lineaments, which may differ for longitudinal and transverse lineaments.

Longitudinal lineaments are approximately parallel to the main elements of relief (such as ridges, basins, elongated intra-mountain depressions, etc.). Usually they form the boundaries between these elements. They are often expressed directly by zones of tectonic faults. Detailed tectonic maps show them as a system of narrow, elongated blocks. However, some parts of longitudinal lineaments are not traced by faults, but by geomorphic evidence: direct contact of contrasting forms of relief such as steep mountain front and flat plain, linear form of valley between ridges, etc.

Transverse lineaments form a large angle with the dominant strike of the main elements of relief and tectonics. They are traced mainly by terminations of these elements, en-echelon patterns, relative displacements of elements, sharp changes in height or strike, sharp changes in types of relief. A regular linear pattern of these features is evidence for a transverse lineament. Transverse lineaments may also show as patterns in the occurrence of volcanoes, the highest peaks or saddle points in parallel ridges, etc. Sometimes the evidence for transverse lineaments may show simply as a straight-line river on the slope of a ridge, which does not flow in the direction of steepest descent of the slope, or distinct, short ranges, or narrow intrusive bodies with transverse strike, or transverse faults of limited extent, shown on tectonic maps. Transverse lineaments are not always marked by the clear traditional evidence of long narrow ridges, valleys and tectonic faults. The evidence for these lineaments may be indirect, subtle and discontinuous. However, the general pattern of evidence allows one to draw transverse lineaments with reasonable certainty, especially with the aid of satellite photography. The name "transverse" could be misleading in that the main distinction is not the strike, but evidence of the kind just described. For example, the Transverse Ranges of California, though perpendicular to the dominant strike of the region, are nevertheless classified as a longitudinal lineament, since the structure carries its own faults, ridges and valleys.

Disjunctive knots. Specific morphostructures of disjunctive knots are formed around the intersections of lineaments. They are zones of especially intensive fracturing and contrasting neo-tectonic movements. Following along a lineament one can notice the appearance of a disjunctive knot by a set of specific phenomena, connected with a general increase of neo-tectonic activity and with manifestations of intersecting lineaments: the types of relief become more diverse (types corresponding to each lineament show up) and appear in new combinations; the heights of ridges and height differences change drastically; boundaries of contrasting forms of relief assume a broken-line form; rivers flow along faults and their valleys straighten; the orientation of valleys corresponds to the direction of both intersecting lineaments; fault activity increases; the lineaments inside the knots are usually expressed by faults; the topographic and geological maps show mosaic patterns; a grid of surface faults expressed tectonically or geomorphically is not uncommon.

As a rule a knot is wider than the lineaments from which it is formed. The linear dimensions of knots depend on the rank of the component lineaments and may vary from 20 to 200 km or more. One knot may include several intersections of lineaments.

It is clear from this description that though the knots have rather distinct boundaries the determination of these boundaries demands quite specialized research – in the field and on detailed maps. In the absence of such research we must consider in our problem "intersections" instead of knots. An intersection is the point of intersection of the axes of lineaments. It is a much more formal object than a knot but it is easier to handle. Each intersection belongs to some knot; but it is unknown a priori which intersections belong to the same knot. Algorithm CLUSTERS was especially devised to deal with this difficulty.

2. The scheme of major lineaments (Fig. 4)

The area of interest is divided into morphostructures of three ranks: countries, megablocks and blocks (see first paragraph of this Appendix). Their boundaries are the lineaments of corresponding rank, shown in Fig. 3. The five following countries are represented:

I = Coast Ranges; *II* = Transverse Ranges; *III* = southern California-northern end of the Peninsula Ranges and the entire Mexican section of the Cordilleras; *IV* = Sierra-Nevada; *V* = western part of the Basin and Range Province.

Country *I* belongs to the Pacific orogenic belt, with active Cenozoic folding. The four other countries belong to the Cordilleran orogenic belt where the folding was mostly finished in the Mesozoic area. These four countries differ in Pliocene–Quaternary orogenic development. Faults of the San Andreas system cross countries, *I, II* and *III.* Its width is about 60 km in the north, near San Francisco Bay and up to 160 km, if not more, in southern California. The details are described elsewhere (Ranzman, in preparation).

References

Atwood, W., 1940. The Physiographic Provinces of North America. Ginn, Boston, Mass., 535 pp.

Bongard, M.M., Vaintsveig, M.I., Guberman, Sh.A., Izvekova, M.L. and Smirnov, M.S., 1966. The use of self-learning programs in the detection of oil containing layers. Geol. Geofiz., 6: 96–105 (in Russian).

Briggs, P. and Press, F., in preparation. Pattern recognition applied to earthquake epicenters in California and Nevada.

Coffman, J.L. and Von Hake, C.A., 1973. Earthquake history of the United States. Natl. Oceanic Atmos. Adm. (U.S.), Publ., 41-1, 208 pp.

Cook, K.L., 1966. Rift system in the basin and range province. In: The World Rift System, Proc. UMC Symp., Ottawa, Ont., September 6–8, 1965, Geol. Surv. Can., Pap., 66-14: 246–275.

Dickinson, W.R. and Grantz, A., 1968. Proc. Conf. on Geological Problems of San Andreas. In: Geological Sciences, Vol. XI, Stanford University, Palo Alto, Calif., 374 pp.

Gelfand, I.M., Guberman, Sh.A., Izvekova, M.L., Keilis-Borok, V.I. and Ranzman, E.Ia., 1972. Criteria of high seismicity, determined by pattern recognition. In: A.R. Ritsema (Editor), The Upper Mantle, Tectonophysics, 13(1/4): 415–422.

Gelfand, I.M., Guberman, Sh.A., Izvekova, M.L., Keilis-Borok, V.I. and Ranzman, E.Ia., 1973a. Recognition of places where strong earthquakes may occur, I. Pamir and Tien-Shan. In: Computational Seismology, Vol. 6 (in Russian).

Gelfand, I.M., Guberman, Sh.A., Kaletskaja, M.S., Keilis-Borok, V.I., Ranzman, E.Ia. and Zhidkov, M.P., 1973b. On the transfer of criteria of high seismicity from Central Asia to Anatolia and adjacent regions. Dokl. Akad. Nauk S.S.S.R., 20, N2 (in Russian).

Gelfand, I.M., Guberman, Sh.A., Kalezkaja, M.S., Keilis-Borok, V.I., Ranzman, E.Ia., Rotwain, I.M. and Zhidkov, M.P., 1974a. Recognition of places where strong earthquakes may occur, II. Four regions of Asia Minor and S-E Europa. In: Computational Seismology, Vol. 7 (in Russian).

Gelfand, I.M., Guberman, Sh.A., Keilis-Borok, V.I., Ranzman, E.Ia., Rotwain, I.M. and Zhidkov, M.P., 1974b. Recognition of places where strong earthquakes may occur, III. The case when the boundaries of disjunctive knots are unknown. In: Computational Seismology, Vol. 7 (in Russian).

Gutenberg, B. and Richter, C.F., 1954. Seismicity of the Earth and Associated Phenomena. Princeton University Press, Princeton, N.J., 310 pp.

Hain, W.Y., 1971. Regional Geotectonics, Vol. 1. Moscow University, Moscow (in Russian).

Hamilton, W. and Myers, W.B., 1966. Cenozoic tectonics of the Western United States. In: The World Rift System, Proc. UMC Symp., Ottawa, Ont., September 6–8, 1965, Geol. Surv. Can., Pap., 66-14: 291–306.

Hill, M.L., 1966. The San Andreas system, California and Mexico. In: The World Rift System, Proc. UMC Symp., Ottawa, Ont., September 6–8, 1965, Geol. Surv. Can., Pap., 66-14: 239–245.

King, F.B., 1969. The tectonics of North America. A discussion to accompany the *Tectonic Map of North America, Scale 1 : 5,000,000.* U.S. Geological Survey, Washington, D.C.

National Oceanic and Atmospheric Administration, 1965–1973. Bulletins of preliminary determination of epicenters. Boulder, Colo.

Ranzman, E.Ia., in preparation. Major morphostructures and lineaments of California.

Richter, C., 1958. Elementary Seismology. Freeman, San Francisco, Calif., 768 pp.

Rothe, J.P., 1969. The seismicity of the Earth. UNESCO, Paris, 336 pp.

Thompson, G.A., 1966. The rift system of the Western United States. In: The World Rift System, Proc. UMC Symp., Ottawa, Ont., September 6–8, 1965, Geol. Surv. Can., Pap., 66-14: 280–289.

USGS, no date. Topographic Map of the United States. Scale 1 : 1,000,000, Washington, D.C.

USGS, 1962. Tectonic Map of the United States. Scale 1 : 2,500,000, Washington, D.C.

USGS, 1965. Geologic Map of North America. Scale 1: 5,000,000, Washington, D.C.

Wright, H. and Frey, D., 1965. The Quaternary of the United States. Princeton University Press, Princeton, N.J., 922 pp.

Table 1 $^{187}Re/^{186}Os$ and $^{187}Os/^{186}Os$ ratio for iron meteorites and the magnetic separate from St Séverin

Sample		Sample weight (mg)	Re (p.p.m.)	Os (p.p.m.)	$^{187}Re/^{186}Os$	$^{187}Os/^{186}Os$
Iron meteorites:						
Tlacotepec	no. 1	187.3	3.28 (3)	57.0 (4)	2.32 (3)	0.980 (4)
(IVB)	no. 2	264.7	3.69 (3)	57.8 (7)	2.56 (5)	0.994 (6)
	no. 3	253.6	3.69 (3)	56.0 (6)	2.65 (5)	1.008 (6)
Canyon Diablo	IA	4075.0	0.236 (1)	2.12 (2)	4.47 (5)	1.122 (8)
Coahuila	IIA	411.9	1.40 (1)	11.11 (6)	5.06 (7)	1.187 (11)
Casas Grandes	IIIA	2446.9	0.731 (6)	9.87 (14)	2.98 (7)	1.035 (5)
Henbury	IIIA	472.0	1.53 (5)	16.9 (2)	3.63 (16)	1.080 (4)
Chondrite:						
St Séverin						
magnetic separate	LL6	636.8	0.939 (8)	8.00 (18)	4.72 (16)	1.164 (15)

The numbers in parentheses are the errors on the results, expressed in last digits: for example, 3.28(3) is equivalent to 3.23 ± 0.03 and 3.63(16) is equivalent to 3.63 ± 0.16.

We can then compute for these different values of (β) the curves showing the variation of $^{187}Os^*/^{187}Re$ with Δt (duration of nucleosynthesis (Fig. 3). Choosing for β the extreme values

$10^{-6}\ yr^{-1}$ for sudden nucleosynthesis

$10^{-12}\ yr^{-1}$ for steady state nucleosynthesis

we obtain for $^{187}Os^*/^{187}Re = 0.152$ two intercepts that give us an interval for the duration of nucleosynthesis: $8{,}700 < \Delta t <$ 17,800 Myr and hence the age of the Galaxy is $13{,}300 < T <$ 22,400 Myr. Note that a ±10% variation on $^{187}Os^*/^{187}Re$ would lead to a ±8% variation on T. Such an uncertainty might come from the cosmic Re/Os ratio or from the factor f or from the neutron capture cross-section ratio.

We thank G. Manhes and J. F. Minster for discussions, A. Pierre for technical assistance and P. Pellas for a magnetic separate from Saint Séverin.

Received 28 September; accepted 20 November 1979.

1. Herr, W., Hoffmeister, W., Hirt, B., Geiss, J. & Houtermans, F. G. *Z. Naturforschung.* **16a**, 1053–1058 (1961).
2. Morgan, J. in *Handbook of Elemental Abundances in Meteorites* (ed. Mason, B.) 435–462 (Gordon and Breach, 1971).
3. Kelly, W. & Larimer, J. *Geochim. cosmochim. Acta* **41**, 93–111 (1977).
4. Birck, J. L. Thesis, Univ. Paris VII (1979).
5. Manhes, G. & Allègre, C. J. *Meteoritics* **11**, 327–328 (1976).
6. Clayton, D. D. *Astrophys. J.* **139**, 637–663 (1964).
7. Hainebach, K. & Schramm, D. *Astrophys. J.* **212**, 347–359 (1977).
8. Watt, D. E. & Glover, R. N. *Phil. Mag.* **7**, 105 (1962).
9. Hirt, B., Herr, W. & Hoffmeister, W. *Radioactive dating*, 35–43, (IAEA, Vienna, 1963).
10. Hirt, B., Tilton, G. R., Herr, W. & Hoffmeister, W. *Earth Science and Meteoritics* (eds Geiss, J. & Goldberg, E.) 273–280 (North Holland, Amsterdam, 1963).
11. Shimizu, N., Semet, M. P. & Allègre, C. J. *Geochim. cosmochim. Acta* **42**, 1321–1334 (1978).
12. Podosek, F. A. *Geochim. cosmochim. Acta* **34**, 341–365 (1970).
13. Niemeyer, S. *Geochim. cosmochim. Acta* **43**, 843–860 (1979).
14. Wasserburg, G. J. & Burnett, D. S. in *Meteorite Research* (ed. Millman), 467–479 (Reidel, Dordrecht, 1969).
15. Rambaldi, E. R., Cendales, M. & Thacker, R. *Earth planet. Sci. Lett.* **40**, 175–186 (1978).
16. Manhes, G., Minster, J. F. & Allègre, C. J. *Earth planet. Sci. Lett.* **39**, 14–24 (1978).
17. Allègre, C. J., Birck, J. L., Fourcade, S. & Semet, M. P. *Science* **187**, 436–438 (1975).
18. Birck, J. L. & Allègre, C. J. *Earth planet. Sci. Lett.* **39**, 37–51 (1978).
19. Minster, J. F. & Allègre, C. J. *Earth planet. Sci. Lett.* **40**, 333–347 (1979); *Earth planet. Sci. Lett.* (submitted).
20. Minster, J. F., Ricard, L. P. & Allègre, C. J. *Earth planet. Sci. Lett.* **44**, 420–440 (1979).
21. Manhes, G. & Allègre, C. J. *Meteoritics* **13**, 543–548 (1978).
22. Turner, G., Enright, M. C. & Cadogan, P. H. *Proc. 9th Lunar Planet. Sci. Conf.* 989–102[illegible] (1978).
23. Browne, J. C. & Berman, B. L. *Nature* **262**, 197–199 (1976).
24. Morgan, J. W. & Lovering, J. F. *Geochim. cosmochim. Acta* **31**, 1893–1909 (1967).
25. Scott, E. R. D. *Geochim. cosmochim. Acta* **36**, 1205–1236 (1972).
26. Lovering, J. F. & Butterfield, D. *Proc. Apollo 11 Lunar Sci. Conf.* 1351–1355 (1970).
27. Allègre, C. J. & Luck, J. M. *Earth planet. Sci. Lett.* (submitted).
28. Lugmair, G. W. *Meteoritics* **9**, 369 (1974).

Bursts of aftershocks, long-term precursors of strong earthquakes

V. I. Keilis-Borok*, L. Knopoff† & I. M. Rotvain*

* Institute of the Physics of the Earth, Academy of Sciences, Moscow, USSR
† Institute of Geophysics and Planetary Physics, University of California, Los Angeles, California 90024

Earthquakes that are followed within a short time by abnormally large numbers of aftershocks are hypothesised to be long-term precursors of stronger earthquakes. In a test of the hypothesis, 18 out of 23 strong earthquakes, in five regions worldwide, were predicted retrospectively. In comparison with a random model the precursory pattern occurs at a satisfactorily high confidence level.

THE hope that patterns of seismicity exist that are premonitory to strong earthquakes relies on the fact that the occurrence of earthquakes is controlled by the distribution of stresses and strengths within the Earth and that each earthquake in turn changes these fields. At present we cannot anticipate what such patterns should be, because the theoretical basis is inadequate. One reason for this is the absence of firmly established empirical regularities in seismicity which would provide a factual basis for theoretical studies of earthquake occurrence.

Searches for premonitory patterns of seismicity have attracted much attention, as these may possibly be established by the analysis of earthquake catalogues which are the longest record of the relevant observations. With regard to precursors of large earthquakes, there have been many purported descriptions of

0028–0836/80/030259—05$01.00

remonitory patterns of seismicity. These patterns are of three asic types—quiescence, activation and migration of epicentres. Although the results from some of these studies support the ope that premonitory patterns of seismicity exist, nevertheless, vith a few exceptions, their definition is not specific enough to stimate the reliability of the pattern. To obtain such estimates ne should be able to detect the pattern by a formal algorithm nd to count how often it is successful as a precursor. Without uch estimates of reliability, any premonitory pattern would be ncertain.

We present here an attempt to derive a formal definition of ome premonitory seismicity patterns from the analysis of an arthquake catalogue, which yields a reasonable success-to-ailure ratio in retrospective long-term earthquake prediction. The definition also provides constraints on the theory of earth-quake occurrence.

One reason why the search for premonitory patterns of eismicity has not been successful is that aftershocks are usually lisregarded as being highly predictable phenomena. The results ve describe below show that, while aftershocks are indeed ighly predictable, some specific traits of aftershock sequences re highly informative. These traits may represent important, lescriptive aspects of the distribution and redistribution of tresses and strengths within the Earth.

Definitions

Let t be the origin time of an earthquake, λ, ϕ the longitude and atitude of its epicentre (west and south are negative), h its focal lepth, and M its magnitude. The subscript j will indicate the equence number of an earthquake in the catalogue, $t_j < t_{j-1}$.

We seek precursors of earthquakes with magnitudes $M \geqslant M_0$. These are called strong earthquakes. Our patterns will be con-tructed from earthquakes in the magnitude range $M_2 \equiv M_0-\mu_2 \leqslant M \leqslant M_0-\mu_1 \equiv M_1$ and in a time interval of duration ; the values of μ_1, μ_2 and s may be different for each of the hree patterns that we will discuss.

Aftershocks are defined as follows. Consider two earthquakes vith sequence numbers i and j, $j > i$. The second earthquake is n aftershock of the first if the following conditions are satisfied: he distance between their epicentres is less than $R(M_i)$; $t_j - t_i \leqslant T(M_i)$; $h_j - h_i \leqslant H(M_i)$; $M_j \leqslant M_i$. Here $T(M)$, $R(M)$ and $H(M)$ re empirical functions. If we apply this definition to an earth-quake catalogue, starting with the first earthquake, we can eparate the catalogue into main shocks and their aftershocks. The first earthquake of the catalogue is identified as a main shock. Each succeeding main shock can be identified after the deletion of aftershocks of the preceding main shocks. Formulas of this type have been developed[1–3] for both the worldwide and the southern California[4] catalogues for use in other aspects of the study of aftershocks and sequences or chains of inter-dependent earthquakes.

Pattern B (burst of aftershocks). Let $b_i(e)$ be the number of aftershocks following the ith main shock within time interval e. We consider only those main shocks with magnitudes between M_2 and M_1 and only aftershocks with magnitudes at or above some threshold $m \equiv M_0 - \mu_3$ We invoke the thresholds M_2 and m for practical reasons as all catalogues are incomplete and inhomogeneous for small magnitudes. However, these thresholds may be of some value in explaining physical rela-tionships, as the details of premonitory patterns may be different in different magnitude ranges.

Pattern B consists of a main shock and its aftershocks for which $b_i \geqslant C$. A more flexible definition of a threshold for b_i can be given[5], but it is not required for the present purposes.

An alarm is declared for a period of τ years after pattern B is diagnosed. A strong earthquake is expected during this period, having been predicted by pattern B. The alarm is terminated either at the time a strong earthquake takes place or at the end of this period, whichever comes first.

Pattern S ('swarms') was introduced elsewhere and its parameters were adjusted to the Italian earthquake catalogue[6]. We count the number of earthquakes in a time-window from $(t-s)$ to t, in the same magnitude range as above and in a restricted depth range. Let $N(t)$ be the total number of such earthquakes in the region. A reduced number of earthquakes $n(t)$ is obtained from $N(t)$ by the elimination of aftershocks of strong earthquakes. We then consider the map of the epicentres that are counted to determine $n(t)$. From this map we determine a third function $r(t)$ which is the maximum number of epicentres that can be surrounded by a small rectangular cell of dimensions $\Delta\phi$ in latitude and $\Delta\lambda$ in longitude.

Pattern S consists of a group of earthquakes such that $n(t) > \max\{C_1, C_2\bar{N}(t)\}$ and $r(t) > C_3 n(t)$. $\bar{N}(t)$ is the average of $N(t)$ over the interval t_0 to t or from $(t-ks)$ to t. The threshold M_1 plays no part in the identification of this pattern.

Pattern Σ has been introduced elsewhere[7] and its parameters were adjusted to some of the strongest earthquakes of the world. It is represented by a peak of the function $\Sigma(t) = \sum_j G(M_j)$ where the summation takes place over the same band of time and magnitude intervals as above. Aftershocks of strong earth-quakes are eliminated from the summation. $G(M)$ is specified[7]

Table 1 Score of patterns B

Regions	1 Southern California	2 N. California–Nevada (excluding Mendocino area)	3 Southern Japan	4 Northern Japan	5 New Zealand	Totals
Catalogue ref.	4, 11, 12	9	10, 14	10, 14	13	
Span of catalogue (yr)	1932–78	1945–76.5	1926–76.5	1940–76.5	1945–77	201.5
M_0	6.5	6.2	8	7.6	6.6	
C	11–13	6–12	4–10	4–6	9–14	
Strong earthquakes						
total	6	4	2	4	7	23
'predicted'	5	3	2	4	4	18
Relative time of alarms	0.46	0.27	0.18	0.36	0.27	0.31
N_m	36	39	9	45	65	194
N_A/N_B	9/14	3/6	2/4	7/13	7/8	28/45
T_τ/T	0.37	0.28	0.10	0.27	0.53	0.30
P	96%	79%	95%	95%	95%	

N_m, total number of main shocks; N_B, number of main shocks which generate pattern B $b_i \geqslant C$, for upper value of C; N_A, number of occurrences of attern B which are followed by strong earthquakes within τ yr; T_τ, sum of times τ before each strong earthquake, counting overlapping parts only nce; T, span of the catalogue (yr); C, threshold for identification of pattern B. A variation of C within the limits indicated in the table will introduce ither zero or one additional error; P, confidence level for upper value of C.

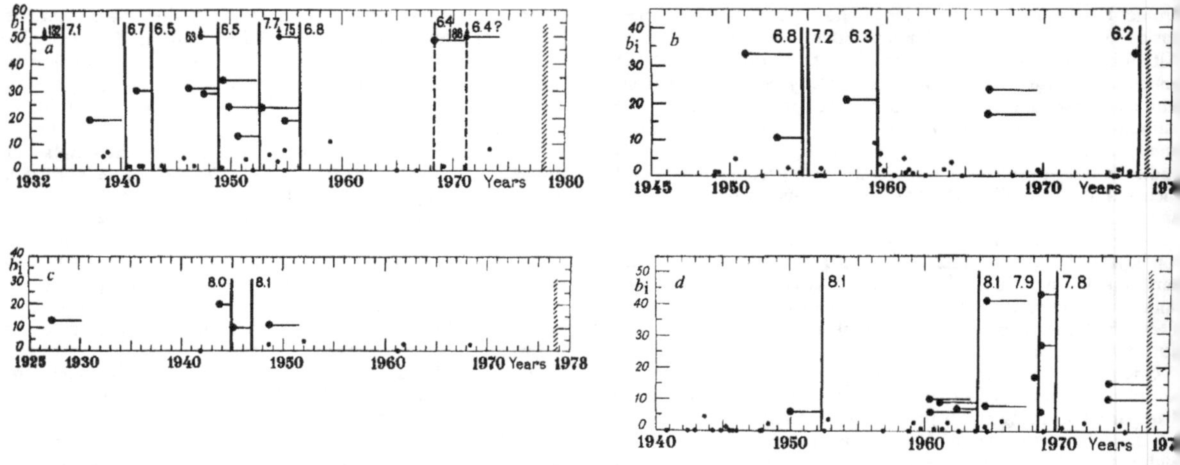

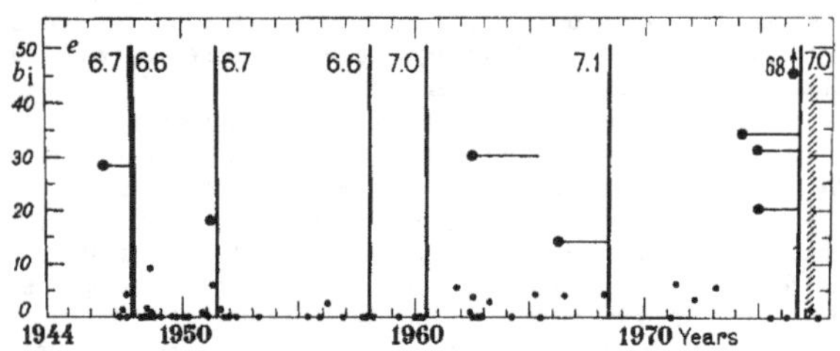

Fig. 1 Pattern B and strong earthquakes for *a*, southern California, $M_0 = 6.5$; *b*, northern California–Nevada, $M_0 = 6.2$; *c*, southern Japan, $M_0 = 8.0$; *d*, northe Japan, $M_0 = 7.6$; *e*, New Zealand, $M_0 = 6.6$. Hatching indicates the end of th catalogue. Vertical lines denote the times of strong earthquakes. Large dots are th values of *b* for main shocks which excited pattern $B(b_i \geq C)$; the subsequent alarm are shown by the horizontal lines. Small dots are the values of *b* for main shock which did not excite pattern $B(b_i < C)$. The two dashed lines denote the times of th Borrego Mountain (1968) and San Fernando (1971) earthquakes; the two fals alarms associated with these events disappear if the magnitude of the San Fernand earthquake is at least 0.1 greater than catalogued. The magnitudes of all events ar body wave magnitudes m_b with the exception of the Kern Country earthquake o July, 1952 which has $M_s = 7.7$ indicated. This latter earthquake has had a body wav magnitude assigned[15] (of $m_b = 7.2$).

to be $G(M) = 10^{d(M-f)}$ where *d* and *f* are numerical parameters. Pattern Σ consists of earthquakes satisfying the condition $\Sigma(t) \geq \bar{\Sigma}$. The values $\bar{\Sigma} = cG(M_0)$, $c = 0.5$, $d = 0.91$ have been assumed elsewhere[7].

Alarms after patterns S and Σ are declared in the same manner as indicated after pattern B.

Data processing

We have tested pattern B in a retrospective 'prediction' of strong earthquakes in the five regions indicated in Table 1. Arbitrary decisions have had to be made regarding the definition of the geographical boundaries of these regions as well as the values of all parameters. No *a priori* theoretical limits can be imposed on these decisions and we have to obtain them by data fitting. This creates a danger of self-deception as the number of these decisions is rather large. To decrease this danger we have tried to make these decisions simple, uniform and, if possible, *a priori*. Also we have accepted, when possible, the arbitrary choices of parameters or thresholds made in previous studies, and we have tested the stability of our results with regard to variations in each of these choices. We have considered only those earthquakes with $h \leq 100$ km.

We have assigned the following values: $\tau = 3$ yr; $e = 2$ days; $\mu_1 = 0.1$; $\mu_2 = 1$; $\mu_3 = 3.5$. These values were obtained by data fitting for southern California[5]. To define aftershocks we took $R = 50$ km; $T(M) = 0.5$ years for $5.0 \leq M \leq 5.4$, $T(M) = 1$ yr for $5.5 \leq M \leq 6.4$ and $T(M) = 2$ yr for $M \geq 6.5$. The values of $R(M)$, $T(M)$ given elsewhere[3] for the southern California catalogue give better results, but these fine gradations are hardly warranted in view of the accuracy of some of the other catalogues used in this study. The values of $R(M)$ and $T(M)$ assumed here have the advantage that they can be used for a fast diagnosis of pattern B even before the epicentres of aftershocks are determined precisely.

The values of M_0 and C listed in Table 1, have been chosen separately for each region due to lack of experience and a satisfactory theory. The fact that these two parameters have been chosen *a posteriori* contributes the single largest difficult in the evaluation of the significance of our results.

We have chosen the other parameters to be the same for a regions, even if they are unreasonably simple. This suffices fo our present purpose which is to test the existence and stability o pattern B. For actual predictions, the parameters could probabl be better adjusted to match the specific geophysical charac teristics of the region as well as its own earthquake catalogue More detailed studies of pattern B in some specific regions wil be published elsewhere[8].

The boundaries of the southern California as well as th northern California–Nevada regions are defined elsewhere[4,5]. We straightened these boundaries slightly from the origina definitions, and have deleted the area around Cape Mendocin from the northern California study as it has a distinctly isolate cluster of epicentres. The boundaries of the other regions ar based on large-scale neotectonics and on the densities of epi centres.

Results

The results are summarised in Table 1. Examples of the detail are documented in Table 2 and displayed in Fig. 1. The tota span of all five catalogues is 202 yr. Twenty-three strong earth quakes occurred in the five regions during these time intervals Eighteen of these earthquakes were retrospectively predicted that is, they took place during the intervals of alarms sounde after each instance of pattern B. The total duration of thes alarms is 60 yr or 30% of the total span of the catalogues. If w exclude the results for southern California, which were highl involved in the data fitting, that is, in the adjustment o parameters, we have 13 out of 17 strong earthquakes predicte for the remaining four regions; the total duration of alarms is 3 out of 155 yr or 25% of the total time.

Although these numerical values illustrate the practica efficiency of pattern B as a tool in prediction, they cannot b used in statistical tests of the existence of pattern B as, by th definition above, an alarm is terminated by the occurrence of a strong earthquake. To perform this statistical test (of the existence of pattern B), the values of the three parameters, N_A, N_B, and T_τ/T were tabulated and listed in Table 1.

The probability that a random binomial process could do as well or better than pattern B in anticipating strong earthquakes is given by the quantity $(1-P)$. This is the probability that N_A events or more will occur out of N_B tries in the fraction T_τ/T of the total catalogue time T according to a random binomial process. The values of P strongly support the hypothesis that B is a premonitory pattern, although these values of probability should only be used in a qualitative sense rather than literally. The fact that some thresholds were chosen *a posteriori* leads to an overestimation of P. On the other hand, the fact that some thresholds are chosen to be the same for all regions, leads to an underestimation of P. For example, the earthquake in southern Japan of 28 June 1948 could be deleted as a cause of a false alarm (Table 2) either by (1) changing e from 2 to 7 days, and/or (2) changing m to 5.0 from 4.5. Both of these changes are not unreasonable: in the first instance because $M_0 = 8$ is especially large and in the second because the catalogue[10] is more reliable for magnitudes greater than 5 than it is for magnitudes between 4.5 and 5. The last two false alarms for the southern California catalogue (Fig. 1) disappear if we assume that the magnitude of the San Fernando earthquake of 9 February 1971 is 6.5 instead of 6.4; this question is discussed elsewhere[5]. Similarly, the values of P could be increased by adjusting some of the parameters to take into account specific traits of the observed seismicity for each region separately. However, these adjustments would be non-unique; for the purpose of testing the existence of pattern B as a precursor, it is better to keep the criteria for all regions as uniform as possible, and that we make them, insofar as possible, *a priori*, and hence avoid tampering with the results by intuitive 'improvements' after inspection of the tabulations.

In the case of the northern California–Nevada catalogue we took $m = 3$, instead of $m = 2.7$ which would have been obtained from a literal application of our uniform definition $\mu_3 = 3.5$; this catalogue is especially inhomogeneous for $m < 3$. The errors in prediction with $m = 2.7$ and $m = 3$ are the same.

Table 2 lists the strong shocks as well as the main shocks that trigger pattern B; the strong shocks are underlined. Thus it is possible to see whether or not a strong shock was preceded by pattern B within 3 yr. Table 2 shows that our results remain essentially unaltered when the value of R is changed to 100 km and/or e to 7 d. Similar results have been obtained for the other catalogues.

We have indicated in Table 2 those events that would have been identified as aftershocks of the event immediately preceding had we used $R = 100$ km; deletion of these events from our listing would have improved our success rate as these invariably result in a reduction in N_B without a corresponding decrease in N_A. There are two examples of entries of this type in the southern California catalogue and none in the northern California and New Zealand catalogues.

A more detailed analysis of the stability of pattern B as well as the choice of the optimum parameters for the diagnosis of pattern B requires a separate study for each region. We have tried here to emphasise that pattern B can score quite satisfactory success-to-failure ratios, even with parameters that are uniform for all regions, despite the fact that each region has widely different tectonics, seismicity and quality of earthquake catalogue from the others.

Table 2 Pattern B and strong earthquakes

Month-day-year	epicentre ϕ	λ	M	b_i $R = 50(100)$ km $e = 2$ d	$e = 7$ d	δT (yr)	n_m
Southern Japan							
Polygon: (30; 130), (33; 139), (39; 136), (33; 127.5)							
3- 7-27	35.6	135.1	7.5	13	18	7.8	
9-10-43	35.5	134.2	7.4	20	25	1.24	
12- 7-44	33.7	136.2	8.0	3	3		3
1-13-45	34.7	137.0	7.1	10	17	1.94	
12-21-46	33.0	135.6	8.1	2	3		1
6-28-48	36.1	136.2	7.3	11	11	≥30	
Northern Japan							
Polygon: (33; 139), (37; 145), (40; 146), (45; 154), (49; 150), (45; 140), (39; 136)							
12-26-49	36.7	139.7	6.7	6	7	2.19	
3- 4-52	42.2	143.9	8.1	6(21)	9(28)		14
3-21-60	39.8	143.5	7.5	10(13)	19(30)	3.56	
3-23-60	39.3	143.8	6.7	6	6	3.56	*
1-16-61	36.0	142.3	6.8	9(13)	9(13)	2.74	
4-12-62	38.0	142.8	6.8	7(8)	10(12)	1.50	
10-13-63	43.8	150.0	8.1	4(8)	9(16)		15
5- 7-64	40.3	139.0	6.9	8	10	4.03	
6-16-64	38.4	139.2	7.5	41(42)	48(52)	3.92	
1-29-68	43.2	147.0	6.9	17(21)	23(33)	0.30	
5-16-68	40.7	143.6	7.9	12(45)	17(65)		6
5-16-68	41.4	142.8	7.5	27	50	1.24	*
5-17-68	39.8	143.5	6.7	6	7	1.24	
6-12-68	39.4	143.1	7.2	43(45)	51(55)	1.17	
8-12-69	42.7	147.6	7.8	7(37)	18(66)		4
6-17-73	43.0	146.0	7.4	15(30)	18(45)	>4	
6-24-73	43.0	146.8	7.1	10(30)	16(45)	>4	*

Values of b_i for $R = 100$ km are indicated only if they are not the same as for $R = 50$ km; δT, time from the given pattern B until the next strong earthquake; n_m, the number of all main shocks between the given and preceding strong earthquakes, or since the beginning of the catalogue in the case of the first strong earthquake; * denotes a main shock, that would have been identified as an aftershock with $R = 100$ km.

A speculation on generalisation

The same catalogues were processed to detect the presence of patterns S and Σ (Fig. 2). The value of M_0 is assumed to be the same as for pattern B. As in the papers where these patterns were introduced[6,7], the parameters were taken to be as follows: $C_1=0$, $C_2=1$, $C_3=0.5$, $\mu_2=2$, $\mu_1=0.5$ for Σ (absent for S), $d=0.91$, $f=4.5$, $s=1$ yr for S. For Σ we varied $\bar{\Sigma}$ and s for each of the regions (Fig. 2). The absence of a pattern signifies that it could not be detected for the given region. We added data for Alaska in view of the striking success of pattern Σ.

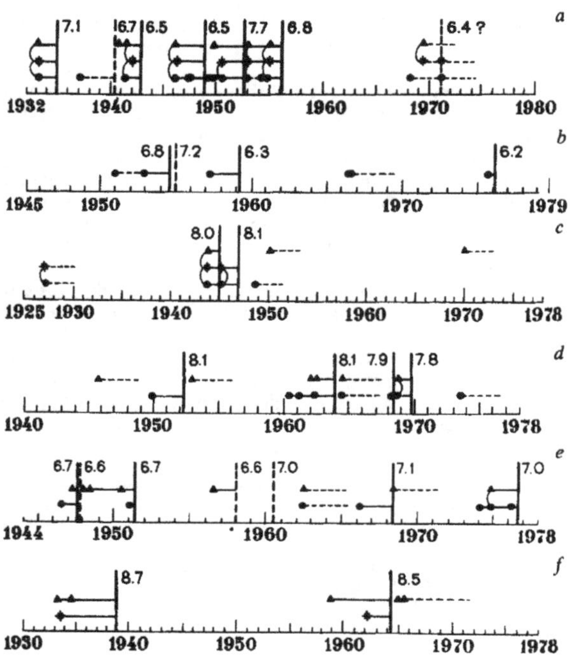

Fig. 2 Occurrence of patterns B (●), S (◆) and Σ (▲).

	$\bar{\Sigma}$	s (yr)	$\Delta\phi$	$\Delta\lambda$
a, Southern California	60	1	0.4°	0.6°
b, Northern California–Nevada				
c, Southern Japan	900	3	0.4	0.6
d, Northern Japan	700	1	1.0	1.4
e, New Zealand	95	0.5	0.4	0.6
f, Alaska	770	2	0.5	1.0

Solid horizontal line is period of alarm; dashed is false alarm. Solid vertical line is time of strong earthquake; dashed is strong earthquake not predicted by pattern B. Brackets denote different patterns generated by the same main shock and the appropriate parts of its aftershock series.

In several cases the same main shock and its aftershocks will give rise to several different patterns from among Σ, S, B. This result is not unreasonable as by definition, a main shock and its aftershocks will generate:

(1) pattern B if the number of aftershocks during the first e days is large enough;

(2) pattern S if the number of aftershocks during the time interval s, or during any part of it, is large enough; we usually take $s=1$ yr;

(3) pattern Σ if a sufficiently large number of events have magnitudes close to the upper limit M_1.

In most cases in southern California the same main event plus its aftershocks generate two or even all three patterns. Moreover, the coincidence of a combination of two or three of the patterns eliminates false alarms without adding any failures-to-predict[5].

Patterns Σ and S also appear in conjunction before extreme strong earthquakes in the Tibet–Himalaya region[8]; in this regi pattern B cannot be diagnosed, as the catalogues that a available do not include aftershocks.

We find that prediction by pattern B is successful when tw conditions are satisfied: not only is a large number of aftershoc b_i required but also the main shock producing it must have large enough magnitude; if the magnitude of a main shock is tc low, but is still followed by an exceedingly large b_i, a false alar will be generated. We have therefore introduced the thresho M_2. On the other hand we find both the numbers of false alarm and failures-to-predict increase significantly if we ignore th number of aftershocks produced by a main shock and mere consider the occurrence of a main shock itself as a possib precursor: for example, these numbers increased when w considered as a premonitor any earthquake with magnitude u to M_0, that is, by removing the gap between M_2 and M_0, b increasing M_1 and ignoring the threshold C.

We can imagine that a hypothetical generalised premonito pattern exists which is an abnormal clustering of earthquakes i a space–time–energy domain. We call this hypothetical pattern 'burst of seismicity'. Our three particular patterns may represe different projections of this generalised pattern. The definitior of the three are neither unique nor optimised; the use of discre magnitude thresholds, especially in regard to large uncertaintie in magnitudes of earthquakes in the catalogues, is an inadequat and troublesome aspect of the work so far. Our results indicat however, that these patterns exist and deserve attention.

We have refrained from providing additional refinement such as geological regionalisations or to assign probabilities t the forecasts as functions of time to the forthcoming event an its magnitude. The results of the present exploration indicat that such refinements are necessary and we hope that they woul improve our success ratios, which are surprisingly high despit the simplicity of the criteria we have used.

This research was performed as part of Project III of th US–USSR cooperative programme in earthquake predictior V.I. K.-B. was a Fairchild Scholar at the Seismological Labora tory, California Institute of Technology, and visiting professor a the Institute of Geophysics and Planetary Physics, University o California, Los Angeles. The following members of the Seis mological Laboratory supported this research: C. Allen, D Anderson, D. Harkrider, H. Kanamori, K. McNally, B. Minste We also thank D. Eckelmann, T. Hearn, C. Johnson, J. Gardne Y. Kagan, B. Bolt, J. Evernden, W. Lee, and R. Lamoreaux fo comments and assistance. Financial support of the Office o Earthquake Studies, US Geological Survey is also acknowl edged. This is publication no. 1882 of the Institute of Geo physics and Planetary Physics, University of California, Lo Angeles and contribution no. 3261 of the Division of Geologica and Planetary Sciences, California Institute of Technology Pasadena.

Received 15 December 1978; accepted 8 November 1979.

1. Keilis-Borok, V. I., Podgayetskaya, V. M. & Prozorov, A. G. *Comput. Seis.* **5**, 55–79 (1971)
2. Knopoff, L. & Gardner, J. K. *Geophys. J. R. astr. Soc.* **28**, 311–313 (1972).
3. Gardner, J. K. & Knopoff, L. *Bull. seis. Soc. Am.* **64**, 1363–1367 (1974).
4. Hileman, J. A., Allen, C. R. & Nordquist, J. M. *Seismicity of the Southern California Region 1 January 1932 to 31 December 1972* (Seismological Laboratory, California Institute o Technology, 1973).
5. Keilis-Borok, V. I., Knopoff, L., Rotvain, I. M. & Sidorenko, T. M. *J. geophys. Res.* (in the press).
6. Caputo, M. *et al. Annali Geofis.* (in the press).
7. Keilis-Borok, V. & Malinovskaya, L. N. *J. geophys. Res.* **69**, 3019–3024 (1964).
8. Keilis-Borok, V., Knopoff, L. & Allen, C. R. *J. geophys. Res.* (in the press).
9. Bolt, B. & Miller, R. D. *Catalog of Earthquakes in Northern California and Adjoining Area 1 January 1910–31 December 1972* (Seismological Stations of the University of Califor nia at Berkeley, 1975).
10. *Catalog of Major Earthquakes which occurred in and near Japan, 1926–52* (Japa Meteorological Agency).
11. Friedman, M. E., Whitcomb, J. H., Allen, C. R. & Hileman, J. A. *Seismicity of the Southern California Region, 1 January 1972 to 31 December 1974* (Seismological Laboratory California Institute of Technology, 1976).
12. *Catalog of the Earthquakes of Southern California, 1974–77* (Seismological Laboratory California Institute of Technology).
13. Smith, W. *N.Z. J. Geol. Geophys.* **19**, 393–4 (1976).
14. *U.S. Geological Survey, Earthquake Hypocenter Data File* (NOAA, 1977).
15. Bolt, B. A. *Bull. seis. Soc. Am.* **68**, 513–515 (1978).

THE LITHOSPHERE OF THE EARTH AS A NONLINEAR SYSTEM WITH IMPLICATIONS FOR EARTHQUAKE PREDICTION

V. I. Keilis-Borok
Institute of Earth Physics
Academy of Sciences
Moscow, USSR

Abstract. The lithosphere of the Earth can be viewed as a hierarchy of volumes, from tectonic plates to grains of rock. Their relative movement against the forces of friction and cohesion is realized to a large extent through earthquakes. The movement is controlled by a wide variety of independent processes, concentrated in the thin boundary zones between the volumes. A boundary zone has a similar hierarchical structure, consisting of volumes, separated by boundary zones, etc. Altogether, these processes transform the lithosphere into a large nonlinear system, featuring instability and deterministic chaos. From this background some integral grossly averaged empirical regularities emerge, indicating a wide range of similarity, collective behavior, and the possibility of intermediate-term earthquake prediction.

The history of science will probably label our time by the sweeping intrusion of the concept of deterministic chaos, with all the advantages and side effects of a bandwagon. Inaugurated by Henri Poincaré near the turn of the century, it keeps infiltrating one science after another, always revolutionizing them. During the last decade the variety of isolated applications started to merge into Weltanschauung, the perception that well-ordered, predictable phenomena present just an archipelago in the ocean of Nature, which is intrinsically nonlinear and chaotic. Luckily, the chaos can be, to some extent, understood. A limited set of universal patterns can be recognized in the chaotic behavior of exceedingly diverse systems: nonlinear pendulums, living matter, economy, burning fuel, brain, ocean, atmosphere; and here we discuss the implications for the lithosphere.

THE LITHOSPHERE

The lithosphere's major trait is so fundamental that it is easy to overlook: it is the hierarchical discreteness of its structure and dynamics. The lithosphere presents a hierarchy of volumes, or blocks, which move in relation to each other. The largest blocks are the major tectonic plates themselves. They are divided into smaller blocks, like shields or mountain chains. After 15–20 divisions we come to the grains of rock of millimeter scale, if not less.

The blocks are separated by less rigid boundary zones, 10–100 times thinner than the corresponding blocks. Each zone presents a similar hierarchical structure: it consists of blocks divided by zones of smaller rank. A lowest rank or two may present an exception. The boundary zones bear different names:

Boundary Zone	*Size of Blocks, km*
Fault zone	10^4–10^2
Fault	10^1–10^{-2}
Crack	10^{-3}–10^{-5}
Microcrack	10^{-6}–10^{-7}
Interface	10^{-8}– ...

The spell of terminology sometimes obscures the fact that all the boundary zones from the San Andreas fault system to a facet of a grain play similar roles in the dynamics of the lithosphere.

Until recently, a boundary zone was regarded as a passive interface, which glues the volumes together by friction and cohesion; it suffers deformation up to fracturing due to the motion of the blocks; and it somehow heals, since it does not turn to dust. That is the truth but not the whole truth. Friction and cohesion are controlled in turn by internal processes confined mainly to the boundary zones: by interaction with fluids, phase and petrochemical transformations, fracturing, buckling, etc. These processes, discovered or rediscovered during the last 5–6

EDITOR'S NOTE: This is an edited version of a paper presented at the International Union of Geodesy and Geophysics in Vancouver in 1987.

Reviews of Geophysics, 28, 1 / *February 1990*
pages 19–34
Paper number 89RG03576

8755-1209/90/89RG-03576 *$05.00*

years, can rapidly and unobtrusively change the friction and cohesion by a factor of 10^5, if not more, and accordingly change the effective strength within the lithosphere. So, while the energy of the motion is stored within and well beneath the lithosphere, the release of the energy and the motion is controlled mainly by the boundary zones.

We shall now briefly review the mechanisms of this control. It will lead to the conclusion that for a wide range of time and space scales it is necessary to regard the lithosphere as an unstable, nonlinear chaotic system.

Rhebinder Effect, or Stress Corrosion

Many solids lose their strength by contact with certain surface-active liquids. A bar of steel may bend under its own weight in this way. The liquid diminishes the surface tension μ, and consequently the strength, which by Griffith's criterion is proportional to $\sqrt{\mu}$. Then cracks may appear under very small stress, liquid penetrates the cracks, and the cracks grow with the drops of liquid propelling forward like tiny knives. This mechanism requires very little energy—like unlocking a door, instead of breaking it. It was first discovered for metals, then for ceramics, and during the last decade such combinations of a solid and sympathetic liquid were recognized among the common ingredients of the lithosphere. An example is basalt in a sulphur solution [*Pertsov and Kogan,* 1981]. When they meet, the basalt will be permeated by a grid of cracks, and its effective strength may drop instantly by an order of magnitude and eventually decrease by 10^{-5} due to this mechanism alone.

The stress field determines the geometry of such cracks, and this brings us to the realm of deterministic chaos. The stress fields in the lithosphere are exceedingly diverse, with all the inhomogeneities and fractures. The geometry of a system of cracks can be equally diverse. The geometry of the weakened areas, where the cracks are concentrated, is strictly limited however. Such areas may be of only a few types [*Gabrielov and Keilis-Borok,* 1983], determined by the theory of singularities. Some examples are shown in Figure 1. The light lines are possible trajectories of cracks. Each heavy line is a separatrix, which separates the areas with different patterns of trajectories.

What does this mean for an observer of the lithosphere? Suppose the source of a fluid appears in the place shown by arrows. The fluid will concentrate in the shaded areas and, out of the blue sky, the strength of this area will plummet. A slight displacement of the source across the separatrix may lead to gross, global change in the geometry of such fatigue—it may be diverted to quite a different place and take quite a different shape, but not an arbitrary one.

A new dimension is brought into this picture by the evolution of this stress field. It evolves incessantly, for many reasons, including feedback from the fatigue itself. Such evolution may change the type of singularity, as

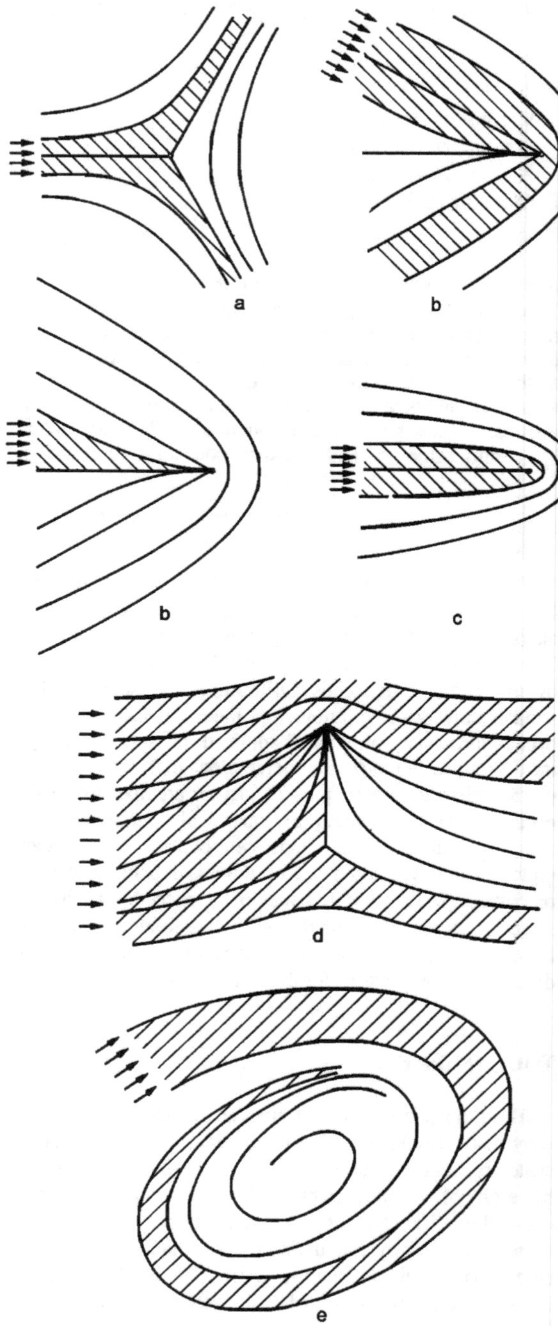

Figure 1. The geometry of fatigue induced by stress corrosion. The sources of surface-active fluids are shown by arrows. The weakened zones are hatched. Figures 1*a*–1*c* correspond to different types of singularities of the principal stress field; Figure 1*d* corresponds to two singularities of the type shown in Figure 1*a*; and Figure 1*e* corresponds to a loop or "limit cycle." After *Gabrielov and Keilis-Borok* [1983].

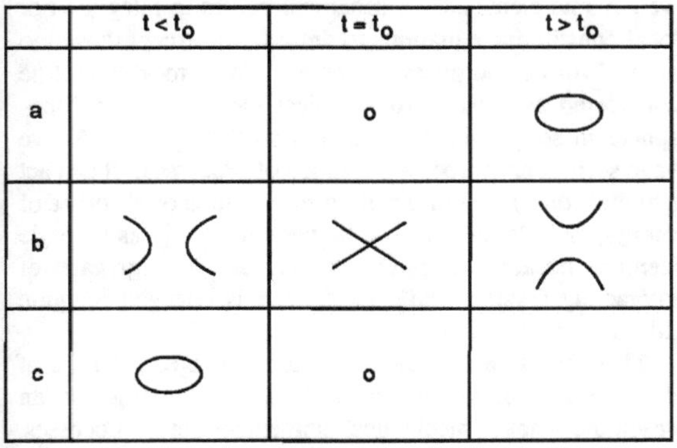

Figure 2. The evolution of singularities in the three-dimensional case: (*a*) birth of a loop; (*b*) bifurcation; and (*c*) annihilation of a loop. After *Gabrielov and Keilis-Borok* [1983].

shown in Figure 2, and the geometry of fatigue will follow suit.

So, the Rhebinder effect brings into the dynamics of the lithosphere a strong and specific instability, controlled by the stress field and by the geochemistry of arriving fluids. Migration of fluids hopefully can be monitored through electroconductivity and by some methods of hydrology. The history of this migration is reflected in mineralogy and isotope geochemistry. In this way, quite a variety of Earth sciences are tied together by the Rhebinder effect. It may possibly suggest an explanation of many mysteries: spontaneous tectonic activations, isostatic anomalies in some volcanic regions, stability of hot spots, etc.

In this way it is easy to explain also the seismicity patterns premonitory to strong earthquakes: seismic gaps, clustering, migration, doughnuts—all of them so far suggested. I dare say that any new pattern can also be explained in this way, since the elementary singularities are sufficiently sophisticated. However, such an explanation would be premature.

First, the singularities are local, and the natural fluctuations of the stress field may prevent their extension to the scale of premonitory patterns, which is tens to hundreds of kilometers. More probably, these singularities are elements which compose a more complicated infrastructure of fatigue. Second, the Rhebinder effect may decay along the trajectories due to diffusion of the fluids. Finally, this effect is not the single major mechanism by which the boundary zones control the dynamics of the lithosphere, generating deterministic chaos for good measure. Even the fluids alone generate other coexisting mechanisms as spectacular as this.

Filtration

One competing mechanism is more conventional filtration of fluids through the pores or cracks along the gradient of pressure [*Barenblatt et al.*, 1985]. A boundary zone between two blocks is modeled as a porous layer. When shear stress exceeds friction, slip begins.

Further development indicates a source of strong instability, illustrated by Figure 3. If porosity is subcritical, the slip, once started, will increase the friction and self-decelerate. At worst, we will have a vacillating creep or a slow earthquake. But if porosity exceeds a critical threshold, the solid frame will be destroyed by the slip, friction will decrease, and the slip will self-accelerate. The porosity can rise above the critical threshold because of the infiltration of a fluid; it will increase the tension and the pores will expand. So, again, just because some fluid filtered in, an instability instantly rises, and the tiny slips, which occur incessantly, self-accelerate, grow, and merge.

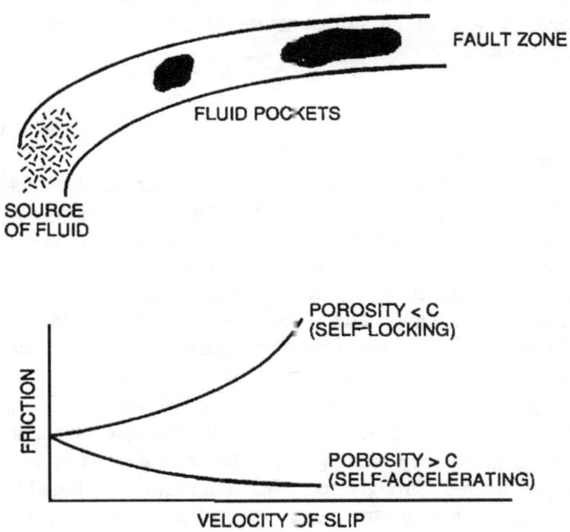

Figure 3. Instability of a fault zone due to filtration of fluids. After *Barenblatt et al.* [1985]. (Top) Fault zone, modeled as a porous layer with residual pockets of fluids between impermeable rigid blocks. (Bottom) The mechanism of instability.

This instability is aggravated by the erratic velocity of filtration. It is described by the nonlinear parabolic equation [*Barenblatt et al.*, 1985]

$$\partial P/\partial t = K P_0 \Delta P^{\alpha} \qquad \alpha > 2$$

Here P is the perturbation of pressure on the background of steady filtration flow; P_0 is the background pressure; $K = k/(m\rho\nu)$, where k and m represent permeability and porosity of the layer and ρ and ν represent the density and viscosity of the fluid. Though this equation is parabolic, P propagates with a final velocity proportional to $\sqrt{P_0}$. The calculated velocity lies in the range of tens to hundreds of kilometers per year [*Barenblatt et al.*, 1985], the same as the observed rate of migration of seismicity along fault zones.

A new model of an earthquake source naturally arises: a residual pocket of fluid, where the background pressure P_0 is large. A front of filtration can quickly cross and destabilize this pocket, turning it into an earthquake source.

This mechanism may explain even more features of real seismicity that can be explained by the Rhebinder effect. The same premonitory seismicity patterns, the foreshocks and aftershocks, and quasi-periodicity in the recurrence of strong earthquakes are explained by this mechanism. The difficulties with spatial scale do not arise here.

But again, there is no reason whatsoever to single out this particular mechanism. First, such instabilities may develop in parallel within boundary zones of different rank and interact along the hierarchy. This model may be only one element of an infrastructure of filtration-generated instabilities.

The Need for a Generalized Model

The boundary zones feature several other mechanisms of instability, potentially as important and certainly as complicated as those discussed above. One is the usual type of lubrication. Another is petrochemical and phase transitions which may generate lubricators, for example, both in formation and decomposition of serpentines. Boundary mechanisms may also create an instant loss of volume, such as in the transformation of calcite into aragonite. This will create a vacuum and unlock a fault, which will at once be closed by hydrostatic pressure, but an instability may be triggered. Most mechanisms of this type depend strongly on the temperature. Finally, traditional mechanisms, such as buckling, clustering of cracks, and viscous flow, remain important, but I will not dwell on them, hoping that the point is made.

The dynamics of the lithosphere can be viewed as the interaction of its blocks across and along the entire hierarchy of sizes. This interaction is realized through a wide variety of mutually dependent mechanisms. Each of them creates instability. However, none can be singled out as a major factor that allows the others to be neglected.

So what we need is the "whole enchilada"—a generalized theory that embraces all of these phenomena on the background of hierarchical structure. To assemble the whole set of corresponding equations is unrealistic. More than that, it may be misleading because of what, at the time of Hegel, was called the Gestalt concept: the whole is more than just a sum of its components. For example, the laws of biology cannot be derived from the evolution of protoplanetary dust, though life is the result of this evolution. Instead, we hope for a model that directly represents the grossly averaged traits of the lithosphere, in particular, the laws of its dynamics.

The problem of averaging is unusual here. Contrary to what would be a typical starting point in such a problem, we have no elementary mechanical bodies to integrate nor local constitutive equations to draw from, except those too general for our purposes. Even a grain of rock cannot be considered as a purely mechanical element of the lithosphere in simple local interaction with its peers. As we have seen, it undergoes a multitude of processes. It can act simultaneously as a material point, a source or absorber of energy, etc. It cannot even be treated exactly as a single element because it may also act as an aggregate of interacting crystals while its surface is engaged in quite different processes.

Therefore, to model the integrated, averaged behavior of the lithosphere, we have to do the averaging for an ensemble—not of mechanical elements—but of processes. The experience of quantum statistical physics shows that this is not easy.

Traditional models are based on the mechanics of continuous media, with different rheologies; sometimes porosity and filtration are included. Such models may be applied beyond their formal limits by expansion of the meaning of parameters. For example, microfracturing is often represented as a drop in elastic modulus, to which a so-called "effective" value is assigned.

Such an approach has a good track record. But, obviously, it is applicable to a limited range of problems: when the blocks of lithosphere are sufficiently homogeneous; when purely mechanical processes dominate; and only for certain time and space scales. Almost none of the phenomena discussed above can be reproduced by such models, even if we present the parameters as some Pickwickian construction (not to be understood literally). Our search for a generalized theory brings us to the concept of deterministic chaos. The qualitative description of chaos is recalled in the following section.

CHAOS

Chaos arises in deterministic systems because of their specific instability. For example, imagine a billiard game (Figure 4). The player sends the ball into the usual array of other balls. The slightest variation in the direction of the original push will send the ball down quite a different path and the difference will not attenuate but will grow with time. Each collision of the balls with each other will further amplify this divergence.

To prolong the motion, let us assume that the loss of energy is small. Newton's laws do determine the trajectory of each ball and the sequence of collisions. But the prediction will be completely wrong after a certain number of collisions even if the initial push is defined with an error as small as the gravitational effect of a single electron on the margin of the galaxy. It is beyond the estimation of necessary precision—it is just a picturesque way of saying that the deviation does not attenuate but grows exponentially in time, so that prediction is impossible at any level of precision of the initial conditions. If the boards are convex ("Sinay billiard") even a single ball reduced to a material point will display the same instability. With a multitude of such turning points, a dynamic system may display erratic, complicated behavior, which looks and is called chaotic. Though deterministic, it will be unpredictable, because prediction would require paradoxical precision of the initial conditions. This is not an abstract extravaganza. On the contrary, chaotic systems can be surprisingly simple, like the nonlinear pendulum, for example.

However, this kind of chaos does contain inherent regularities. These regularities can be understood, and some integral traits of chaotic behavior can even be predicted. Quoting Shakespeare, "though this be madness, there is method in't."

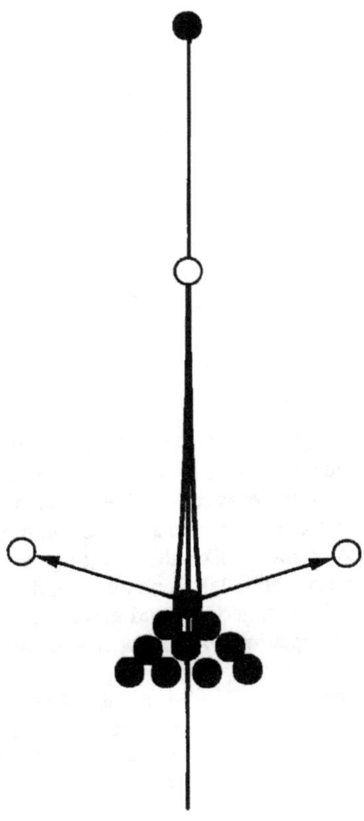

Figure 4. An illustration of the origin of deterministic chaos.

One fundamental regularity in chaotic behavior was discovered by E. Lorenz in 1963; Poincaré apparently suspected its existence. Lorenz studied thermal convection in the atmosphere, which has the astonishing ability to self-organize into a honeycomb of convection cells. He introduced a grossly simplified model for a single cell, defined by ordinary differential equations:

$$\partial x/\partial t = -\sigma x + \sigma y$$
$$\partial y/\partial t = Rx - y - xz$$
$$\partial z/\partial t = -bz + xy$$

Here the functions x, y, and z characterize the intensity of the convection stream and horizontal and vertical temperature gradients, respectively: σ, b, and R are numerical parameters.

Figure 5 shows a phase space for this system [*Crutchfield et al.*, 1986]. Its coordinates are these three functions, so that a point completely defines the state of the system at some moment of time; the evolution in time is defined by a trajectory. There is a strong divergence. The dots in Figure 5 show 10,000 different states at some moment of time. They evolved from 10,000 initial states, which were so close that they are all merged into one dot in Figure 5 (somewhere in the top right corner). In other words, microscopic initial perturbation leads to macroscopic divergence, and prediction is impossible. However, we see an inherent regularity: all states eventually congregate around the configuration represented by the lines. These lines are the asymptotic trajectories. The evolution of the system will gradually be attracted to them. They occupy the subspace called the chaotic or strange attractor. The subspace has smaller dimensionality than the whole phase space.

A nonchaotic system may have two rather prosaic stable attractors: a fixed state and a harmonic oscillation; in phase space it will be a point and a loop, called a limit cycle, respectively. (One may add quasi-periodic oscillation, formed by two incommensurate frequencies, that is, a torus in phase space, but it is not stable to the change of parameters of the system.)

Strange attractors are more exotic. They have fractal structure. Actually, the one in Figure 5 has the geometry of the Cantor set. Any small part of this attractor, when examined through a magnifying glass, will show a similar uneven density of lines. If we cut off a small part of this enlarged sample and enlarge it again, we will still see a similar pattern, and so on indefinitely.

The probability that the system will cross any point specified in advance is zero. But the system will certainly pass within an arbitrarily small distance of an attractor. So, we now know of a very restrictive global regularity: the system will eventually move along this attractor. However, we cannot go into much more detail, we cannot predict specific trajectories, and the times of transition from one branch to another are completely random. The future cannot be determined from the present, no matter how many observations and supercomputers are engaged.

The global traits of a chaotic system cannot be derived from the interaction of its components: it is a Gestalt phenomenon. The natural tendency to understand such a system by breaking it down into small elements is futile. A large zoo of diverse strange attractors has been encountered, but their general theory has not been discovered so far.

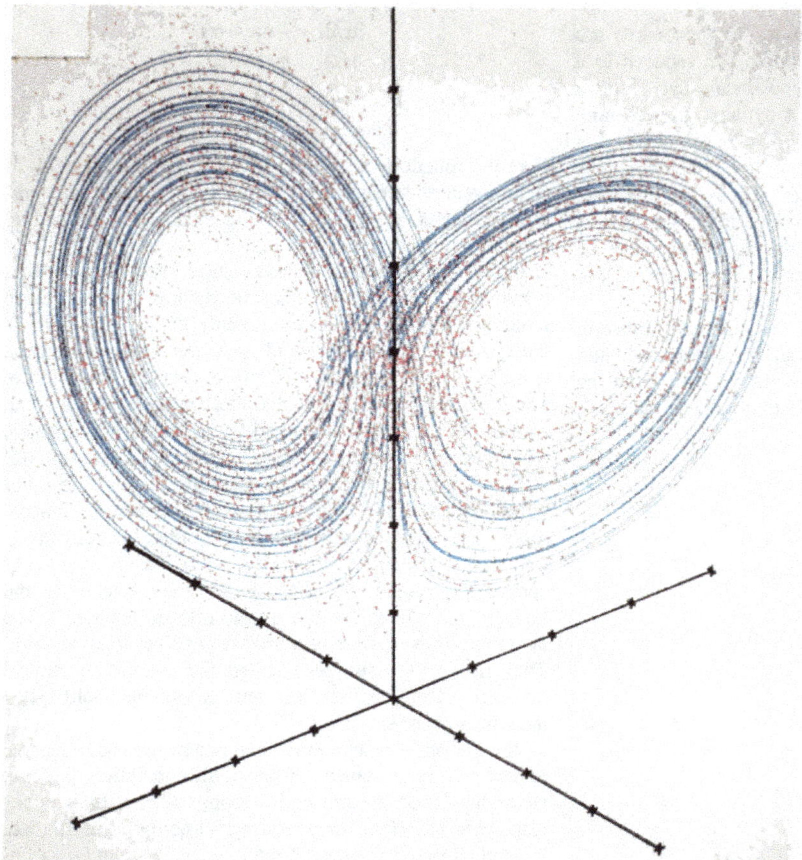

Figure 5. The Lorenz strange att After *Crutchfield et al.* [1986].

Nonlinear systems show many other behavior patterns that are incomprehensible for linearly oriented intuition: intermittency, i.e., alternation of different types of motion; transitions between chaotic and orderly behavior with the change of a "control parameter" (Figure 6); etc. There are also other kinds of self-organization in the lithosphere, of a statistical physics type; they are not considered here.

Chaotic patterns generate a hypnotic fascination, like a waterfall or burning wood. Accordingly, nonlinear scien has a dramatic vernacular; for example, the turn of a cur to a vertical asymptote is called blue-sky catastrophe. T temptation is not unknown to stampede into new applic tions of these patterns like the avant-garde of a lootir army, sticking romantic labels on all complicated phenon ena in sight. So far, geodesy and geophysics have wisel escaped this temptation. Actually, it is difficult to find

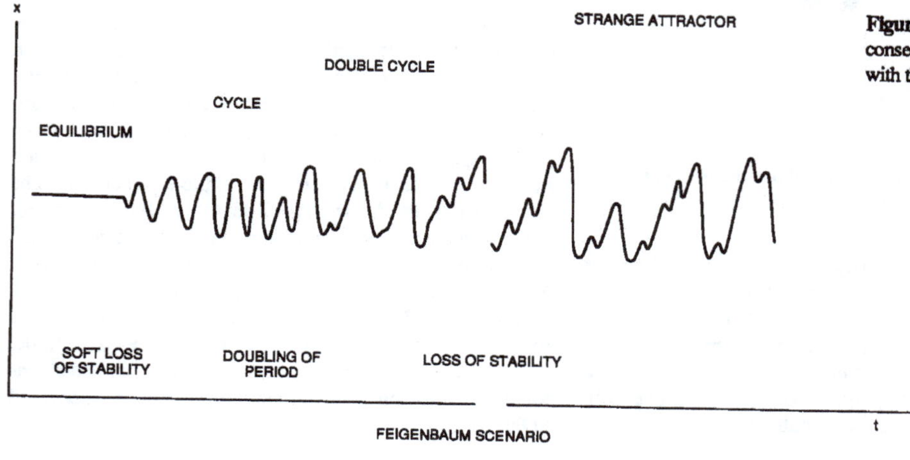

Figure 6. Feigenbaum scenar consecutive doubling of the peri with the change of control parame

useful model for the lithosphere. The difficulty of this kind for the modeling of climate is brilliantly described by *Puppi and Speranza* [1988, pp. 23, 25].

> . . . The reactions and the attitudes with respect to such complexity are linked between two extremes: on the one hand is he who identifies as the sole possible solution to the problem a meticulous treatment of every process operating on the climatic system, on the other hand is he who considers as the only hope that of "guessing" the right equations.

For the lithosphere, the problem seems even more difficult since the lithosphere, with its fractures and fixed inhomogeneities, does not mix as easily as the atmosphere. For some time and space scales, even the purely mechanical description of the lithosphere would probably require mechanics not of continuous but of fractal media.

Predictability

It may seem unexpected from the previous description, but some features of deterministic chaos can be predicted, though with limited accuracy and lead time. For example, transition to another branch in a Lorenz system (Figure 5) can be predicted with sufficiently small lead time, because it is always preceded by a sharp bend of a trajectory. This does not contradict the fact that the time of transition is completely random. Rather, the transition itself is not an instant jump but a continuous, though short, process, starting with a bend. One may hope that an earthquake is similarly a part of a more extended scenario.

Contrary to the Lorenz system, the lithosphere is hierarchical. The hope of predicting a feature of a hierarchical system with deterministic chaos lies in averaging (smoothing); the more averaged a feature is, the larger are the time and space scales in which prediction may be possible. Accordingly, in the dynamics of the lithosphere, and particularly in the problem of earthquake prediction, we first have to find, for a given time and space scale, an adequate smoothed description, without losing the phenomena we are interested in. A "smoothed description" is the reduction of the system to a small number of integral characteristics along with the relations between them (a low-dimensional phase space). The first steps have to be empirical—this field is in the same prehistoric stage as the theory of gravitation was before discovery of Kepler's laws. The next section considers empirical regularities relevant to the approach of a strong earthquake.

EARTHQUAKE PREDICTION

The above considerations are not confined to a specific scale. They seem consequential on many scales: the interaction between core, mantle, and crust; unexplained correlations of geochemical, tectonic, and geophysical fields over the Earth's surface; the evolution of the Earth—its whole tectonic history possibly may be rewritten as a sequence of instability episodes, like an intermittency.

We discuss here only the results relevant to the time scale of 1–10 years, which is intrinsic for earthquake prediction. This is particularly challenging, since the bulk of the lithosphere is inaccessible for the direct measurement of earthquake-related stress and strength fields. The lithosphere is virtually a black box. For each strong earthquake, it generates a multitude of smaller ones, and we shall try to use this multitude to diagnose the approach of a strong earthquake.

The occurrence of a particular earthquake cannot be entirely isolated from the dynamics of the whole lithosphere. That is why the problem of earthquake prediction consists of consecutive step-by-step narrowing of the time-space domain where strong earthquakes are expected; this process is in accordance with the hierarchical nature of the lithosphere.

Here, however, the formulation of the problem is truncated: we consider only one step, which is the prediction of the strongest earthquakes in a fixed territory with an accuracy of a few years and a few hundred kilometers. This is classified sometimes as intermediate-term and sometimes as long-term forecasting. The second classification is used in the papers referred to, but I will use the first one, which is apparently more common.

General Scheme

The symptoms of an incipient strong earthquake may be different from case to case in the same area and may vary even more from region to region. Nevertheless, we have to look for a uniform diagnosis, applicable at all times in different regions and magnitude ranges, otherwise the test of prediction algorithms would require hundreds of years. We try to overcome this contradiction by considering an integral representation of earthquake activity where the diversity of circumstances is smoothed while the premonitory phenomena are hopefully not smoothed away. The uniformity of diagnosis is made possible by normalization of earthquake flow as described by *Allen et al.* [1986] and *Gabrielov et al.* [1986*b*].

The scheme of diagnosis is outlined in Figure 7. It shows a sequence of earthquakes in a region. A vertical dashed line indicates a sliding moment of time. At each moment we look back in time and define several traits of the earthquake sequence within sliding time windows, shown by horizontal lines. We hope that variation of these traits will indicate the time of increased probability (TIP) of a strong earthquake.

The scheme is open for inclusion of any other hypothetically predictive phenomena, but so far only the traits of the five types listed in Figure 7 have been considered. They were selected for the following reasons.

The intensity of earthquake flow ("activity") was selected because abnormal activity and/or quiescence has

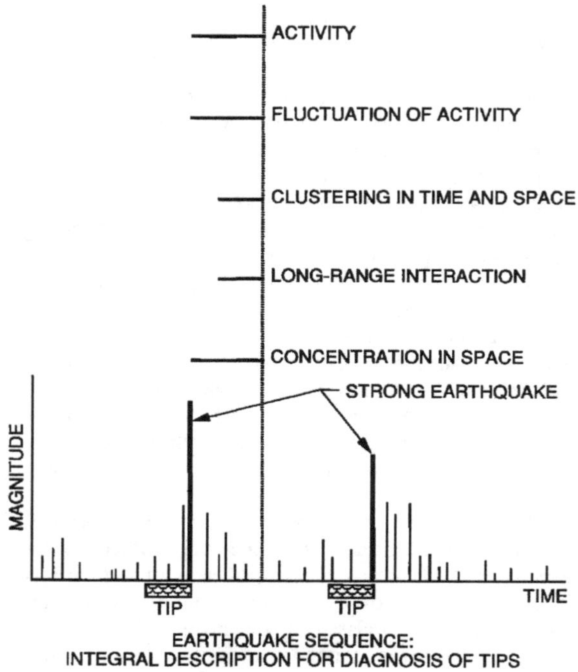

Figure 7. Scheme of description of earthquake sequence for diagnosis of times of increased probability (TIPs). After *Allen et al.* [1986] and *Gabrielov et al.* [1986*b*]. The earthquake sequence is described by several integral traits. Horizontal lines show the sliding time windows, on which different traits are defined. All windows end at the common moment of time, shown by the vertical line. The traits are attributed to this moment.

been reported before many strong earthquakes, separately or in combination. Among different kinds of quiescence, seismic gaps are most popular.

The next three traits represent phenomena common to many nonlinear systems: when instability approaches, the fluctuations increase as well as the response to excitation. An earthquake itself is a source of excitation. Therefore the response to it may be expressed in clustering and in an increase of the distance within which the earthquakes are not independent. Also, clustering is the only promonitory pattern for which statistical significance is strictly established at present [*Molchan et al.*, 1988].

Concentration is suggested by laboratory experiments with rocks; when the density of microfractures exceeds some rather universal threshold, the failure of a whole sample occurs [*Zhurkov et al.*, 1978].

The smoothing of the earthquake flow is achieved here in several steps:

1. The traits are defined for large areas and time windows.

2. Each trait is represented by several not independent functionals; for example, the intensity of the earthquake flow is estimated in different overlapping magnitude ranges.

3. We distinguish only large, medium, and small values of each functional, so that their definition is sufficiently robust.

4. Finally, the diagnosis of TIPs is based on the whole set of traits, admitting that each one by itself may be insufficient.

To diagnose the TIPs is a common problem of pattern recognition.

A Soviet-American team considered the seismic history of California and Nevada [*Allen et al.*, 1986; *Keilis-Borok et al.*, 1988]. These traits were compared at moments in the past either preceding or not preceding earthquakes with $M > 6.4$. By this comparison we found the rule for recognizing the TIPs. Roughly quantitatively, this rule is as follows: most of the traits, listed in Figure 7, simultaneously became more prominent during a period of 3–4 years preceding a strong earthquake. Before this period a relative quiescence often occurs. A similar analysis was made for the strongest earthquakes (magnitude 8 or more) over the whole world [*Keilis-Borok and Kossobokov*, 1986]. The corresponding algorithms are named CN (for "California and Nevada") and M8 (for "magnitude 8"). The sample of such retrospective diagnosis by algorithm CN is shown at the top of Figure 8.

Worldwide Tests

If reliable, these algorithms would have significant implications: that the blocks of lithosphere do show a collective behavior even in the erratic process of generating earthquakes; that our traits are promising candidates for basic integral characteristics of this behavior, so far somewhat clumsily formulated; and that we may predict strong earthquakes over the intermediate term. However, these conclusions are merely hypothetical, because in lieu of a theory we had to derive the algorithms by retrospective data fitting, and we improved the success-to-failure score by adaptation of many free parameters such as the length of time windows and the magnitude intervals. To test the prediction on independent data, we had to introduce an additional hypothesis: that the phenomena indicating the approach of a strong earthquake are similar in different regions, independent of the tectonic environment and the level of seismicity. This allows us to test the algorithms on independent data for other regions not involved in the data fitting. Technically, this is possible because the definition of all traits is normalized, so that looking at our representation of earthquake activity we would not distinguish a subduction zone from a mildly active platform, Kamchatka from Belgium.

So far, algorithms CN and M8 have been applied to the regions listed in Tables 1 and 2 [*Gabrielov et al.*, 1986*b*; *Keilis-Borok and Kossobokov*, 1986; V. I. Keilis-Borok and I. M. Rotwain, Diagnosis of time of increased probability of strong earthquakes in different regions of the

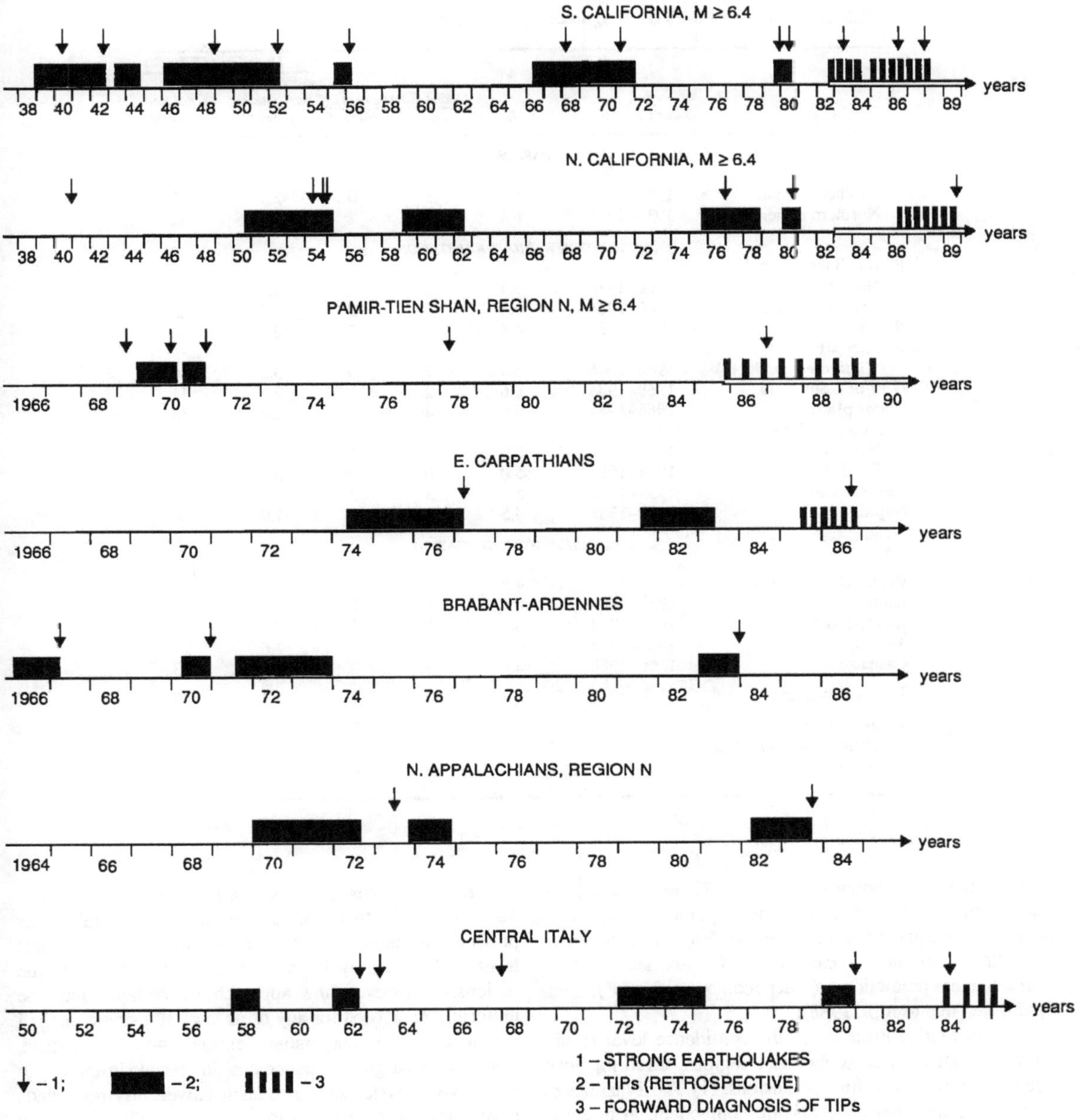

Figure 8. Test of the algorithm CN on independent data. After *Gabrielov et al.* [1986*b*] and V. I. Keilis-Borok and I. M. Rotwain (unpublished manuscript, 1988).

world: Algorithm CN, unpublished manuscript, 1988]. The results are summarized in those tables; for some specific regions they are shown in Figures 8 and 9. So far, TIPs were diagnosed in advance for six strong earthquakes: four in California-Nevada (including that on October 18, 1989), one in Pamir-Tien Shan, and one in Armenia (December 7, 1987).

It is encouraging that the success-to-failure score for these regions is, on the average, not much worse than it was for the regions used for the data fitting (compare, for example, the regions in Figure 8).

The practical value of such predictions is limited, though by no means annulled, by the fact that TIPs extended over areas 5–10 times larger in diameter than the

TABLE 1. Summary of the TIPs Diagnosed by Algorithm CN

Region	Time Considered	M_0	Strong Earthquakes "Predicted"	Strong Earthquakes Missed	Duration of TIPs: Average per Earthquake, years	Duration of TIPs: Total per Region, %
		"LEARNING"				
California-Nevada						
Southern region	1938–1983	6.4	9	0	2.2	42
Northern region	1938–1983	6.4	4	2	2.6	22
		APPLICATION WITHOUT ADAPTATION				
Pamir–Tien Shan						
Northern region	1966–1986	6.4	2	2	1.5	14
Southern region	1966–1986	6.4	5	0	1.9	45
Baikal	1966–1984	6.4	0	0	0	0
E. Carpathians						
(intermediate depth)	1966–1986	6.4	2	0	3.3	30
Gulf of California	1968–1984	6.6	2	1	2.6	31
Cocos plate	1968–1984	6.5	4	0	1.6	38
N. Appalachians						
Northern region	1964–1985	5.0	1	1	3.2	28
Southern region	1964–1985	5.0	0	0	0	0
Central Italy	1954–1986	5.6	3	2	2.0	18
Nepal	1970–1988	6.5	3	0	1.9	30
		APPLICATION WITH ADAPTATION				
Brabant-Ardennes	1966–1987	4.5	3	0	1.9	30
Kangra	1968–1988	6.4	1	0	0.3	2
Gaval-Kumaon	1968–1988	6.4	1	1	3.4	18
Assam	1968–1988	6.4	2	1	2.6	27
Caucasus	1966–1988	6.4	3	2	1.0	13
Total or average			45	12	1.9	23
Total or average, without "learning" and adaptation			22	6	1.8	23

source of the upcoming earthquake. There is some hope indicating, however, that we can pinpoint the place of a coming earthquake within about two lengths of its source (S. Smith et al., Localization of intermediate term earthquake prediction ("Mendocino scenario"), unpublished manuscript, 1988).

A quantitative estimate of the confidence level is still difficult, but the results for these regions, taken together, do support the following conclusions: (1) The dynamics of the lithosphere, having a chaotic component, also shows collective behavior. (2) The traits of the earthquake activity, considered here, provide an integral description of this behavior, which is similar in a wide range of regions, and are intrinsically relevant to the approach of a strong earthquake. About 80% of strong earthquakes may be predicted on the basis of these traits, with alarms occupying 10–20% of space-time.

Though within the spirit of ideas about deterministic chaos, this similarity seems very controversial so far. I have heard exclamations (not simply questions): how could it be possible that the same diagnostics are uniformly applicable in such different seismotectonic environments while each single region is eminently inhomogeneous and the stress field changes from earthquake to earthquake? The answer is that the choice of integral traits was probably fortunate. For example, the readers of this lecture, if any, may be at least as different as seismic regions. However, the approach of collapse may be uniformly diagnosed for any of us, e.g., if a single integral parameter, body temperature, exceeds +44°C. Indeed, some universal global phenomena do precede instability in very diverse systems. Since such universality transcends from brain to atmosphere to nonlinear oscillator, it may well transcend the mere differences between seismic regions. But it is difficult to accept this similarity if we are traditionally focused on a more detailed scale.

Two simple models of the lithosphere may illustrate both the individuality of earthquake activity and the universality of integral laws. Figure 10 shows a computer simulation of the interaction of lithospheric blocks [*Gabrielov et al.*, 1986*a*]. Energy is provided by the shear movement of boundaries, indicated by the hatched rectangles and arrows. The model generates earthquakes, in stick-slip fashion. In spite of its extremely simple geometry, the earthquake activity is chaotic. Not only is it unstable to the change of structure, but it has bifurcations:

TABLE 2. Summary of the TIPs Diagnosed by Algorithm M8

Region	M_0	*Time Considered*	*Strong Earthquakes "Predicted"*	*Strong Earthquakes Missed*	*Space-Time Volume, % of Total: TIPs*	*Space-Time Volume, % of Total: Tes*
		"LEARNING"				
1. The world	8.0	1967–1982	5	2	5	3
		APPLICATION WITHOUT ADAPTATION				
2. Central America	8.0	1977–1986	1	--	16	16
3. The Kurils and Kamchatka	7.5	1975–1987	2	--	17	7
4. Japan and Taiwan	7.5	1975–1987	5	1	20	8
5. Southern America	7.5	1975–1986	3	--	18	13
6. Western United States	7.5	1975–1987	--	--	5	5
7. Southern California	7.5	1947–1987	1	--	12	1
8. Western United States	7.0	1975–1987	2	--	24	10
9. Baikal and Stanovoy Range	6.7	1975–1986	--	--	0	0
10. The Caucasus	6.5	1975–1987	2	1	12	7
11. East central Asia	6.5	1975–1987	4	1	24	11
12. Eastern Tien Shan	6.5	1963–1987	4	--	27	15
13. Western Turkmenia	6.5	1979–1986	--	--	0	0
14. Apennines	6.5	1970–1986	1	--	10	1
15. Koyna reservoir	4.9	1975–1986	1	--	42	28
		APPLICATION WITH ADAPTATION				
16. Greece	7.0	1973–1987	3	--	18	10
17. The Himalayas with surroundings	7.0	1970–1987	2	--	8	2
18. Vrancea	6.5	1975–1986	2	--	58	29
19. Vancouver Island	6.0	1957–1985	4	--	20	14
Total or average			42	5	18	10
Total or average, without "learning" and adaptation			26	3	16	9

the sequence of earthquakes changes drastically when initial conditions are changed just by the lowest few digits in the computer memory. But many integral traits remain the same, e.g., the linear frequency-energy relation, migration, clustering, and average annual energy release. Even the same algorithms for the diagnosis of TIPs are applicable here, though the success-to-failure score is lower, of course.

It seems suggestive that the observed regularities are reproduced by such a simple model, as in Figure 10. They may be intrinsic to some general type of deterministic chaos and not to an exclusive feature of the lithosphere.

Another model, more of a statistical physics type, similar to the well-known "life game," is shown in Figure 11. It illustrates how small displacements of elementary blocks can self-organize into a strong earthquake. Elements are elastic discs which can move and rotate, while general compression holds them together. When the stress between two discs exceeds friction, slip occurs, releasing energy; one elementary slip may trigger another. The discs show prominent collective behavior. The most stressed (darkest) discs are self-organized in these linear configurations which eventually become unstable; this triggers the "strong earthquakes" in the model. Again, a slight initial perturbation may drastically change specific patterns; there is no answer as to why this delineation is here and a minimum of stress is there. But, again, the model displays some integral traits of real earthquake activity.

A single disk (if having intellect) would hardly believe that his erratic pressure history from one slip to another, is averaged to some universal regularities. Predictable collective behavior is not uncommon, however, especially in systems with complicated interactions on an elementary level. This can be illustrated by a system of interactive electorate blocks in the American midterm Senatorial elections [*Lichtman and Keilis-Borok*, 1989]. The situation before past elections was described by averaged functionals (Table 3)—the "yes" or "no" answers to questions referring to the state as a whole. The questions and the pattern recognition prediction rule are the same for all states and all election years, while both are indeed diverse. Nevertheless, the prediction, published in advance, was correct for 30 out of 34 elections in 1986 (Table 4). The confidence level is above 99.9%. It remains above 97% if we allow only for the outcomes

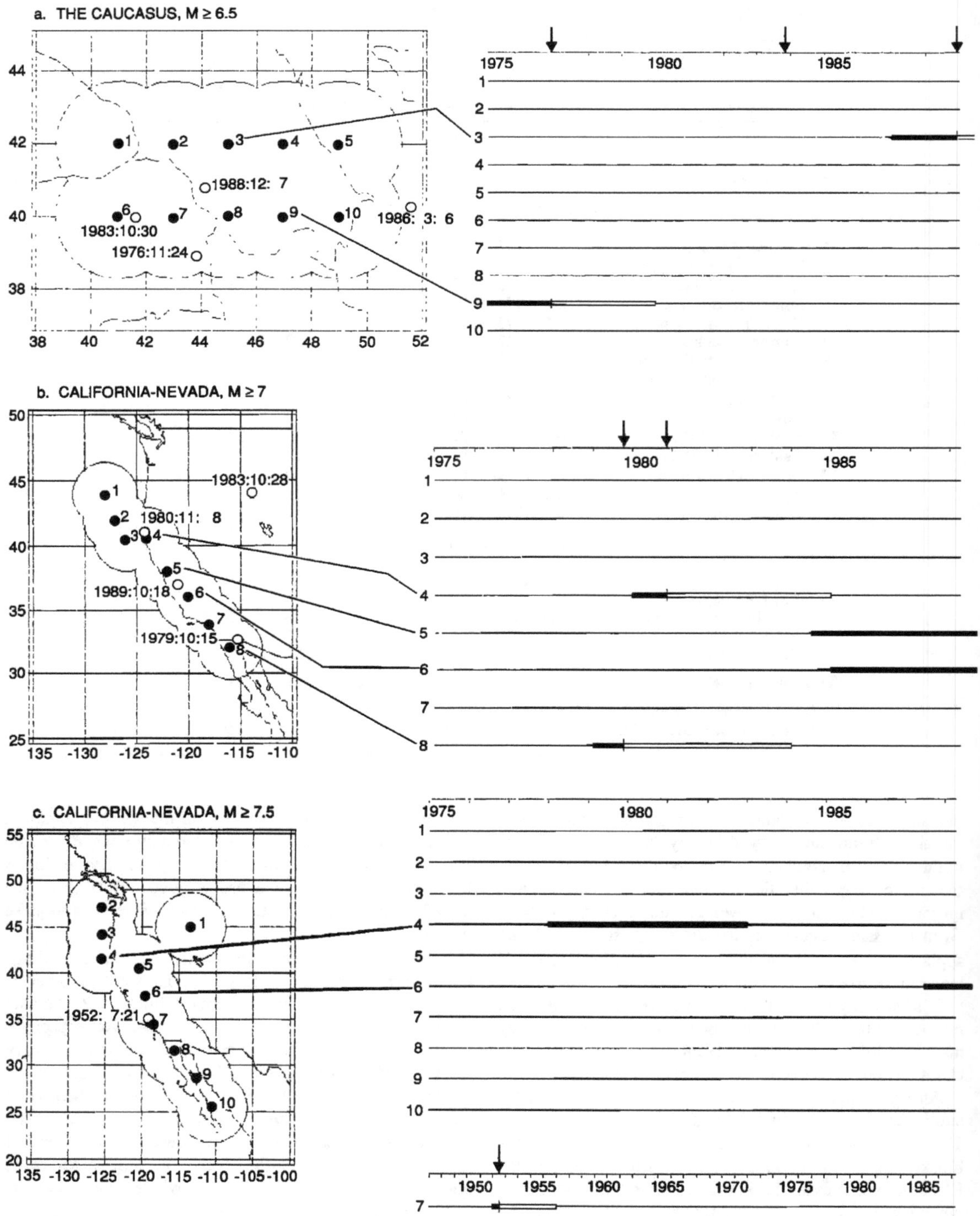

Figure 9. Test of the algorithm M8 on independent data. After *Keilis-Borok and Kossobokov* [1986]. Each line is a time scale for an area, centered around a point indicated on the map. The heavy lines indicate times of increased probability (TIPs). Arrows show the moment of an earthquake; its magnitude is indicated at the upper time scale. (*a*) Caucasus, $M \geq 6.5$. (*b*) California-Nevada, $M \geq 7$. (*c*) California-Nevada, $M \geq 7.5$. The TIPs preceding the Armenian earthquake on December 7, 1988, and the Santa Cruz earthquake on October 18, 1989, were reported in advance.

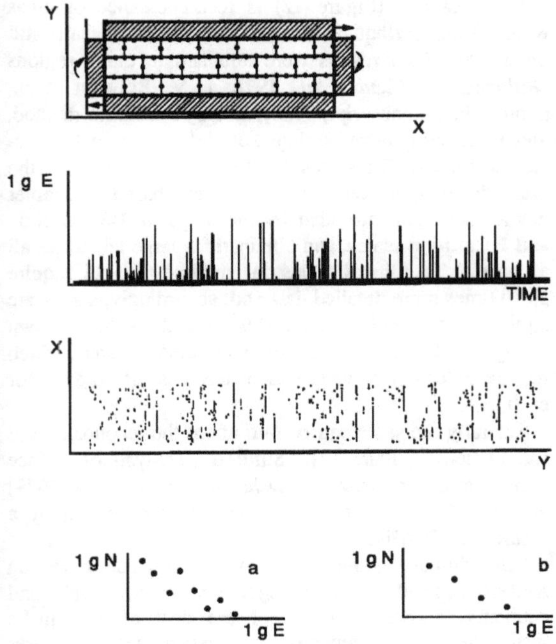

Figure 10. Unstable earthquake activity generated by a simple model of interacting blocks [after *Gabrielov et al.*, 1986*a*]. (Top) Geometry of the model. (Middle) Time sequence of earthquakes, *T* being the time and *E* energy; earthquake sources on (*T*, *x*) plane. (Bottom) Magnitude-frequency relation of occurrence relation for (*a*) small and (*b*) large energy intervals.

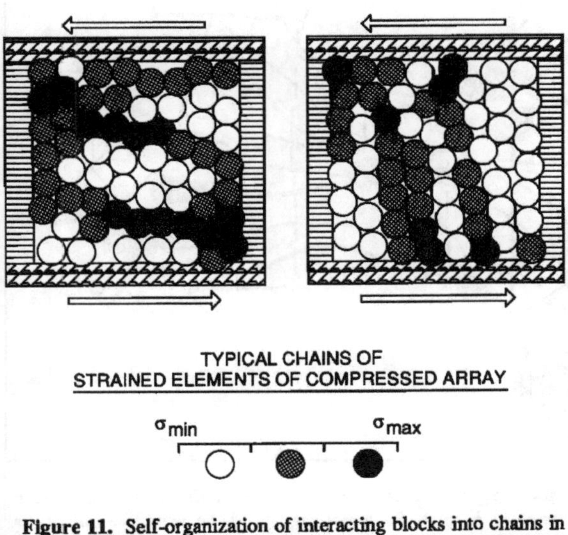

Figure 11. Self-organization of interacting blocks into chains in a simple model (after V. F. Pisarenko et al., unpublished manuscript, 1989).

TABLE 3. The Integral Parameters Depicting a Situation Before the American Midterm Senatorial Elections

Incumbent party candidate is sitting senator?
Incumbent party candidate is major national figure?
No serious contest for incumbent-party nomination?
Incumbent party got 60%+ in the previous election?
Challenging party candidate is not a national figure or past or present governor or member of Congress?
No serious contest for challenging party nomination?
Incumbent party candidate not of the same party as the President?
Incumbent party candidate outspends challenger by at least 10%?

After A. J. Lichtman and V. I. Keilis-Borok (Pattern recognition applied to America mid-term Senatorial elections, unpublished manuscript, 1988).

TABLE 4. Prediction of the Outcome of the 1986 Senatorial Elections From Advance Publication in *Washingtonian* (November 1986, p. 144).

State	*Keys Against*	*Incumbent-Party Candidate*	*Challenger*
		REPUBLICAN SEATS	
Alabama	5	☐ Jeremiah Denton	☑ Richard Shelby
Alaska	3	☑ Frank Murkowski	☐ Glenn Olds
Arizona	4 *	☑ John McCain	☐ Richard Kimball
Florida	5 *	☐ Paula Hawkins	☑ Bob Graham
Georgia	5	☐ Mack Mattingly	☑ Wyche Fowler
Idaho	4 *	☑ Steve Symms	☐ John Evans
Indiana	3	☑ Dan Quayle	☐ Jill Long
Iowa	3	☑ Charles Grassley	☐ John Roehrick
Kansas	2	☑ Robert Dole	☐ Guy MacDonald
Maryland	6	☐ Linda Chavez	☑ Barbara Mikulski
Nevada	6	☐ James Santini	☑ Harry Reid
New Hampshire	3	☑ Warren Rudman	☐ Endicott Peabody
New York	4 *	☑ Alphonse D'Amato	☐ Mark Green
• North Carolina	4 *	☑ James Broyhill	☐ Terry Sanford
• North Dakota	2	☑ Mark Andrews	☐ Kent Conrad
Oklahoma	4 *	☑ Don Nickles	☐ James Jones
Oregon	3	☑ Robert Packwood	☐ Rick Bauman
• Pennsylvania	5	☐ Arlen Specter	☑ Robert Edgar
South Dakota	6	☐ James Abdnor	☑ Thomas Daschle
Utah	1	☑ Jake Garn	☐ Craig Oliver
• Washington	4 *	☑ Slade Gorton	☐ Brock Adams
Wisconsin	4 *	☑ Robert Kasten	☐ Ed Garvey
		DEMOCRATIC SEATS	
Arkansas	2	☑ Dale Bumpers	☐ Asa Hutchinson
California	3	☑ Alan Cranston	☐ Ed Zschau
Colorado	4 *	☑ Timothy Wirth	☐ Ken Kramer
Connecticut	2	☑ Christopher Dodd	☐ Roger Eddy
Hawaii	1	☑ Daniel Inouye	☐ Frank Hutchinson
Illinois	3	☑ Alan Dixon	☐ Judy Koehler
Kentucky	3	☑ Wendell Ford	☐ Jackson Andrews
Louisiana	4	☑ John Breaux	☐ Henson Moore
Missouri	5 *	☐ Harriet Woods	☑ Kit Bond
Ohio	1	☑ John Glenn	☐ Thomas Kindness
South Carolina	1	☑ Ernest Hollings	☐ Henry McMaster
Vermont	3	☑ Pat Leahy	☐ Richard Snelling

If four or fewer keys fall, the incumbent-party candidate wins; if five or more fall, the challenger wins.
☑ Winners ☐ Losers.
* Prediction would change if the spending key changed.
• Incorrect prediction.

which the experts regarded as uncertain. A similar prediction rule was formulated for the Presidential elections [*Lichtman and Keilis-Borok,* 1981]. This was based on other but equally simple integral parameters. This rule cannot claim statistical significance, since only two elections, 1984 and 1988, happened after the rule was published. It is encouraging, nevertheless, that correct predictions could be made about a year in advance.

The differences of subject matter notwithstanding, these examples illustrate how the behavior of a chaotic system may become predictable after proper integration. Accordingly, the existence of uniform diagnostics of TIPs in diverse regions is not so paradoxical as it may seem from a too detailed perception of the lithosphere.

CONCLUSION

The integral approach to other studies of the lithosphere, besides its dynamics on short time scales, deserves attention. It is a major and, I believe, the major workhorse in the present perestroika of the solid-Earth sciences: their integration and globalization, integration of basic and applied problems, change of theoretical base, etc.

Following are a few more examples. Since most of the studies in this direction are also in similar preliminary stages—an empirical search for decisive integral parameters—I will discourse once more from the spirit of a Union Lecture, which should be an ivory tower exaltation. After the circus of elections, the next examples belong to the marketplace: industrial applications of geodesy and geophysics.

One example (Figure 12) is reconnaissance of areas where strong earthquakes, $M \geq 7$, are possible even if still unknown. Such results were obtained in many regions worldwide [*Gelfand et al.*, 1976; I. M. Rotwain et al., unpublished manuscript, 1988], also by a uniform method, and most subsequent earthquakes did appear in the prescribed places. These results illustrate the power of the hierarchical approach, since they are based on rather nondetailed data, provided by the maps of 1:2.5 million and 1:1 million scales, and obtained for large territories all at once. Traditional industrial methods would require 5–10 times more detailed data and, accordingly, a separate survey of each building site. Results such as those shown in Figure 12 drastically reduce the number of areas which require industrial surveys which may cost up to $\$10^7$ for each spot.

Literally the same can be said about the reconnaissance for oil fields (Figure 13). Similarly, analysis of surface waves from earthquakes [*Levshin and Berteussen,* 1979] allows reduction of expensive seismic prospecting by a factor of 5–10 (Figure 14).

Also similarly, game theory analysis of existing data on neotectonics and seismicity, together with demography and economy, allows reduction of the cost of earthquake insurance and optimization of preparedness measures [*Keilis-Borok et al.*, 1984].

All this calls for reconsideration of the strategy of research, especially, but not only, in developing countries. It may be better to start with intense analysis of existing data, rather than with expansion of local observations and hardware. In this way one may save a good part of the cost of industrial surveys. It may even be the only possible way when dealing with chaotic phenomena.

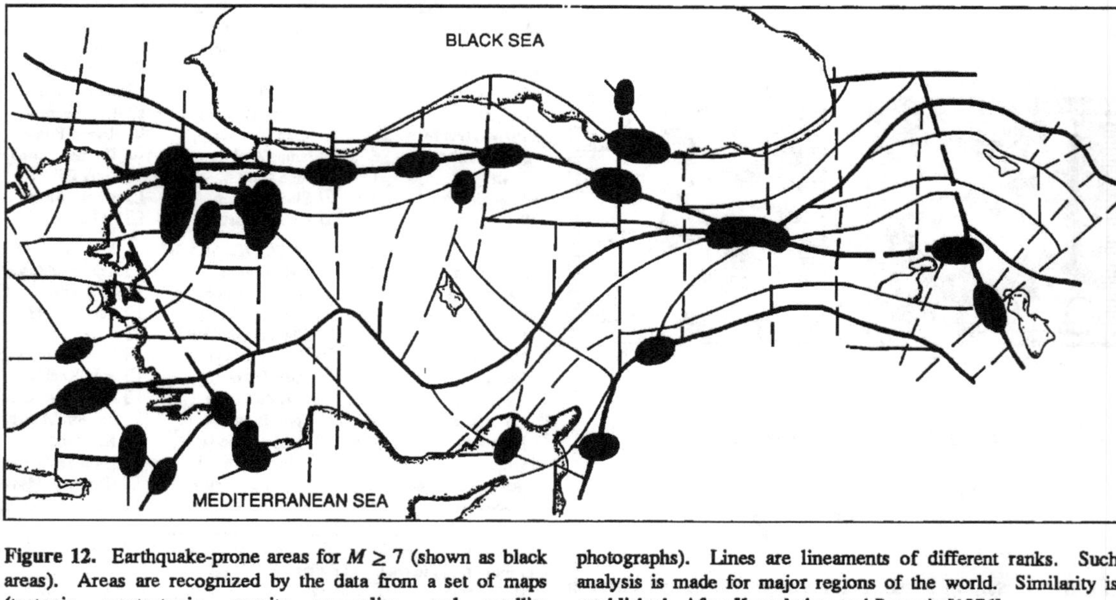

Figure 12. Earthquake-prone areas for $M \geq 7$ (shown as black areas). Areas are recognized by the data from a set of maps (tectonic, neotectonic, gravity anomalies, and satellite photographs). Lines are lineaments of different ranks. Such analysis is made for major regions of the world. Similarity is established. After *Kossobokov and Rotwain* [1976].

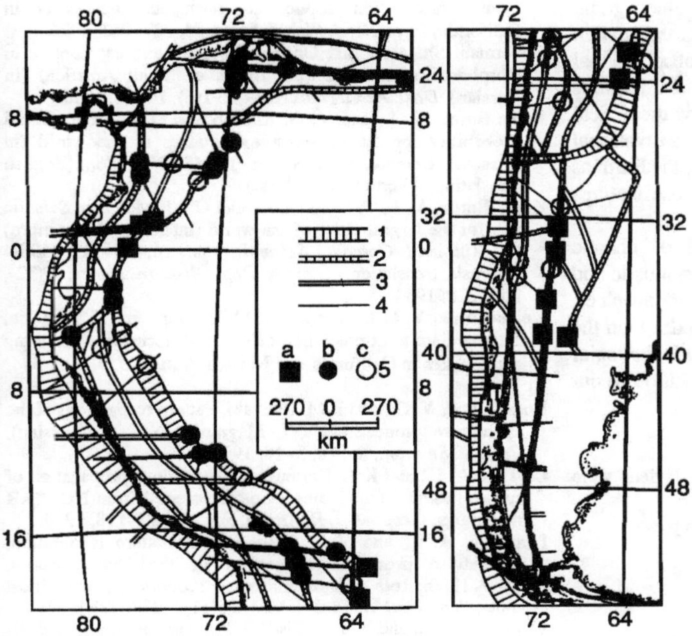

Figure 13. Recognition of areas promising for super large oil deposits. After *Guberman et al.* [1986]. The data used were taken from a set of maps (see caption to Figure 12). The method is tested on independent data.

NOTATIONS

1 – OCEANIC TRENCHES AND CONTINENTAL SLOPES;
2 – MAJOR STRIKE-SLIP FAULTS;
3 AND 4 – LINEAMENTS OF THE 2ND AND THE 3RD RANK;
5 – LARGE DEPOSITS
a – known, b – probable, c – possible

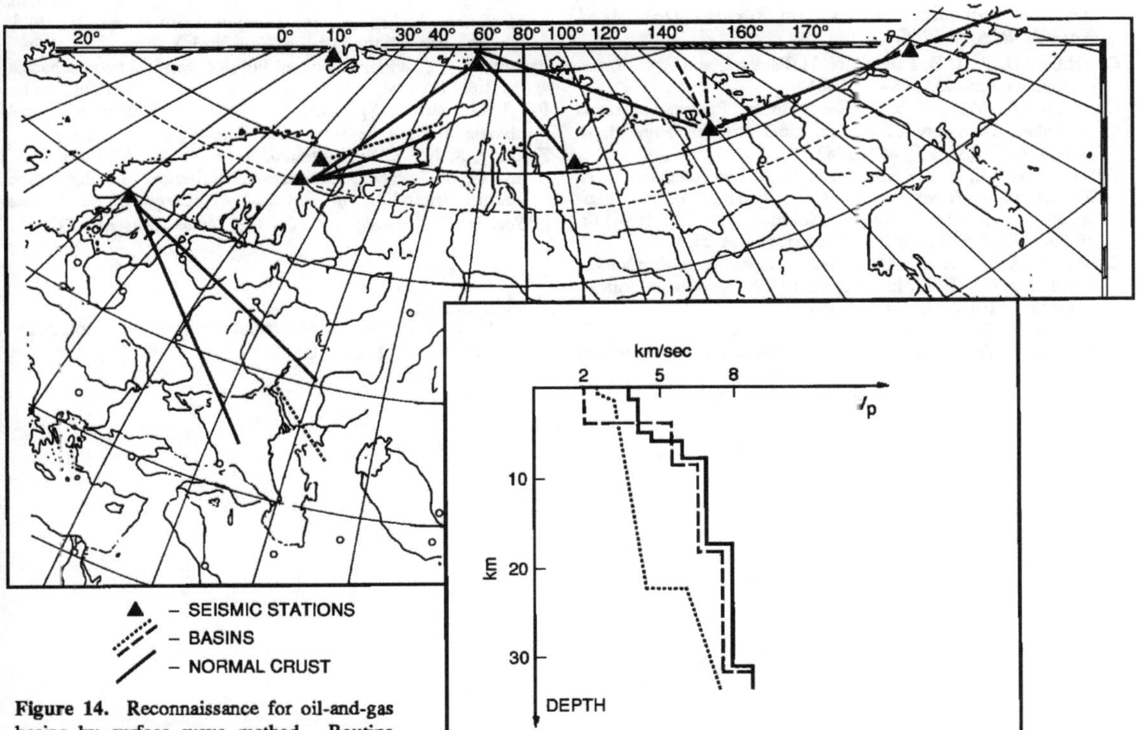

Figure 14. Reconnaissance for oil-and-gas basins by surface wave method. Routine seismological observations are used. After *Levshin and Berteussen* [1979].

In summary, I believe that among the other grand possibilities for better understanding of the Earth, modern nonlinear science in combination with global observational networks does deserve attention. It promises and requires further consolidation of Earth sciences on a new theoretical basis, consolidation and trimming of our observational base and our software, and better perception of the Earth as a whole with sufficient resolution to depict those features important for the economy.

We may even expect that a new kind of chaotic behavior, probably intermediate between deterministic and statistical types, will be discovered under our auspices. This would reconfirm the tradition which is older than the present civilization: that the study of the Earth is the major source of basic milestone concepts which revolutionize our perception of Nature.

ACKNOWLEDGMENTS. Discussions with E. Nyland were inspiring and helpful.

H. Jay Melosh was the editor responsible for this paper.

REFERENCES

Allen, C., V. I. Keilis-Borok, I. M. Rotwain, and K. Hutton, A set of long-term seismological precursors; California and some other regions (in Russian), *Comput. Seismol.*, no. 19, 24–34, 1986. (English translation, Allerton, New York.)

Barenblatt, G. M., V. I. Keilis-Borok, and A. S. Monin, Filtration model of earthquakes sequence (in Russian), *Dokl. Akad. Nauk SSSR*, April 1985.

Crutchfield, J. P., J. D. Farmer, N. H. Packard, and R. S. Shaw, Chaos, *Sci. Am.*, *255*, 46–57, 1986.

Gabrielov, A. M., and V. I. Keilis-Borok, Patterns of stress corrosion: Geometry of the principal stresses, *Pure Appl. Geophys.*, *121*(3), 477–494, 1983.

Gabrielov, A. M., V. I. Keilis-Borok, T. A. Levshina, and V. A. Shaposhnikov, Block model of the dynamics of the lithosphere (in Russian), *Comput. Seismol.*, no. 19, 168–178, 1986*a*. (English translation, Allerton, New York.)

Gabrielov, A. M., et al., *Algorithms of Long-Term Earthquake Prediction*, CERESIS, Lima, Peru, 1986*b*. (Available from A. Giesecke, CERESIS, Lima 14, Peru.)

Gelfand, I. M., Sh. A. Guberman, V. I. Keilis-Borok, L. Knopoff, F. Press, I. Ya. Ranzman, I. M. Rotwain, and I. M. Sadovsky, Pattern recognition applied to earthquake epicenters in California, *Phys. Earth Planet. Inter.*, *11*, 227–283, 1976.

Guberman, Sh. A., et al., On criteria of oil and gas content in morphostructural knots (the Andes of South America) (in Russian), *Dokl. Akad. Nauk SSSR*, *291*(6), 14–36, 1986.

Keilis-Borok, V. I., and V. G. Kossobokov, Time of Increased Probability for the strongest earthquakes of the world (in Russian), *Comput. Seismol.*, no. 19, 48–57, 1986. (English translation, Allerton, New York.)

Keilis-Borok, V. I., T. L. Kronrod, and G. M. Molchan, Seismic risk for the largest cities of the world (intensity VIII or more) (in Russian), *Comput. Seismol.*, no., 16, 99–117, 1984. (English translation, *Geneva Pap. Risk Insurance*, *9*(32), 255–270, 1984.

Keilis-Borok, V. I., L. Knopoff, I. M. Rotwain, and C. R. Allen, Intermediate-term prediction of times of occurrence of strong earthquakes in California and Nevada, *Nature*, *335*, 690–694, 1988.

Kossobokov, V. G., and I. M. Rotwain, Pattern recognition of the earthquake prone areas, VI, Magnitude ≥ 7 (in Russian), *Comput. Seismol.*, no. 10, 3–18, 1976.

Levshin, A. L., and K.A. Berteussen, Anomalous propagation of surface waves in the Barents Sea as inferred from NORSAR recordings, *Geophys. J. R. Astron. Soc.*, *56*, 97–118, 1979.

Lichtman, A. J., and V. I. Keilis-Borok, Pattern recognition, applied to presidential elections in the United States, 1860–1980: Role of integral social, economic, and political traits, *Proc. Natl. Acad. Sci. USA*, *78*(11), 7230–7234, 1981.

Lichtman, A. J., and V. I. Keilis-Borok, Aggregate-level analysis and prediction of midterm senatorial elections in the United States, 1974–1986, *Proc. Natl. Acad. Sci. USA*, *86*, 10,176–10,180, 1989.

Molchan, G. M., O. E. Dmitrieva, and I. M. Rotwain, Statistical analysis of the results of long-term prediction, based on bursts of aftershocks (in Russian), *Comput. Seismol.*, no. 21, 1988. (English translation, Allerton, New York.)

Pertsov, N. V., and B. S. Kogan, Physicochemical geomechanics (in Russian), *Dokl. Akad. Nauk SSSR*, *235*, 1375–1378, 1981.

Puppi, G., and A. Speranza, The physical problem of global climate on Earth, *Bologna Alma Mater Stud.*, no. 1, 1988.

Zhurkov, S. N., V. S. Kuksenko, V. A. Petrov, V. I. Saveljev, and U. S. Sultanov, Concentration criterion of volume-type fracture in solids, in *Physical Processes in Earthquake Sources* (in Russian), pp. 101–116, Nauka, Moscow, 1978.

V. I. Keilis-Borok, Geophysical Committee, Molodezhnaya 3, Moscow 117296, USSR.

Kantorovich L.V, *Keilis-Borok V.I.*

EARTHQUAKE PREDICTION AND DECISION-MAKING: SOCIAL, ECONOMIC AND CIVIL PROTECTION ASPECTS.

"Of course, things are complicated ... But, in the end every situation can be reduced to a simple decision: do we act, or not? If yes, in what way?"

E.Burdick

"Far better is an approximate answer to the right question, which is often vague, than an exact answer to the wrong question which can always be made precise."

J.Tukey

This review outlines the framework of decision-making, concerning time-dependent seismic risk: models, data base and modes of necessary actions. Hopefully it may also help to get oriented in the vast literature; major topical reviews are referred to. Main conclusion can be formulated as follows:

The earthquake predictions, which are feasible at the present, that is limited both in accuracy and reliability, can be used nevertheless to reduce the casualties and economic losses. The key to this is diversity and preparedness of safety measures, arranged into an escalating scale and combined in flexible scenaria of response to predictions.

1

THE PROBLEM

A wide set of different, rather diverse and often costly safety measures (SM) may be undertaken to reduce the damage from the earthquakes. The problem is to select the realistic and most efficient set of SM, given the current estimation - when and where a destructive earthquake may occur and what damage it may cause; such estimations are provided by earthquake prediction.

Decision-making requires the estimation of *total damage*, for a time-interval, territory and safety measures considered.

The problem is intrinsically probabilistic for two reasons:

- the limited knowledge of all relevant phenomena and
- probabilistic nature of these phenomena.

Let us outline the basic components involved.

THE MECHANISMS by which the damage from the earthquakes are inflicted, are exceedingly diverse. They include partial or total destruction of buildings, lifelines and other constructions; damage to the objects inside the buildings; release of dangerous materials - radioactive, toxic, genetically active etc; fires; triggering of the natural phenomena, such as floods, avalanches, landslides, tsunamis, liquefaction of the ground etc.

As consequential are *socio-economic mechanisms:* disruption of vital services - supply, medical, financial, law enforcement etc; epidemics; disruptive anxiety of population; profiteering and crime; interruption of normal functioning of economy; destabilization of economy industrial production, prices, credit, commerce, stock-market, employment etc.

These consequences are developing *in different time-scales,* from immediate damage to long-term chain reaction, spreading over large regions.

THE DAMAGE, inflicted by these calamities, may be described for the purposes of decision-making by three integral estimations:

- possible casualties,
- economic losses and
- damage to ecology.

The number of casualties is a vector, differentiating the death toll and the wounds of different degrees.

Also useful can be *indirect measures* of the damage, which show just the score of vulnerable objects, affected by strong motion of a given intensity during a given time interval. This score may represent the number of people, and/or economic valuables, and/or production capacity, and/or the area affected, and/or number of specific constructions etc.

All measures are essentially time-dependent, due to such factors as population growth, rate of interest, cost dynamics, dynamics of constructions and ground, development of infrastructures etc.

The past experience concerns mainly the damage in a locality, to a construction, ground etc. *The integrated damage,* affecting socio-economic infrastructures and ecology in epicentral zone and eventually spreading beyond it, has been studied in fewer cases and it is more difficult to foresee: the infrastructures are changing and ecology becomes more fragile annulling past experience.

It is clear only, that the gross vulnerability of mankind to the earthquakes is critically growing due to:

- population growth;

- proliferation of high-risk constructions such as nuclear power plants, high dams, toxic waste disposals, etc;
- destabilization of ground in densely populated areas.

For example: hundreds of thousands instead of thousands people live on the shores, which had been covered by tsunami in Lima, Lisbon, Messina; an earthquake in E.Carpatheans can affect several nuclear power plants simultaneously; seismically vulnerable became the cities in low-seismicity areas, including New York, Moscow, Kiev, Rome ... .

Today a single earthquake may take up to a million lives; cause material damage up to trillion USD with chain reaction expanding to economic depression; destabilize social order in a country.

Ecological catastrophe can be triggered by the earthquakes e.g. due to release of toxic materials.

"Surely these problems ... are the moral equivalent of the war; they are a threat to civilization's survival as great as any posed by Hitler, Stalin and the atom bomb... Only an international effort will be able to cope (with them)." (J.Wiesener)

THE DIVERSITY OF SAFETY MEASURES, which may reduce the damage, can be outlined as follows.

Permanent SM maintained during the decades.

- restriction of land - use, especially for high-risk objects and earthquake - inducing activities;
- building code, demanding reinforcement of constructions;
- tightening of safety regulations;
- enforcement of public safety services;
- insurance and special taxation;
- observations and data-processing to estimate seismic risk and

to monitor earthquake precursors;

- preparation of the response to time-prediction, and of post-disaster activities: *planning; legal background; accumulation of the stand-by resources; simulation alarms; training of population* etc.

Temporary safety measures activated in response to time prediction:

- enhancement of permanent measures;
- emergency (up to martial) legislation, to facilitate the rational response to prediction;
- mandatory regulation of economy;
- neutralization of the sources of high risk: life-lines; nuclear power plants; chemical plants; unsafe buildings etc, up to suspension of operation, partial disassembling, demolition...;
- evacuation of population and highly vulnerable objects (e.g. schools and hospitals);
- mobilization of post-disaster emergency services;
- preparation of measures for long-term post-disaster relief (restoration of dwellings, jobs, production, credit etc.);
- monitoring of socio-economic changes, and prevention of prediction - induced hazards.

These measures are in different forms applicable on international, national, provincial and local levels.

EARTHQUAKE PREDICTION which would allow to activate the temporary safety measures,consists of gradual, step-by-step reduction of {time-space-magnitude} domain, where a strong earthquake has to be expected.

The following stages are distinguished (time-interval, for which prediction is made are indicated for each stage):

Background *(100 years):* mapping of long-term characteristics of seismic risk, most often represented by seismic zonation.

Long-term: tens of year.

Intermediate-term: years.

Short-term: weeks or days.

Immediate: hours or less.

In other words the estimations of probability of the destructive earthquakes are changing with time, depending on the current dynamics of earthquake-phone zones, and of our current understanding of it. The safety measures has to be escalated or deescalated accordingly.

The area, where the earthquake is predicted depends on the size of expected source. It is usually assumed, that the linear dimension of this area is proportional to this side.

As we know from other sections of this Symposium, only the background stage is an established routine.

Long-term prediction is based mostly on the concept of seismic gaps, with its track record (success and failures score) recently disputed.

Intermediate-term prediction, based mainly on premonitory seismicity patterns, is relatively wider tested; the large rate of false alarms is its major handicap.

Prediction on other stages remains so far occasional, without a systematic track record.

Only for one precursor, intermediate-term one, a statistical significance is established.

All consecutive stages of prediction are accomplished only in exceptional cases; we are aware of one case - Haicheng catastrophic earthquake in PRC, 1975.

THE MODELS

To estimate the total damage one need the models of:

- earthquake occurrence;
- strong ground motion, generated by a single earthquake;
- damage, caused by a single episode of strong motion in a given place;
- time-dependence of this damage.
- territorial distribution of vulnerable objects

These models have to be defined as probabilistic distribution functions.

Then the distribution of total damage represents the following process:

- destructive earthquakes occur according to the first model;
- each earthquake causes strong motion, in the area around the epicenter according to the second model;
- if vulnerable objects are situated in this area, the damage is inflicted, according to the third and fourth models.
- the damage from different earthquakes is accumulated, according to the two last models, with time-dependent weights, in a total damage.

The data of Geophysics, Geology, Engineering, Economy and Demography have to be this integrated.

Geophysical and geological data determine the models of the earthquake occurrence and together with engineering data - the models of strong motion.

The data of Engineering, Economy and Demography determine the damage caused by an episode of strong motion and, in another time-scale, the integrated damage over the years to come.

Only approximate and/or simplified models are realistic so far. Nevertheless they can provide a valuable quidance for reduction of the damage, if the formulation of the problem is simplified according to the way, in which estimation will be used.

Earthquake occurrence.

For the background seismicity, the most comprehensive model would be given by the number of earthquakes N, within a time-interval T, hypocentral region G and magnitude range M: F(N/T,G,M).

Poison or branching Poison distributions are mainly used so far, with intensity constant or slowly changing in time.

Of independent importance are two auxiliary models, which are involved in the estimation of N:

- The territorial distribution of the earthquakes: the maximal magnitude of the earthquakes $M_{max}(G)$, which may occur in the area G.
- The areas, where M_{max} is above certain threshold; this estimation is often called "recognition of earthquake-prone areas."

Strictly speaking, M_{max} has to be attributed to certain time - interval T, during which it may be exceeded only with sufficiently small probability $\leq$ p.

On the consecutive stages of earthquake prediction this model is replaced by the probability, that prediction will come true.

Strong motion

An earthquake with hypocenter G and magnitude m is causing in a point s the strong motion

$$I = I(s,g,m).$$

Here I is the measure of strong motion: maximal acceleration, or maximal velocity, or macroseismic intensity. I may be a vector, indicating the values of I for different frequencies.

Instrumental data may seem "superior" to macroseismic ones and they are, by and large. If should be noted, however, that the impact on constructions depends on the whole time-function, representing a strong motion. The fundamental problem remains open: which functional defined on this function is best correlated with the damage to a construction, allowing for the ground - construction interaction.

Until this problem is to some extent solved, the macroseismic data may have some advantage too, not mentioning their larger volume. Often, a regression is introduced

$$I = I_c(\Delta,h,A)f(m) + l(s) + \delta$$

Here Δ and h are epicentral distance and, focal depth; A is the direction from epicenter to a point, counted from dominant strike of the faults; l is correction for the ground in point s and δ - random deviation.

Localized damage

Its complete representation is P(D/I) - conditional probability of damage being $\leq$ D if strong motion is I. Practically this distribution is discredized into damage probability matrix.

9

Integrated damage

Its definition is the following. Consider the earthquake - phone territory. Each earthquake may generate a strong motion of the ground, causing damage to population, property and natural environment.

The time-sequence of these episodes may be represented by a flow of vectors (d_i, t_i, g_i, m_i). Here i is the sequence number of an earthquake, t_i, g_i and m_i are its moment, hypocenter and magnitude, d_i is the damage, which it inflicted.

The total damage from the earthquakes on the territory G during time interval from T_1 to T_2 is

$$D = \sum_i d_i, \quad T_1 < t_i < T_2, \quad g_i \in G.$$

The damage can be partly prevented by different SM.

The general base for selection of SM is the probabilistic distribution of the total damage:

$$F(D/T_1, T_2, EP, SM).$$

Here EP symbolizes the earthquake prediction, i.e. the estimation of probability of a occurrence of a destructive earthquake during the period T_1, T_2; SM symbolizes the safety measures.

Analytical representation for F(D) is obtained under the hypothesis, that the effects d_i are statistically independent. In that case the effect of the earthquakes during the time-interval T can be represented trough two distributions:

- of earthquake sequence N(T,G,M) ;
- of the effect of a single earthquake d(t,g,m).

If $\quad F_N[\varphi] = M\ \{\iiint_{TGM}(t,g,m)N(dt,dg,dm)\},$

10

$$F_d[\theta/t,g,m] = \int_0^\infty e^{\theta d} dF_d(d/t,g,m),$$

$\rho(t,g)$ - time-dependence of effect d,

then

$$F_D(\theta) = F_N[\ln F_d(\theta/\rho(t,g)/t,g,m)].$$

The estimations of damage may be attributed to the whole territory, or to some set of objects, to which the SM have to be applied. This can be:

- a set of "vulnerable points" e.g. small towns, nuclear power plants.
- a system of life-lines, e.g.roads, communication networks power supply lines etc;
- a vulnerable area, e.g. a megacity.

Accordingly the necessary data include maps of the towns, life-lines, density of population, density of production etc.

The attribution to specific objects is more adequate for long-term decision making and also makes the estimation of damage more realistic. However for selection of temporary safety measures one usually have to estimate the damage to all potentially vulnerable objects on the territory considered.

DECISION-MAKING

We are returning now to the crux of the problem: given is an estimation of the seismic risk that is of the potential damage from the earthquakes within certain territory and time-interval. What set of safety measures to choose?

The choice requires comparison of cost vrs prevented damage for different sets of SM. Prevented damage is estimated from the distributions of the damage with and without SM considered.

So far this problem was explicitly posed for only for two of the background SM, namely for the building codes and insurance. Building codes are based on seismic zonation: structures should be safe against strong motion of intensity

$$I \le I_{max} + D_C.$$

Here I_{max} is the maximal intensity of strong motions, in a zone of construction, with average recurrence time below a fixed threshold; D_C is correction for the risk, associated with the damage to construction.

This is a staple of building codes; among its advantages is relative simplicity of estimation of I_{max}.

Next essential addition is the estimation of average recurrence time for strong motions of different intensity. This may be necessary since the areas with the same I_{max} may differ in frequency of occurrence of strong motion episodes by factor up to 10^4. This addition allows more comprehensive decision-making, but still on qualitative level. Its major drawback is that it allows to estimate only the average expected damage, which may be misleading due to the large dispersion, which often exceeds the average.

The estimation of the whole distribution of the damage F(D/T,SM) would allow much more efficient and flexible regulation of SM. They could be divided into compulsory and optional ones.

Compulsory measures - should secure two relationships:

$F_D(0/T,SM) > I - \varepsilon$ for population (probability of *no* casualties is large enough)

$F_D(X/T,SM) > X$ for economy (probability of economic losses being less, than X, is large enough)

Optional measures should be regulated economically, by insurance rates, tax deductions, subsidies and loans etc.

Some examples of damage distributions, estimated to help the decision making are given in fig.1 and tables 1-2. Fig.1 concerns reinforcement of the railroad tracks on Baikal-Amur railroad in Eastern Siberia.

About 1000 km of the railroad track had to be reinforced, to prevent purely economic losses (public safety was not threatened). Fig.1 shows the distribution function for prevented damage. The probability that at least 60% of investments into reinforcement will return as damage prevented, is below 1%, and 80% will just never return.

Such a definite conclusion was reached in spite of very limited data base, by considering the upper estimation of the damage. Since the possibility not to reinforce was considered all uncertainties were resoled by assuming the worst

Other example (table 2) concerns the regional insurance, for the village dwellings in Georgia, Caucasus. The distribution of damage was used to determine the fair insurance premium.

Last example (table 3) concerns worldwide reinsurance; it gives the base for qualitative decision of necessary capacity of reinsurance system.

Decision-making in response to time-predictions is currently based mainly on experience and intuition. Optimization of scenaria of response requires more complete description of predictions, including the tradeoff between the total duration of alarms and the rate of failures to predict.

The choice of scenario obviously depends on three factors:

- Quality of predictions: rates of false alarms and failures-to-predict; total volume of alarms in percents of monitoring time-space.
- Cost of switching on the SM, in response to an alarm and the cost of maintaining SM during the alarms.
- Damage, prevented by SM.

Basic Publications

Series: Seismic Design Decision Analysis. School of Engineering, Massachusetts, USA, 1972- ... .

Series: Computational Seismology, iss 5 - Ac. Sci. USSR, Moscow 1970- ... . English Translation of relevant papers in reference above.

G.M.Molchan l.al. Papers in Physics of the Earth and Planetary Interior, v.61 NN 1,2 May 1990.

Series: J. Evernden(ed) National Earthquake Hazards Reduction Program, Technical Reports, USGS, Menlo Parc, California, USA. Summarized in Report 86-383, 1986.

A.Kreimer and M.Minasinghe (ed-s). Managing Natural Disasters and Environment. Publication of the World Bank, 1991 Washington D.C.USA.

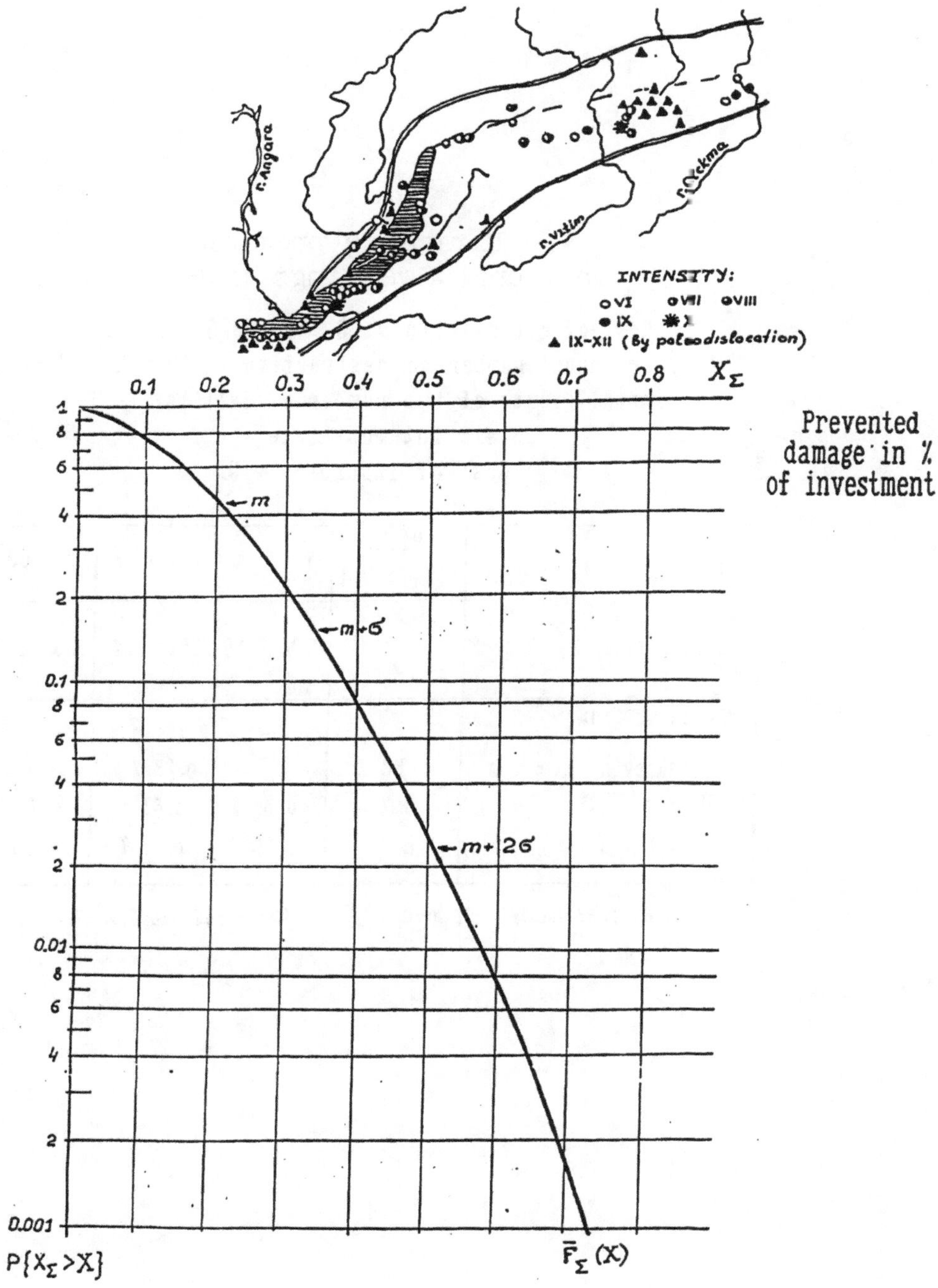

Fig. 1.

Distribution function for the damage, prevented by reenforcement of the railroad track, Baikal-Amur railroad (shown by dashed line).

Table 1

Insurance premium
for rural dwellings in Georgia

Cost of all dwellings $S = 7\ 908.6$ mln roub.
Average annual number of destructive earthquakes $\lambda = 1.9$
Annual growth of the number of dwellings - 1 %
Basic interest rate - 8 %
Cost of insurance - U_T

	Risk, $P\{X_T > U_T\}$	T = 5 y.	T = 30 y.	T = 50 y.
Total damage, X_T, % of S		0.98±0.89	2.91±1.24	3.32±1.25
Profit margin:			Rates:	
break even, $U_T = \lambda \cdot m$	> 50 %		0.23 %	
$U_T = m + 2\sigma$	6 %	0.65 %	0.43 %	0.41 %
$U_T = m + 3\sigma$	2 %	0.86 %	0.53 %	0.50 %

m - average damage per event, σ - its dispersion

Table 2

: Estimated number D of inhabitants of large cities affected by strong earthquakes (≥ VIII degrees on ≥ 100 km²) during 1971-2000

Category (n	Total population, mlns	D, mlns			Probability of no events	A per 100 years
		$m_D \pm \delta_D$	$x_D(p)$			
			$p = 5\%$	$p = 1\%$		
1. Exceptional risk (Tokyo group, 7 cities)	24	20 ± 15.5	48	65	8 %	8
2. Very high risk (17)	47	17 ± 15	46	65	8 %	8.6
3. High risk (52)	76	2.9 ± 3.3	9	15	28 %	4.3
All cities (76)	147	40.1 ± 21.6	79.8	101.7	0.2%	21

Note. A is the average number of the earthquakes, which generate $J \geq$ VIII MMS at least in one of the cities, while the area of VIII-degree isoseist is ≥ 100 km².

: Estimated number N of large cities affected by strong earthquakes (≥ VIII degrees on ≥ 100 km²) in 30 years

Category	Number of cities	N		
		$m_N \pm \delta_N$	$x_N(p)$	
			$p = 5\%$	$p = 1\%$
1. Exceptional risk (Tokyo group)	7	4.1 ± 2.8	10	13
2. Very high risk	17	2,6 ± 1.7	6	8
3. High risk	52	1.3 ± 1.2	4	6
All cities	76	8.0 ± 3.5	15	18

m_N — average ; δ_N — dispersion ; $x_N(p)$ will not be exceeded with probability $(1-p)$.

9. The Self-Organization of American Society in Presidential and Senatorial Elections

V. I. Keilis-Borok and A. J. Lichtman

9.1 Historical Background

The American electorate is commonly regarded as a hierarchic system of electoral groups that differ in their responses to the issues and tactics of electoral campaigns. Taken together, the groups comprise the entire electorate; they are normally represented in social or political organizations, but do not coincide with them. Cultural, social, economic, territorial, ethnic, and other factors divide voters into electoral groups. Each group, in turn, can be subdivided into smaller units, down to the indivisible, though variable element of the individual voter.

American elections are often summarized as follows:

- The candidate's task is to optimize his platform and campaign strategy to win over as many electoral groups as possible, without antagonizing too many other groups.
- At the peak of electoral campaigns, the quest for such optimization goes out of proportion, which makes the campaign shallow and separated from the real interests of society.
- The outcome of the election to a large extent depends on the ability of each candidate to manipulate electoral groups through advertising, rhetoric, and stage-managed events.

This formulation appears plausible, and it has consistently guided coverage of American elections. But it leaves a number of questions unanswered. Though bifurcations do occur in nature, there needs to be specific evidence for the belief that a huge electorate (10^8 for a presidential election and 10^6–10^7 for senatorial ones) easily changes its collective decision under the influence of transient factors that are irrelevant to the governing of the country.

Besides, this account shows little respect for American democracy or the American voter. It portrays voters as excitable simpletons who are easily swayed by television commercials and emotional appeals.

The problem of elections in the U.S. deserves an alternative hypothesis. In our papers [9.3, 9.5, 9.6] we have proposed the hypothesis that the collective decision of the American electorate (on the national scale when the president is elected, or in each state during senatorial elections) is determined by long-term factors that can usually be gauged well in advance of an electoral campaign. This means, first, that debates, advertising, nightly news reports, and other campaign events have little to do with the outcomes of elections and, second, that election results can be predicted reliably many months in advance without recourse to polls.

In [9.5] we have tried to find parameters which describe the political environment prior to an election and predict the outcome of American presidential elections. In [9.6] we examine the same problem for senatorial elections in all 50 American states.

Given a lack of adequate theory, both problems were tackled phenomenologically through the analysis of pre-election situations in the past. The results are relevant for understanding the electoral process, developing theoretical models, and for predicting electoral results.

Our work shows that presidential elections are referenda on the performance of the party in power. Rather than choosing between individual candidates or between Republicans and Democrats, the electorate decides whether to retain or reject the party that controls the White House. If the nation has fared well during the term, the executive party is re-elected; otherwise the presidency changes hands. This pattern has remained unchanged at least since the election of Abraham Lincoln in 1860 despite enormous changes in the electorate and the society during the subsequent 130 years.

Although other investigators have also probed the conditions under which incumbent administrations rise or fall, their work has focused narrowly on the state of the economy. Strictly economic models, however, fail to capture the dynamics of many American elections (e.g., 1860, 1912, 1968, and 1976). Our work in [9.5] gauges not just the economy, but such measures of incumbent performance as policy change, foreign successes and failures, scandal, and social unrest.

We found that the choice of a senator is fundamentally different from the election of a president. Senate elections are not primarily referenda on the performance of the previous term. They are to some extent contests between parties and candidates, although in fact their outcome depends more upon the relative strengths of the competitors going into the campaign than on what they say or do during the campaign itself. That is why we found that senatorial elections, like presidential elections, could also be forecast in advance of campaigns.

The logic and algorithms of the present analysis follow Gelfand's school of pattern recognition analysis. Specifically, we draw upon the experience of earthquake prediction research [9.1]. For both presidential and senatorial elections we sought to predict the winners of elections, but not their percentage of the vote. Likewise, we eliminated certain information through the discretization of parameters to the lowest level of resolution, 'yes' or 'no'. Similar 'robust' methods are widely used in the heuristic analysis of complicated data, especially when dealing with small samples. The apparent loss of information involved has often produced the stable results that elude more 'detailed' analyses that are subject to fluctuations in the values of particular variables.

9.2 The American Presidential Election: Formal Analysis

Initial Data. The historical study of American elections suggested the choice of a set of integral parameters that may affect the outcome of presidential elections. We were guided by the hypothesis that elections turned on the strength and performance of the incumbent party. These parameters were determined at the minimum level of resolution, as answers to certain 'yes' or 'no' questions given in Table 9.1. Each question can be answered prior to an upcoming election, usually by the time that parties have nominated their candidates. The questions are formulated so that an affirmative answer favors victory of the party in power. The unexpected finding is that this set of questions is sufficient to predict the outcome of elections.

The answers to the questions for pre-election situations from 1860–1980 are given in Table 9.2. This material forms our initial data.

The analysis follows a procedure that is common for pattern recognition. Let us divide the pre-election situations into two types: *I* (incumbent will win) and *C* (challenger will win). Victory is defined as a popular vote plurality regardless of the verdict in the electoral college.

The problem is formulated as follows. We are given 'learning material' that consists of examples of elections *I* and *C*, with the answers to the questionnaire (Table 9.1) for each election. From this data

Table 9.1. Questionnaire for presidential elections

A 'yes' answer favors the incumbent party. When five or more answers are 'yes' a victory for the incumbent party is predicted. Otherwise a victory for the challenging party is predicted.

1 **Party mandate.** After the midterm elections, the incumbent party holds more seats in the U.S. House of Representatives than it did after the midterm elections.

2 **Contest.** There is no serious contest for the incumbent party nomination.

3 **Incumbency.** The incumbent party candidate is the sitting President.

4 **Third party.** There is no significant third party or independent campaign.

5 **Short-term economy.** The economy is not in recession during the election campaign.

6 **Long-term economy.** Real per-capita economic growth during the term equals or exceeds mean growth during the previous two terms.

7 **Policy change.** The incumbent administration effects major changes in national policy.

8 **Social unrest.** There is no sustained social unrest during the term.

9 **Scandal.** The incumbent administration is untainted by major scandal.

10 **Foreign military failure.** The incumbent administration suffers no major failure in foreign military affairs.

11 **Foreign military success.** The incumbent administration achieves a major success in foreign or military affairs.

12 **Incumbent charisma.** The incumbent candidate is charismatic or a national hero.

13 **Challenger charisma.** The challenger party candidate is not charismatic or a national hero.

Table 9.2. Presidential elections 1860–1988 classified by questionnaire

Question numbers 1–13 correspond to Table 9.1. Answer 'yes' = 0 and 'no' = 1.

Incumbent victories

Year	1	2	3	4	5	6	7	8	9	10	11	12	13	Total
1864	0	0	0	0	0	1	0	1	0	0	0	1	0	3
1868	0	0	1	0	0	0	0	1	0	0	0	0	0	2
1872	1	0	0	0	0	0	1	1	0	0	0	0	0	3
1880	0	1	1	0	0	0	0	0	0	0	1	1	0	4
1888	1	0	0	0	0	0	1	1	0	0	1	1	0	5*
1900	1	0	0	0	0	0	0	0	0	0	0	1	1	3
1904	0	0	0	0	0	0	0	0	0	0	0	0	0	0
1908	0	0	1	0	0	1	0	0	0	0	0	1	0	3
1916	1	0	0	0	0	1	0	0	0	0	0	1	0	3
1924	1	0	0	1	0	0	0	0	1	0	0	1	0	4
1928	0	0	1	0	0	0	1	0	0	0	0	1	0	3
1936	0	0	0	0	0	0	0	0	0	0	1	0	0	1
1940	1	0	0	0	0	0	0	0	0	0	1	0	0	2
1944	1	0	0	0	0	0	0	0	0	1	0	0	0	2
1948	1	0	0	1	0	1	0	0	0	1	0	1	0	5
1956	0	0	0	0	0	0	1	0	0	0	0	0	0	1
1964	1	0	0	0	0	0	0	0	0	1	0	1	0	3
1972	1	0	0	0	0	1	1	0	0	0	0	1	0	4
1984	0	0	0	0	0	1	0	0	0	0	1	0	0	2
1988	0	0	1	0	0	0	1	0	0	0	0	1	0	3

Challenger victories

Year	1	2	3	4	5	6	7	8	9	10	11	12	13	Total
1860	0	1	1	1	0	0	1	1	0	0	1	1	0	7
1876	1	1	1	0	1	1	1	0	1	0	1	1	0	9*
1884	1	1	1	0	1	1	1	0	0	0	1	0	0	7
1892	1	1	0	1	0	0	0	1	0	0	1	1	0	6
1896	1	1	1	0	1	1	1	1	0	0	1	0	0	8
1912	1	1	0	1	0	0	1	0	0	0	1	1	0	6
1920	1	1	1	0	1	1	0	1	0	1	0	1	0	8
1932	1	0	0	0	1	1	1	1	0	0	1	1	1	8
1952	0	1	1	0	0	1	1	0	1	1	0	1	1	8
1960	1	0	1	0	1	1	1	0	0	1	1	1	1	9
1968	1	1	1	1	0	0	0	1	0	1	1	1	0	8
1976	1	1	0	0	0	1	1	0	1	1	1	1	0	8
1980	1	1	0	1	1	0	1	0	0	1	0	1	1	8

* Electoral vote did not coincide with popular vote results.

we attempt to find the 'recognition rule' that predicts the outcome of a specific election from answers to the same questionnaire.

Algorithm. Given that the learning material consists of only 31 elections, we have chosen the simplest algorithm, 'recognition by the Hamming distance' [9.5]. It may be described briefly as follows.

First, the 'kernel' is determined. This is the set of answers which are encountered more frequently in elections I than in elections C.

To find the kernel the following numbers are calculated:

$$K(i) = \frac{n(i/I)}{n(I)} - \frac{n(i/C)}{n(C)},$$

where $i = 1,2,\ldots$ is the sequence number of a question from Table 9.1, $n(i/I)$ and $n(i/C)$ show how many elections I and C have the answer 'yes' to question i, and $n(I)$ and $n(C)$ show how many elections I and C are in the learning material.

The answer to question i in the kernel is 'yes' if $K(i) \geq k$ and 'no' if $K(i) \leq -k$, where k is a numerical threshold. If $|K(i)| < k$ the question is not used for recognition.

Second, the distance D between the election and the kernel is calculated as follows:

$$D = \sum_i W(i),$$

where $W(i)$ is the 'weight' of question i:

$$W(i) = \frac{|K(i)|}{\max_i |K(i)|}.$$

Only the answers which are contrary to those in the kernel are included in this summation. If all $W(i)$ are replaced by 1 then D is simply the number of answers which are unfavorable to the incumbent party (the number of 'no' answers).

Third, a rule of recognition is formulated as follows. If $D \leq L$, an election is recognized as I, a win for the incumbent party, and if $D > L$, an election is recognized as C, a win for the challenging party, where L is a certain threshold.

Data Analysis. For purposes of analysis we assumed $W(i) = 1$ and $k = 0.1$. Including all elections from 1860 to 1980 in the questionnaire we obtain a kernel of all zeroes (all answers 'yes'). This follows the intuitive formulation of the questionnaire. The division of elections acording to that kernel is shown in the last column of Table 9.2, which sets the threshold *a posteriori* as $L = 5$.

A central question for pattern recognition is the reliability of the rule of recognition. The rule derived for presidential elections correctly divides past elections into incumbent and challenger victories. Successful forecasts of the 1984, 1988, and 1992 elections provide additional evidence of the rule's reliability. For its statistical evaluation more forecasts are required.

The stability of our results is also established by exploring variations in the questionnaire. We performed a logical exercise similar to the 'seismic history' experiment in [9.1]. Beginning in 1900 we determined the outcome of each successive election using information from all previous elections. Thus we have recreated the position of a forecaster anticipating every election from 1900 through 1980 on the basis of data from the previous elections dating back to 1860. This experiment gives correct results for 19 of 21 elections, indicating that the situations leading to incumbent or challenger victories has changed little during the past 130 years.

A number of additional tests were performed. We found, for example, that results were unchanged when we excluded the five elections in which the number of votes cast for the two main parties differed by 1% or less (1880, 1884, 1888, 1960, 1968) or when we included only twentieth-century elections in the questionnaire.

We also used the algorithm Cora III [9.1] which can formulate composite questions from combinations of the original set. This failed to produce better or additional results. What makes the Hamming distance algorithm preferable is that it is less vulnerable to accidental combinations that have little substantive meaning.

9.3 Midterm Senatorial Elections: Formal Analysis

Analysis of senatorial elections, held between presidential election years, provides a sufficient sample size to obtain statistically verifiable results. The logic and algorithms of the analysis were the same as those used for presidential elections. The work for [9.6] was performed in 1985 and used to forecast the outcomes of the 1986 elections.

The basic data were taken from the history of the three previous midterm elections, those of 1974, 1978, and 1982. Each of the 50 states was included in at least one of these contests. For the analysis, we selected the same set of integral parameters for all states. They are given in Table 9.3.

The answers to the questions from Table 9.3 are given in [9.6]. As before, the wording of the questions implies that 'yes' is preferable for *I* elections. Accordingly, the kernel again consists of all zeroes.

Figure 9.1 shows the results of applying pattern recognition to the data. These results establish the following decision rule: the incumbent party retains its senatorial seat (I) if $D<5$ and loses it to the opposition party (C) if $D \geq 5$.

The forecast of the 1986 election published a week in advance provided a test of this decision rule. This forecast is found in Table 9.4, together with the evaluations of predictions by political authorities and the actual results. Our forecast was correct in 30 states of the 34.

We subjected the 1986 results to statistical analysis, considering (a) the correlation between questionnaire results and election outcomes, and (b) the predictive power of the questionnaire.

			WY82I				
			UT82I	VT82I			
			TX82I	NJ82I			
			TN82I	NE82I			
			RI82I	MO82I			
			PA82I	IN82I			
	WV82I		MN82I	MT78I			
	WI82I		MI82I	KY78I			
	WA82I	OH82I	MA82I	KS78I			
	ND82I	NY82I	HI82I	GA78I			
	MT82I	ME82I	DE82I	DE78I			
	MS82I	MD82I	CT82I	AR78I			
	TN78I	FL82I	AZ82I	AL78I	WY78I		
	OR78I	NM78I	WV78I	PA74I	NM82C		
	WI74I	NC78I	VA78I	OR74I	NJ78C		
	WA74I	ID78I	TX78I	OK74I	NH78C		
SC78I	SC74I	AK78I	RI78I	MD74I	NE78C	CA82I	
IL78I	MO74I	SD74I	LA78I	AR74I	ME78C	UT74I	
LA74I	IN74I	KS74I	NC74I	AK74I	CO78C	OK78C	SD78C
ID74I	IL74I	IA74I	NY74I	MI78C	VT74C	MN78C	MS78C
HI74I	CT74I	CA74I	ND74I	NV82C	KY74C	NV74C	OH74C
GA74I	AL74I	AZ74I	MA78C	IA78C	CO74C	NH74C	FL74C
0	1	2	3	4	5	6	7

Fig. 9.1. Division of off-year election returns (1974–1982) by the Hamming distance *D* from the kernel (i.e., by the number of 'no' answers to the questionnaire in Table 9.3). Each election is represented by the two-letter state abbreviation, the year, and the outcome: *I* for an incumbent-party victory, *C* for a challenger-party victory

To test condition (a) we examined the results for all 34 elections. To test condition (b) we excluded elections whose outcome, according to expert opinion, was easily predictable. Specifically, we excluded all elections identified as 'probably secure' in Congressional Review (Table 9.4).

A detailed analysis, presented in [9.6], established the statistical significance of our results both for the 34 elections and the smaller set of less predictable contests.

Our forecast of the 1990 midterm elections, likewise published in advance [9.4], proved correct in 32 of 35 states. Thus 62 of 69 predictions (90%) for the midterm elections of 1986 and 1990 were correct. In the 1990 cycle, moreover, the forecast correctly anticipated the victory of incumbent party candidates in virtually every state, despite widespread press reports that 'anti-incumbent' sentiment was sweeping the nation.

The reliability of the rule of recognition for midterm senatorial elections was also established through the same numerical experiments applied to presidential elections. Detailed results can be found in [9.6]. The sole source of instability lies in the threshold $L = 5$ that divides elections into I and C. Changing the threshold by 1 leads to numerous mistakes, as shown in Fig. 9.1. With only eight parameters this instability is hardly avoidable; it remains when weights $W(i)$ are introduced into the algorithm.

That our prediction was successful despite this instability implies that each parameter is an essential indicator of electoral outcomes. It may also suggest that the parameters are mutually dependent so that a change in one is likely to be accompanied by a change in others. There may yet be tendencies within the electorate that we have not been able to diagnose through additional questions.

We also used the same methods to analyze senatorial elections held in presidential years [9.5]. Although we did not publish forecasts of the 1988 elections ahead of time, we review presidential year elections to make the picture complete.

The first six questions from Table 9.3 remained unchanged for presidential election years. The last two were replaced with the questions 7A and 8A shown in Table 9.3.

The use of this questionnaire with $D = 5$ correctly predicted the results of 32 of 35 senatorial elections in 1988.

Table 9.3. Eight questions for midterm senatorial elections

Senators are elected for a term of six years. Approximately one-third of the senators are re-elected every two years, in some cases in the middle of the presidency, and in others simultaneously with the presidential election.

A 'yes' answer favors the incumbent party. If no more than four answers are 'no' a victory for the incumbent party is predicted. Otherwise a victory for the challenging party is predicted.

Incumbent party candidate:

1 The incumbent party candidate is the sitting Senator.

2 The incumbent party candidate is a major national figure.

3 There was no serious contest for the incumbent party nomination (the candidate collected no less than two-thirds of the vote in the first round)

Incumbent party:

4 The incumbent party won 60% or more of the vote in the previous election.

5 The incumbent party raised at least 10% more money for the campaign than the opposition.

Challenger candidate:

6 The challenger candidate is not a national figure or a past or present Governor or Member of Congress.

7 The challenger candidate is from the same party as the current President.

8 There was no serious contest for the challenger party nomination.

If the senatorial and presidential elections take place in the same year, the last two questions are replaced with:

7A The questionnaire for the presidential election predicts victory for the incumbent party.

8A The incumbent party has a majority in the lower house of the state legislature.

Table 9.4. 1986 senatorial election forecast and returns

State	Answer[1]									CR[2]	Fore-	Actual
	1	2	3	4	5	6	7	8	*D*		cast[3]	result
HI	0	0	0	0	0	1	0	0	1	PS	*I*	*I*
OH	0	0	0	0	1	0	0	0	1	PS	*I*	*I*
SC	0	0	0	0	0	1	0	0	1	PS	*I*	*I*
UT	0	0	0	0	0	0	1	0	1	PS	*I*	*I*
AK	0	1	0	1	0	0	0	0	2	PS	*I*	*I*
CT	0	1	0	1	0	0	0	0	2	PS	*I*	*I*
KS	0	0	0	0	0	1	1	0	2	PS	*I*	*I*
KY	0	1	0	0	0	1	0	0	2	PS	*I*	*I*
ND	0	1	0	0	0	0	1	0	2	PV	*I*	*C* *
AR	0	1	0	1	0	0	1	0	3	PS	*I*	*I*
CA	0	0	0	1	1	1	0	0	3	V	*I*	*I*
IL	0	1	0	1	0	1	0	0	3	PS	*I*	*I*
IN	0	0	1	1	0	0	1	0	3	PS	*I*	*I*
IA	0	1	0	1	0	0	1	0	3	PS	*I*	*I*
NH	0	1	0	1	0	0	1	0	3	PS	*I*	*I*
OR	0	0	1	1	0	0	1	0	3	PV	*I*	*I*
VT	0	1	0	1	1	0	0	0	3	PV	*I*	*I*
AZ	1	1	0	1	0	0	1	0	4	PS	*I*	*I*
CO	1	1	0	1	1	0	0	0	4	V	*I*	*I*
ID	0	1	0	1	1	0	1	0	4	V	*I*	*I*
LA	1	1	0	0	1	0	0	1	4	HV	*I*	*I*
NY	0	1	0	1	0	1	1	0	4	PS	*I*	*I*
NC	0	1	0	1	1	0	1	0	4	PV	*I*	*C* *
OK	0	1	0	1	1	0	1	0	4	V	*I*	*I*
WA	0	1	0	1	1	0	1	0	4	PV	*I*	*C* *
WI	0	1	0	1	0	1	1	0	4	V	*I*	*I*
AL	0	1	0	1	1	1	1	0	5	V	*C*	*C*
FL	0	1	0	1	1	0	1	1	5	HV	*C*	*C*
GA	0	1	0	1	1	1	1	0	5	PV	*C*	*C*
MO	1	1	0	1	1	0	0	1	5	HV	*C*	*C*
PA	0	1	0	1	1	1	1	0	5	V	*C*	*I* *
MD	1	1	0	0	1	1	1	1	6	HV	*C*	*C*
NV	1	1	0	1	1	0	1	1	6	HV	*C*	*C*
SD	0	1	1	1	1	0	1	1	6	HV	*C*	*C*

[1] Answer 0 = 'yes' and 1 = 'no'.

[2] CR gives Congressional Review assessment of the incumbent party:

- PS probably secure
- PV potentially vulnerable
- V vulnerable
- HV highly vulnerable

[3] Result *I* is incumbent party victory and *C* is challenger party victory.

* Forecast happened to be wrong.

9.4 Discussion

Self-organization and Predictability. In the natural world, intricate chaotic systems, after appropriate smoothing, often display stable regularities, including predictability. These regularities are difficult, if not impossible, to derive from the behavior of the system's elementary components. Our results for the American political system suggest that American society comprises such a system during presidential and senatorial elections. The hierarchic system of American electoral groups has stable and predictable aggregate-level behavior with a high degree of integration for the entire nation and even for individual states.

This integration occurs despite the contradictory interests and outlooks of electoral groups. The laws governing the outcome of elections have remained stable at the aggregate level from 1860 through 1988, even though three-fourths of today's voters – women, 18 to 20-year-olds, African-Americans, and the great majority of descendants from Latin America, Asia, and Eastern and Southern Europe – were not part of the nineteenth-century electorate. Electoral systems may thus display features similar to large-scale physical systems that likewise exhibit collective behavior comprehensible only at the level of the system as a whole.

What determines collective choice in American presidential elections is the enduring, pragmatic nature of the American electorate. Contrary to the conventional wisdom, our results suggest that issues and ideology, party affiliation, speeches, debates, and advertising count for little or nothing on election day. What matters is the electorate's assessment of how well an incumbent administration has governed the nation. And that assessment is usually clear before the general election campaign even begins. The analysis thus restores the unity between politics and governing that is torn apart in conventional accounts of how elections turn on the strategy and tactics of campaigns.

The parameters developed for presidential elections probe the multiple dimensions of incumbent-party power and performance. The first four parameters primarily gauge conditions that reflect the strength and unity of the party in power. The next seven parameters measure incumbent achievements and failures across a wide range of public concerns. The final two parameters recognize that personality can make a difference in presidential politics, but only when a candidate is either unusually compelling or of heroic stature.

Only one of the thirteen parameters (13) can be influenced directly by the opposition party. This indicates that an incumbent party largely holds its fate in its own hands. Still, many of the parameters may not be within the administrations's control.

Our findings, for example, suggest an explanation of George Bush's 1988 victory that is radically different from the generally accepted version of events. According to the conventional wisdom, after trailing by as many as 17 percentage points in the polls, Bush began a remarkable 'comeback' with his eloquent convention speech (primarily crafted by master speechwriter Peggy Noonan). He then launched a devastating barrage of negative attacks on Mike Dukakis, orchestrated by political adviser Lee Atwater and designed by advertising expert Roger Ailes. When Dukakis failed to respond to charges that he furloughed dangerous criminals and fouled Boston Harbor, Bush surged permanently ahead. Thus, a brilliantly designed – if shallow and vicious – campaign allegedly changed the minds of the voters.

Our conclusions compel a different version of what actually happened in 1988. Based on the record of the previous four years, as measured by the 13 parameters, a Bush victory was apparent long before the public ever heard of speechwriter Peggy Noonan or furloughed rapist Willie Horton.

Six months prior to the election and three months before Bush's alleged comeback, the following forecast was published: "Barring a suddenly stalled economy and a major disaster between now and election day, George Bush is a shoo-in for the presidency, no matter who winds up as the Democratic nominee." [9.3]

The ability to forecast elections prior to campaigns, however, does not mean that candidates can cease campaigning. Campaigns are integral to the political system in which the parameters operate. If one or both parties decided not to take part in the campaign, the nature of the electoral system might change, with consequences that are as yet unclear.

Still, our results do raise the question of what a party should do when historical factors show that it is likely to lose an election. What certainly will not help are the usual attempts to manipulate voters. No master stroke of strategy, advertising trick, or campaign event has ever reversed an unfavorable situation for either the incumbent or the challenging party.

But the disadvantaged party can at least attempt to change the rules of the political game. A party that has the historical odds stacked against it has nothing to lose and perhaps much to gain by running a campaign of candor and substance. For the opposition, it may be the only chance to recover the White House, short of waiting for disaster to befall the president and the country. To run a new kind of campaign, an opposition candidate would have to fire the ad men, cancel the TV spots, and talk straight to the public about how they would govern the country, revealing at least:

- Prospective cabinet appointments,
- Specific legislative changes,
- Concrete taxing and spending plans,
- Drafts of international treaties,
- Alternative solutions to national crises.

If circumstances were against the party in power, the incentive to run an honest and substantive campaign would be reversed. But the results would still be to elevate the shallow and trivial campaigns that have become routine in America's quadrennial contests for president.

As democracy has spread throughout the world, the contagion of a politics based on attack strategies, sound bites, and stage-managed events has followed. A more substantive and honest presidential campaign in the United States might remove this sand from the bearings of democracy worldwide.

The pragmatic electorate also emerges in senatorial elections. Once again, issues, ideology, and campaign events play no role in prediction. But the analysis of senatorial elections also includes no direct performance measures comparable to those included in the presidential analysis. What is most surprising is that neither the national nor the state economy affects the results of senatorial elections.

Voters thus have more diffuse expectations of senators than of presidents. For Senate elections, perceptions of incumbent performance are captured only indirectly through questions about nomination contests, the political stature of opposition-party candidates, and the competition for financial support.

Taken together, our results illustrate the potential to analyze social systems with the same aggregate-level methods used for studying intricate chaotic systems in the natural world. For presidential elections, extremely smoothed parameters describing the social, economic, and political situation in an election year are averaged for the United States as a whole. The same set of parameters is correlated with the outcomes of elections at least since 1860. For senatorial elections, the same set of similarly smoothed parameters are applied to each of the 50 states. At least since 1874, in all states, for each election year, the outcomes of elections for the Senate follow the same basic pattern.

Our conclusions are generally confirmed by early, published forecasts of senatorial elections in 1986 and 1990 and of presidential elections in 1984, 1988, and 1992. The finding that elections can be predicted without reference to campaigns, issues, or ideologies points to the need for a radically revised understanding of how presidential and senatorial elections really work in the United States.

References

9.1 I.M. Gelfand, Sh.A. Guberman, V.I. Keilis-Borok, L. Knopov, F. Press, Ye. Ratsman, I.M. Rotvain, A.M. Sadovski: The conditions of occurrence of strong earthquakes in California and some other regions. Vychyslitelnaya Seismologiya [Computer Seismology] 9, 3–90 (1976) (in Russian)

9.2 K. DeCell: Interview. The Washingtonian, pp. 142–145 (November 1986)

9.3 K. DeCell, A.J. Lichtman: The thirteen keys to the presidency. Madison Books 1990

9.4 K. DeCell, A.J. Lichtman: Interview. The Washingtonian (October 1990)

9.5 A.J. Lichtman, V.I. Keilis-Borok: Pattern recognition applied to presidential elections in the United States 1860–1980: role of integral, social, economic and political traits. Proc. Nat. Acad. Sci. USA 78, 7230–7234 (1981)

9.6 A.J. Lichtman, V.I. Keilis-Borok: Aggregate-level analysis and prediction of midterm senatorial elections in the United States 1974–1986. Proc. Nat. Acad. Sci. USA 86, 1076–1080 (1989)

Proc. Natl. Acad. Sci. USA
Vol. 93, pp. 3838–3842, April 1996
Geophysics

Geometric incompatibility in a fault system

ANDREI GABRIELOV*†, VLADIMIR KEILIS-BOROK*, AND DAVID D. JACKSON‡

*International Institute of Earthquake Prediction Theory and Mathematical Geophysics, Russian Academy of Sciences, Warshavskoye sh.79, k.2, Moscow 113556, Russia; †Departments of Mathematics and Earth and Atmospheric Sciences, Purdue University, West Lafayette, IN 47907-1395; and ‡Institute of Geophysics and Planetary Physics, University of California, Los Angeles, 405 Hilgard Avenue, Los Angeles, CA 90024-1567

Contributed by Vladimir Keilis-Borok, August 7, 1995

ABSTRACT Interdependence between geometry of a fault system, its kinematics, and seismicity is investigated. Quantitative measure is introduced for inconsistency between a fixed configuration of faults and the slip rates on each fault. This measure, named geometric incompatibility (*G*), depicts summarily the instability near the fault junctions: their divergence or convergence ("unlocking" or "locking up") and accumulation of stress and deformations. Accordingly, the changes in *G* are connected with dynamics of seismicity. Apart from geometric incompatibility, we consider deviation $\vec{K}$ from well-known Saint Venant condition of kinematic compatibility. This deviation depicts summarily unaccounted stress and strain accumulation in the region and/or internal inconsistencies in a reconstruction of block- and fault system (its geometry and movements). The estimates of *G* and $\vec{K}$ provide a useful tool for bringing together the data on different types of movement in a fault system. An analog of Stokes formula is found that allows determination of the total values of *G* and $\vec{K}$ in a region from the data on its boundary. The phenomenon of geometric incompatibility implies that nucleation of strong earthquakes is to large extent controlled by processes near fault junctions. The junctions that have been locked up may act as transient asperities, and unlocked junctions may act as transient weakest links. Tentative estimates of $\vec{K}$ and *G* are made for each end of the Big Bend of the San Andreas fault system in Southern California. Recent strong earthquakes Landers (1992, $M = 7.3$) and Northridge (1994, $M = 6.7$) both reduced $\vec{K}$ but had opposite impact on *G*: Landers unlocked the area, whereas Northridge locked it up again.

The geometry of a fault system imposes certain limitations on the movements within it; a wide class of movements would require a change of geometry. A simple illustration of such limitations is the intersection ("junction") of strike-slip faults (Fig. 1*A*). If their movement could follow the arrows, the corners A and C would penetrate each other as shown in Fig. 1*B*. Obviously, this is not possible, so that the movement along the faults has to be accommodated by additional fracturing and/or deformation, changing fault geometry near the junction point.

The concept of such phenomena was first introduced by McKenzie and Morgan (1). They considered subduction zones, mid-ocean ridges, and transform faults under a formulated condition in which triple junctions of the plate boundaries were "stable"—that is, "could retain their geometry as the plates moved"; this was expressed as a relation between the rates of relative plate movement at each plate boundary. Systematic description of 125 possible types of triple junctions is given in ref. 2.

If the stability condition were not satisfied, the plate movements would lead to stress accumulation around the junction and formation of new faults of a smaller scale (3, 4). Generally speaking, this situation will not prevent the accumulation of stress but only redistribute it among newly formed fault junctions. King (3) suggested that such redistribution will lead to generation of new faults of progressively smaller and smaller scale, so that a hierarchical system of faults is formed around the initial junction. This conclusion is in good accordance with neotectonic data on specific structures that are formed around the junctions of active faults (5). These structures, called "morphostructural nodes" or "knots," are wider than the intersecting fault zones. They are characterized by particularly intensive fracturing and contrasting neotectonic movements with a mosaic pattern of structure and topography resulting. Their formalized definition is described in ref. 6. It was demonstrated in a series of studies (e.g., refs. 7, 8) that epicenters of strong earthquakes are situated only within some of the nodes, identified by pattern recognition.

In this paper we introduce a quantitative measure of incompatibility between the geometry and kinematics of a fault system. We call this measure "geometric incompatibility."

Let us outline basic definitions. Consider a system of blocks separated by faults. At each intersection of the faults (fault junction) the slip rates $\vec{v}_i$ on faults at the junction point should satisfy the condition of Eq. 1

$$\vec{K} \equiv \sum \vec{v}_i = \vec{0}, \quad [1]$$

which is a discrete analogue of a well-known Saint Venant compatibility condition in continuum mechanics. Here the sum is taken over all faults entering the junction. This condition ensures that the relative movements on the faults can be realized through the absolute movements of the volumes (blocks) separated by the faults. The value of $\vec{K}$ computed from observations may differ from zero because of observational errors or because some deformations are not allowed for. Accordingly, we call $\vec{K}$ the Saint Venant incompatibility.

However, the Saint Venant condition is not sufficient to ensure that faults movements are possible without a change in the fault system geometry around the junction. For example, if the rate of movement along each fault in Fig. 1*a* is the same for both sides of the junction point, the Saint Venant condition is satisfied. Nevertheless, the movement requires a geometric change; otherwise the junction point would split into a parallelogram (Fig. 1*b*). After a time interval *t* the (oriented) area of this unrealizable parallelogram would be $[\vec{v}_1, \vec{v}_2]t^2$. In general case, a junction point would split into a polygon with area

$$S(t) = Gt^2/2. \quad [2]$$

The coefficient *G* is our measure of geometric incompatibility. It depends on all the slip rates $\vec{v}_i$ at the junction point. In our example, $G = 2[\vec{v}_1, \vec{v}_2]$. Note that in this case *G* is negative because the corners A and C overlap. In such cases, compression concentrates at the junction, locking it up. Fig. 2 shows an example of a junction with all corners diverging; then *G* is positive, and the junction is unlocked by tension.

Below we derive the formulas for geometric incompatibility in a fault system with many junctions. In such a system, Saint Venant and geometric incompatibilities depend also on rota-

The publication costs of this article were defrayed in part by page charge payment. This article must therefore be hereby marked "*advertisement*" in accordance with 18 U.S.C. §1734 solely to indicate this fact.

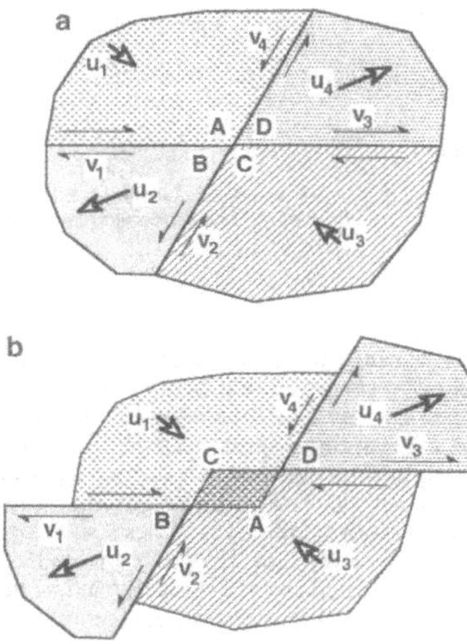

FIG. 1. Example of geometric incompatibility near fault junction. 1, Faults; 2, horizontal component of absolute movement of a block, $\vec{u}_i$; 3, horizontal component of slip rate on a fault, $\vec{v}_i$; A, B, C, and D, corners of the blocks; a, initial position of the blocks; b, extrapolation of initial movement. Corners A and C are converging and would overlap; this indicates that the movement cannot be realized without the change of the fault geometry.

tion and deformation of blocks separated by the faults. Both measures of incompatibility have the following useful property, reminiscent of the Stokes formula: To estimate their values in a whole region, it is sufficient to know only the movements on the faults that cross the boundary of this region.

Analytical expressions for G are derived below first for a single junction and then for a system of faults with many junctions within a given contour. The expressions for a single junction are the same for rotating and nonrotating blocks. For a fault system with many junctions these expressions are

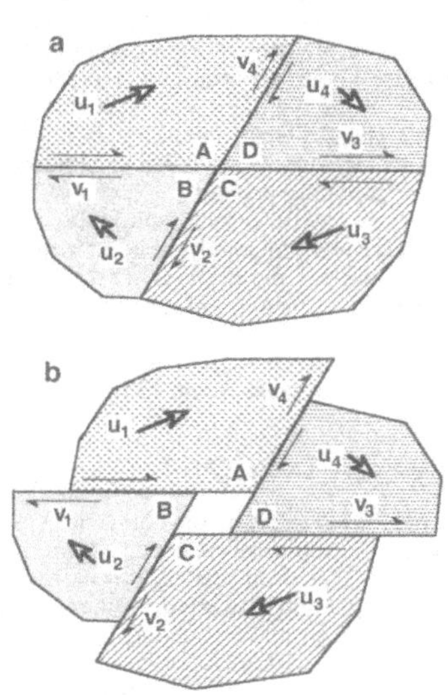

FIG. 2. Example of geometric incompatibility. Notations are the same as in Fig. 1. Geometric incompatibility leads to divergence of the block corners.

different for deformable/rotating blocks; we consider this case in the *Appendix*.

Geometric Incompatibility for a Single Junction

Consider n fault segments with a common junction point; a fault crossing this point is regarded as two segments. We choose some segment as the first one, and assign the numbers from 2 to n to the other segments counterclockwise. The n blocks separated by these faults meet at the junction. We number them in such a way that the ith fault separates the blocks i and $i + 1$.

Let $\vec{u}_i$ be the rate of movement of the ith block at the junction point, and $\vec{v}_i = \vec{u}_{i+1} - \vec{u}_i$ the rate of relative movement on the ith fault. We always set $\vec{u}_{n+1} = \vec{u}_1$ and $\vec{u}_{-1} = \vec{u}_n$. Because we consider only horizontal components of the movements, the vectors $\vec{u}_i$ and $\vec{v}_i$ are two-dimensional. For two such vectors, $\vec{a} = (a_1, a_2)$ and $\vec{b} = (b_1, b_2)$, we define the cross-product $[\vec{a}, \vec{b}] = a_1b_2 - a_2b_1$. After a time period t the junction, which we place at the origin, becomes, due to block movement, a polygon with the vertices $\vec{u}_1t, \ldots, \vec{u}_nt$. Computing the (oriented) area of this polygon, we obtain the following expression for the geometric incompatibility G defined in ref. 2:

$$G = [\vec{u}_1, \vec{u}_2] + [\vec{u}_2, \vec{u}_3] + \ldots + [\vec{u}_n, \vec{u}_1]. \qquad [3]$$

We easily verify that this sum does not change when we add a common vector to all the rate vectors $\vec{u}_i$. Replacing $\vec{u}_i$ by $\vec{u}_i - \vec{u}_1$ and substituting $\vec{u}_i - \vec{u}_1 = \vec{v}_1 + \ldots + \vec{v}_{i-1}$, we can rewrite G in terms of the relative movement rates $\vec{v}_i$:

$$G = [\vec{v}_1, \vec{v}_2] + [\vec{v}_1 + \vec{v}_2, \vec{v}_3] + \ldots + [\vec{v}_1 + \ldots + \vec{v}_{n-2}, \vec{v}_{n-1}]. \qquad [4]$$

This formula depends on the arbitrary choice of the first segment. However, it will lead to the same values of G, for any choice, if $\vec{K} = 0$—i.e., if the observed values of $\vec{v}$ satisfy the Saint Venant condition Eq. **1**. If this condition is not satisfied—e.g., due to observational errors or block deformation, we have to satisfy it by modification of $\vec{v}_i$. This can be done in a nonunique way. In the absence of additional information we simply subtract from each of $\vec{v}_i$ a fraction of $\vec{K}$ proportional to the absolute value of v_i:

$$\vec{v}_i' = \vec{v}_i - \frac{\vec{K}|\vec{v}_i|}{|\vec{v}_1| + \ldots + |\vec{v}_n|}. \qquad [5]$$

The following properties of G may be of interest: (*i*) The value of G does not change when all slip rates reverse directions. (*ii*) For intersections of strike-slip faults, a negative sign of G indicates the tendency of the blocks to "penetrate" each other at the junction (Fig. 1), so that the movement along one fault locks up another fault. On the contrary, if G were positive, the movement on one fault unlocks another one (Fig. 2).

Geometric Incompatibility for a System of Faults with Multiple Junctions

We consider here the case of rigid nonrotating blocks, which is analytically simpler than the general case and can be of independent interest. Generalization for the case of deformable and/or rotating blocks is given in the *Appendix*.

Suppose we have a map with a system of blocks divided by faults. Consider on this map a simply connected region D surrounded by a boundary L. Because the blocks do not rotate, the rates of block movement are the same at all points of a block, and the rates of relative fault movements are the same between fault junctions.

As before, we number the blocks crossing the boundary L of the region D counterclockwise, and number the faults crossing

L in such a way that the ith fault separates the blocks i and $i + 1$. Let $\vec{u}_i$ be the rate of movement of the ith block crossed by L, and $\vec{v}_i$ be the slip rate along the ith fault crossing L. These rates do not change when we deform the boundary L of the region, as long as it does not cross any fault junctions and no intersection points of L with faults appear or disappear.

The Saint Venant condition is again expressed by Eq. **1** with the new definition of the sequence $\vec{v}_i$.

Let us define now geometric incompatibility G. Consider the contour L drawn at a time moment t_0. At a time $t > t_0$, this contour will be deformed into a contour L_t. For example, if L is the external boundary in Figs. 1*A* and 2*A* then L_t is the boundary in Figs. 1*B* and 2*B*. Generally, L_t is defined as follows. The segments of L within blocks move together with blocks. Each crossing of L with a fault splits into two points. To form a closed contour L_t, we connect these points by a segment of a straight line. Let $S(t)$ be the area of the region bounded by L_t. We define the geometric incompatibility as

$$G = \left.\frac{d^2S(t)}{dt^2}\right|_{t=t_0}. \qquad [6]$$

It can be shown that (*i*) G can be computed from slip rates $\vec{v}_i$ by the same formula (Eq. **4**) as for a single junction, but with the new meaning of $\vec{v}_i$, and (*ii*) the value G for the region D equals the sum of the values G for all fault junctions that exist inside D at the time moment t_0.

Application to Southern California Fault Junctions

To illustrate the method suggested here, we considered a schematic representation of two nodes with multiple fault junctions around the ends of the Big Bend segment of the San Andreas fault. Only a few major faults out of the much more complicated fault system are allowed for in our analysis. Such a fault may actually represent a set of faults with a common dominant orientation. With the data available, we could only get an insight into the range of the values of geometric incompatibility.

We used the slip rate data from comprehensive summaries (9) and (10).

The values of $\vec{K}$ and G were computed from formulas (Eq. **1**) and (Eq. **4**), which do not allow for rotation, little known so far, or for blocks deformation.

Fig. 3 shows a simplified scheme of the Big Bend, after (ref. 9, Fig. 3). We analyze separately two nodes of the Big Bend shown in Fig. 3 by dashed ellipses.

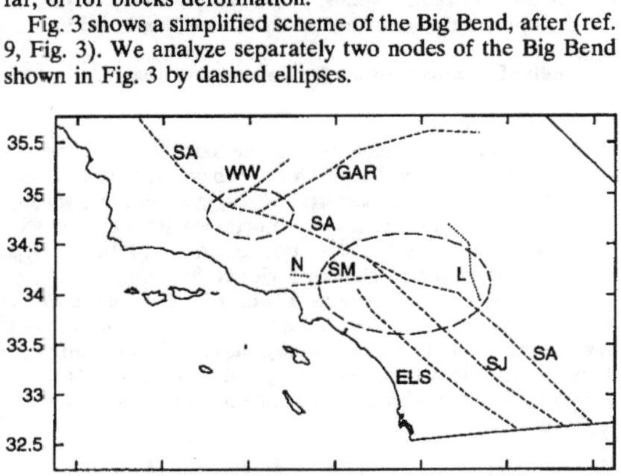

FIG. 3. Scheme of the Big Bend in San Andreas fault system, Southern California (9, 10). Dotted lines, fault breaks during Landers, 1992 (L), and Northridge, 1994 (N), earthquakes; dashed ellipses, contours within which geometric and kinematic incompatibilities are estimated.

North-West Node. This area encompasses the junction of San Andreas and Garlock faults. One of the two strongest earthquakes of this century in Southern California (Kern County, $M = 7.5$, 1952) occurred on the adjacent White Wolf fault. Ironically, it is sometimes not regarded as a major fault, probably due to the small slip rate, which is 2 mm/yr.

The slip rate vectors are given in Table 1. The values of $\vec{K}$ and G are given in the first two lines of Table 2. Line 1 corresponds to the case when only the San Andreas and Garlock faults are considered. The Saint Venant incompatibility lies within 12% of the slip rate on the San Andreas fault. We can reduce $\vec{K}$ by taking into account the White Wolf fault (line 2 in Table 2). The value of G does not change much. Its negative sign indicates that this node is "locked up".

South-East Node. This area encompasses the branching of the San Andreas fault into at least three faults: San Andreas proper, San Jacinto, and Elsinore. The San Andreas fault is joined here from the west by a complicated series of bilateral thrusts represented in Fig. 3 by the Sierra Madre fault. The slip rate vectors are given in Table 1. The values of $\vec{K}$ and G are given in lines 3–6 of Table 2.

Line 3 corresponds to four major faults only, after ref. 9. We see that $\vec{K}$ is rather large, of the same order of magnitude as slip values on some faults. To reduce $\vec{K}$, we take into account 3 mm/yr horizontal shortening across the Sierra Madre fault, and change the slip rate on the Mojave segment of San Andreas fault from 34 to 30 mm/yr, as suggested in ref. 10, table 5. The results are given in line 4 in Table 2. We see that only the lateral component of $\vec{K}$ is noticeably reduced by these changes.

Finally, we can reduce $\vec{K}$ to acceptable limits by increasing the slip rate on Sierra Madre fault from 3 to 8 mm/yr, which is within the difference between estimates by different authors (see ref. 10); the results are given in line 5 of Table 2. We see that this junction is also locked up by geometric incompatibility.

Note that G cannot be reduced to zero in such a configuration of faults without reversing the movement on some faults or introducing new faults.

Impact of Strong Earthquakes. Qualitatively, an earthquake is equivalent to additional slip rate directed as the slip vector in the earthquake source. Let us discuss now how the values of $\vec{K}$ and G may be influenced by two strongest recent earthquakes in this region, Landers (1992, $M = 7.3$) and Northridge

Table 1. Slip rates along the faults in the Big Bend junctions

Fault name	Azimuth	Slip rate, mm/yr Tangent	Normal
North–West node			
San Andreas N	318	34	0
San Andreas S	118	34	0
Garlock	57	−10	0
White Wolf	55	0	−2
South–East node			
San Andreas N			
After ref. 9	298	34	0
After ref. 10	298	30	0
Elsinore	125	5	0
San Jacinto	132	10	0
San Andreas S	133	19	0
Sierra Madre			
After ref. 10	270	0	−3
Modified	270	0	−8*
"Landers"	0	5	0

Data are taken from ref. 10 unless otherwise indicated. Tangent component is positive for right-lateral strike-slip; the normal component is positive for extension.
*Closer to the upper estimate indicated in ref. 10.

Table 2. Saint Venant ($\vec{K}$) and geometric (G) incompatibilities in the Big Bend junctions

No.	$\vec{K}$, East, mm/yr	$\vec{K}$, North, mm/yr	G, mm^2/yr^2
	North-West end of Big Bend		
1	−1.1	3.9	−326.7
2	0.0	2.2	−324.3
	South-East end of Big Bend		
3	−4.6	−6.6	−22.4
4	−1.1	−5.4	−106.9
5	−1.1	−0.4	−234.1
6	−1.1	−0.4	30.0

(1994, $M = 6.7$). The corresponding fault breaks are indicated on Fig. 3 by dotted lines.

To analyze the impact of the Landers earthquake, we included in our computation 5 mm/yr of right-lateral slip along the meridional part of its fault break. The rest of the parameters were the same as for line 4 of Table 2. The results are shown in line 6. Comparing these lines, we see that the Landers earthquake leads to a reduction of both incompatibilities and, because G becomes positive, unlocks the junction.

At the same time, this earthquake created a new junction and therefore new incompatibility on the bend of its own fault break.

The Northridge earthquake occurred on a bilateral thrust fault in the above-mentioned compression zone represented in our scheme (Fig. 3) by the Sierra Madre fault. The Landers earthquake, although more complicated, started as a sub-meridional right-lateral strike-slip. Though different in mechanism and location, both earthquakes reduce $\vec{K}$ similarly. However, their contribution to G is opposite.

The Northridge earthquake is equivalent to an increase of the thrust rate on Sierra Madre fault. Comparing lines 4 and 5 in Table 2, we see that, due to such an increase, the meridional component of $\vec{K}$ would be reduced. However, $|G|$ becomes larger, so that this earthquake locks up the region.

Coarse quantitative estimates can be done as suggested in ref. 11. Consider a fault with length L and width H. An earthquake on this fault causes a fault break of length l and width h and an average slip vector $\vec{d}$. We spread this slip over the whole fault and over the recurrence time T. The additional slip rate on the fault is roughly estimated as

$$\vec{v}_e = \vec{d}\frac{lh}{LHT}.$$

Because this relation is very coarse, we can use, at the best, only the orders of magnitude for all parameters involved, taking the values indicated in ref. 10.

Assuming $l \approx 10^2$ km, $h \approx 10^1$ km, $d \approx 10^3$ mm, $L \approx 10^2$ km, $H \approx 10^1$ km, $T \approx 10^2$–10^3 yr, we get for additional slip rate an estimate $v_e \approx 10^0$–10^1 mm/yr. Comparing these numbers with numerical experiments in Table 1, we see that the contribution of strong earthquakes to geometric and kinematic incompatibilities is indeed essential. For example, Landers earthquake alone could well unlock the region completely, reversing the sign of G.

Discussion

(*i*) This work is a continuation of the quest for integral parameters that control the dynamics of seismicity. Geometric G and kinematic $\vec{K}$ incompatibilities in an active fault system seem to be highly relevant for this purpose, representing the tendency of a system to stress and strain accumulation and fracturing. This interconnection may determine some of the features of seismicity, regardless of the physical processes involved.

Geometric incompatibility seems to be the only known parameter that gives a comprehensive description of fault junctions.

The analogs of the Stokes formula established here allow one to estimate G and $\vec{K}$ in a complicated area from observations on its boundary, thus reducing the requirements for observations.

(*ii*) Because geometric incompatibility depicts the locking up or unlocking of the fault junctions, it may be one of the major factors controlling nucleation (triggering) of strong earthquakes. It is generally accepted that earthquakes occur when the stress exceeds the static friction (strength). Traditionally, such triggering is assigned to "strong" fault segments. Actually, this condition will be first reached rather near a fault junction where both factors are more favorable: The strength may be much smaller due to intense fracturing. And the stress itself may rise faster due to geometric incompatibility.

This conjecture is in accordance with the well-confirmed conclusion that epicenters of strong earthquakes—that is, their nucleation areas, are, as a rule, confined to the vicinity of fault junctions (7, 8).

(*iii*) Existing observations allow, at best, only coarse estimates of G and $\vec{K}$. Still, they provide important insight into how different faults and earthquakes may affect the general equilibrium of a region. In particular, the impact of a strong earthquake cannot be neglected: it is comparable with the role of long-term ("geodetic") movements and, moreover, it may reverse the geometric incompatibility, locking the junction up or unlocking it. A junction that is locked up may act as a transient asperity until it is broken by a strong earthquake or unlocked by subsequent impacts. A junction that is unlocked may act as a "weakest link" in the earthquake triggering or, alternatively, as a relaxation barrier.

In this way, geometric incompatibility and, therefore, seismicity may migrate within a fault system.

(*iv*) Simplifications in our analysis may be summarized as follows: we considered rigid blocks and horizontal movements; in computations possible rotation, being unknown, was not allowed for. When more data are available, these simplifications may have to be removed.

(*v*) Summing up, geometric incompatibility deserves attention as an important integral characteristic of the tectonic development of fault systems. In particular, it may be one of the parameters that controls the dynamics of seismicity.

Appendix: Incompatibilities for Deformable and Rotating Blocks

For a system of deformable blocks, the Saint Venant compatibility condition, in the integral form, can be written as follows (ref. 12, see also ref. 13). Let $\vec{u}(\mathbf{x})$ be displacement rate vector. Its components $u_i(\mathbf{x})$ are smooth functions within each block, with discontinuities at the block boundaries. Let L be a closed contour crossing the block boundaries at the points $\mathbf{x}_1, \ldots, \mathbf{x}_n$ counterclockwise, and let x_0 be a fixed reference point. For $\nu = 1, \ldots, n$, let $\vec{u}^-(\mathbf{x}_\nu)$ and $\vec{u}^+(\mathbf{x}_\nu)$ be the rate of movement $\vec{u}(\mathbf{x}_\nu)$ in the blocks that contain the contour L before and after it crosses the block boundary at $\mathbf{x}_\nu$, and let $\vec{v}(\mathbf{x}_\nu) = \vec{u}^+(\mathbf{x}_\nu) - \vec{u}^-(\mathbf{x}_\nu)$ be the rate of relative movement of the two blocks at $\mathbf{x}_\nu$. Then

$$\oint_L du_i(\mathbf{x}) = \oint_L (x_0^k - x^k)\left(\frac{\partial e_{ik}}{\partial x^l} - \frac{\partial e_{kl}}{\partial x^i} + \frac{\partial e_{il}}{\partial x^k}\right)dx^l + \sum_\nu \left[v_i(\mathbf{x}_\nu) + (x_0^k - x_\nu^k)\frac{\partial v_i}{\partial x^k}(\mathbf{x}_\nu)\right] \quad \textbf{[A1]}$$

should be zero, for all i. Here $e_{ik}(x) = \frac{1}{2}(\partial u_i/\partial x^k + \partial u_k/\partial x^i)$ is the strain tensor.

Linearization of $\vec{u}(\mathbf{x})$ defines affine operators

$$U_{\mathbf{x}}(\mathbf{y}) = \vec{u}(\mathbf{x}) + (y^k - x^k)\frac{\partial \vec{u}}{\partial x^k}(\mathbf{x}). \quad \textbf{[A2]}$$

For $\mathbf{x} = \mathbf{x}_\nu$, let $U_\nu^\pm$ be the operators corresponding to $\vec{u}^\pm(\mathbf{x}_\nu)$. The operator $V_\nu = U_\nu^+ - U_\nu^-$, linearization of the relative rate of movement of two blocks at $\mathbf{x}_\nu$, does not depend on the rate of movement of the coordinate system. Expression in brackets in **A1** can be rewritten as the ith component of the vector $\Sigma_\nu V_\nu(\mathbf{x}_0)$.

When the blocks are rigid, the integral in **A1** vanishes, and the Saint Venant compatibility condition is satisfied if and only if the affine operator

$$\mathbf{K} = \sum_\nu V_\nu \quad \textbf{[A3]}$$

is zero. This is an element of the Lie algebra $\mathcal{A}$ of the orthogonal affine group. This operator is a natural generalization of the Saint Venant incompatibility $\vec{K}$. Note that **K** satisfies an analog of the Stokes formula: its value on any contour is equal to the sum of its values over the junctions within the contour. This can be easily deduced from **A1**.

Geometric incompatibility for a system of deformable blocks can be defined as

$$\mathbf{G} = \oint_L U_{\mathbf{x}} \wedge dU_{\mathbf{x}} + \sum_\nu U_{\mathbf{x}_\nu}^- \wedge U_{\mathbf{x}_\nu}^+. \quad \textbf{[A4]}$$

It is easy to check that the value of **G** does not depend on the rate of movement of the coordinate system. Thus **G** is an invariant of the rate of deformation.

For rigid blocks, the integral in **A4** vanishes, and the geometric incompatibility becomes

$$\mathbf{G} = \sum_\nu U_\nu^- \wedge U_\nu^+, \quad \textbf{[A5]}$$

similar to Eq. **3**. This is an element of $\mathcal{A} \wedge \mathcal{A}$, the external square of the Lie algebra $\mathcal{A}$ of the orthogonal affine group. If $\mathbf{K} = 0$, this can be rewritten as

$$\mathbf{G} = V_1 \wedge V_2 + (V_1 + V_2 \wedge V_3) + \ldots + (V_1 + \ldots + V_{n-2} \wedge V_{n-1}), \quad \textbf{[A6]}$$

similar to (Eq. **4**). Otherwise, V_ν should be modified as in (Eq. **5**) to make $\mathbf{K} = 0$.

For a system of rigid blocks, an analog of the Stokes formula is valid for **G**: its value for a contour is equal to the sum of its values over the junctions within the contour, when the Saint Venant incompatibility is zero for all these junctions.

For a system of two-dimensional rigid blocks, let $\mathbf{x} = (x, y)$, and $V_\nu(x, y) = (X_\nu - y\omega_\nu, \mathbf{Y}_\nu + x\omega_\nu)$. Here $(\mathbf{X}_\nu, \mathbf{Y}_\nu)$ is the relative displacement rate at the origin, and ω_ν is the relative rotation rate of the two blocks adjacent at $\mathbf{x}_\nu$. In this case, $\mathcal{A}$ is three-dimensional, and the Saint Venant incompatibility has three components:

$$\mathbf{K} = \left(\sum_\nu X_\nu, \sum_\nu Y_\nu, \sum_\nu \omega_\nu\right). \quad \textbf{[A7]}$$

The space $\mathcal{A} \wedge \mathcal{A}$ is also three-dimensional. Accordingly, the geometric incompatibility **G** has three components:

$$\mathbf{G} = \left(\sum_\nu (X_\nu^- Y_\nu^+ - Y_\nu^- X_\nu^+), \sum_\nu (X_\nu^- \omega_\nu^+ - \omega_\nu^- X_\nu^+), \sum_\nu (Y_\nu^- \omega_\nu^+ - \omega_\nu^- Y_\nu^+)\right). \quad \textbf{[A8]}$$

Here $(X_\nu^\pm, Y_\nu^\pm, \omega_\nu^\pm)$, define the operators $U_\nu^\pm$ in Eq. **A5**. Expression **A6** can be rewritten similarly in terms of X_ν, Y_ν, and ω_ν.

This work was supported by National Science Foundation Grant no. 931650, and P.O. no. 56659 from University of Southern California, on behalf of the Southern California Earthquake Center. The first author (A.G.) was partly supported by the U.S. Army Research Office through the Mathematical Sciences Institute of Cornell University, Contract DAAL03-91-C-0027, and Department of Geological Sciences at Cornell University, under National Science Foundation Grant no. EAR-94-23818.

1. McKenzie, D. P. & Morgan, W. J. (1969) *Nature (London)* **224,** 125–133.
2. Cronin, V. S. (1992) *Tectonophysics* **207,** 287–301.
3. King, G. C. P. (1983) *Pure Appl. Geophys.* **121,** 761–815.
4. King, G. C. P. (1986) *Pure Appl. Geophys.* **124,** 567–583.
5. Gerasimov, I. & Rantsman, E. (1973) *Geomorphol. Res.* **1,** 1–13.
6. Alekseevskaya, M., Gabrielov, A., Gel'fand, I., Gvishiani, A. & Rantsman, E. (1977) *J. Geophys.* **43,** 227–233.
7. Gelfand, I. M., Guberman, Sh. A., Keilis-Borok, V. I., Knopoff, L., Press, F., Ranzman, E. Y., Rotwain, I. M. & Sadovsky, A. M. (1976) *Phys. Earth Planet. Int.* 11, 227–283.
8. Guberman, Sh.I. & Rotwain, I. M. (1986) *Izvestia Akad. Nauk SSSR* **12,** 72–74.
9. Feigl, K. L., Agnew, D. C., Bock, Y., Dong, D., Donnellan, A., Hager, B. H., Herring, T. A., Jackson, D. D., Jordan, T. H., King, R. W., Larsen, S., Larson, K. M., Murray, M. H., Shen, Z. & Webb, F. H. (1993) *J. Geophys. Res.* **98,** 21677–21712.
10. Working Group on California Earthquake Probabilities (1995) *Bull. Seis. Soc. Am.* **85,** 379–439.
11. Brune, J. (1968) *J. Geophys. Res.* **73,** 777–784.
12. Fung, Y. C. (1965) *Foundation of Solid Mechanics* (Prentice-Hall, Englewood Cliffs, NJ), Chap. 4.
13. Minster, J. B. & Jordan, T. H. (1984) in *Tectonics and Sedimentation Along the California Margin: Pacific Section*, eds. Crouch, J. K. & Bachman, S. B. (Soc. Econ. Paleontol. Mineral., Los Angeles), Vol. 38, pp. 1–6

Journal of Forecasting
J. Forecast. **19**, 65–80 (2000)

Pre-recession Pattern of Six Economic Indicators in the USA

V. KEILIS-BOROK,[1]* J. H. STOCK,[2] A. SOLOVIEV[1] and P. MIKHALEV[3]

[1]*International Institute for Earthquake Prediction Theory and Mathematical Geophysics, Moscow, Russia*
[2]*Kennedy School of Government, Harvard University, USA*
[3]*McKinsey & Co., Los Angeles, USA*

ABSTRACT

This paper applies a tightly parameterized pattern recognition algorithm, previously applied to earthquake prediction, to the problem of predicting recessions. Monthly data from 1962 to 1996 on six leading and coincident economic indicators for the USA are used. In the full sample, the model performs better than benchmark linear and non-linear models with the same number of parameters. Subsample and recursive analysis indicates that the algorithm is stable and produces reasonably accurate forecasts even when estimated using a small number of recessions. Copyright © 2000 John Wiley & Sons, Ltd.

KEY WORDS leading indicators; pattern recognition; business cycle forecasting

INTRODUCTION

A central problem of macroeconomic forecasting has been the prediction of the onset and end of recessions using time series data on economic indicators; recent contributions include Diebold and Rudebusch (1989), Fair (1993), Stock and Watson (1989, 1993), Sims (1993), Berk and Bikker (1995), and Mostaghimi and Rezayat (1996). At least four features of this problem make this particularly difficult. First, recessions are rare; in the United States, for example, since 1959 (the earliest date for which many monthly series are available) to date, there have been only five completed expansions. Second, recessions are discrete events, so forecast of recessions naturally entail some degree of non-linearity. Third, recessions are complicated events and their accurate prediction arguably requires the use of many series at once. Indeed, the presence of an economy-wide decline is an important factor in the *ex-post* identification of recessions, and statistical models that focus on a single series miss this aspect. Fourth, economic institutions, demographics, and economic policy are constantly evolving, and it is natural to expect that

* Correspondence to: V. Keilis-Borok, International Institute of Earthquake Prediction Theory and Mathematical Geophysics, Warshavskoye sh., 79, kor.2, Moscow 113556 Russia.

CCC 0277–6693/2000/010065–16$17.50

Revised September 1998

reduced-form relations will also evolve over time. In short, the problem requires fitting non-linear, high-dimensional models to a handful of observations generated by a possibly non-stationary economic environment.

In this paper we address these problems by considering recession forecasts produced by a pattern-recognition algorithm that was previously developed for the prediction of infrequent events (e.g. Gelfand *et al.*, 1976; Press and Allen, 1995, and the references therein). The general methodology was developed by the artificial intelligence school of I. M. Gelfand for analysis of rare phenomena of complex origin. The specific algorithm used here has previously been applied to earthquake prediction (e.g. Keilis-Borok and Kossobokov, 1990; Vorobieva and Levshina, 1994) and to prediction of the results of American elections (Lichtman and Keilis-Borok, 1989; Keilis-Borok and Lichtman, 1990). The motivation for this application to recession forecasting is that, from a time series perspective, recessions and large earthquakes bear similarities: both are discrete, relatively rare events, and both have a rich set of observable indicators that can potentially help in their prediction. This raises the question of whether the empirical methods developed for handling many geophysical series to predict earthquakes can be useful for the analogous task of predicting recessions.

Applied to this macroeconomic problem, this algorithm warns of a recession when negative signals are widespread across a group of economic indicators. In the work presented here, six indicators (listed in the next section) are used; if four or more signal a recession, then a recession warning is signalled. The result is a binary warning signal, with a value of one indicating that the economy is in a time of increased probability of a recession. Formally, the pattern recognition algorithm uses a tightly parameterized non-linear model that produces a 0/1 alarm, A_t. Let $\{y_{kt}\}$, $t = 1, \ldots, T$, denote the time series of observations on the kth of N economic indicators. With this notation, the model can be written as

$$A_t = \begin{cases} 1 & \text{if} \sum_{k=1}^{N} \Psi_{kt} \geqslant N - b \\ 0 & \text{otherwise} \end{cases} \tag{1}$$

where $\Psi_{kt} = 1$ if $y_{kt} \geqslant c_k$ and $=0$ otherwise where $(b, c_1, \ldots, c_N)$ are free parameters. The time series $\{y_{kt}\}$ typically are transformations of the original economic indicators, for example a series might be transformed by taking quarterly growth rates, and it might be 'inverted' (that is, the series is multiplied by minus one) so that large values of the series are associated with an impending recession. (Details of such transformations are discussed in the next section.) The pattern-recognition interpretation of the model stems from noting that

$$N - \sum_{k=1}^{N} \Psi_{kt}$$

is the (Hamming) distance from the ideal prerecessionary pattern of $\Psi_{kt} = 1$, $k = 1, \ldots, N$.

From the perspective of the problems listed in the introductory paragraph, this model has several advantages. It uses multiple economic indicators (in our application, $N = 6$); it is highly non-linear and by construction produces 0/1 warnings; and it is tightly parameterized, with only $N + 1$ parameters (given the choice of $\{y_{kt}\}$). Because the economic indicator series are transformed to binary variables, the model also has the desirable feature of being robust to outliers in the data. Arguably, this transformation to binary variables could in addition make the model less sensitive to drift in the relationship between the individual indicators and the overall economy, so that this model would be robust to some types of structural instability.

Three variants of model (1) are explored in this paper. The first follows the earthquake prediction methodology and involves judgemental choice of the free parameters $(c_1, \ldots, c_N, b)$ and series transformations that are not usually found in econometric applications. This variant is interesting because of its foundation in the geophysical applications but it has the drawback of requiring significant user judgement to implement. The other two variants incorporate more conventional econometric methods: the parameters are estimated by non-linear least squares, and in one variant standard linear filters are applied to the series. We also explore a modification of the alarm in model (1). Our analysis uses monthly data, and because model (1) is based on data only from the current month, the alarm could have high-frequency noise. We therefore also consider a smoothed alarm which issues a recession warning at date t if A_t signals an alarm in the current month or in any of the three preceding months.

Interestingly, it turns out that the predictions made by these three variants are quantitatively similar. For the full sample of cyclical peaks since 1961, we find that this model provides a recession warning in the months before each of the five cyclical peaks from 1961 to 1990, and, depending on the specific model, it provides few or no false alarms. This pattern of reliable performance holds up in a recursive (pseudo out-of-sample) forecasting experiment, and in this sense the model is found to be robust.

The data and preliminary transformations are described in the next section. Estimation methods and results based on the full sample are presented in the third section. The judgemental model is examined in more detail in the fourth section. The fifth section focuses on the two pattern-recognition models with econometrically estimated parameters, and subjects them to a simulated out-of-sample comparison with some more conventional multivariate models for recession probabilities based on the same leading indicators. The final section presents conclusions.

THE DATA AND PRELIMINARY TRANSFORMATIONS

We use monthly macroeconomic data for the United States for the period from January 1959 to April 1996. Over this period, there were six complete recessions (cyclical peaks followed by cyclical troughs). The cyclical peaks and troughs (NBER dates) are listed in Table I. The

Table I. NBER-dated cyclical turning points

Episode	Peak	Trough
1	1960:4	1961:2
2	1969:12	1970:11
3	1973:11	1975:3
4	1980:1	1980:7
5	1981:7	1982:11
6	1990:7	1991:3

Source: National Bureau of Economic Research.

focus of this exercise is predicting recessions, and we therefore consider the five complete expansionary episodes: W_1: 1961: 08–1969:12 (101 months); W_2: 1971: 05–1973:11 (31 months); W_3: 1975: 09–1980:01 (53 months); W_4: 1981:01–1981:07 (7 months); and W_5: 1983:05–1990:07 (87 months). The beginning dates of these episodes were set at six months after the previous cyclical trough. The six-month lag was chosen to allow for recognition lags in *ex post* identification of peaks and troughs, and because by convention turning points are dated so that expansions and recessions have a minimum length of six months. In all, these episodes contain 279 months.

The forecasting models use five leading economic indicators and a coincident index. The purpose of this paper is to explore model (1), not to compare the forecasting properties of different leading indicators. The indicators used here were selected prior to performing any of the calculations in this project. In particular no further variable selection was performed to improve the performance of a particular model. The series were selected judgementally based on their coverage of different aspects of the economy, the consistency of data definitions and availability, and their past performance as leading indicators. The five leading indicators used are:

G10FF. Difference between the interest rate on 10-year US Treasury bond and the federal funds interest rate, on an annual basis.
FYGM3. Interest rate on 90-day US Treasury bills at an annual rate (in per cent).
INVMTQ. Total manufacturing and trade inventories, in real (1987) dollars.
LUINC. Average weekly initial claims for state unemployment insurance.
LHEL. Index of help wanted advertising in newspapers.[1]

Considerable research has demonstrated the historical role of the term structure spread as a leading indicator (see, for example, Stock and Watson, 1989; Estrella and Mishkin, 1996). The specific long/short spread used here, the 10-year Treasury bond/federal funds spread, is the spread currently used in the Composite Index of Leading Indicators (LEI) maintained by The Conference Board (this index was formerly maintained by the US Department of Commerce). The interest rate plays an important role in many macro models and also generally leads economic activity (cf. Stock and Watson, 1998). Inventories were included to provide a leading indicator from the production side of the economy. Average weekly unemployment insurance claims are included in The Conference Board's LEI. Help wanted advertising is another leading indicator that measures labor market tightness, which also leads the business cycle (Stock and Watson, 1998).

The final variable included in the model is a measure of overall economic activity, the Stock–Watson (1989) coincident index of overall monthly economic activity (a weighted average current and past values of non-agricultural employment, industrial production, manufacturing and trade sales, and real personal income), here referred to as the *XCI*. The *XCI* is close numerically to The Conference Board's Composite Index of Coincident Indicators. The *XCI* is included because past values of the *XCI* are useful in predicting its future values and turning points in the *XCI* are close to NBER-dated turning points.

Many of these series exhibit high-frequency noise and/or trends, either stochastic or deterministic. These series were therefore subject to further transformation prior to inclusion in the model. Two alternative transformations were therefore used. The first entails transformation to quarterly

[1] Data sources: *XCI*: authors' calculations. The other series were obtained from CITIBASE. *IVMTQ*, *LHEL*, *LUINC*, *FYGM3* are CITIBASE mnemonics, and *G10FF* = *FYGT10* − *FYFF*.

growth rates (*QG*); this is a procedure based on standard linear filtering theory and as such is familiar in economic applications. The second is a procedure which we will refer to as local linear (*LL*) detrending which entails interactive user judgement and was developed for the original geophysical applications of the model.

For the quarterly growth rate transformation, y_{kt} is computed as $(1 - L^3)x_{kt}$, where x_{kt} denotes the natural logarithm of the original series. Because $(1 - L^3) = (1 + L + L^2)(1 - L)$, this has the interpretation as a three-quarter moving average of the first differences of the series. This constitutes a bandpass filter which filters out both linear deterministic trends and stochastic trends that are integrated of order one. It also filters out high-frequency noise (albeit imperfectly because of a side lobe in the transfer function of this filter at the highest frequencies). For results based on *QG* detrending, this transformation was applied to *XCI*, *LHEL*, *LUINC*, and *IVMTQ*. The term spread, *G10FF*, has little high-frequency noise and little or no trend. The 90-day Treasury Bill rate, *FYGM3*, does not exhibit a deterministic trend, and unit root tests are ambiguous about whether *FYGM3* exhibits a stochastic trend. In levels, it has little power at high frequencies. For results based on *QG* detrending, these two series were therefore not further transformed.

The local linear detrending methods were developed for use in earthquake prediction and were applied here using that methodology. Applying these transformations requires user judgement. The series *XCI*, *IVMTQ*, and *FYGM3* were transformed to be deviations from their trend value over the current expansion. Specifically, the series y_t was regressed on $(1, t)$ by recursive least squares over the period $t = \tau_s + 1, \ldots, s - 1$, where τ_s is the date of the most recent cyclical trough as of month s. This yields the estimated intercept and trend coefficients $\hat{\alpha}_s$ and $\hat{\beta}_s$. The transformed series, $\tilde{y}_s$, is the deviation from this local trend, $\tilde{y}_s = y_s - \hat{\alpha}_s - \hat{\beta}_s s$. The series so transformed are denoted *XCIR*, *INVR*, and *FYG3R*. The series *LHEL* and *LUINC* were transformed to be weighted averages of their changes, specifically, they were transformed to be the estimated slope coefficients of a local linear least squares regression of the series on $(1, t)$, for $t = s - m, \ldots, s$. For *LHEL*, the value $m = 5$ was used, and for *LUINC*, the value $m = 10$ was used. The transformed values of these two series are denoted *LHK5* and *LUK10*. No transformation was applied to *G10FF*. Because these transformations are recursive they maintain the information set of the original variables, augmented by the trough dates.[2] A representative locally linearly detrended series, *INVR*, is plotted in Figure 1 (the values of *INVR* are arbitrarily set to zero during recessions; these observations are not used).

The signs of the series are set so that, after series transformation, a one is indicated by values of *XCI*, *IVMTQ*, *LHEL*, and *G10FF* below the threshold and by values of *LUINC* and *FYGM3* above the threshold. For example, a recession is signalled by: a value of the coincident indicator series *XCI* sufficiently below its trend over the current expansion; a low value of the spread between the long- and short-term interest rates, *G10FF*; and a value of the 90-day Treasury bill rate series *FYGM3* that is far above its trend value over the current expansion.

[2] The NBER business cycle dating committee announces cyclical peaks and troughs with a variable lag, which can be a year or longer. Thus the most recent trough date is not known during the initial part of an expansion. This concern is mitigated here somewhat by starting the W_i episodes six months into the recession. An alternative would to be define these functionals on a sufficiently long sliding time window.

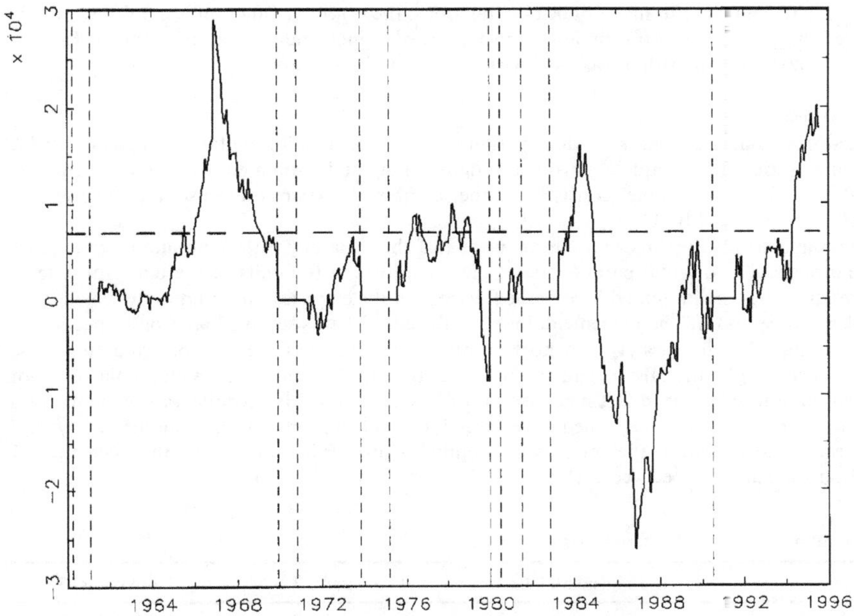

Figure 1. Locally linearly detrended investment (INVR, solid line), discretization threshold (horizontal dashed line), and NBER-dated cyclical turning points (vertical dashed lines), 1960–95

FULL-SAMPLE RESULTS

Parameter selection

The model depends on $N+1$ free parameters $(c_1, \ldots, c_N, b)$. In this paper we consider two different approaches to the specification of these parameters. The first uses the methods developed in the original application of this model to earthquake prediction, and employs algorithms developed at the International Institute of Earthquake Prediction Theory and Mathematical Geophysics in Moscow. This entails judgemental selection of the parameters over a restricted set using interactive software. Because of its history in earthquake prediction, it is intriguing to apply it to the problem of recession prediction. This methodology is, however, difficult for other researchers to apply to new applications.

The second approach requires less user judgement because parameters are estimated by non-linear least squares. Specifically, the objective function is $\Sigma_t(I_t - A_t)^2$, where $I_t = 1$ if an NBER-dated cyclical peak will occur within nine months, and $= 0$ otherwise. The nine-month horizon was chosen arbitrarily as representative of short-term recession forecasting horizons.[3] Because

[3] Results have also been computed for the estimated models at 6- and 12-month horizons. The results are generally similar, although of course the accuracy of all the models is better at the 6-month horizon, and worse at the 12-month horizon, relative to the 9-month horizon reported here. To save space, only results for the 9-month horizon are reported.

the alarms are defined in terms of indicator functions, the objective function is not differentiable in any of the parameters, so estimation must proceed using random search methods. Here, a simulated annealing algorithm was used with 10,000 random draws.

Empirical results

This section summarizes results for three variants of the model: judgemental parameters with *LL* detrending; estimated parameters with *LL* detrending; and estimated parameters with *QG* detrending. The full sample consists of the combined expansion episodes, denoted by $W = W_1 \cup W_2 \cup W_3 \cup W_4 \cup W_5$.

The parameters for the three variants are given in Table II, along with some summary statistics for the models. The original parameters $\{c_k\}$ are quantiles in the units of the data; for ease of interpretation they are presented here as the fractions of the empirical distribution to which these quantiles correspond. In the judgemental model, the cutoff b was set to 2 based on inspection of the individual 0/1 indicators $\{y_{kt}\}$. A horizon of nine months was selected for evaluating these models. Accordingly, the false negative rate is the fraction of months in which an alarm is not signalled but a recession is in fact nine months or less away; the false positive rate is the fraction of months in which an alarm is issued but a recession is in fact more than nine months away; and the overall accuracy rate is the fraction of months in the prediction episode in which the 0/1 prediction was, in retrospect, correct.

Table II. Summary results, pattern-recognition models

Parameter	Judgemental/*LL*	Estimated/*LL*	Estimated/*QG*
Estimated parameters			
p_{XCI}	0.25 (1:3)	0.266	0.873
p_{IVMTQ}	0.75 (3:1)	0.764	0.839
p_{GIOFF}	0.10 (1:9)	0.125	0.123
p_{LHEL}	0.333 (1:2)	0.845	0.019
p_{LUINC}	0.833 (5:1)	0.007	0.999
p_{FYGM3}	0.75 (3:1)	0.514	0.404
b	2	1	2
Summary statistics for prediction of a recession within 9 months:			
False neg. rate	0.311	0.111	0.222
False pos. rate	0.027	0.023	0.008
Accuracy	0.931	0.964	0.961

The entries in parentheses are the probabilities for the judgemental model expressed as odds ratios. The estimated parameters are non-linear least squares estimates computed by simulated annealing with 10,000 random draws.

For the models with *LL* detrending, the estimated and judgemental parameters are quite close for the *XCI*, inventories and the term spread, but differ for the other three variables. In the estimated model, the quantile for *LUINC* is effectively driven to zero, so that the *LUINC* always signals a recession which has the effect of dropping this variable from the model. For the *QG* model, *LUINC* is also effectively dropped, although in the opposite way: the cutoff is such that *LUINC* never delivers a recession signal, but the number of series for which a signal is required to sound an alarm drops from five to four in the estimated *QG* model. In both the estimated models, the term spread is not transformed, and essentially the same parameter values are estimated for

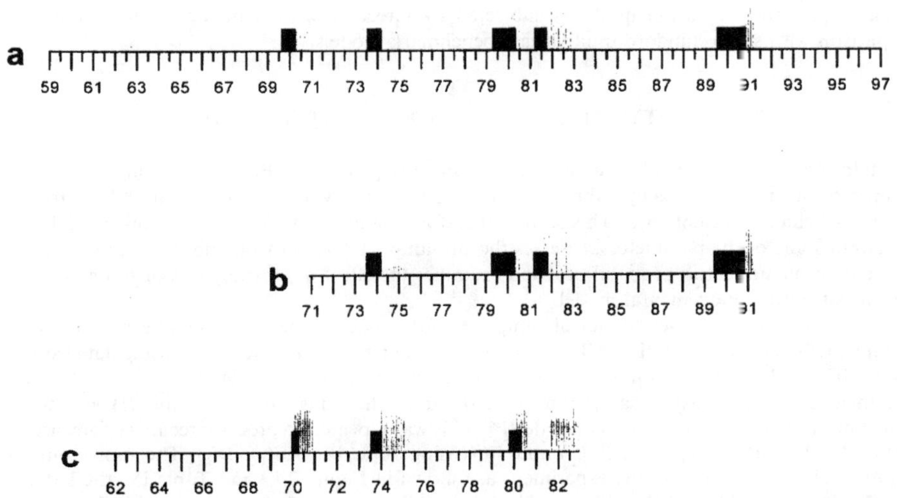

Figure 2. Alarms (shown by black bars): (a) main version; (b) prediction history; (c) reverse prediction history

its cutoff. As expected, when measured by overall accuracy the two estimated models have somewhat better performance than the judgemental model.

The four-month smoothed alarms are shown in Figure 2(a) by black bars. The periods covered by the alarms inside the set *W* are listed in Table III. Note that each alarm extends into the first two or three months of a recession. Evidently all five recessions were preceded by continuous alarms. The longest alarm lasted 13 months, one alarm lasted 10 months, and three alarms lasted 5 months. There were no continuous alarms not ending in a recession. The total duration of the alarms was 38 months, 13.6% of the time covered by the analysis (*W* set). There was no recession from 1991:3 to 1996:4, and the application of this algorithm to this period gives no alarms.

Although the pattern-recognition model is tightly parameterized relative to alternative non-linear models, nonetheless it has several free parameters that have been selected, either informally or formally, using a small number of recessions. We therefore turn to an analysis of the sensitivity of these models to the parameters. This is done in two parts. The next section performs a sensitivity analysis on the judgemental pattern-recognition model. In the fifth section, the

Table III. Alarms within the set *W*

#	Period	Duration (months)
1	1969:08–1969:12	5
2	1973:07–1973:11	5
3	1979:04–1980:01	10
4	1981:03–1981:07	5
5	1989:07–1990:07	13

estimated pattern-recognition model is subjected to a recursive, or simulated out-of-sample, comparison with some standard multivariate benchmark models.

SENSITIVITY ANALYSIS OF JUDGEMENTAL MODEL

The judgemental model involves considerable data fitting. (Dr W. Press, discussing an early example of such an approach, reminded one author of the words of E. Fermi: 'With four exponents I can fit an elephant.') This section therefore examines the sensitivity of this model to changes in some of the parameters, assesses the subsample stability of the model using the full-sample judgemental cutoff probabilities, and examines the effect of omitting one or more of the indicators from the judgemental model.

We first consider the pseudo out-of-sample performance of the algorithm by a recursive experiment. Given the odds ratios in Table II, the thresholds were first determined using data from episode W_1, and the prediction algorithm was then applied to the episode W_2. The thresholds were then recomputed using data from $W_1 \cup W_2$ (and the odds ratios in Table II) and the algorithm was then used to forecast episode W_3. This was repeated to produce recursive forecasts of episodes W_4 (using $W_1 \cup W_2 \cup W_3$) and W_5 (using $W_1 \cup W_2 \cup W_3 \cup W_4$). The four-month smoothed alarms obtained in the experiment are shown in Figure 2(b) and Table IV; the latter contains the thresholds and the alarms within the prediction set. The agreement with the main version is practically complete. The only difference is that the last alarm became a month longer.

The recursive experiments should not be interpreted literally as what would have happened had this approach been applied in real time, in particular the odds ratios $p{:}q$ are fixed here at their full-sample judgemental values. A recursive (simulated real-time) evaluation of the models with estimated parameters is presented in the next section.

We also performed a reverse recursive experiment, in which the data set was expanded in reverse chronological order: given the odds ratios, the thresholds were estimated on W_5, $W_4 \cup W_5$, $W_3 \cup W_4 \cup W_5$, and $W_2 \cup W_3 W_4 \cup W_5$ and the resulting model was then used to make predictions over W_4, W_3, W_2, and W_1 respectively. The alarms obtained in this experiment are shown in Figure 2(c). Table V lists the thresholds and alarms. Most of the results from this reverse recursive exercise are the same as from the full sample. However, the recession in 1981–2 was missed and the alarms before the recessions in 1969–70 and in 1980 are shorter than in the full-sample version. Technically this is due to the differences in the thresholds: comparing to the main version they are smaller for the functions *XCIR*, *INVR*, and *LHK5*, and larger for the functions *LUK10* and *FYG3R*.

In another set of experiments, we altered the odds ratio $p{:}q$, as indicated in Table VI; the table also shows the corresponding change in the alarms within the set W. Again, the results are similar to those in the main version. Note that the second part of the alarm before the 1973–5 recession (line 8) starts in the same month as the recession and has no intersection with the set W. Of course, this alarm still has some predictive value because recessions are formally identified only after a lag of at least several months.

We also performed experiments in which we dropped, one at a time, each of the six economic indicators, and computed forecasts using the thresholds estimated over the full sample. The alarms are declared using $b = 1$ instead of $b = 2$ in the main version. The major change is failure to predict the 1981–2 recession after the elimination of *XCIR*, *INVR*, *G10FF*, or *LHK5*. The smallest changes are caused by elimination of *LUK10*. When *INVR*, *G10FF*, *LHK5*, or

Table IV. Results of the historical experiment

#	Set for definition of thresholds	Thresholds for functions						Results		
		XCIR	*INVR*	*G10FF*	*LHK5*	*LUK10*	*FYG3R*	Examined set	Alarms within the set W	Duration of alarms (months)
1	W_1	−0.77	11531	−0.61	1.38	12.15	0.26	W_2	1973:07–1973:11	5
2	$W_1 \cup W_2$	−0.74	8548	−1.32	2.47	11.17	0.51	W_3	1979:04–1980:01	10
3	$W_1 \cup W_2 \cup W_3$	−0.78	8036	−1.59	3.04	20.41	0.48	W_4	1981:03–1981:07	5
4	$W_1 \cup W_2 \cup W_3 \cup W_4$	−0.86	7991	−1.99	2.78	20.16	0.92	W_5	1989:06–1990:07	14

Table V. Results of the reverse historical experiment

#	Set for definition of thresholds	Thresholds for functions						Results		
		XCIR	*INVR*	*G10FF*	*LHK5*	*LUK10*	*FYG3R*	Examined set	Alarms within the set W	Duration of alarms (months)
1	W_5	−3.65	5305	−0.61	−0.69	29.87	1.27	W_4	No	0
2	$W_5 \cup W_4$	−3.61	4511	−0.95	−1.03	27.89	1.18	W_3	1979:10–1980:01	4
3	$W_5 \cup W_4 \cup W_3$	−3.30	7024	−1.38	0.97	44.47	1.29	W_2	1973:07–1973:11	5
4	$W_5 \cup W_4 \cup W_3 \cup W_2$	−3.04	6320	−1.59	2.33	29.87	1.29	W_1	1969:12	1

Table VI. Results of changing $p:q$

#	Function	New $p:q$	Threshold	Alarms, different from those in the main version (Table III)	Total duration of alarms in % of total time
1	*XCIR*	1:2	−0.98	1989:06–1990:07	14.0
2		1:4	−2.20	1981:04–1981:07	13.3
3	*INVR*	2:1	6167	1969:08–1969:10, 1969:12	13.3
4		4:1	8870	1979:03–1908:01	14.0
5	*G10FF*	1:7	−0.99	No	13:6
6		2:23	−1.90	1969:09–1969:12, 1973:10–1973:11	12.2
7	*LHK5*	1:1	5.38	1985:03–1985:08 (added)	15.8
8		1:3	−0.33	1973:07–1973:09, alarm after 1973:11	12.9
9	*LUK10*	4:1	19.78	No	13:6
10		6:1	29.87	1979:05–1980:01	13:3
11	*FYG3R*	2:1	0.54	1969:07–1969:12	14:0
12		4:1	1.17	1969:08–1969:10, 1969:12	13:3

FYGR3 is eliminated, the alarm before the recession in 1973–5 starts in the same month as the recession.

We also considered prediction using a single indicator, in which case an alarm is declared if that indicator passes its full-sample threshold. The performance of these single-indicator models was worse than the multivariate models, in some cases because recessions were missed, in others, because predicted recessions did not occur.

Performance of the different variations of the prediction algorithm is juxtaposed in the error diagram (Molchan, 1994), shown in Figure 3.

COMPARISON WITH BENCHMARK MODELS

This section compares the pattern-recognition forecasts to those of two conventional, simple models in a recursive, or simulated out-of-sample, experiment. All three versions of the pattern recognition model in Table II are considered here. The benchmark comparison models are a linear probability model and a probit model, which predict the probability of a recession beginning some time in the next nine months, using current values of the six economic indicators. Like the pattern-recognition model, the linear probability model and the probit model each have seven free parameters. For both the linear probability and probit models, an alarm was indicated if a recession probability exceeded 50%. These models were estimated using both *LL* and *QG* detrended data.[4]

[4] Attempts were also made to estimate Hamilton (1989) switching models using these indicators. However, when more than two indicators were included as predictors, these models failed to converge because the parameters were essentially unidentified. This is not surprising because these models contain a large number of parameters to be estimated and there are only six recessions in this data set. This underscores the problem raised in the Introduction faced by empirical models generally, and non-linear models in particular, of predicting recessions using many leading indicator and only a handful of recessions.

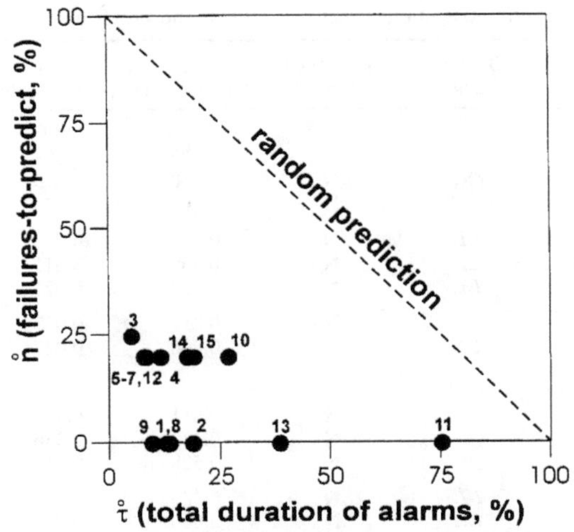

Figure 3. Error diagram: (1) main version; (2) prediction history; (3) reverse prediction history; (4)–(9) exclusion of variables: (4) *XCIR*, (5) *INVR*, (6) *G10FF*, (7) *LHK5*, (8) *LUK10*, (9) *GYG3R*, (10)–(15), prediction using a single variable: (10) *XCIR*, (11) *INVR*, (12) *G10FF*, (13) *LHK5*, (14) *LUK10*, (15) *GYG3R*

The performance of these models in predicting recessions at the nine-month horizon is summarized in Table VII, panel A for the full sample. For comparison purposes, summary measures of the performance of the judgemental model are repeated there as well. All models have overall accuracy rates that exceed 92%. The estimated pattern-recognition models outperform the linear probability and probit models based on this criterion. Interestingly, the choice of the non-standard *LL* or the more conventional *QG* detrending method does not have a large qualitative impact on the results or on the rankings across models.

An interesting question is whether the non-linear transformations $\{\Psi_{kt}\}$ contain information not present in the linear regression. To address this question $\{\Psi_{kt}\}$ were included as six additional regressors in the linear probability model. For the *QG* detrended regression, the *F*-statistic for these six non-linear variables is 20.18; for the *LL* detrended regression, the *F*-statistic is 25.33. The conventional *F* distribution is suspect here because the $\{\Psi_{kt}\}$ variables contain estimated parameters (the cutoffs) and because of overlapping horizons in the dependent variable, and we do not provide any distribution theory for these statistics. Nonetheless, the values of these *F*-statistics far exceed standard critical values, which suggests that these non-linear transformations have important marginal predictive content beyond the linear terms.

Panels B–E report the results of four recursive experiments. The experiments differ by the estimation and evaluation data sets. For example, in panel C, the parameters were estimated on the sample $W_1 \cup W_2$, but the reported accuracy rates refer to the episodes $W_3 \cup W_4 \cup W_5$. In each experiment, the parameters of the linear model were estimated by ordinary least squares; the parameters of the probit model were estimated by maximum likelihood; and the seven parameters

Table VII. Comparison of pattern recognition, linear probability, and probit models

Algorithm	Parameters	Trans-formation	Smoothed alarms?	False negative rate	False positive rate	Overall accuracy rate
(A) Full sample						
Pattern rec.	Estimated	*QG*	N	0.222	0.008	0.961
Linear prob.	Estimated	*QG*	N	0.400	0.019	0.924
Probit	Estimated	*QG*	N	0.267	0.039	0.928
Pattern rec.	Estimated	*LL*	N	0.111	0.023	0.964
Linear prob.	Estimated	*LL*	N	0.333	0.015	0.938
Probit	Estimated	*LL*	N	0.222	0.027	0.944
Pattern rec.	Judgemental	*LL*	N	0.311	0.027	0.931
Pattern rec.	Judgemental	*LL*	Y	0.200	0.027	0.947
(B) Estimation: W_1, prediction: $W_2 \cup W_3 \cup W_4 \cup W_5$						
Pattern rec.	Estimated	*QG*	N	0.611	0.074	0.828
Linear prob.	Estimated	*QG*	N	0.333	0.056	0.894
Probit	Estimated	*QG*	N	0.528	0.012	0.894
Pattern rec.	Estimated	*LL*	N	0.389	0.086	0.859
Linear prob.	Estimated	*LL*	N	0.056	0.086	0.919
Probit	Estimated	*LL*	N	0.111	0.198	0.818
(C) Estimation: $W_1 \cup W_2$, prediction: $W_3 \cup W_4 \cup W_5$						
Pattern rec.	Estimated	*QG*	N	0.185	0.185	0.815
Linear prob.	Estimated	*QG*	N	0.000	0.178	0.852
Probit	Estimated	*QG*	N	0.000	0.489	0.593
Pattern rec.	Estimated	*LL*	N	0.296	0.089	0.877
Linear prob.	Estimated	*LL*	N	0.278	0.160	0.827
Probit	Estimated	*LL*	N	0.037	0.267	0.772
(D) Estimation: $W_1 \cup W_2 \cup W_3$, prediction: $W_4 \cup W_5$						
Pattern rec.	Estimated	*QG*	N	0.333	0.174	0.798
Linear prob.	Estimated	*QG*	N	0.444	0.105	0.837
Probit	Estimated	*QG*	N	0.500	0.070	0.856
Pattern rec.	Estimated	*LL*	N	0.778	0.163	0.731
Linear prob.	Estimated	*LL*	N	0.389	0.140	0.817
Probit	Estimated	*LL*	N	0.278	0.360	0.654

of the pattern-recognition model were estimated by non-linear least squares using simulated annealing (10,000 random draws).

Several results are apparent. The other non-linear model, the probit model, is unstable, and produces forecasts that have quite low accuracy in several of the experiments. The ranking between the linear probability model and the estimated pattern-recognition model is less clear; generally speaking, the estimated pattern-recognition model does better than the linear model when the estimation data set is longer. The estimated pattern-recognition model appears to be more robust than the probit model in the sense of being less prone to significant deteriorations in accuracy.

CONCLUSIONS

Taken together, the evidence presented here suggests that these simple binary transformations of these economic indicators have significant predictive content for recessions. The pattern-recognition model has the interesting feature of being highly non-linear yet only having the same number of free parameters as a linear regression model. The sensitivity analysis of the judgemental pattern-recognition model and the simulated out-of-sample analysis of the estimated pattern-recognition model suggests that this model is stable to outliers and changes in regimes. It is striking that these models, in which the information in the data is reduced to binary indicators, has predictive content comparable to or, in many cases, better than that of more conventional models.

Several caveats should be emphasized. First, despite the subsample and pseudo-out-of-sample analysis, the only true test of this algorithm is to track the performance of the forecasts from the pattern-recognition models during the current expansion and beyond.[5] Second, only two methods for parameter selection (judgemental and full non-linear least squares) have been investigated here, and it might be that estimation methods could be refined to further reduce the effective number of parameters of the pattern-recognition model. Third, although the six economic indicators used here were selected because of their quality (documented elsewhere) as coincident or leading economic indicators, there is no guarantee that these indicators will continue to be good predictors in the future. Fourth, the assessment here has relied on pseudo-out-of-sample forecasting exercises to assess the validity of this algorithm, and it would be useful (but difficult) to augment this with formal statistical measures of sampling uncertainty. Despite these cautionary notes, however, we consider these results encouraging and supportive of additional investigation of this approach.

ACKNOWLEDGEMENTS

The authors are grateful to two anonymous referees and the editor for helpful suggestions.

REFERENCES

Berk, J. M. and Bikker, J. A., 'International interdependence of business cycles in the manufacturing industry: the use of leading indicators for forecasting and analysis', *Journal of Forecasting*, **14** (1995), 1–23.

Diebold, F. X. and Rudebusch, G. D., 'Scoring the leading indicators', *Journal of Business*, **62** (1989), 369–92.

Estrella, A. and Mishkin, F. S., 'The yield curve as a predictor of U.S. recessions', *Current Issues in Economics and Finance*, Federal Reserve Bank of New York, **2** (1996) No. 7, June.

Fair, R., 'Estimating event probabilities from macroeconomic models using stochastic simulation', in Stock, J. H. and Watson, M. W. (eds), *Business Cycles, Indicators and Forecasting*, (NBER Studies in Business Cycles, Vol. 28) 1993, pp. 157–78.

[5] We can provide some evidence on this. The analysis in this paper used data through 1996:4. In a late revision of this paper we had the opportunity to update the signals using data through 1997:3. This constitutes a brief, truly out-of-sample experiment using the judgemental algorithm. During the ten months from May 1996 through March 1997 (inclusive), no recession alarm has been sounded by the judgemental model. As of May 1998, no cyclical turning point had been declared by the NBER, nor is one suspected to have occurred, during this period. This brief episode provides a test of the judgemental model, which it passes.

Gelfand, I., Keilis-Borok, V. I., Knopoff, L., Press, F., Rantsman, E., Rotwain, I. and Sadovsky, A., 'Pattern recognition applied to earthquake epicenters in California', *Phys. Earth Planet. Inter.*, **11** (1976), 227–83.

Hamilton, J. D., 'A new approach to the economic analysis of nonstationary time series and the business cycle', *Econometrica*, **57** (1989), 357–84.

Keilis-Borok, V. I. and Kossobokov, V. G., 'Premonitory activation of seismic flow: algorithm M8', *Physics of the Earth and Planetary Interiors*, **61** (1990), 73–83.

Keilis-Borok, V. I. and Lichtman, A., 'Understanding and prediction of large and complex unstable systems in the absence of basic equations: concepts and selforganization and similarity', in Costa, G., Calucci, G. and Giorgi, M. (eds), *Proceedings of the First International Symposium on Conceptual Tools for Understanding Nature*, Trieste 26–28 September, 1990, p. 145.

Lichtman, A. and Keilis-Borok, V. I., 'Aggregate-level analysis and prediction of midterm senatorial elections in the United States', *Proc. Natl. Acad. Sci. USA*, 86, December 1989, pp. 10176–80.

Molchan, G. M., 'Models for optimisation of earthquake prediction', *Computational Seismology and Geodynamics*, (AGU, Washington, DC) **2** (1994) 1–10.

Mostaghimi, M. and Rezayat, F., 'Probability forecast of downturn in U.S. economy using classical statistical decision theory', *Empirical Economics*, **21** (1996), 255–79.

Press, F. and Allen, C., 'Pattern of seismic release in the Southern California region', *Journal of Geophysical Research*, **100** (1995), 6421–30.

Sims, C. A., 'A nine-variable probabilistic macroeconomic forecasting model', in Stock, J. H. and Watson, M. W. (eds), *Business Cycles, Indicators, and Forecasting*, NBER Studies in Business Cycles, Vol. 28, 1993, pp. 179–212.

Stock, J. H. and Watson, M. W., 'New indexes of leading and coincident economic indicators', *NBER Macroeconomics Annual*, 1989, pp. 351–94.

Stock, J. H. and Watson, M. W., 'A procedure for predicting recessions with leading indicators', in Stock, J. H. and Watson, M. W. (eds), *Business Cycles, Indicators, and Forecasting*, NBER Studies in Business Cycles, Vol. 28, 1993, pp. 95–156.

Stock, J. H. and Watson, M. W., 'Business cycle fluctuations in U.S. macroeconomic time series', in Taylor, J. and Woodford, M. (eds), *Handbook of Macroeconomics*, forthcoming, 1998.

Vorobieva, I. A. and Levshina, T. A., 'Prediction of a second large earthquake based on aftershock sequence', *Computational Seismology and Geodynamics*, (AGU, Washington, DC), **2** (1994), 27–36.

Authors' biographies:

V. Keilis-Borok is the Research Group Leader in the International Institute of Earthquake Prediction Theory and Mathematical Geophysics, Russian Ac. Sci., and the Chairman of several international projects aimed at modeling and prediction of critical phenomena. His recent publications are in the areas of non-linear dynamics, pattern recognition, Geophysics, and socio-economic sciences.

J. H. Stock is the Roy Larsen Professor of Political Economy at the Kennedy School of Government, Harvard University. His recent publications are in the area of time series analysis, economic forecasting, econometric methodology, and monetary economics.

A. Soloviev is the Director of the International Institute of Earthquake Prediction Theory and Mathematical Geophysics, Russian Ac. Sci. His recent publications are in the area of mathematical geophysics, particularly, application of pattern recognition methods to geophysical problems and modelling of block structure dynamics and seismicity.

P. Mikhalev is an Associate in the Los Angeles Office of McKinsey and Company. His recent work concerns consulting in marketing and manufacturing strategies in different industries. Previously he worked also with IBM East Europe/Asia and Harris Corporation. He received an M.Sc. in Chemical Engineering from Moscow Chemical-Technological University and an M.B.A. from the University of California at Berkeley.

Authors' addresses:
V. Keilis-Borok, and **A. Soloviev**, International Institute of Earthquake Prediction Theory and Mathematical Geophysics, Warshavskoye sh., 79, kor. 2, Moscow 113556, Russia.

J. H. Stock, Kennedy School of Government, Harvard University, Cambridge, MA 02138, USA.

P. Mikhalev, McKinsey & Co., 400 South Hope Street, Suite 700, Los Angeles, CA 90071-2890, USA.

PHYSICAL REVIEW E VOLUME 62, NUMBER 1 JULY 2000

Critical transitions in colliding cascades

A. Gabrielov
Department of Earth and Atmospheric Sciences and Department of Mathematics, Purdue University, West Lafayette, Indiana 47907-1395

V. Keilis-Borok
International Institute of Earthquake Prediction Theory and Mathematical Geophysics, Russian Academy of Sciences, Warshavskoe shosse, 79, korpos 2, 113556, Moscow, Russia
and Department of Earth and Space Sciences and Institute of Geophysics and Planetary Physics, University of California, Los Angeles, California 90095-1567

I. Zaliapin
International Institute of Earthquake Prediction Theory and Mathematical Geophysics, Russian Academy of Sciences, Warshavskoe shosse, 79, korpos 2, 113556, Moscow, Russia

W. I. Newman
Department of Earth and Space Sciences, Department of Physics and Astronomy, and Department of Mathematics, University of California, Los Angeles, California 90095-1567
(Received 9 November 1999)

We consider here the interaction of direct and inverse cascades in a hierarchical nonlinear system that is continuously loaded by external forces. The load is applied to the largest element and is transferred down the hierarchy to consecutively smaller elements, thereby forming a direct cascade. The elements of the system fail (i.e., break down) under the load. The smallest elements fail first The failures gradually expand up the hierarchy to the larger elements, thus forming an inverse cascade. Eventually the failures heal, ensuring that the system will function indefinitely The direct and inverse cascades collide and interact. Loading triggers the failures, while failures release and redistribute the load. Notwithstanding its relative simplicity, this model reproduces the major dynamical features observed in seismicity, including the seismic cycle, intermittence of seismic regime, power-law energy distribution, clustering in space and time, long-range correlations, and a set of seismicity patterns premonitory to a strong earthquake. In this context, the hierarchical structure of the model crudely imitates a system of tectonic blocks spread by a network of faults (note that the behavior of such a network is different from that of a single fault). Loading mimics the impact of tectonic forces, and failures simulate earthquakes. The model exhibits three basic types of premonitory pattern reflectin seismic activity, clustering of earthquakes in space and time, and the range of correlation between the earthquakes. The colliding-cascade model seemingly exhibits regularities that are common in a wide class of complex hierarchical systems, not necessarily Earth specific

PACS number(s): 05.65.+b, 91.30.Px, 91.30.Dk, 64.60.Ht

I. INTRODUCTION

We synthesize here three phenomena that play an important role in many complex systems. First, the system has a *hierarchical structure*, with the smallest elements merging in turn to form larger and larger ones, the largest element being the entire system. Second, the *system is continuously loaded* or driven by external sources. Finally, the *elements of the system fail* (break down) under the load, causing redistribution of the load throughout the system. Eventually the failed elements "heal" and regain their structural integrity, thereby facilitating the continuous operation of the system.

The load is transferred from the top of the hierarchy to the bottom, thus forming a *direct cascade* from the largest to the smallest scales. Failures are initiated at the lowest level of the hierarchy or tree, and gradually propagate upward, thereby forming an *inverse cascade*. The interaction of direct and inverse cascades establishes the dynamics of the system.

Direct cascades are well known, for example, in the theory of three-dimensional turbulent flo where energy and/or momentum is transferred from large eddies to small ones, eventually dissipating through viscosity [1]. Another example is plate tectonics—the influenc of mantle convection is transferred to consecutively smaller structures [2]. Examples of inverse cascades [3] include the escalating sequence of earthquakes; avalanches in rock, sand, and snow; the consecutive coalescence of fractures in a solid body; forest fire and epidemics; the clustering of animals into flocks herds, schools, and so on; and chain reactions in physics, chemistry, and economics. In many systems both direct and inverse cascades coexist and interact. Loading increases instability and causes an inverse cascade of failures, while failures release and redirect loading. In this study we model such an interaction.

The hierarchical structure of the model imitates the active lithosphere, composed of a hierarchy of volumes ("blocks") separated by faults. Loading in the model imitates the influ ence of tectonic forces, while failures imitate earthquakes.

We focus here on the *collective* behavior of multiscale failures, which correspond to the seismicity of a fault net-

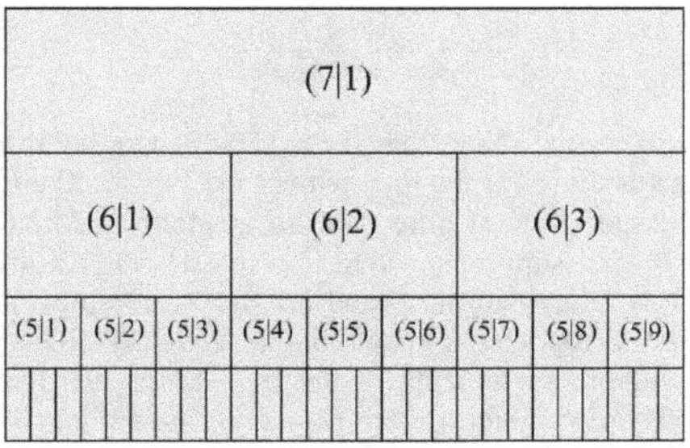

FIG. 1. Structure of the three-branched or ternary tree model. Figure shows four highest levels of the seven-level tree, used in simulation.

work, and not on the dynamics of a single failure. Moreover, we explore the major robust features of the behavior of colliding cascades. Accordingly, as is usually the case in such models, we employ the basic condition of failure specifie in Sec. II D below.

Heuristic or empirical constraints are derived from studies of seismicity. We make the model as simple as we can, providing a skeletal representation of a real lithosphere with its immense complexity. Nevertheless, the model reproduces the major regularities in the observed dynamics of seismicity, namely, the seismic cycle, intermittency in the seismic regime, power-law energy distribution, clustering in space and time, long-range spatial correlations, and a wide variety of seismicity patterns premonitory to a strong earthquake.

We consider here three basic types of premonitory pattern. Two patterns of the firs type reflec the rise of seismic activity. The pattern of the second type reflect the rise in the clustering of earthquakes in space and time; and two patterns of the third type mirror the rise in the correlation distance between the earthquakes. Each pattern was define separately for different magnitude ranges.

Although the model of colliding cascades reproduces so well the major features of earthquake sequences, the design of the model is not specifi to seismicity, or even to the more general phenomenon of multiple fracturing in solids. Our model probably exhibits regularities that are common to a wide class of complex hierarchical systems. Turcotte *et al.* [4] provide a theoretical basis for such universality, outlining the connection between seismicity and other processes with power-law scaling, such as those studied in statistical physics.

II. COLLIDING-CASCADE MODEL

We begin by providing an overview of the structure and dynamics of the simple dynamical system that we are constructing. We then proceed to describe the processes of loading, failure, and healing as well as the governing differential equations. Finally, we describe a synthetic sequence of failures (''earthquakes'') and demonstrate how it is generated by the interaction of direct and inverse cascades.

A. Structure

We consider a dynamical system acting on the ternary tree shown in Fig. 1. Nodes of the tree are called the *elements* of the system, and each element has its own *index*, consisting of two numbers $i=(m|g)$. Here, m is the level where an element is situated and is enumerated from the bottom to the top of the hierarchy. The second integer g identifie the position of an element within its level m, counting from left to right, namely, from 1 to $3^{\bar{m}-m}$, where $\bar{m}$ identifie the top level in the hierarchy. Indexing of elements is illustrated in Fig. 1. It is convenient to describe the taxonomy of this system using the imagery of a family tree. The top element $(\bar{m}|1)$ has three ''children''—the elements $(\bar{m}-1|1)$, $(\bar{m}-1|2)$, and $(\bar{m}-1|3)$. They are referred to as ''siblings'' while the element $(\bar{m}|1)$ identifie their ''parent.'' For example, the elements $(\bar{m}-2|4)$, $(\bar{m}-2|5)$, and $(\bar{m}-2|6)$ are the children of the parent element $(\bar{m}-1|2)$ and the siblings of each other. Conceptually, this structure is similar to that of a wavelet, where the complementary dimensions, crudely, the position and the (logarithm) of the wavelength, have a direct physical meaning and significance

B. Dynamics

The behavior of an arbitrary element i is described by two functions, namely, a continuous positive-valued function $\sigma_i(t)$ and a Boolean function $f_i(t)$. We think of $\sigma_i(t)$ as the ''load'' supported by an element and of $f_i(t)$ as its ''state.'' An element i is ''whole'' or intact when $f_i(t)=0$, and ''broken'' or failed when $f_i(t)=1$. The direct cascade of loading is described by the set of functions $\{\sigma_i(t)\}$ while the inverse cascade of fracturing is described by the set of functions $\{f_i(t)\}$. The dynamics of the system is described by interaction of direct and inverse cascades. The functions $\sigma_i(t)$ satisfy a system of ordinary differential equations with the right sides depending upon the functions $\{f_j(t)\}$. The functions $f_i(t)$ change their values according to certain logical rules or conditions that depend upon $\sigma_i(t)$ and $\{f_j(t)\}$.

C. Loading

First, we introduce equations for the top element $(\bar{m}|1)$, namely.

$$\dot{\sigma}_{\mathrm{top}}(t)=\begin{cases} v-\dfrac{\beta\sigma_{\mathrm{top}}(t)}{[\theta-\sigma_{\mathrm{top}}(t)]} & \text{if} \quad f_{\mathrm{top}}(t)=0 \\ v-\alpha\sigma_{\mathrm{top}}(t) & \text{if} \quad f_{\mathrm{top}}(t)=1, \end{cases} \tag{2.1}$$

where $v>0$, $\beta>0$, $\theta>0$, and $\alpha>0$. Here v is a constant describing the application of the load to the top element of the system. Note that, in the present realization of the model, this is the only method for introducing a load into the system. The load is transferred in this ternary model to the three elements $(\bar{m}-1|1)$, $(\bar{m}-1|2)$, and $(\bar{m}-1|3)$. The rate of load outflo in Eq. (2.1) depends on the state of the top element. If this element is whole, i.e., $f_{\mathrm{top}}(t)=0$, then the rate is equal to

$$\frac{\beta\sigma_{\mathrm{top}}(t)}{[\theta-\sigma_{\mathrm{top}}(t)]},$$

where β is a constant. If this element is broken, i.e., $f_{\mathrm{top}}(t)$

$=1$, then the rate is proportional to the load accumulated by the element at that instant of time.

In the stationary or steady state case, when the time derivatives in Eqs. (2.1) vanish, we have

$$\sigma^{\mathrm{w}}_{\mathrm{top,ss}} \equiv \frac{\theta v}{v+\beta} \quad (2.2)$$

for a whole element, and

$$\sigma^{\mathrm{b}}_{\mathrm{top,ss}} \equiv \frac{v}{\alpha} \quad (2.3)$$

for a broken one. The load of the top element $(\overline{m}|1)$ tends to approach the values (2.2) or (2.3), depending on the state of the element.

It is clear from Eq. (2.1) that the load applied to the top element in the whole state can never exceed θ. We shall call θ a "critical threshold" for the load.

Equations for $\sigma_i(t)$ for all $i \neq (\overline{m}|1)$ are constructed in a manner analogous to those for $(\overline{m}|1)$. The only difference is that each element $i \neq (\overline{m}|1)$ receives a load not only from its parent but from its siblings as well. Furthermore, the rate of load transfer depends on the load accumulated by the parent and the siblings at that time, namely,

$$\dot{\sigma}_i(t) = U_i(t) - W_i(t). \quad (2.4)$$

Here

$$W_i(t) = \begin{cases} \dfrac{\beta\sigma_i(t)}{[\theta-\sigma_i(t)]} & \text{if} \quad f_i(t)=0 \\ \alpha\sigma_i(t) & \text{if} \quad f_i(t)=1 \end{cases}$$

and

$$U_i(t) = CW_p(t) + \frac{1-C}{2}W_{s1}(t) + \frac{1-C}{2}W_{s2}(t), \quad 0 \leqslant C \leqslant 1.$$

The subindex p refers to the parent of the ith element, while subindices $s1$ and $s2$ refer to its two siblings. In order to maintain parallelism, we defin for the top element

$$W_{\mathrm{top}}(t) = \begin{cases} \dfrac{\beta\sigma_{\mathrm{top}}(t)}{[\theta-\sigma_{\mathrm{top}}(t)]} & \text{if} \quad f_{\mathrm{top}}(t)=0 \\ \alpha\sigma_{\mathrm{top}}(t) & \text{if} \quad f_{\mathrm{top}}(t)=1, \end{cases}$$

$$U_{\mathrm{top}}(t) = v.$$

With this notation, the loading is described by the system (2.4).

As before, the load supported by an intact element can never exceed the critical threshold θ. In the stationary case, when the time derivatives in Eqs. (2.4) vanish, we have the steady state solutions

$$\sigma^{\mathrm{w}}_{\mathrm{i,ss}} \equiv \frac{\theta v}{v+\beta} \quad (2.5)$$

for a whole element i, and

$$\sigma^{\mathrm{b}}_{\mathrm{i,ss}} \equiv \frac{v}{\alpha} \quad (2.6)$$

for a broken one. Note that these steady state solutions are the same as those for the top element via Eqs. (2.2) and (2.3).

We assume that at time $t=0$ all elements are intact, i.e., $f_i(0)=0$, and support no load, i.e., $\sigma_i(0)=0$. As given by Eqs. (2.4), the load is added to the hierarchical system through the top element and is subsequently redistributed among all other elements in the tree. Since all dynamical equations are symmetric with respect to the siblings' indices $(i,s1,s2)$, all of the elements on any level retain the same load until at least one element fails.

D. Failure

A whole element $i \neq (\overline{m}|1)$ fails when the following condition is satisfied

$$\sigma_i(t) \geqslant \theta q^{-[f_{c1}(t)+f_{c2}(t)+f_{c3}(t)]} \times s^{-[f_{s1}(t)+f_{s2}(t)]},$$
$$q \geqslant 1, \quad s \geqslant 1. \quad (2.7)$$

Here, the subindices $c1$, $c2$, and $c3$ refer to the three children of the ith element while $s1$ and $s2$ refer to its two siblings. The exponents of q and of s indicate the number of broken children and siblings, respectively. If all children and siblings of the ith element are intact, then this condition reduces to

$$\sigma_i(t) \geqslant \theta.$$

If some of the siblings or children are broken, the ith element is weakened, that is, the threshold for failure is reduced. The parameters q and s in Eqs. (2.7) quantitatively determine this weakening. Equation (2.7) describing the top element reduces to

$$\sigma_{\mathrm{top}}(t) \geqslant \theta q^{-[f_{c1}(t)+f_{c2}(t)+f_{c3}(t)]}$$

due to the absence of siblings. As we have mentioned above, the load applied to an intact element can never exceed θ. Therefore, an element cannot fail until at least one of its siblings or children fails. Accordingly, the failures propagate upward and thereby form an inverse cascade.

At the bottom level of the tree where the elements have no children, we introduce random failures with a rate proportional to the intensity of the direct cascade. This mimics "juvenile cracking" in earthquake phenomenology. Let t_s be the time when the load of an element rises close to the stationary value, $\sigma_i(t) \geqslant \sigma^{\mathrm{w}}_{\mathrm{i,ss}} - \epsilon$ for ϵ small and positive. This element fails at a later time $t_s + \chi$, where χ is a random variable, distributed exponentially with a decay time λ. This randomness ensures that the dynamics of our model shows a degree of inhomogeneity in spite of the above mentioned symmetry.

E. Healing

In order to ensure the perpetual operation of our system, we introduce the effect of "healing," i.e., the restoration to an unbroken state of a previously broken element. Otherwise the system will cease to function once all elements have failed. We assume that a broken element heals when the following two conditions hold during the ensuing exponentially distributed time interval with a decay time L. At least n children of the ith element are intact, and

$$\sigma_i(t) < \theta q^{-[f_{c1}(t)+f_{c2}(t)+f_{c3}(t)]} s^{-[f_{s1}(t)+f_{s2}(t)]}. \tag{2.8}$$

Finally, at the bottom level we replace the latter by

$$\sigma_i(t) < \theta. \tag{2.9}$$

Having formulated our model, we now provide it in dimensionless form.

F. Dimensionless equations

Equations (2.4) contain values of $\sigma_i(t)$ measured in units $[u]$, and time measured in units $[t]$. Meanwhile, the variables v and β are measured in units $[u/t]$, variable θ in $[u]$, α in $[t^{-1}]$, and λ and L in $[t]$.

We now defin a time scale $t_0 = \theta/v$. Let us introduce the following dimensionless variables:

$$\tau \equiv t/t_0, \quad \tilde{\sigma}_i(\tau) \equiv \frac{\sigma_i(t)}{\theta},$$

and dimensionless parameters

$$\gamma \equiv \beta/v,$$

$$\tilde{\alpha} \equiv \frac{\alpha\theta}{v}.$$

We then obtain the dimensionless equations

$$\dot{\tilde{\sigma}}_i(\tau) = \tilde{U}_i(\tau) - \tilde{W}_i(\tau). \tag{2.10}$$

Here

$$\tilde{W}_i(\tau) = \begin{cases} \dfrac{\gamma\tilde{\sigma}_i(\tau)}{[1-\tilde{\sigma}_i(\tau)]} & \text{if} \quad f_i(\tau)=0 \\ \tilde{\alpha}\tilde{\sigma}_i(\tau), & \text{if} \quad f_i(\tau)=1, \end{cases}$$

$$\tilde{U}_i(\tau) = C\tilde{W}_p(\tau) + \frac{1-C}{2}\tilde{W}_{s1}(\tau) + \frac{1-C}{2}\tilde{W}_{s2}(\tau)$$

$$i \neq (\overline{m}|1),$$

$$\tilde{U}_i(\tau) = 1, \quad i = (\overline{m}|1).$$

As before, the subindex p refers to the parent of the ith element, while the subindices $s1, s2$ refer to the two siblings of ith element.

G. Sequence of failures

The colliding-cascade model determines for each element both the load $\sigma_i(t)$ and the state $f_i(t)$—whether the element is broken or whole. In many relevant processes, including seismicity, the data for failure events are especially complete since it is easier to both measure and catalog the failures than the load. We represent the sequence of failures as

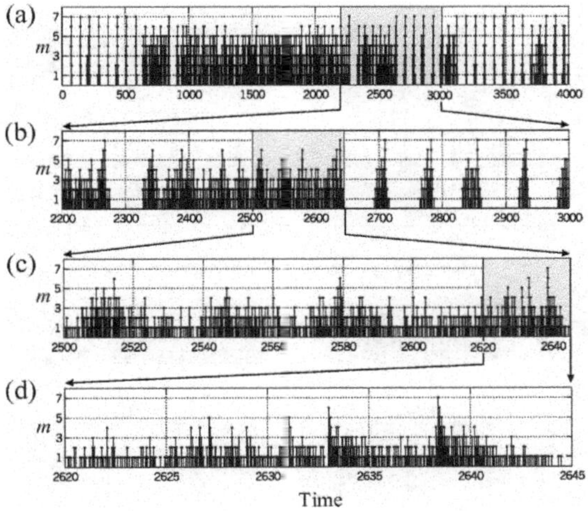

FIG. 2. Synthetic earthquake sequence. The complete sequence is shown on the top panel followed by exploded views in the following three panels. Note that all times given, in this and subsequent figures are in dimensionless units.

$$(t_k, m_k, g_k), \quad k=1,2,\ldots, \quad t_k \leq t_{k+1}. \tag{2.11}$$

Here, t_k is the time of failure for an element, whereas m_k and g_k indicate its position within the model (see Fig. 1). The usual basic representation of an observed earthquake sequence is very similar to Eq. (2.11), where the vector g_k constitutes the coordinates of the hypocenter, i.e., the actual point of origin of an earthquake event (the epicenter is its projection on the Earth's surface). Often overlooked, however, is the fact that the region of rupture, which we will later call L, is 10^1-10^2 km in extent for earthquakes with magnitudes between 7 and 8. Earthquakes are *not* point source events. An earthquake starts with a localized rupture that then spreads on a complex surface of finit dimensions and has distinctly different near-field intermediate zone, and far-fiel effects.

Strictly speaking, this model has no three-dimensional space of hypocenters, but we regard g to be a coarse analog of the hypocenter, since it identifie the position of a failed element in relation to other ones (see Fig. 1). Finally, we regard the value m in synthetic earthquake sequences as an analog of the earthquake magnitude—the latter is a logarithmic measure of energy released by an earthquake.

An example of the sequence generated by the colliding-cascades model is shown in Fig. 2. It was computed using the numerical parameters provided in Table I. The entire sequence is shown in the top panel, followed by a sequence of exploded views. Each event is formed by simultaneous failure of one or more elements. In a case when several elements fail at the same time, only the events at the highest

TABLE I. Values of parameters used in computations.

γ	$\tilde{\alpha}$	C	q	s	ϵ	λ	L	n
0.2	27.2	0.9	1.096	1.03	0.1	3.0	3.4	2

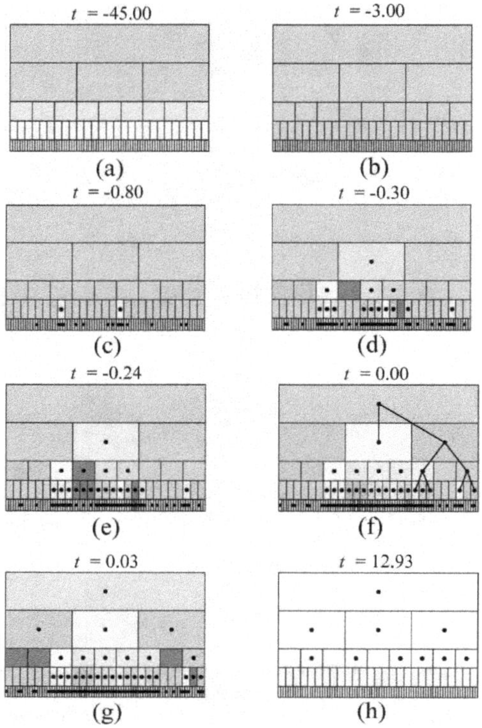

FIG. 3. Cascades in the fiv highest levels of the model. Figure portrays the case history of a cycle. Shading is employed to describe proximity in time to failure—the darker the shade, the closer an element to the critical threshold θ. Black dots identify broken elements. The time remaining to the major event is indicated at each frame.

level in the hierarchy are indicated. The vertical scale shows their maximum level, referred to as a ''magnitude'' m.

Figure 3 shows the interaction between direct and inverse cascades prior to a major failure, i.e., $m=7$, and following its aftermath. Until some time (i.e., the initiation of rupture), we can see only a direct cascade moving downward (a). After a while (b), the cascade has reached the lowest level and triggers the inverse cascade that is shown in (c). The inverse cascade triggers the rupture that in turn triggers secondary direct cascades. Panel (d) describes the instant when inverse cascade has reached the second-highest level, namely, level, 6. The elements with the darkest shading identify a secondary direct cascade, triggered by a failure at that level. Meanwhile, the secondary direct cascade is evolving (e) and triggers an aftershock at level 5 (the darkest broken element). Panel (f) demonstrates another inverse cascade that is initiated in the right hand branch of the model. It triggered instantaneous failure that propagated to top level. These failures are connected in the figure Panel (g) shows that yet another secondary direct cascade has started. Finally, the inverse cascade is weakening (h). Healing prevails until the subsequent direct cascade reaches level 1.

In subsequent sections, we investigate whether synthetic seismicity generated by the colliding-cascade model reproduces the basic regular features of real seismicity.

III. HEURISTIC CONSTRAINTS: REGULARITIES IN OBSERVED SEISMICITY

Real earthquake sequences exhibit a high degree of complexity but, upon averaging, manifest a remarkable degree of regularity—such is the usual case in complex systems. We summarize these features in this section and compare them with synthetic seismicity in Secs. IV and V.

A. Seismic cycle

The level of seismic activity or seismicity in an area goes through three different phases [5]. First, there is a ''preseismic'' rise culminating in one or several major earthquakes. This is followed by a period of ''postseismic'' activity, which gradually declines with time. Finally, there emerges a long period of relatively low activity that ultimately returns to another rise, and so on. Such transitions take place in different time and space scales. The characteristic time interval between the strongest earthquakes in active regions, such as southern California, is of the order of 10^1-10^2 years. Usually, however, there is no real periodicity. Observed intervals between major earthquakes depart considerably from the mean, and the character of each phase in the ''seismic cycle'' varies strongly from case to case. In different epochs, seismic cycles may culminate in earthquakes of different magnitude. Generally, a seismic regime exhibits large intermittency.

B. Power-law energy distribution

Otherwise known as the *Gutenberg-Richter* law [6–11], the energy distribution of earthquakes in a fault system may be approximately described as

$$\log_{10} N(m) = a - bm. \tag{3.1}$$

Here, $N(m)$ is the average annual number of earthquakes with magnitude above m in a specifie spatial region. This law is valid for sufficientl large fault systems and time intervals and for a certain magnitude range $[m_1, m_2]$ [8]. It is important to note that it does *not* describe the frequency of events on a small-scale, individual fault. Typically, $b \approx 1$, not changing much with variation of the geometry of the fault system, the local tectonics, and so on, thereby hinting at a universal mechanism for the dynamics of seismicity.

The power-law energy distribution (3.1) is at the heart of many models of seismicity. It is noteworthy, therefore, that Eq. (3.1) is only an approximation to real seismicity. The most frequently observed deviations from Eq. (3.1) that are not attributable to inaccurate observations are summarized below.

(a) $\log_{10} N(m)$ bends over or changes slope at the largest magnitudes [6,8]. In the case of a downward bend, the Kolmogoroff log-normal relation [12], which describes the distribution of the size of the rocks in a fractured massive region, sometimes happens to fi the earthquake energy distribution better than Eq. (3.1). In some cases, however, there is an upward bend [13], or the values of $N(m)$ for the largest m are scattered above the power-law approximation

(3.1). Such magnitudes are often attributed to what is sometimes called a "characteristic earthquake," an event having the maximum possible size in the region considered [7,14].

(b) $\log_{10} N(m)$ may be represented by a collection of overlapping power laws with statistically significan differences in the slope b of the different segments [8,15]. (This refers to the whole magnitude range, as well as to what was described above for its ends.)

(c) The shape and parameters of the energy distribution change with time. This includes two kinds of changes premonitory to a major earthquake. (1) The upward bend of distribution for relatively large m [13,16]; it reflect premonitory increase of seismic activity. (2) A possible increase in the b value for smaller magnitudes. In [15] such changes are found both in the observed seismicity and in fracturing of steel samples.

(d) For the physical interpretation of the Gutenberg-Richter law (3.1) its multiscale nature is primal [10,8]. Statistics of earthquakes with magnitude m have to be established in spatial regions of linear dimension much larger than $L(m)$, the characteristic dimension of the earthquake source. Accordingly, the parameters a and b have to be determined for increasingly larger areas, when larger magnitude ranges are considered. The above deviations from the power law (3.1) depend on the spatial scale considered. For larger regions some of them may disappear.

We now return to other forms of regularity observed in seismicity.

C. Clustering

Earthquakes are observed to cluster in both space and time [9,17,18]. A typical cluster is a main shock, namely, the earthquake, followed by a string of weaker shocks, called aftershocks. Some earthquakes are also preceded by foreshocks of a smaller magnitude. Typically, about 30% of main shocks have foreshocks. The number of foreshocks is usually small. Clusters of another kind, known as "swarms," are formed by main shocks of comparable magnitude that occur in proximity (in both space and time) to each other, with their own overlapping sets of aftershocks and foreshocks [19].

D. Premonitory seismicity patterns

Often, a strong earthquake in a region is preceded by a set of unusual patterns of seismicity ([20–22] and references therein). This anticipatory behavior was summarized by Keilis-Borok [23] as follows. Typically, a few years in advance of a strong earthquake, a sequence of seismic events in a medium magnitude range becomes increasingly intense and irregular. These events become more clustered in both space and time, and their correlation distance probably increases. All of these phenomena may be caused by an increased response of the lithosphere to excitation.

The scaling of these phenomena depends on the scale of the "strong" earthquake they foretell. Let M be its magnitude, and $L(M)$ be the linear size of its source. These phenomena are defined then, in the magnitude range down to about $M-4$, and in a region of linear dimension of $5L-10L$. They precede a strong earthquake by a few years.

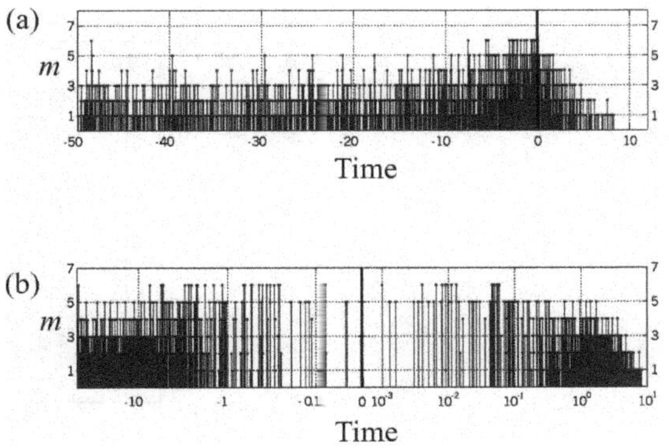

FIG. 4. Composite catalog. All cycles are combined, with each major event placed at time $t=0$

Such premonitory phenomena, formally defined have been used in algorithmic earthquake prediction—see [22,23] and references cited therein. We apply here the same definitions of premonitory behavior to synthetic seismicity generated by our model.

In addition, we consider accelerating Benioff stress release—a precursor described in [24–27]. It consists of power-law escalation of seismicity possibly including an accelerating and, possibly, log-periodic oscillatory component.

IV. SYNTHETIC SEISMICITY: SEISMIC CYCLE, ENERGY DISTRIBUTION, AND CLUSTERING

Here, and in the next section, we demonstrate that the colliding-cascade model reproduces many of the regular features of observed seismicity, described in the previous section. The major advantage provided by our synthetic sequence is the relative simplicity of its temporal structure. That facilitates the establishment of connections between different forms of regularity, and the possible identificatio of yet undiscovered regularities or patterns, which can then be validated by analysis of observations.

Seismic cycles. About half of the time, the sequence shown in Fig. 2 consists of well-separated cycles, culminating in a "major event" of magnitude 7. There is clear intermittency in the seismic regime—seismicity is very different during the time interval 900–2200 time units—the magnitudes there do not exceed 6, and separation of the record into individual time cycles is not clear. Possibly, this interval includes shorter seismic cycles culminating in earthquakes with $m=6$. Such seismic intermittency is also typical in real seismicity. A composite portrait of the seismic cycle is shown in Fig. 4. All seismic cycles are superimposed in this figur with each major event centered at the same time origin, i.e., $t=0$. The top panel shows all earthquakes, in a linear time scale, while the bottom panel shows mainshocks and aftershocks in logarithmic time scale.

Energy distribution. Figure 5 shows the energy distribution in our synthetic earthquake sequence. We see that it fit well the Gutenberg-Richter law (3.1) without strong deviations.

Clustering. We have identifie the aftershocks by the same method as is used in analysis of observations; it is

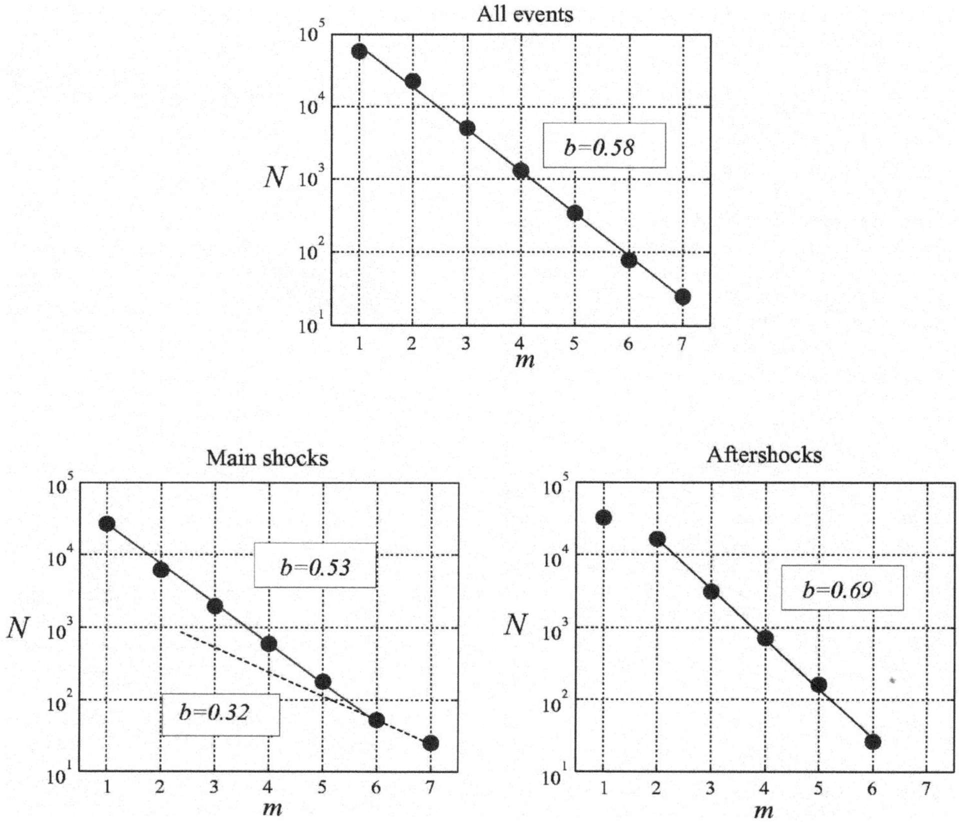

FIG. 5. Energy distribution function, $\log_{10} N = a - bm$ where N is the number of events with magnitude m. Note that the magnitude is given discrete integer values from 1 to 7.

described in the Appendix. Statistics of aftershocks are shown in Fig. 5. We see here the features in common with real seismicity—aftershocks constitute about half of the earthquake events, and the b parameter of the Gutenberg-Richter law for aftershocks is slightly higher than for main shocks.

V. SYNTHETIC SEISMICITY: PREMONITORY SEISMICITY PATTERNS

Here we apply to the analysis of earthquake precursors an approach known as "pattern recognition of infrequent events." It was developed by Gelfand co-workers over two decades ago for the study of rare phenomena of highly complex origin, a situation where classical statistical methods are inapplicable. The methodology of pattern recognition is very robust and its essence will be clear from the way synthetic data are analyzed here. In our problem, the goal of pattern recognition is the identificatio of behavior that almost always occurs before a main shock, yet almost never occurs otherwise. A more detailed description can be found in [28].

In this approach to the search for earthquake precursors, we conventionally divide the time period considered into intervals of three kinds, which are designated by the symbols D, N, and X. Figure 6 illustrates this division. An interval D precedes a major earthquake—that letter is chosen to mnemonically represent "danger." An interval X follows it and is characterized by post-major-earthquake activity, including waning aftershock sequences. The latter have a characteristically inverse time rate of occurrence sometimes known as Omori's law. The rest of the time between the major earthquakes comprises the (null) intervals N; thus, they are distanced from major earthquakes. This division of time allows

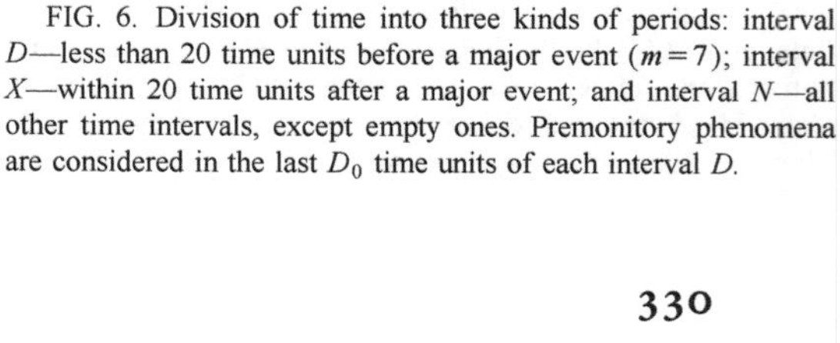

FIG. 6. Division of time into three kinds of periods: interval D—less than 20 time units before a major event ($m=7$); interval X—within 20 time units after a major event; and interval N—all other time intervals, except empty ones. Premonitory phenomena are considered in the last D_0 time units of each interval D.

us to explore precursors by the pattern recognition approach. A precursor is recognized if it seems to emerge in the intervals D more frequently in a statistically significan way than in intervals N. (Ideal precursors emerging only in the intervals D have yet to be found.) The application of such an approach to the observed seismicity is described in [23,29] and references therein. Several precursors considered in the next section have been identifie using the above condition. Now, we explore whether they exist in our synthetic seismicity.

First, we eliminate the intervals X from the search for precursors, since these intervals are dominated by specifi features of post-major-earthquake activity, even if we exclude aftershocks. (These intervals, however, are considered separately for prediction of a second strong earthquake, in a pair [30]. This problem is beyond the scope of the present study.)

The exact duration of D intervals is not known *a priori*. To ensure sufficien separation between intervals D and N, we consider in each interval D only the fina D_0 time units preceding a major event. Empty intervals between the cycles are disregarded.

We assume the duration 20 time units for the intervals D and X, independently of what precursor is explored. D_0 is chosen to be 3 or 5 time units, depending on the precursor.

A. Statistically significan precursors validated by prediction of real earthquakes

The precursors considered here have been identifie by the analysis of observations of real earthquakes, with the pattern recognition approach described above ([21,22] and references therein). They are define on the sequence of main shocks. Elimination of aftershocks from the sequence is necessary for the following reason. On average, the total number of aftershocks grows with the magnitude of the main shock. Therefore, unless aftershocks are eliminated, the relatively strong main shocks would be represented in premonitory patterns twice—by themselves and by their aftershocks. However, for each main shock, we retain the number of its aftershocks; these numbers are used in the definitio of premonitory clustering.

1. Clustering

We consider as a measure of clustering the function $B_m(t_k|\tau)$ which identifie the number of aftershocks within a time interval τ following a main shock [31–33]. Here, m is the magnitude of the main shock, while k identifie its position in the sequence of main shocks. Thus, t_1 denotes the time of the firs main shock, t_2 the second, and so on. In earthquake prediction studies, B_m is assessed over a very short time interval, namely, $\tau=2$ days, while the ensuing aftershock sequence may last a year or more. The application of this measure to observed seismicity is described in detail in [31,34]. In our analysis of synthetic events, we assume $\tau=0.05$ time units.

Let $P_D(B_m)$ be the density distribution of the values of B_m taken collectively from all intervals D. Similarly, $P_N(B_m)$ is the density distribution corresponding with the intervals N. Precursory behavior in the variable B_m is characterized, according to a familiar Bayesian approach, by the difference of these densities, namely,

$$\Delta(B_m)=P_D(B_m)-P_N(B_m).$$

If large values of B_m preferentially occur prior to a major event, then the function Δ should be positive for these values of B_m, and negative for smaller ones.

To make our analysis robust, we divide all the values of B into three groups, namely, "small," "medium," and "large." They are separated by the percentiles of the levels 33.3% and 66.6%. In other words, this division is made using the condition that each group contains equal numbers of events and, therefore, equal numbers of values B. This is done for the intervals D and N taken together so that *a priori* knowledge of major events is not used in discretization of the values of B. The functions $\Delta(B_6)$ and $\Delta(B_5)$, thus discretized, are shown in panels (a) and (b) of Fig. 7 . The last $D_0=5$ time units are considered in the intervals D. Note that Δ is given as a percentage. We see clearly that the generation of aftershocks tends to be larger in D. Similar results were obtained for $m=2$, 3, and 4.

2. Premonitory rise of seismic activity

Following [29,35] and references therein, we consider two functions reflectin the premonitory rise of seismic activity. These functions are $N_m(t|s)$, define as the number of main shocks with magnitude m, and

$$\Sigma_m(t|s)=\sum S_k(t|s).$$

Here, S_k is the area of rupture in the earthquake source, where k again is its number in the sequence of main shocks. Summation is taken over all the main shocks with magnitude less than or equal to m. In the analysis of real observations, this area is crudely estimated from the earthquake magnitude. For synthetic seismicity, we assume that $S_i=3^{\bar{m}-m}$, where $\bar{m}$ is the index of the top level; this value is proportional to the number of elements at the lowest or firs level that are the offspring of the element on level m (Fig. 1). The contribution of the top element, namely, $(\bar{m}|1)$, is unity. Both functions $N_m(t|s)$ and $\Sigma_m(t|s)$ are counted in a sliding time window $(t-s,t)$, where $s=2$. Note that their values are attributed to the *end* of this window so that information from the future is not used. Panels (c)–(e) of Fig. 7 display the functions $\Delta(N_1)$, $\Delta(N_2)$, and $\Delta(\Sigma_6)$. The fina $D_0=3$ time units are considered in the intervals D. We clearly see that each measure tends to be larger in D. Similar results were obtained for N_m, $m=3,4,5$, and for Σ_m with m varying from 2 to 5.

B. A precursor explored by retrospective analysis of observations: "Accelerated Benioff strain release"

In the studies by Bufe and Varnes, [24], Sornette and Sammis [25], Newman *et al.* [26], and Johansen *et al.* [27] (see also references therein), premonitory escalation of seismicity is represented by the empirically based function

$$\epsilon(t)=\epsilon_0-B(t_f-t)^\alpha.$$

Here, ϵ is the cumulative Benioff stress release

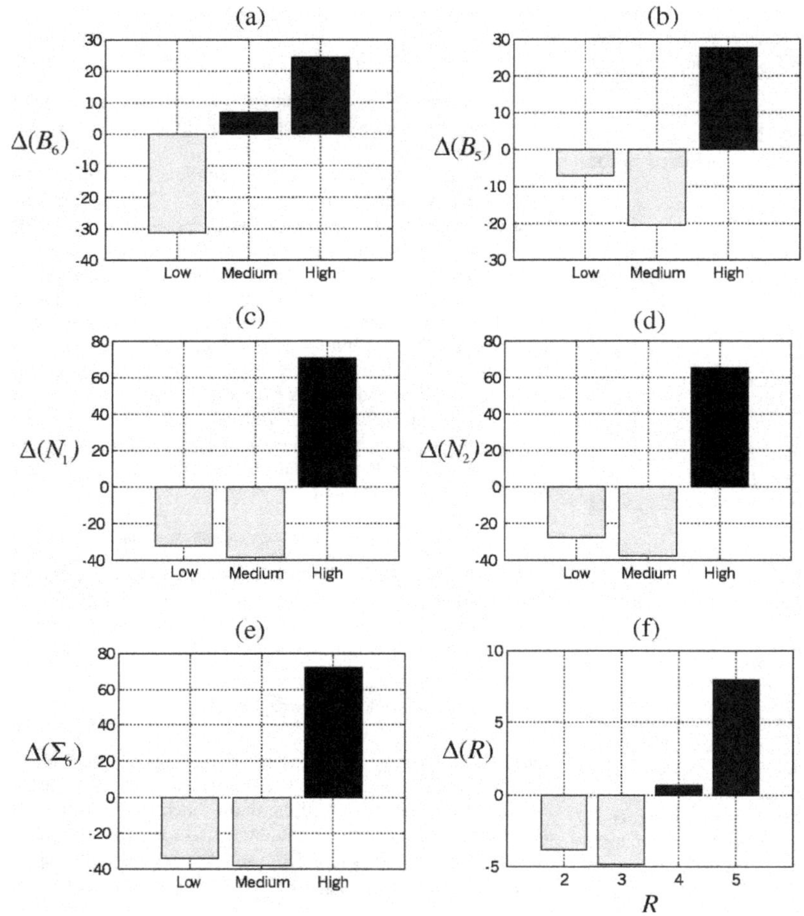

FIG. 7. Premonitory phenomena depicted by the difference of distributions within intervals D and N. $\Delta(F)$ is the difference between distributions of values of premonitory function $F(t)$ within intervals D and N. Concentration of high values of the function $F(t)$ within the intervals D is depicted by positive differences $\Delta(F)$, which can be (a) $\Delta(B_6)$; (b) $\Delta(B_5)$; (c) $\Delta(N_1)$; (d) $\Delta(N_2)$; (e) $\Delta(\Sigma_6)$; or (f) $\Delta(R)$.

$$\epsilon=\sum E_k^{1/2}$$

as a function of time over the interval (t_0,t), E_k is earthquake energy estimated from the magnitude with the summation taken over all earthquakes in the region without the elimination of aftershocks, and t_f is the actual time of a strong earthquake. Figure 8 demonstrates this phenomenon on the composite seismic cycle shown in Fig. 4. We see that ϵ indeed is rising steeply starting at about 10 time units before the major earthquake. Panel (a) in Fig. 8 gives the usual representation of that function while panel (b) shows that, during a time interval from -5 to -1 time units, the function ϵ is rising according to a power law, with $\alpha=0.4$. However, unlike the reported real observations, the growth of synthetic seismicity terminates about 0.5 time units before a major earthquake.

C. Potential precursors not yet identifie in observations

We consider the premonitory increase in the range of spatial correlation between the earthquakes. Qualitatively, this phenomenon was hypothesized by Keilis-Borok [20]. However, specifi precursors of that type are introduced in this paper.

1. Range of correlation (precursor ROC)

Consider two main shocks (t_k,g_1,m_1) and (t_l,g_2,m_2) generated by the failure of elements $i=(m_1|g_1)$ and $j=(m_2|g_2)$. The pairwise (ultrametric) distance $r(i,j)$ between these elements along the ternary tree (see Fig. 1) is define as

$$\min_{m_p}(m_p-m_1,m_p-m_2),$$

where m_p is the level of an element from which both i and j descend. Let $R(\tau)$ be the pairwise distance $r(i,j)$ between the main shocks, which occurred within an interval τ from each other. The function $\Delta(R)$ is shown in panel (f) of Fig. 7. The last $D_0=3$ time units were considered in the intervals D and a threshold $\tau=0.1$ was used. We clearly see that values of R tend to be larger in D. A related effect was introduced into studies of the dynamics of seismicity by Prozorov

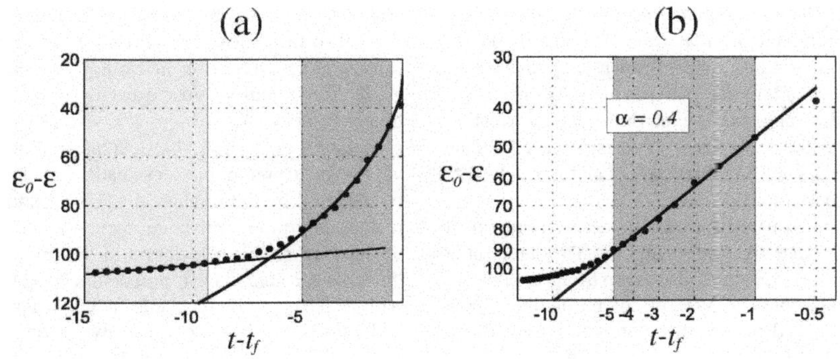

FIG. 8. Premonitory escalation of the failure sequence—the power-law rise of the cumulative Benioff stress release ϵ. The function $\epsilon(t)$ is determined for the composite sequence (see Fig. 4) with aftershocks included. (a) Linear scale. Dark zone at the right marks the interval $[-5,-1]$, where power-law behavior is observed. Dark zone at the left marks interval with linear behavior of ϵ. (b) Bilogarithmic scale. Line corresponds to the power law $\epsilon(t)=\epsilon_0-B(t_f-t)^\alpha$ with parameters $B=47.3$, $\alpha=0.4$ estimated by the least squares method within the interval $[-5,-1]$. Parameter ϵ_0 equals $\epsilon(0)=117.1$. Note that values of $(\epsilon_0-\epsilon)$ are shown in the reverse direction.

in the early 1970s [36]. He considered "long-range aftershocks," i.e., intermediate magnitude earthquakes, which sometimes follow a major earthquake within a short time interval. According to these studies of real data, such aftershocks mark the location where another major earthquake is going to occur.

2. Accord

This precursor depicts simultaneous rise of activity in the three major branches of the model that descend from the elements on the second highest level, $m=6$. The intuitive significanc of this precursor is that several branches of the tree are undergoing simultaneous activation. Hence, the term "accord" is appropriate since it means (i) to be in agreement, and (ii) in music, to strike several chords at once.

We consider the function $\Sigma_6(t|s)$, define in Sec. V A 2, individually for each branch. Our precursor is depicted by the function $A(t)$ which is the number of branches where $\Sigma_6(t|s)$ simultaneously exceeds a common threshold, say C_A. By definition $A(t)$ may assume only integer values from 0 to 3. The distribution of $A(t)$ in the intervals D and N is shown in panel (a) of Fig. 9. Their difference $\Delta(A)$ is shown in panel (b). The last $D_0=3$ time units are considered in the intervals D. Computations are made with the threshold $C_A=1/9$. We clearly see that values of A tend to be larger in D. In application to real observations, the different branches of fault zones, which are known to be organized hierarchically, have to be considered instead of branches of the model.

VI. DISCUSSION

We enumerate below how the colliding-cascade model compares and contrasts with real seismicity.

(1) The colliding-cascade model indeed reproduces the basic features of observed earthquake sequences, although the latter are more complicated and irregular. Seismicity generated by this model satisfie the basic empirical pattern derived from observations—the seismic cycle, intermittency of the seismic regime, power-law energy distribution, earthquake clustering in space and time, and a set of seismicity patterns premonitory to a strong earthquake. These seismic-

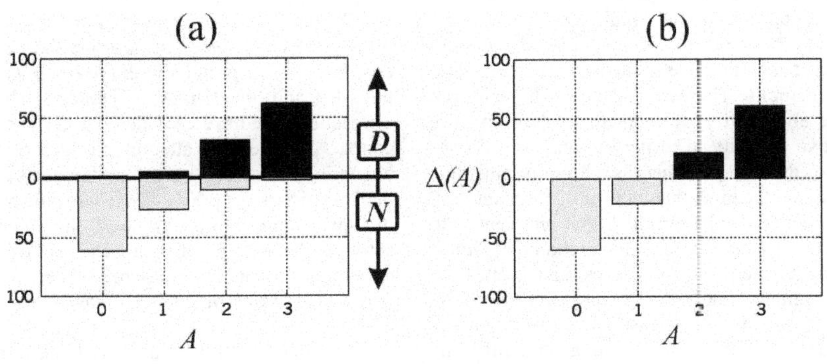

FIG. 9. Premonitory increase of the range of correlation—simultaneous rise of activity in the major branches of the model (precursor called *Accord*). $A(t)$ may assume integer values from 0 to 3. The last $D_0=3$ time units are considered in the intervals D. Threshold $C_A=1/9$ is used. (a) Density distributions, in percent, of $A(t)$ in the intervals D (upward dark bars) and N (downward light bars). (b) $\Delta(A)$—difference of these density distributions.

ity patterns include the three patterns established by statistically significan prediction of real (observed) earthquakes. They reflec the premonitory rise of seismicity and earthquake clustering. Two additional patterns reflectin premonitory increase of the range of the earthquakes' correlation have been found in the model and have become candidate premonitors for real events, which we plan to test by the analysis of real observations.

We have reproduced using a single model a broad set of major features of the dynamics of seismicity. This agreement between model and observation provides strong support to the colliding-cascade concept. We fin of particular importance the reproduction of the three major types of premonitory seismicity patterns. Note, also, that we have explored here the collective behavior of the fault network. Due to the scale invariance of such networks, this behavior may be manifest on different scales. However, this similarity may be limited [23,30] and is probably not applicable to a single fault of a nonhierarchical structure.

Here we detail the differences between observed and synthetic sequences (a) Real seismic cycles do not have the same degree of periodicity; (b) during real seismic cycles, periods of low seismicity are not as ''silent;'' (c) the energy distribution in the observed seismicity has considerable deviations from the Gutenberg-Richter power law at the largest and smallest magnitudes; and (d) a distinction between the intervals D and N is much clearer in synthetic seismicity than in the observed one, or in other words, premonitory phenomena (not unexpectedly) are more prominent in the model.

We believe that these quantitative differences are a manifestation of the relative simplicity of our model. On the whole, the model provides an acceptable description of the dynamics of seismicity. In particular, no other existing model to our knowledge has shown such a wide set of spatiotemporal structures characteristic for real seismicity preceding major earthquakes, albeit several precursors have been shown to be reproducible in the models (see [13,37]). Indeed, simplicity is an essential virtue of the colliding-cascade model, since it facilitates the model's exploration, testing, and application. Moreover, it facilitates understanding of the processes considered, while reproducing their high complexity. The basic characteristics of seismicity considered here are robust. Extensive numerical experiments have demonstrated that they are stable to initial conditions.

(2) In Sec. V, we established that the precursory phenomena associated with real earthquake sequences are a hallmark of our synthetic seismicity. The essential value of such precursors is that the ability to identify them could lead to a significan predictive capacity. The performance of precursors is a measure of their overall success. Not only must the presence of a precursor establish that an event will occur, but the absence of a precursor must establish that an event will *not* occur. The performance of a precursor, therefore, must involve the ratio of its successes to failures, and it must be applied to the prediction of individual strong events in both real and synthetic seismicity. We must learn how to avoid ''false alarms,'' yet maintain confidenc that no significan event will go unpredicted. This will be a topic of a separate study.

(3) The ultimate measure of success of a model is the discovery of previously unknown phenomena, which are confirme by real observations. Our modeling has suggested several so far unobserved precursors depicting an increase of the range of correlation in earthquake sequences. These include the simultaneous escalation of activity in major branches of a fault system—this would be depicted by the precursor ''*Accord*;'' the increased number of faults involved in a premonitory escalation of activity—this would be depicted by the precursor ''*ROC*;'' and the more general realization of the same phenomenon expressed in a spreading of the premonitory escalation of seismicity over an increasingly larger area. These precursors may be used in parallel with other ones, or they may help to discriminate the false and confirme alarms obtained with other precursors.

(4) The model allows us to explore a broad variety of other relevant phenomena. For example, different measures of loading may be considered, ranging from straightforward ones (e.g., stress or energy) to geometric incompatibility in the fault system [21] (this is due to the driven motion of disjoint and disparate fragments of crustal material). Loading can be applied to more than one level of the model, and not necessarily to the top one. We have considered thus far *time-independent* loading. It is certainly worth exploring the case when the loading is changing in time. The relevance of such changes to formation of earthquake precursors is discussed in [2,13]. Moreover, the model may be truncated from above so that the top level will consist of several elements.

(5) We have considered, thus far, the dynamics of the model with fixe parameters. It remains to be seen how the occurrence of major events is influence also by change of the parameters, e.g., loading, strength, time of healing, and so on.

(6) The colliding-cascade model is not particularly earthquake specific At the same time, it reproduces the precursors that were firs discovered in observed earthquake sequences. This confirm the previous conclusion, as in [23,38] and references therein, that these precursors are not earthquake specifi but symptomatic of more general features of critical transitions in hierarchical nonlinear systems. (The features of the seismically active lithosphere may be reflected of course, in the values of the parameters of the model.) One can hardly doubt that earthquake-specifi precursors also exist, but they have to be reproduced on models of a different kind (e.g., [21]).

(7) The colliding-cascade model is more complex than other models advanced to describe the earthquake phenomena. It lacks the grand simplicity of slider block [39], sandpile [40], and fibe bundle [26] models. This is largely due to the fact that we have considered both inverse and direct cascades. Also, we have attempted to describe the full scope of phenomena observed in real earthquake sequences—namely, the seismic cycle, power-law energy distribution, clustering in space and time, as well as premonitory seismic activity—by incorporating a richer set of interactions. Nevertheless, it would be a worthwhile endeavor to explore whether the same phenomena may be reproduced by a simpler model.

We remain optimistic that the richer array of phenomena described by this model, and its overall agreement with the basic premonitory characteristics evident in real earthquakes, will ultimately facilitate the design of better algorithms for the prediction of real earthquakes.

ACKNOWLEDGMENTS

This study was partly supported by the National Science Foundation (Grant No. EAR-9804859) and the International Science and Technology Center (Project No. 1293-99). This study was partially supported by a subcontract with Cornell University, Geological Sciences, under Grant No. EAR-9804859 from the National Science Foundation and administered by the CRDF. This study was also partially supported by NSF Grant No. ATM95-23787.

APPENDIX: DEFINITION OF AFTERSHOCKS

In the analysis of precursory behavior in observed seismicity, we employed [31] the following definitio of aftershocks:

$$M_j \leqslant M_i;$$

$$0 < t_j - t_i \leqslant T(M_i);$$

$$d_{ij} \leqslant D(M_i);$$

$$|h_i - h_j| \leqslant H(M_i).$$

Here, the indices i and j refer to a main shock and any of its aftershocks, respectively. The distance between epicenters is given by d_{ij}, while h is the focal depth, i.e., the depth of the hypocenter. The windows T, D, and H are empirical functions of the main shock's magnitude M_i. This definitio is robust—it is coarse yet stable. A more subtle definitio is given by Molchan and Dmitrieva [18].

We employed a similar definitio for synthetic seismicity. An aftershock is a child or a sibling of a main shock of magnitude m that occurred within a time interval $T(m)$ from the main shock. For convenience, we assumed time thresholds $T(m) = m$, $m = 1,2,\ldots,7$. The restriction of aftershocks to the children and siblings of a main shock is a coarse analog of the proximity of epicenters in real seismicity.

[1] A.N. Kolmogorov, Dokl. Akad. Nauk SSSR **30**, 299 (1941); U. Frish, *Turbulence: The Legacy of Kolmogorov* (Cambridge University Press, Cambridge, 1995); A. S. Monin and A. M. Yaglom, *Mechanics of Turbulence Vol. 2 of Statistical Fluid Mechanics* (MIT Press, Cambridge, 1975); G.M. Molchan, Commun. Math. Phys. **179**, 681 (1996).

[2] V.I. Keilis-Borok, Rev. Geophys. **28**, 19 (1990); F. Press and C. Allen, J. Geophys. Res. **100**, 6421 (1995); B. Romanowicz, Science **260**, 1923 (1993); A.T. Ismail-Zadeh, V.I. Keilis-Borok, and A.A. Soloviev, Phys. Earth Planet. Inter. **111**, 267 (1999).

[3] G.I. Barenblatt, *Similarity, Self-Similarity and Intermediate Asymptotics* (Consultants Bureau, New York, 1982).

[4] D.L. Turcotte, W.I. Newman, and A. Gabrielov, in *Geocomplexity and the Physics of Earthquakes*, edited by J.B. Rundle, D.L. Turcotte, and W. Klein (American Geophysical Union, Washington, DC, 2000).

[5] C.H. Scholz, *The Mechanics of Earthquakes and Faulting* (Cambridge University Press, Cambridge, 1990).

[6] B. Gutenberg and C.F. Richter, *Seismicity of the Earth and Associated Phenomena* (Princeton University Press, Princeton, 1954); G.M. Molchan and V.M. Podgaetskaya, Comput. Seism. **6**, 44 (1973).

[7] D.L. Turcotte, *Fractals and Chaos in Geology and Geophysics*, 2nd ed. (Cambridge University Press, Cambridge, 1997).

[8] G.M. Molchan, T.L. Kronrod, and G.F. Panza, Bull. Seismol. Soc. Am. **87**, 1220 (1997).

[9] G.M. Molchan and O.E. Dmitrieva, Phys. Earth Planet. Inter. **61**, 99 (1990).

[10] E.A. Okal and B.A. Romanovicz, Phys. Earth Planet. Inter. **87**, 55 (1994).

[11] J.B. Rundle, J. Geophys. Res. **98**, 21 943 (1993).

[12] A.N. Kolmogoroff, Dokl. Akad. Nauk SSSR **31**, 99 (1941).

[13] G.S. Narkunskaya and M.G. Schnirman, Phys. Earth Planet. Inter. **61**, 29 (1990).

[14] D.P. Schwartz and K.J. Coppersmith, J. Geophys. Res. **89**, 5681 (1984).

[15] I.M. Rotwain, V.I. Keilis-Borok, and L. Botvina, Phys. Earth Planet. Inter. **101**, 61 (1997).

[16] G.M. Molchan, T.L. Kronrod, and A.K. Nekrasova, Phys. Earth Planet. Inter. **111**, 229 (1999).

[17] Y.Y. Kagan and L. Knopoff, Geophys. J. R. Astron. Soc. **62**, 303 (1980); Y. Ogata, J. Geophys. Res. **97**, 19 845 (1992); Y. Ogata, T. Utsu, and K. Katsura, Geophys. J. Int. **121**, 233 (1995).

[18] G.M. Molchan and O.E. Dmitrieva, Geophys. J. Int. **109**, 501 (1992).

[19] V.I. Keilis-Borok, R. Lamoreaux, C. Johnson, and B. Minster, in *Earthquake Prediction Research*, edited by T. Rikitake (Elsevier, Amsterdam, 1982).

[20] V.I. Keilis-Borok, Physica D **77**, 193 (1994).

[21] A. Gabrielov, O.E. Dmitrieva, V.I. Keilis-Borok, V.G. Kossobokov, I.V. Kuznetsov, T.A. Levshina, K.M. Mirzoev, G.M. Molchan, S.Kh. Negmatullaev, V.F. Pisarenko, A.G. Prozoroff, W. Rinehart, I.M. Rotwain, P.N. Shebalin, M.G. Shnirman, and S.Yu. Shreider (unpublished).

[22] Phys. Earth Planet. Inter. **111**, (3-4) (1999) special issue, edited by V.I. Keilis-Borok and P.N. Shebalin.

[23] V.I. Keilis-Borok, Proc. Natl. Acad. Sci. USA **93**, 3748 (1996).

[24] C.G. Bufe and D.J. Varnes, J. Geophys. Res. **98**, 9871 (1993).

[25] D. Sornette and C.G. Sammis, J. Phys. I **5**, 607 (1995).

[26] W.I. Newman, D.L. Turcotte, and A. Gabrielov, Phys. Rev. E **52**, 4827 (1995).

[27] A. Johansen, D. Sornette, H. Wakita, U. Tsunogai, W.I. Newman, and H. Saleur, J. Phys. I **6**, 1391 (1996).

[28] I.M. Gelfand, Sh.A. Guberman, V.I. Keilis-Borok, L. Knopoff, F. Press, E.Ya. Ranzman, I.M. Rotwain, and A.M. Sadovsky, Phys. Earth Planet. Inter. **1**_, 227 (1976).

[29] Phys. Earth Planet. Inter. **61**, (1-2) (1990), special issue, edited by V.I. Keilis-Borok.

[30] I.A. Vorobieva, Phys. Earth Planet. Inter. **111**, 197 (1999); L. Knopoff, T. Levshina, V.I. Keilis-Borok, and C. Mattoni, J. Geophys. Res. **101**, 5779 (1996).

[31] V.I. Keilis-Borok, L. Knopoff, and I.M. Rotwain, Nature (London) **283**, 258 (1980).

[32] M. Caputo, R. Console, A.M. Gabrielov, V.I. Keilis-Borok,

and T.V. Sidorenko, Geophys. J. R. Astron. Soc. **75**, 71 (1983); M. Caputo, P. Gasperini, V.I. Keilis-Borok, L. Marcelli, and I.M. Rotwain, Ann. Geofis **30**, 269 (1977).

[33] V.I. Keilis-Borok, L. Knopoff, and I.M. Rotwain, Nature (London) **283**, 258 (1980).

[34] G.M. Molchan, O.E. Dmitrieva, I.M. Rotwain, and J. Dewey, Phys. Earth Planet. Inter. **61**, 128 (1990).

[35] V.I. Keilis-Borok and L.N. Malinovskaya, J. Geophys. Res. **69**, 3019 (1964); V.I. Keilis-Borok and V.G. Kossobokov, Phys. Earth Planet. Inter. **61**, 73 (1990).

[36] A.G. Prozorov, Vychisl. Seismol. **8**, 71 (1975) (in Russian).

[37] B.E. Shaw, J.M. Carlson, and J.S. Langer, J. Geophys. Res. **97**, 479 (1992); Y. Huang, H. Saleur, C. Sammis, and D. Sornette, Europhys. Lett. **41**, 43 (1998).

[38] V.I. Keilis-Borok, Phys. Earth Planet. Inter. **111**, 179 (1999).

[39] R. Burridge and L. Knopoff, Bull. Seismol. Soc. Am. **57**, 341 (1967).

[40] P. Bak, C. Tang, and K. Wiesenfeld, Phys. Rev. A **38**, 364 (1988).

JOURNAL OF GEOPHYSICAL RESEARCH, VOL. 107, NO. B10, 2221, doi:10.1029/2000JB000034, 2002

Premonitory spreading of seismicity over the faults' network in southern California: Precursor Accord

I. Zaliapin[1] and V. Keilis-Borok[2]

International Institute of Earthquake Prediction Theory and Mathematical Geophysics, Russian Academy of Sciences, Moscow, Russia

G. Axen

Department of Earth and Space Sciences, University of California, Los Angeles, California, USA

Received 31 October 2000; revised 16 March 2002; accepted 21 March 2002; published 10 October 2002.

[1] We establish a connection between geometry of the faults' network in a region and seismicity preceding the strong earthquakes in the intermediate-term timescale of years. Previous studies of observed and computer-simulated seismicity demonstrated that strong earthquakes are preceded in that timescale by the rise of seismic activity in a lower magnitude range. Here, we explore a complementary phenomenon: spreading of activity over the fault network. This phenomenon is depicted by the seismicity pattern Accord, defined as a simultaneous rise of seismicity in a sufficiently large number of fault zones (hence its name). Pattern Accord has been found recently in synthetic seismicity generated by the colliding cascades (CC) model. The present study demonstrates this pattern in observed seismicity of southern California. We applied to observations a prediction algorithm based on the pattern Accord. Scaled to the target magnitude 7.5, the pattern Accord emerges within a few years before each of the three largest earthquakes in southern California (Kern County, 1952; Landers, 1992; and Hector Mine, 1999) and at no other time. Scaled to the target magnitude interval from 6.5 to 7.4, the pattern Accord precedes six out of nine earthquakes. The duration of alarms in both cases is about 30% of the time considered. We repeated prediction with different combinations of adjustable numerical parameters of the algorithm, summing up the success and error scores on an error diagram. This numerical experiment shows that the prediction results are stable under moderate variations of adjustable parameters. The final test would be, as always, an advance prediction. The pattern Accord has a simple physical explanation connected with equilibrium of the crustal blocks' system under the impact of tectonic driving forces. At the same time, Accord is a manifestation of a broader phenomenon: premonitory increase of earthquakes' correlation range. The latter phenomenon was found for the first time in the CC model, although it was hypothesized by one of the authors much earlier. *INDEX TERMS:* 0910 Exploration Geophysics: Data processing; 3220 Mathematical Geophysics: Nonlinear dynamics; 7223 Seismology: Seismic hazard assessment and prediction; *KEYWORDS:* earthquake prediction, long-range earthquake correlations, nonlinear dynamics

Citation: Zaliapin, I., V. Keilis-Borok, and G. Axen, Premonitory spreading of seismicity over the faults' network in southern California: Precursor Accord, *J. Geophys. Res.*, *107*(B10), 2221, doi:10.1029/2000JB000034, 2002.

1. Introduction

1.1. Long-Range Correlations in Seismicity

[2] One of the prominent features of seismicity is the correlation in the earthquakes' occurrence at long distances, greatly exceeding the source dimensions. That correlation is expressed in many ways: simultaneous change of seismic activity within large regions [*Mogi*, 1968]; migration of earthquakes along fault zones [*Mogi*, 1968; *Vilkovich and Schnirman*, 1982; *Ma et al.*, 1990]; alternate rise of seismicity in distant areas [*Press and Allen*, 1995] and even in distant tectonic plates [*Romanowicz*, 1993]; and seismicity patterns, premonitory to strong earthquakes (see section 1.2 below). Global correlations have been found also between seismicity and other geophysical phenomena, such as Chandler wobble, variations of magnetic field, and velocity of Earth's rotation [*Press and Allen*, 1995; *Romanowicz*, 1993; *Press and Briggs*, 1975]. To explain these

[1]Now at Institute of Geophysics and Planetary Physics, University of California, Los Angeles, California, USA.

[2]Now at Institute of Geophysics and Planetary Physics and Department of Earth and Space Sciences, University of California, Los Angeles, California, USA.

0148-0227/02/2000JB000034$09.00

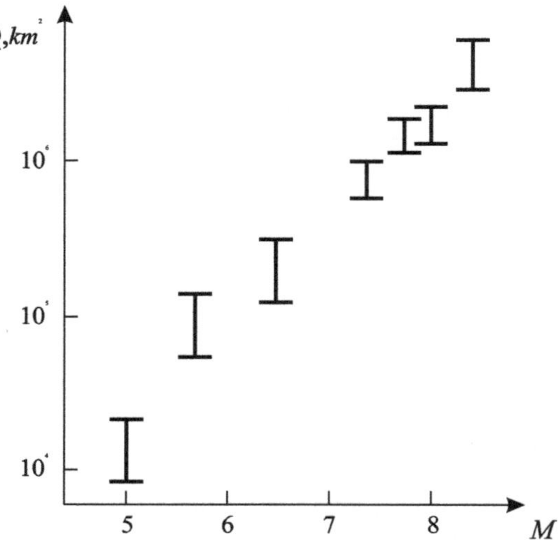

Figure 1. Size Q (in km^2) of the regions where pattern Sigma was observed prior to the earthquakes of magnitude M. After *Keilis-Borok and Malinovskaya* [1964].

phenomena, several mechanisms (not mutually exclusive) have been suggested. They may be divided into two groups.

1. Some authors attribute long-range correlations to a specific large-scale process, controlling stress and strength in the lithosphere. Among such processes are microrotation of tectonic plates [*Press and Allen*, 1995]; interaction of crustal blocks within the fault zones [*Gabrielov and Keilis-Borok*, 1983; *Rundkvist and Rotwain*, 1994; *Soloviev et al.*, 1999]; microfluctuations in the direction of mantle currents [*Ismail-Zadeh et al.*, 1999]; migration of fluids in fault systems [*Barenblatt et al.*, 1983; *Yamashita*, 1998, 1999]; hydrodynamic waves in the upper mantle, triggering the strong earthquakes [*Romanowicz*, 1993]; and perturbations of the ductile layer beneath the seismically active zone [*Aki*, 1996].

2. In another approach, long-range correlations are not attributed to any specific mechanical process. The lithosphere is regarded as a complex system with complexity contributed by a multitude of instability mechanisms [*Allegre et al.*, 1982; *Keilis-Borok*, 1990; *Turcotte*, 1997; *Newman et al.*, 1994; *Bak and Tang*, 1989; *Keilis-Borok*, 1992; *Mandelbrot*, 1983; *Rundle et al.*, 2000b; *Ben-Zion*, 2001]. The long-range correlation between earthquakes is considered then as a general feature of chaotic systems in a near-critical state [*Turcotte et al.*, 2000; *Smalley et al.*, 1985; *Sornette and Sammis*, 1995; *Bowman et al.*, 1998; *Rundle et al.*, 1997, 2000a; *Zoeller et al.*, 2001].

1.2. Premonitory Seismicity Patterns

[3] We discuss here the intermediate-term patterns preceding a strong earthquake with characteristic lead time of years. Studies of observed and computer-simulated seismicity demonstrated that an earthquake of magnitude M can often be preceded by certain spatiotemporal patterns of seismicity in the magnitude range (M-δM, M), with δM varying from 2 to 4 for different patterns [*Keilis-Borok and Shebalin*, 1999]. Relatively better validated among them are the patterns reflecting premonitory rise of seismic activity and earthquake clustering. They are used, separately or jointly, in the earthquake prediction algorithms, which have been successfully tested by advance prediction; statistical significance of predictions is above 95% [*Molchan et al.*, 1990; *Keilis-Borok and Rotwain*, 1990; *Kossobokov et al.*, 1999a; *Vorobieva*, 1999].

[4] A prominent feature of these patterns is that they are nonlocal. The patterns preceding an earthquake of magnitude M with linear source dimension $L(M)$ are observed within the region of linear dimension $5L(M)-10L(M)$ [*Keilis-Borok and Rotwain*, 1990; *Keilis-Borok et al.*, 1980; *Kossobokov et al.*, 1999a, 1999b; *Vorobieva*, 1999; *Bowman et al.*, 1998].

[5] An early (and probably the first) evaluation of the size of that region, still valid, has been given by *Keilis-Borok and Malinovskaya* [1964]. That study introduced a premonitory pattern Sigma: the measure $\Sigma(t)$ of seismic activity is the sum of the areas of earthquake sources (fault breaks) in the medium magnitude range. The areas are coarsely estimated from the magnitudes. Essential for our study is the large size of the regions where pattern Sigma was observed. This size grows with magnitude, approximately by a power law, from 10^4 km^2 for $M_0 = 5$, to 10^6 km^2 for $M_0 = 8$ (Figure 1). Similar estimations for other precursors are summed up in Table 1. It shows linear dimension

Table 1. Linear Size of the Regions, Where Premonitory Phenomena Are Observed[a]

Size $d(M)$	Measure/Algorithm	Reference
up to $10L$	total area of fault breaks	*Keilis-Borok and Malinovskaya* [1964]
$\sim 10L$	distant aftershocks	*Prozorov* [1975] and *Prozorov et al.* [1990]
$3L-5L$	Algorithm CN	*Keilis-Borok and Rotwain* [1990]
$5L-10L$	Algorithm M8	*Kossobokov et al.* [1999a]
L-$3L$	Algorithms M8 and MSc	*Kossobokov et al.* [1990]
$\sim 5L$	number of earthquakes	*Knopoff et al.* [1996]
$\sim 5L$	Benioff strain release	*Bufe and Varnes* [1993] and *Bowman et al.* [1998]
$\sim 100L$[b]	seismic activity	*Romanowicz* [1993]
$50L-100L$[b]	seismic activity	*Press and Allen* [1995]
$\sim 3L$	near-simultaneous pairs of earthquakes	*Shebalin et al.* [2000]
$\sim 5L$	correlation range via Single Link Cluster analysis	*Zöller et al.* [2001]

[a] L is the linear size of rupture during an approaching earthquake.

[b] The works of *Romanowicz* [1993] and *Press and Allen* [1995] are related to the timescale of tens of years.

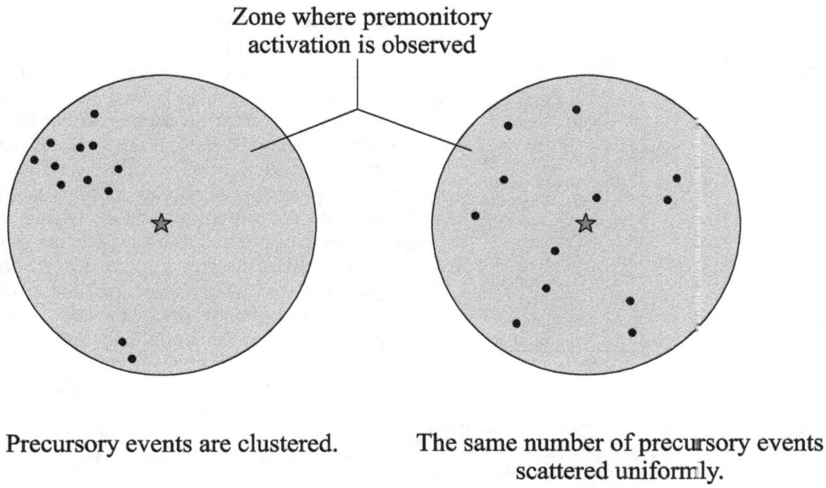

Figure 2. Two types of the spatial distribution of premonitory seismic activity. Stars depict the epicenter of approaching event. The number of precursory events is the same in both cases.

$d(M)$ of a region where premonitory patterns were observed. On the one hand, the dimension $d(M)$ may be reduced down to $3L$–L [*Kossobokov et al.*, 1990]. On the other hand, an even larger dimension, up to $100L$, was observed in the longer timescale, tens of years. According to *Press and Allen* [1995, p. 6428] the Parkfield (California) earthquake, with M about 6, $L \approx 10$ km, "is not likely to occur until activity picks up in the Great Basin or the Gulf of California," hundreds of kilometers away.

2. Pattern Accord

[6] Previous studies impose no limitations on territorial distribution of precursory activity. The earthquakes forming a premonitory pattern may be either spread over the whole region or concentrated in a smaller area not necessarily close to the epicenter of an incipient strong earthquake [*Keilis-Borok and Malinovskaya*, 1964; *Keilis-Borok et al.*, 1980; *Caputo et al.*, 1983; *Keilis-Borok et al.*, 1990; *Keilis-Borok and Kossobokov*, 1990; *Keilis-Borok and Rotwain*, 1990; *Bufe and Varnes*, 1993; *Knopoff et al.*, 1996; *Shaw et al.*, 1997; *Bowman et al.*, 1998; *Kossobokov et al.*, 1999a, 1999b]. The difference between these situations is schematically illustrated in Figure 2. Pattern Accord is designed to reflect premonitory spreading of precursory activity over the faults network.

2.1. Definitions

[7] Pattern Accord consists, qualitatively, of simultaneous rise of seismic activity in several branches of the fault network. This pattern has been found recently in synthetic seismicity generated by the colliding cascades (CC) model [*Gabrielov et al.*, 2000a, 2000b]. It is defined as follows.

[8] Consider a network of major faults $F = \{F_i,\ i = 1,2,\ldots,n\}$ in a region R. The latter is divided into subregions R_i, $R = \{R_i,\ i = 1,2,\ldots,n\}$ such that a subregion R_i corresponds to the fault (or, more generally, to subnetwork) F_i, $i = 1,\ldots,n$.

[9] For each subregion R_i we define the measure $\Sigma_i(t)$ of seismic activity as follows:

$$\Sigma_i(t) = \sum_{k:M_1 \le m_k \le M_2; E_k \epsilon R_i} H(t - t_k, s, \lambda)10^{Bm_k};\ i = 1, \cdots, n. \quad (1)$$

The summation in equation (1) is taken over all the main shocks E_k within ith subregion with magnitudes m_k within the range $M_1 \le m_k \le M_2$; t_k is the occurrence time of kth earthquake. The function H is defined as follows:

$$H(t, s, \lambda) = \begin{cases} 10^{t/\lambda}, & 0 \le t < s \\ 0, & \text{otherwise} \end{cases}. \quad (2)$$

In this definition, M_1, M_2, B, s, and λ are numerical parameters.

[10] With finite positive λ we introduce an attenuating "memory," giving a larger weight to the more recent earthquakes. Similar memory was introduced in the "seismic flux," a smoothed description of seismicity, suggested by *Khokhlov and Kossobokov* [1994].

[11] The pattern Accord is depicted by the function

$$A(t) = \sharp\{i : \Sigma_i(t) \ge \Sigma_0, i = 1, \cdots, n\}. \quad (3)$$

The value of $A(t)$ is the number of subregions where seismic activity measured by the functions $\Sigma_i(t)$ exceeds the common threshold Σ_0. By definition, $A(t)$ may assume only positive integer values from 1 to n.

2.2. Prediction Algorithm

[12] The general scheme of prediction is taken from the past studies [*Keilis-Borok*, 1990; *Keilis-Borok and Shebalin*, 1999]:

1. We consider the earthquake sequence with aftershocks eliminated.

2. The target of prediction is the "strong" earthquakes, defined by the condition $M \ge M_0$; M_0 is a parameter of the

Table 2. Earthquakes Analyzed

Date	Magnitude	Longitude	Latitude	Area
30 Dec. 1934	6.5	−115.5	32.3	Laguna Salada
31 Dec. 1934	7.0	−114.8	32.0	Colorado River delta
19 May 1940	6.7	−115.5	32.7	El Centro
21 Oct. 1942	6.5	−116.0	33.0	Borrego Valley
4 Dec. 1948	6.5	−116.4	33.9	Desert Hot Springs
21 July 1952	7.7	−119.0	35.0	Kern County
9 Feb. 1956	6.8	−115.9	31.8	San Miguel
15 Oct. 1979	7.0	−115.3	32.6	Imperial Valley
2 May 1983	6.7	−120.3	36.2	Coalinga
24 Nov. 1987	6.5	−115.8	33.1	Elmore Ranch
24 Nov. 1987	6.7	−115.8	33.0	Superstition Hills
28 June 1992	7.6	−116.4	34.2	Landers
17 Jan. 1994	6.8	−118.5	34.2	Northridge
16 Oct. 1999	7.6	−116.2	34.4	Hector Mine

algorithm. The target magnitude may be truncated from above as well: $M_0 > M \geq M_0$. Note that the threshold Σ_0 in equation (3) is taken as a specific fraction of contribution to $\Sigma_i(t)$ from an earthquake of magnitude M_0: $\Sigma_0 = 10^{BM_0}/R$, R being a parameter of the algorithm.

3. An alarm is declared for the time period $(t, t + \Delta)$ whenever $A(t) \geq C_A$. The threshold C_A is defined as $q\%$ percentile of the function A (that is, from condition that $A(t)$ is lower that C_A during $q\%$ of the time). Δ and q are numerical parameters of the algorithm.

4. An alarm is terminated after a major earthquake occurs or the time Δ expires, whichever comes first. The former case is called a confirmed prediction, while the latter is called a false alarm. A failure to predict is the case when a major earthquake occurs outside an alarm period.

3. Applications for Southern California

[13] Here we explore the pattern Accord in the observed seismicity. We apply the prediction algorithm retrospectively, in the same way as it would be applied in actual (advance) prediction: at any moment t the algorithm uses only information on seismicity prior to t and not on subsequent seismicity. First, we try to predict the three strongest ($M > 7.5$) southern California earthquakes. Then we turn to the earthquakes with magnitudes from 6.5 to 7.4. Table 2 provides information about the earthquakes analyzed.

3.1. Division of the Region

[14] By the definition developed in the CC model we have to divide our system into major subregions.

[15] We use the traditional criteria of division based upon fault orientation, subjective grouping, and sense of slip, along with tectonic setting. In general this subdivision was not a difficult task because southern California geology is naturally divided. Northern regions are separated from southern ones by the Transverse Ranges, where east-west striking left-lateral faults dominate. The relatively stable Sierra Nevada-Great Valley block and the Peninsular Ranges separate the eastern areas, which have been extended and thinned, from western areas that are mostly underlain by diverse oceanic rock types.

[16] Let us clarify some essential details of this division. Some individual faults form the geological boundaries between blocks. However, we do not wish to divide earthquakes associated with a given fault into two separate subregions, so faults must lie within subregions, not along their borders. The Garlock fault is an example of this situation; it separates the tectonically and morphologically distinct Mojave Desert from the western Great Basin and southern Sierra Nevada. We outline the Garlock fault with its own narrow subregion. Similarly, the San Andreas fault subnetwork is generally continuous through southern California, transecting the entire southern California fault network, but is too large to be put into its own subregion as was done for the Garlock fault. Thus, we split the San Andreas among subregions that were otherwise defined. Some of the geologically most obvious subnetworks are too large. Examples of this are the faults west of the central San Andreas and in the northeastern part of southern California. These two subnetworks were divided into southern and northern parts; the northern parts are not considered further.

[17] Finally, given these considerations, we defined seven subregions shown in Figure 3a. The faults shown there are taken from *Jennings* [1977, 1994].

[18] The southwestern subnetwork includes mainly northwest striking dextral faults in the offshore continental borderland. The southeastern subnetwork also includes northwest striking dextral faults that cut through the Peninsular Ranges or lie within the strongly extended Salton Trough-northern Gulf of California region, where smaller sinistral faults and normal faults are present. The northwestern subnetwork includes mostly northwest striking dextral faults north of the Western Transverse Ranges, which separate it from the southwestern subnetwork. The Western Transverse Ranges are dominated by east striking sinistral and thrust faults. Between the southeastern and Garlock subnetworks is a complicated subnetwork that includes the Eastern Transverse Ranges and Mojave Desert. The Eastern Transverse Ranges are also dominated by east striking sinistral and thrust faults. The Mojave Desert is largely underforming in its westernmost part, but includes a broad array of northwest striking dextral faults in its central part and a small array of east striking sinistral faults in its northeastern corner. The seventh subnetwork includes north to northwest striking dextral and normal faults in the western Great Basin.

[19] What determines the number of subregions in our division? The basic factor is how the fault network is organized into a hierarchy. The nature of premonitory pattern Accord, as it was originally defined during the analysis of the CC model, demands to consider first only subregions of the highest rank. In case of observed fault network this corresponds to considering first only the major fault zones. The quantitative approach to such a division was suggested by *Alekseevskaya et al.* [1977]. Here we followed qualitative criteria, described above, which appear to be sufficient for the subregions at the scale we considered.

3.2. Pattern Accord and the Three Largest Earthquakes

[20] Here, we juxtapose the pattern Accord and the three largest earthquakes in southern California during the time considered: Kern County, $M = 7.7$, 21 July 1952; Landers, $M = 7.6$, 28 June 1992; and Hector Mine, $M = 7.6$, 16 October 1999. We explore whether the pattern Accord could

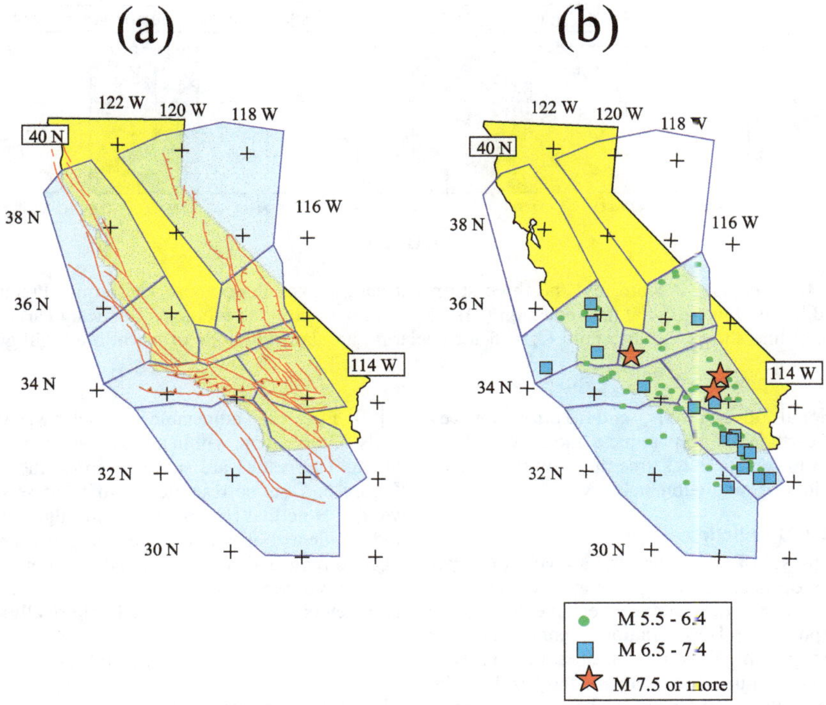

Figure 3. The fault network of southern California. (a) Faults and subregions (fault zones). (b) Subregions and seismicity. The epicenters are shown for the period 1928–2000. See sections 3.1. and 3.2.

predict them. With our choice of magnitude this would correspond to $M_0 = 7.5$.

3.2.1. Data Source

[21] We analyze the data from the U.S. Geological Survey-National Earthquake Information Center (USGS-NEIC) earthquake catalog for the time period 1 January 1928 to 30 January 2000. The data for the period from 1 January 1928 to 31 December 1989 are from the USGS-NEIC Global Hypocenter Data-Base (1990, available at http://neic.usgs.gov/products_and_services.html) with duplicates removed as described by *Shebalin* [1992]. The data for the period from 1 January 1990 to 30 January 2000 are from the NEIC/Preliminary Determination of Epicenters reports (monthly reports at http://gldfs.cr.usgs.gov/pde; weekly reports and QED at http://gldfs.cr.usgs.gov/weekly). The resulting catalog is preprocessed by the same rules as in other prediction algorithms of that kind [*Keilis-Borok and Shebalin*, 1999]: (1) The catalog contains magnitudes in different scales. For each earthquake we use the largest value; for the region and magnitude range considered this is practically equivalent to using Ms scale. (2) The aftershocks are eliminated from the catalog. We choose here an intentionally crude definition of aftershocks that has been used for other premonitory patterns in this region [*Molchan et al.*, 1990; *Keilis-Borok and Rotwain*, 1990]. More precise definitions of aftershocks are described by *Molchan and Dmitrieva* [1991], *Keilis-Borok et al.* [1980], *Gardner and Knopoff* [1974], and *Knopoff et al.* [1982]. Distribution of main shocks among the subregions is illustrated in Figure 3b.

3.2.2. An Example of Prediction

[22] To apply the prediction algorithm, we need to choose the values of its numerical parameters, defined in section 2. Let us consider first the values given in the first row of Table 3. Function $A(t)$ computed with these parameters is shown in Figure 4. We see that $A(t)$ exceeds 5 only in the vicinity of the earthquakes targeted for prediction. This suggest the threshold $C_A = 5$; the corresponding value of q is 80%. With this threshold each strong earthquake is preceded by an alarm, and there are no false alarms. The total duration of alarms is 29 years or 41% of the time considered. Figure 4 may sound not very impressive. Note, however, that we consider here the three strongest earthquakes of the century. More than twofold reduction of the time might be useful in different ways: (1) as initial approximation for more accurate prediction, (2) for instal-

Table 3. Numerical Parameters of the Algorithm[a]

M_0	M^0	M_1	M_2	s, years	λ, years	Δ, years	B	R	q,%
7.5	∞	5.0	7.4	12	∞	3	0.9	21	80
6.5	7.5	5.0	6.4	2	∞	2	0.9	25	90

[a] See sections 3.2.2., 3.3.1, and 3.5 and Figures 4, 6, and 10.

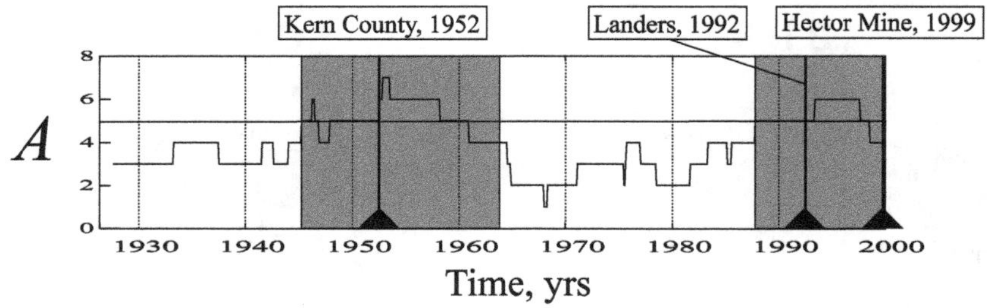

Figure 4. Precursor Accord and the three strongest earthquakes in southern California. Precursor is depicted by the function $A(t)$ (see equation (3)). Vertical lines show the moments of strong earthquakes. Horizontal line shows the threshold $C_A = 5$ for declaring an alarm. Periods of alarm are highlighted.

lation of additional observations, and (3) for low-key enhancement of earthquake preparedness measures. Moreover, as is shown in section 3.2.3, this time can be reduced to 25% without loss of prediction stability.

3.2.3. Stability of Prediction

[23] For a single set of parameters the above scores, per se, are of little relevance, because parameters of the algorithm are retrospectively data fitted. We have to explore how stable such prediction is to variation of parameters. For that purpose we repeat prediction using different combinations of parameters' values and compare the results. The pivotal tool for such an analysis is the error diagram introduced in earthquake prediction studies by *Molchan* [1997]. Its definition follows.

[24] Consider prediction made during a time interval T. Within that time N strong events occurred and N_f of them were not predicted; altogether A alarms were declared and A_f of them were false; the total duration of alarms was D time units. The error diagram brings together the three major scores, characterizing the quality of prediction: the relative duration of alarms $\tau = D{:}T$; the rate of failures to predict $n = N_f{:}N$; and the rate of false alarms $f = A_f{:}A$. A prediction algorithm, with adjustable parameters fixed, corresponds to a point in the (τ, n, f)-space. We plot an error diagram in two projections: (τ, n) and (f, n). The random prediction, when we randomly pick up p% of the time to be covered by alarms, corresponds to the diagonal $n + \tau = 1$.

[25] First, each adjustable parameter was varied independently of the others, within a broad range indicated in Table 4, test a. The results are summed up on the error diagram in Figure 5. Dots correspond to different sets of parameters' values. Note that parameters of our algorithm might be not independent. For example, reducing the "memory" λ, we have to reduce also the threshold Σ_0 in order to produce the same prediction scores. The following relation between parameters λ and R is obtained empirically:

$$R = \lambda^{0.55} 10^{3.5}. \quad (4)$$

The stars in Figure 5 correspond to predictions with R calculated by this relation while the other parameters are varied independently; specific values of parameters are given in Table 5; 36 variants are considered altogether. The obvious improvement of the results stresses the necessity to explore the connections among the adjustable parameters.

[26] Could the parameters of our algorithm be interdependent? Equation (4) above illustrates that this is indeed possible. Could our results be artificially improved by other correlations that are overlooked? Quite to the contrary: as clearly illustrated by Figure 5, the results are greatly improved when the correlation of parameters is known. Not knowing the interdependence of parameters, one can only increase the number of unacceptable versions of an algorithm (like those lying close to the diagonal of the error diagram in Figure 5).

Table 4. Test of Stability: Independent Variation of Parameters

		M_1	M_2	s, years	B	λ, years	Δ,	q, %	R	Sections and Figures in Text
Test A	From	4.5	7.0	5	0.9	10	2	80	15	section 3.2.3
	To	5.1	7.4	15	0.9	∞	3	90	40	Figure 5
Test B	From	4.5	5.5	2	0.9	5	0.5	70	5	section 3.3.2
	To	4.5	6.1	4	0.9	10	0.5	70	7	Figure 7
	To	5.1	6.4	10	0.9	∞	2	90	15	
	To	5.5	7.0	12	0.9	∞	2	90	50	
Test C	From	4.0	7.4	10	0.9	5	5	80	10	section 3.4.3
	To	5.5		1	0.9	∞	6	90	10	Figure 9a
Test D	From	4.5	6.4	2	0.8	5	2	90	10	section 3.4.3
	To	5.2	7.0	7	1	∞	5	95	10	Figure 9b

Table 5. Test of Stability: Parameter R is Determined by λ[a]

M_1	M_2	s, years	B	λ, years	Δ, years	q
4.5	7.0	10	0.9	10	3	80
5.0	7.4	12		15		
5.1				20		

[a] See section 3.2.3 and Figure 5.

3.3. Pattern Accord and Earthquakes With Magnitude 6.5–7.4

[27] Here we explore in a similar way the emergence of the pattern Accord before earthquakes with magnitudes from 6.5 to 7.4. Eleven such earthquakes occurred during the time period considered. Two of them, Colorado River Delta of 31 December 1934, $M = 7.0$, and Superstition Hills of the 24 November 1987, $M = 6.7$, are eliminated from consideration, because they follow the preceding strong earthquakes within a too-short time interval, 29 and 12 hours respectively. Such intervals are incompatible with characteristic resolution time of our analysis, which is of order of months.

3.3.1. An Example of Prediction

[28] We consider for a start the values of adjustable parameters given in the second row of the Table 3. All the parameters are readjusted to the new target magnitude in about the same way as was done for other premonitory patterns [*Keilis-Borok et al.*, 1980]. Figure 6 shows the function $A(t)$ and alarms determined with these parameters. Six out of nine strong earthquakes are "predicted" with no false alarms; three earthquakes are missed. The total duration of alarms is 20 years or 28% of the time considered.

[29] The question naturally arises: Why is this value less than that obtained above in section 3.2.2 for the large and presumably easier to predict earthquakes? In fact, 28% correspond to the case when three earthquakes out of nine were not predicted, while for the strongest earthquakes we have no failures to predict. As shown in section 3.3.2 below, the duration of alarms increases up to 62% if parameters of the algorithm are readjusted to predict all nine earthquakes.

3.3.2. Stability of Prediction

[30] We varied adjustable parameters of the algorithm within the limits indicated in Table 4, test b (lines 1 and 4). Acceptable prediction score ($\tau < 0.4$; $n < 0.4$; $f < 0.5$) is obtained within more narrow limits indicated in lines 2 and 3 of Table 4, test b. The corresponding error diagram is shown in Figure 7. Again, we see that the quality of prediction remains acceptable not only for some special combinations of parameters but for a reasonably wide domain in parameter space. Thus the prediction is stable. In particular, parameters s, λ, and R can be varied much more than one could dare to hope. It is encouraging also that combinations leading to good scores are not scattered randomly over parameter space but form a connected cluster. The overall quality of prediction is slightly lower than that for the three largest earthquakes (see section 3.2).

3.4. Variation of Basic Definitions

[31] In sections 3.2 and 3.3 above we varied adjustable parameters of the algorithm to check its stability. However, we also have freedom of choice in determination of subregions and in the choice of the function that depicts our pattern. In this section we introduce an alternative regionalization based entirely on the territorial distribution of seismicity and explore to what extent the seismic regionalization is different from the geological one and how this difference affects the result. We consider also an alternative measure of correlation of seismic activity. The whole previous analysis, including the stability tests, is repeated here with these two alternative choices.

3.4.1. Alternative Subregions

[32] Consider the case of zoning, based solely on territorial distribution of seismicity. Figure 8 shows sub-

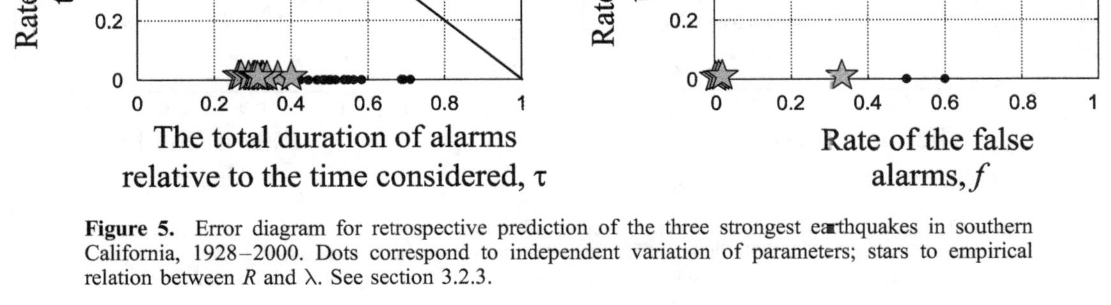

Figure 5. Error diagram for retrospective prediction of the three strongest earthquakes in southern California, 1928–2000. Dots correspond to independent variation of parameters; stars to empirical relation between R and λ. See section 3.2.3.

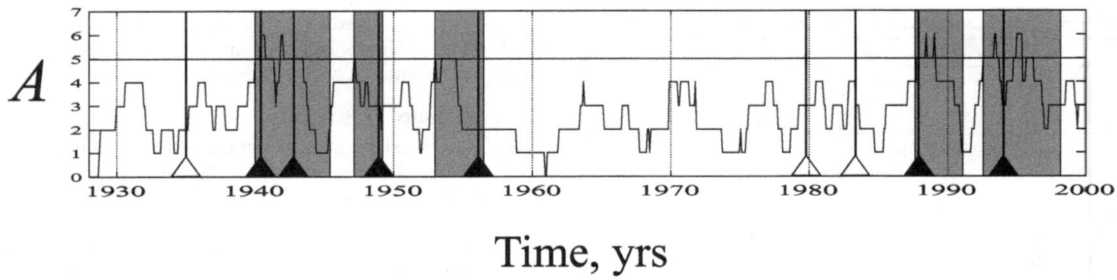

Figure 6. Precursor Accord and earthquakes with magnitudes from 6.5 to 7.4 in southern California. "Predicted" earthquakes are marked with solid triangles; "unpredicted" are marked with empty triangles. Other notations are the same as in Figure 4.

regions obtained this way; note that there are only five of them, compared to seven in Figure 3. The seismological regionalization is less detailed because the clouds of epicenters associated with different faults merge or even overlap.

3.4.2. Alternative Measure of Correlation

[33] It is defined as follows:

$$A^*(t) = (r_1^2 + r_2^2 + \cdots + r_n^2)^{-1}; r_i = \Sigma_i(t)/\Sigma_{sum}(t);$$
$$i = 1, \cdots, n; \Sigma_{sum}(t) = \Sigma_1(t) + \Sigma_2(t) + \cdots + \Sigma_n(t). \quad (5)$$

[34] Function $A^*(t)$ shows how uniform the distribution of seismicity is among the subregions. It is easy to see that $A^*(t)$ takes the maximal value of n when seismic activity $\Sigma_i(t)$ is the same in all subregions and decreases to the minimal value of unity when activity is totally concentrated within a single subregion.

3.4.3. Stability of Prediction

[35] With two alternative sets of boundaries (Figures 3 and 8) and two measures of correlation (A and A^*) we have four versions of the prediction algorithm. One of them, which uses geological zoning and the measure $A(t)$, was considered above.

[36] Figure 9 shows the error diagrams for prediction by the other three versions. The ranges of parameters' variation are shown in Table 4, tests c and d. Table 4, test c, is related to parameters scaled for prediction of the three largest earthquakes, while Table 4, test d, is related to parameters scaled for prediction of earthquakes with magnitudes greater than 6.5.

[37] The following results are noteworthy: (1) All four versions give almost the same errors' scatter; this confirms the robustness of the algorithm. (2) The points on the error diagram are well separated from the diagonal, that is, from random prediction. (3) The rate of false alarms is small. These results are highly encouraging for further exploration of the pattern Accord.

3.5. What Does Work in the Pattern Accord?

[38] Here we compare the pattern Accord with the activity of the whole southern California measured by the function $\Sigma(t)$ described above (section 1.2.2). The functions

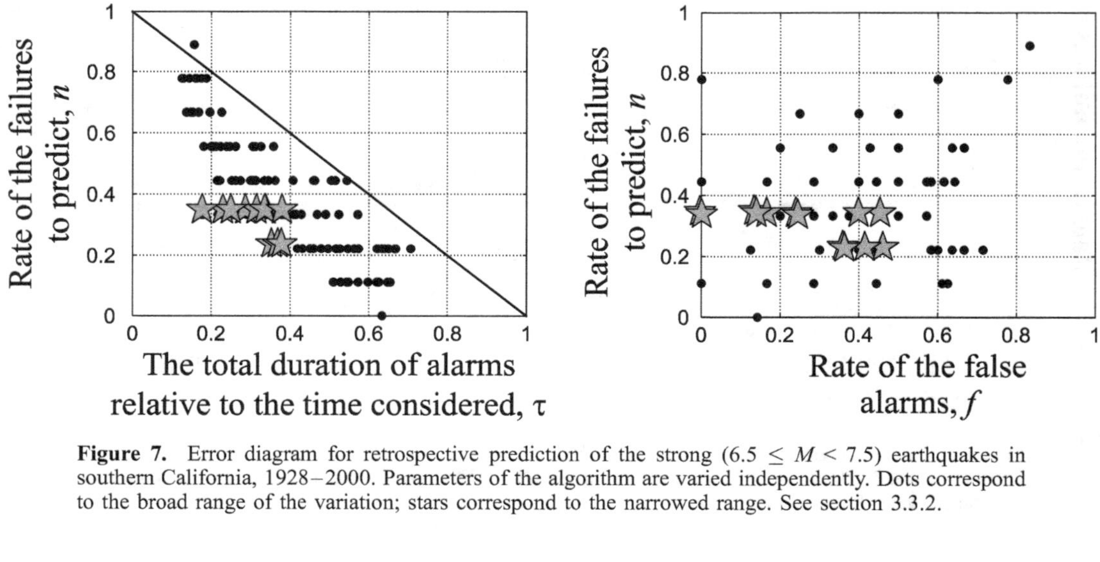

Figure 7. Error diagram for retrospective prediction of the strong ($6.5 \leq M < 7.5$) earthquakes in southern California, 1928–2000. Parameters of the algorithm are varied independently. Dots correspond to the broad range of the variation; stars correspond to the narrowed range. See section 3.3.2.

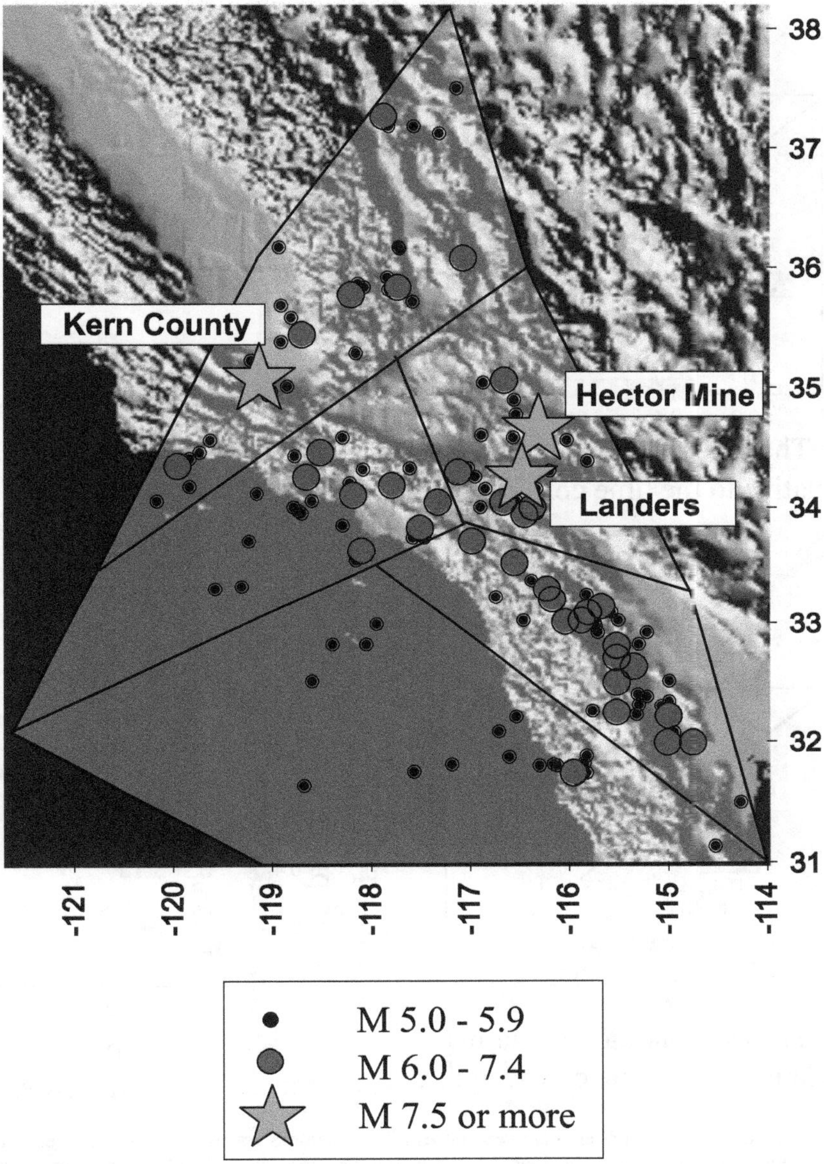

Figure 8. Alternative zoning of southern California based solely on seismicity. The epicenters are shown for the period 1928–2000.

$\Sigma(t)$ and $A(t)$ are juxtaposed in Figure 10. Both functions are calculated with the same values of numerical parameters, given in Table 3. Figure 10a and the first line of Table 3 correspond to prediction of the three strongest earthquakes ($M \geq 7.5$), while Figure10b and the second line of Table 3 correspond to prediction of earthquakes with $6.5 \leq M \leq 7.4$. Alarms declared by the pattern Accord are shown at the top panel of each figure. Figure 10a demonstrates that both functions $\Sigma(t)$ and $A(t)$ have two explicit peaks over the considered time interval: one of these peaks precedes Kern County earthquake, and another precedes Landers and Hector Mine. Figure 10b demonstrates a more diverse behavior. It shows that the pattern Accord is not equivalent to the total activity measured by $\Sigma(t)$. There are periods where $A(t)$ is abnormally high while $\Sigma(t)$ remains within its usual boundaries (e.g., 1940–1944), and vice versa (e.g., 1981–1983). At the same time, there are periods where both the functions are abnormally high (e.g., 1994–1995) or low (e.g., 1959–1961) simultaneously. Similar results are obtained with different sets of numerical parameters. There-

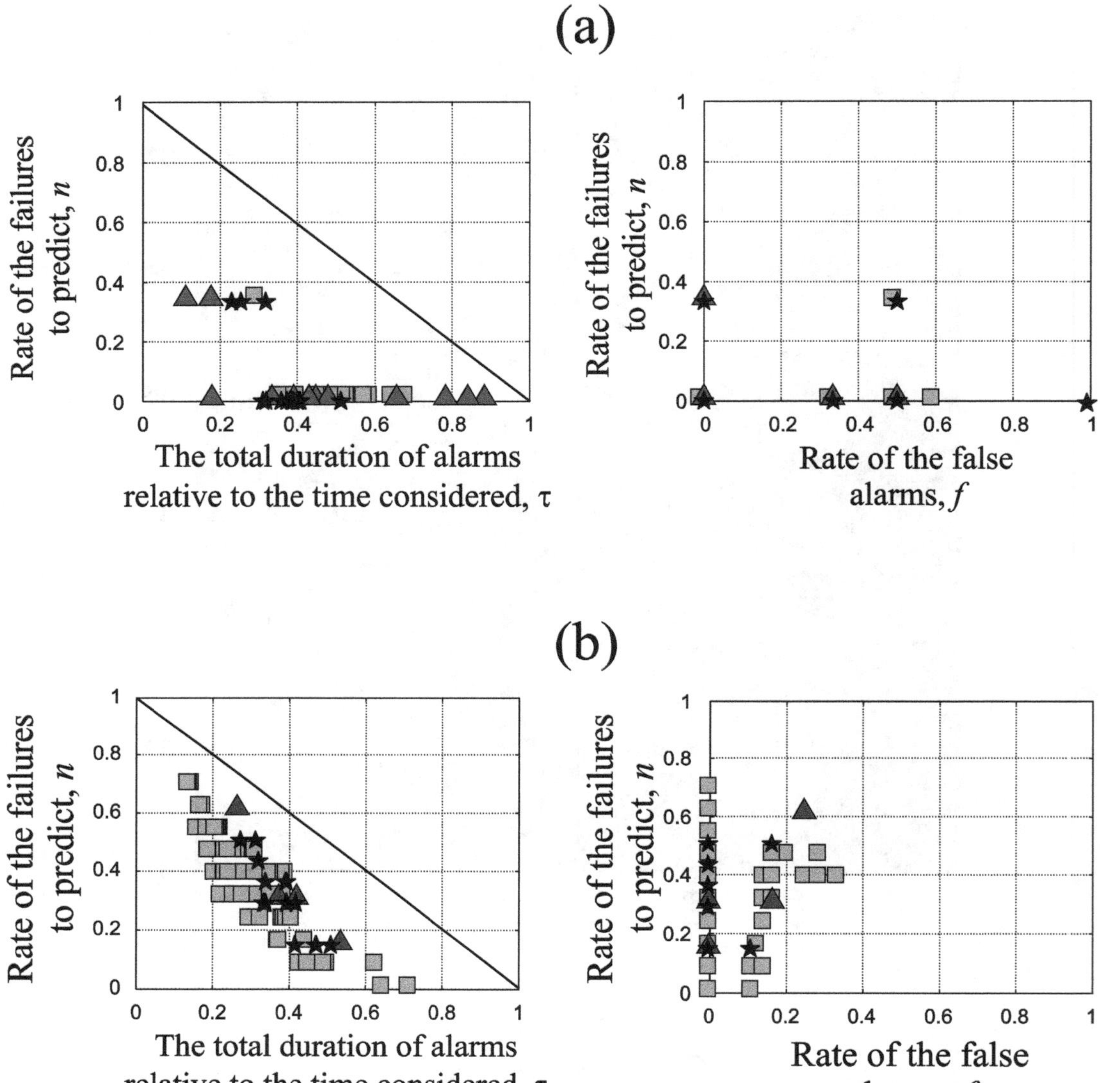

Figure 9. Error diagram for alternative versions of the precursor. Stars correspond to the version which employs geological zoning and the function $A^*(t)$; triangles correspond to alternative zoning and the function $A(t)$; and rectangles correspond to alternative zoning and the function $A^*(t)$. (a) Retrospective prediction of the three strongest earthquakes ($M \geq 7.5$). (b) Retrospective prediction of earthquakes with $M \geq 6.5$.

fore what does work in the pattern Accord is not only the general rise of activity but its spreading over the faults' network as well.

[39] As one can see in Figure 10b, this version of the pattern Accord fails to predict three earthquakes: Colorado River delta of 31 December 1934, Imperial Valley of 15 October 1979, and Coalinga of 2 May 1983. The natural question for future study is: What is the reason for these errors (particularly the two recent ones that are relatively better studied)?

3.6. Alarm From a Decision Maker's Point of View

[40] A prediction algorithm has errors; that is certainly the case for currently known algorithms and possibly is inevitable [*Holland*, 1995; *Gell-Mann*, 1994; *Turcotte*, 1997; *Rundle et al.*, 2000b]. There is, however, a freedom to

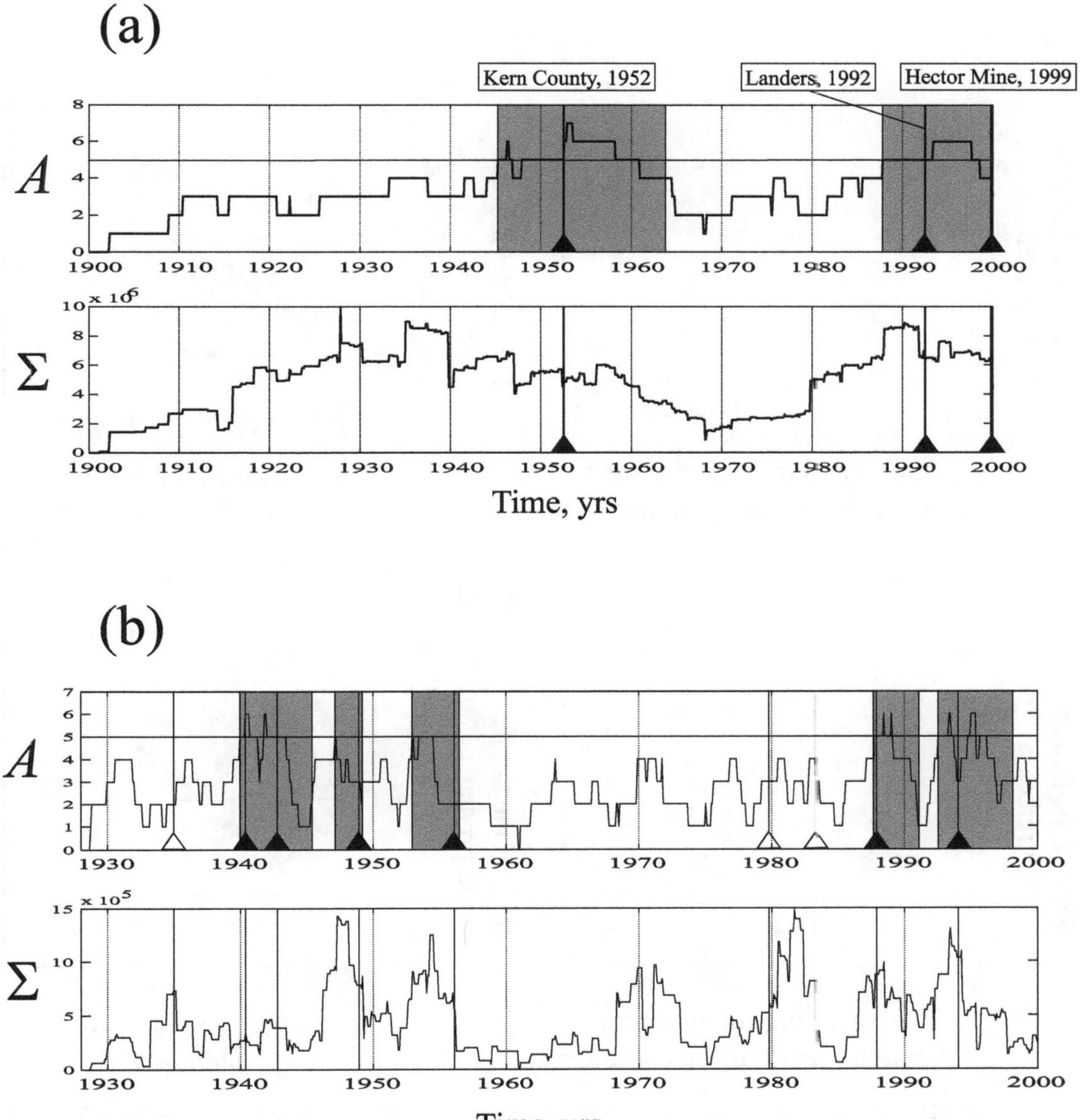

Figure 10. Comparison of precursors Accord and Sigma. (a) Parameters scaled to prediction of the three largest earthquakes. (b) Parameters scaled to prediction of the earthquakes with magnitudes from 6.5 to 7.4. Notations are the same as in Figure 4. See section 3.5.

choose a trade-off between different kinds of errors as illustrated in the error diagrams of Figures 5, 7, and 9. What version of a prediction algorithm is "the best"? Let us consider that question from the point of view of a decision maker who has to choose the safety measures undertaken in response to a prediction. The question arises because earthquake preparedness requires a wide variety of safety measures; see their overview in *Kantorovich and Keilis-Borok* [1991]. The choice of safety measures depends on two factors: (1) expected damage and (2) reliability of the prediction: probabilities of false alarm and failure to predict and duration of the alarm.

[41] Different measures are justified in different cases. Some critical objects (e.g., high dams) have to be protected even at high cost, and expensive safety measures are justified even if the probability of false alarm is high. In other cases

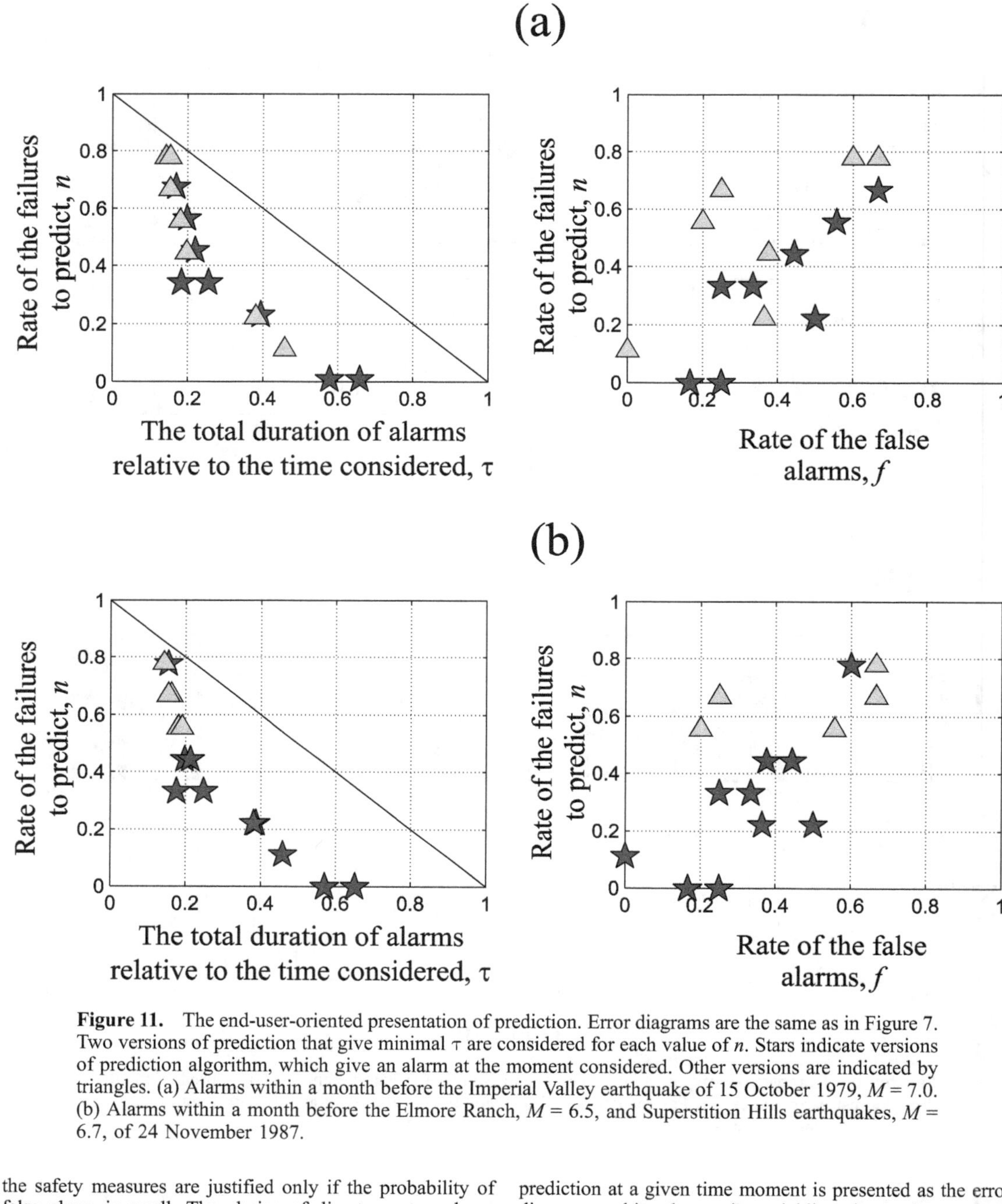

Figure 11. The end-user-oriented presentation of prediction. Error diagrams are the same as in Figure 7. Two versions of prediction that give minimal τ are considered for each value of n. Stars indicate versions of prediction algorithm, which give an alarm at the moment considered. Other versions are indicated by triangles. (a) Alarms within a month before the Imperial Valley earthquake of 15 October 1979, $M = 7.0$. (b) Alarms within a month before the Elmore Ranch, $M = 6.5$, and Superstition Hills earthquakes, $M = 6.7$, of 24 November 1987.

the safety measures are justified only if the probability of false alarm is small. The choice of disaster preparedness measures belongs to a decision maker responsible for disaster management; *Molchan* [1997] suggested a theoretical framework for optimization of a response to prediction. Accordingly, contrary to the established tradition, we propose here not to select a single version of the algorithm, that is, a "best" set of adjustable parameters, but provide a decision maker with the whole available information. Then prediction at a given time moment is presented as the error diagram marking the versions yielding alarms. An example of such a presentation is shown in Figure 11. It shows performance of several versions of the prediction algorithm from Figure 7. Comparing Figure 11 to Figure 7, we retained only the versions that give minimal duration of alarm for the same rate of failures to predict. Panels refer to different time moments: panel a refers to a moment one month before the Imperial Valley earthquake of 15 October 1979; panel b refers to one month before the Elmore Ranch and Superstition Hills earthquakes of 24 November 1987.

[42] A new element on these diagrams is the following: we show which version of prediction algorithm produces an alarm at the given moment (stars) and which versions do not (triangles). In these particular cases predictions are very stable, being generated by a majority of the versions considered. Implications for the decision maker are beyond the scope of the present paper.

4. Discussion

[43] 1. The definition of the algorithm Accord, developed first for the abstract model of colliding cascades [*Gabrielov et al.*, 2000a], is adapted here to analysis of observations. The retrospective application of the algorithm to seismicity of southern California is highly encouraging. Predictions yield acceptable success-to-failures scores in a reasonably wide range of adjustable parameters and choice of subregions. Thus this paper establishes a hypothesis that may be tested on independent data. In support of our hypothesis is the fact that it was first found in a model [*Gabrielov et al.*, 2000a], then in a shorter timescale in real seismicity [*Shebalin et al.*, 2000], and finally in an intermediate timescale considered here. The final test would be the advance prediction.

[44] 2. It is encouraging also that the pattern Accord is directly connected with geometry of fault networks. Two mechanical explanations of that pattern may be hypothetically suggested: (1) strength drop (weakening) at the fault's segments broken by the earthquakes and (2) stress redistribution after the earthquakes. If either of them occurs in several subregions, it is likely to affect the equilibrium of a whole system more than the equivalent change in a single subregion.

[45] 3. There are further as-yet unexplored possibilities to improve the prediction algorithm described here. (1) The size of the fault network considered probably has to be scaled by the size $L(M)$ of the source of the earthquakes which we are trying to predict. (2) One may consider instead of Σ the pattern Active Zone Size (AZS). Its difference from Σ is that overlapping parts of several fault breaks are counted only once. This pattern was suggested by *Kossobokov and Carlson* [1995], who found that AZS yields better predictions than Σ. (3) It would be most interesting to consider instead of fault subregions the "nodes": fractured structures, formed around faults' intersections and junctions [*Gabrielov et al.*, 1996, and references therein]. As was demonstrated by *Gelfand et al.* [1976] and *Gorshkov et al.* [2001, and references therein], strong earthquakes are nucleated mainly in such nodes. Moreover, the tendency of the fault system to fracturing and stress accumulation is concentrated at fault intersections [*Gabrielov et al.*, 1996; *Minster and Jordan*, 1984].

[46] 4. The pattern Accord is a manifestation of a broader phenomenon: premonitory increase of earthquake correlation range. This phenomenon has been formally defined and statistically analyzed for the first time during the analysis of the CC model, although it was previously hypothesized by *Keilis-Borok* [1994, 1996]. Its qualitative explanation by analogy with phase transitions in statistical physics was suggested by *Turcotte et al.* [2000] and *Bowman et al.* [1998]. *Zoeller et al.* [2001] have reported the growth of correlation range prior to the strong earthquakes for observed seismicity of southern California. The CC model suggests also a similar phenomenon in the shorter timescale of months; its first test on observations [*Shebalin et al.*, 2000] is encouraging too.

[47] 5. Although long-range correlations are well known in seismology, they are sometimes regarded as counter-intuitive in earthquake prediction studies. This is probably due to the fact that they cannot be explained by the linear elastic model of stress redistribution after a rupture. In such models redistribution is confined to the distances comparable with the rupture dimension (the "Saint Venant principle"). However, many other mechanisms, such as mentioned in section 1.1, do explain long-range correlations. It is interesting to note in this context that Charles Richter, who was generally skeptical about feasibility of earthquake prediction, made an exception to the pattern Sigma specifically because it was based on long-range correlations. [*Richter*, 1964, p. 3025] wrote: "It is important that [the authors] confirm the necessity of considering a very extensive region including the center of the approaching event. It is very rarely true that the major event is preceded by increasing activity in its immediate vicinity".

[48] 6. More evidence of long-range correlations comes from statistical modeling. It was shown recently by *Liu et al.* [1999] that the Stress Release model, introduced by *Vere-Jones* [1978], does fit seismicity of North China much better if long-range correlations are introduced.

[49] 7. Contrary to the established tradition, it seems better not to choose for prediction the "best" version of an algorithm, but to use, in parallel, different acceptable versions. Then one may present a prediction by the error diagram marking the versions yielding the current alarm (Figure 11). Such representation is important for a decision maker, who has to choose preparedness measures to be undertaken in response to prediction.

[50] **Acknowledgments.** We are grateful to M. Harrison, P. Bird, A. Gabrielov, and W. Newman for inspiring discussions and constructive criticism. This study was partially supported by a subcontract with Cornell University, Geological Sciences, under grant EAR-9804859 from the National Sciences Foundation. This study was also partially supported by grants from the International Science and Technology Center (Project 1293-99), USGS/SCEC, INTA 0748, and James S. McDonnell Foundation.

References

Aki, K., Scale dependence in earthquake phenomena and its relevance to earthquake prediction, *Proc. Natl. Acad. Sci. U.S.A.*, *93*, 3740–3747, 1996.

Alekseevskaya, M. A., A. M. Gabrielov, A. D. Gvishiani, I. M. Gelfand, and E. Y. Rantsman, Formal morphostructural zoning of mountain territories, *J. Geophys.*, *43*, 227–233, 1977.

Allegre, C. J., J. L. Le-Mouel, and A. Provost, Scaling rules in rock fracture and possible implications for earthquake prediction, *Nature*, *297*, 47–49, 1982.

Bak, P., and C. Tang, Earthquakes as a self-organized critical phenomenon, *J. Geophys. Res.*, *94*, 15,635–15,637, 1989.

Barenblatt, G. M., V. I. Keilis-Borok, and A. S. Monin, Filtration model of earthquake sequence, *Dokl. Akad. Nauk SSSR*, *269*, 831–834, 1983.

Ben-Zion, Y., Dynamic ruptures in recent models of earthquake faults, *J. Mech. Phys. Solids*, *49*, 2209–2244, 2001.

Bowman, D. D., G. Ouillon, C. G. Sammis, A. Sornette, and D. Sornette, An observational test of the critical earthquake concept, *J. Geophys. Res.*, *103*, 24,359–24,372, 1998.

Bufe, C. G., and D. J. Varnes, Predictive modeling of the seismic cycle of the Great San Francisco Bay Region, *J. Geophys. Res.*, *98*, 9871–9883, 1993.

Caputo, M., R. Console, A. M. Gabrielov, V. I. Keilis-Borok, and T. V. Sidorenko, Long-term premonitory seismicity patterns in Italy, *Geophys. J. R. Astron. Soc.*, *75*, 71–75, 1983.

Gabrielov, A. M., and V. I. Keilis-Borok, Patterns of stress corrosion: Geometry of the principal stresses, *Pure Appl. Geophys.*, *121*, 477–494, 1983.

Gabrielov, A. M., V. I. Keilis-Borok, and D. D. Jackson, Geometric incompatibility in a fault system, *Proc. Natl. Acad. Sci. U.S.A.*, *93*, 3838–3842, 1996.

Gabrielov, A. M., V. I. Keilis-Borok, I. V. Zaliapin, and W. I. Newman, Critical transitions in colliding cascades, *Phys. Rev. E*, *62*, 237–249, 2000a.

Gabrielov, A. M., I. V. Zaliapin, W. I. Newman, and V. I. Keilis-Borok, Colliding cascades model for earthquake prediction, *Geophys. J. Int.*, *143*, 427–437, 2000b.

Gardner, J., and L. Knopoff, Is the sequence of earthquakes in S. California with aftershocks removed Poissonian?, *Bull. Seismol. Soc. Am.*, *64*(5), 1363–1367, 1974.

Gelfand, I. M., S. A. Guberman, V. I. Keilis-Borok, L. Knopoff, F. Press, E. Y. Ranzman, I. M. Rotwain, and A. M. Sadovsky, Pattern recognition applied to earthquake epicenters in California, *Phys. Earth Planet. Inter.*, *11*, 227–283, 1976.

Gell-Mann, M., *The Quark and the Jaguar: Adventures in the Simple and the Complex*, W. H. Freeman, New York, 1994.

Gorshkov, A. I., V. G. Kossobokov, E. Ya. Rantsman, and A. Soloviev, Recognition of earthquake prone areas: Validity of results obtained from 1972 to 2000, *Comput. Seismol.*, *32*, 48–57, 2001.

Holland, J. H., *Hidden Order: How Adaptation Builds Complexity*, Addison-Wesley-Longman, Reading, Mass., 1995.

Ismail-Zadeh, A. T., V. I. Keilis-Borok, and A. A. Soloviev, Numerical modeling of earthquake flow in the southeastern Carpathians (Vrancea): Effect of a sinking slab, *Phys. Earth Planet. Inter.*, *111*, 267, 1999.

Jennings, C. W., Geological map of California, 1:750,000 scale, Calif. Div. of Mines and Geol., Sacramento, 1977.

Jennings, C. W., Fault activity map of California and adjacent areas, scale 1:750:000, Calif. Div. of Mines and Geol., Sacramento, 1994.

Kantorovich, L. V., and V. I. Keilis-Borok, Earthquake prediction and decision-making: Social, economic and civil protection aspects, in *International Conference on Earthquake Prediction: State-of-the-Art, Sci.-Tech. Contrib. CSEM-EMSC*, pp. 586–593, Strasbourg, France, 1991.

Keilis-Borok, V. I., (Ed.), Intermediate-term earthquake prediction: Models, algorithms, worldwide tests, *Phys. Earth Planet. Inter.*, *61*, special issue, 1990.

Keilis-Borok, V., Chaos in solid earth, in *Proceedings of Conference on Chaos*, Univ. of Tokyo, Tokyo, 1992.

Keilis-Borok, V. I., Symptoms of instability in a system of earthquake-prone faults, *Physica D*, *77*, 193–199, 1994.

Keilis-Borok, V. I., Intermediate-term earthquake prediction, *Proc. Natl. Acad. Sci. U.S.A.*, *93*, 3748–3755, 1996.

Keilis-Borok, V. I., and V. G. Kossobokov, Premonitory activation of earthquake flow: Algorithm M8, *Phys. Earth Planet. Inter.*, *61*, 73–83, 1990.

Keilis-Borok, V. I., and L. N. Malinovskaya, One regularity in the occurrence of strong earthquakes, *J. Geophys. Res.*, *69*, 3019–3024, 1964.

Keilis-Borok, V. I., and I. M. Rotwain, Diagnosis of time of increased probability of strong earthquakes in different regions of the world: Algorithm CN, *Phys. Earth Planet. Inter.*, *61*, 57–72, 1990.

Keilis-Borok, V. I., and P. N. Shebalin (Eds.), Dynamics of lithosphere and earthquake prediction, special, *Phys. Earth Planet. Inter.*, *111*, 179–330, 1999.

Keilis-Borok, V. I., L. Knopoff, and I. M. Rotwain, Bursts of aftershocks, long-term precursors of strong earthquakes, *Nature*, *283*, 259–263, 1980.

Keilis-Borok, V. I., L. Knopoff, V. Kossobokov, and I. M. Rotwain, Intermediate-term prediction in advance of the Loma Prieta earthquake, *Geophys. Res. Lett.*, *17*, 1461–1464, 1990.

Khokhlov, A. V., and V. G. Kossobokov, Seismic flux and major earthquakes in the northwestern Pacific (in Russian), *Comput Seismol*, *26*, 3–8, 1994.

Knopoff, L., Y. Kagan, and R. Knopoff, b-values for foreshocks and aftershocks in real and simulated earthquake sequences, *Bull. Seismol. Soc. Am.*, *72*(5), 1663–1675, 1982.

Knopoff, L., T. Levshina, V. I. Keilis-Borok, and C. Mattoni, Increased long-range intermediate-magnitude earthquake activity prior to strong earthquakes in California, *J. Geophys. Res.*, *101*, 5779–5796, 1996.

Kossobokov, V. G., and J. M. Carlson, Active zone size vs. activity: A study of different seismicity patterns in the context of the prediction algorithm M8, *J. Geophys. Res.*, *100*, 6431–6441, 1995.

Kossobokov, V. G., V. I. Keilis-Borok, and S. W. Smith, Localization of intermediate-term earthquake prediction, *J. Geophys. Res.*, *95*, 19,763–19,772, 1990.

Kossobokov, V. G., K. Maeda, and S. Uyeda, Precursory activation of seismicity in advance of Kobe, 1995, M = 7.2 earthquake, *Pure Appl. Geophys.*, *155*, 409–423, 1999a.

Kossobokov, V. G., L. L. Romashkova, V. I. Keilis-Borok, and J. H. Healy, Testing earthquake prediction algorithms: Statistically significant real-time prediction of the largest earthquakes in the Circum-Pacific, 1992–1997, *Phys. Earth Planet. Inter.*, *111*, 187–196, 1999b.

Liu, J., Y. Chen, Y. Shi, and D. Vere-Jones, Coupled stress release model for time-dependent seismicity, *Pure. Appl. Geophys.*, *155*, 649–667, 1999.

Ma, Z., Z. Fu, Y. Zhang, C. Wang, G. Zhang, and D. Liu, *Earthquake Prediction: Nine Major Earthquakes in China*, Springer-Verlag, New York, 1990.

Mandelbrot, B., *The Fractal Geometry of Nature*, W.H. Freeman, New York, 1983.

Minster, J. B., and T. H. Jordan, Vector constraints on Quaternary deformation of the Western United States east and west of the San Andreas fault, in *Tectonics and Sedimentation Along the California Margin*, Publ. 38, edited by J. K. Crouch and S. B. Bachman, pp. 1–16, Pac. Sect., Soc. of Econ. Paleontol. And Mineral., Los Angeles, Calif., 1984.

Mogi, K., Migration of seismic activity, *Bull. Earthquake Res. Inst. Univ. Tokyo*, *46*(1), 53–74, 1968.

Molchan, G. M., Earthquake prediction as a decision-making problem, *Pure Appl. Geophys.*, *149*, 233–247, 1997.

Molchan, G. M., and O. E. Dmitrieva, Identification of aftershocks: Methods and new approach, *Comput. Seismol.*, *24*, 19–50, 1991.

Molchan, G. M., O. E. Dmitrieva, I. M. Rotwain, and J. Dewey, Statistical analysis of the results of earthquake prediction, based on bursts of aftershocks, *Phys. Earth Planet. Inter.*, *61*, 128–139, 1990.

Newman, W., A. Gabrielov, and D. L. Turcotte, *Nonlinear Dynamics and Predictability of Geophysical Phenomena*, *Geophys. Monogr. Ser.*, vol. 83, AGU, Washington, D.C., 1994.

Press, F., and C. Allen, Patterns of seismic release in the southern California region, *J. Geophys. Res.*, *100*, 6421–6430, 1995.

Press, F., and P. Briggs, Chandler wobble, earthquakes, rotation and geomagnetic changes, *Nature*, *256*, 270–273, 1975.

Prozorov, A. G., Variations of seismic activity related to locations of strong earthquakes, *Vychislitel'naya Seismolog.*, *8*, 71–72, 1975.

Prozorov, A. G., Schreider, and S. Yu, Real time test of the long-range aftershock algorithm as a tool for mid-term earthquake prediction in southern California, *Pure Appl. Geophys.*, *133*, 329–347, 1990.

Richter, C., Comment on "One regularity in the occurrence of strong earthquakes" by V. I. Keilis-Borok and L. N. Malinovskaya, *J. Geophys. Res.*, *69*, 3025, 1964.

Romanowicz, B., Spatiotemporal patterns in the energy-release of great earthquakes, *Science*, *260*, 1923–1926, 1993.

Rundkvist, D. V., and I. M. Rotwain, Present-day geodynamics and seismicity of Asia Minor (in Russian), *Comput. Seismol.*, *27*, 201–244, 1994.

Rundle, J. B., S. Gross, W. Klein, C. D. Ferguson, and D. L. Turcotte, The statistical mechanics of earthquakes, *Tectonophysics*, *277*, 147–164, 1997.

Rundle, J. B., W. Klein, and K. Tiampo, Linear pattern dynamics in nonlinear threshold systems, *Phys. Rev. E*, *61*, 2418–2431, 2000a.

Rundle, J. B., D. L. Turcotte, W. Klein (Eds.), *Geocomplexity and the Physics of Earthquakes*, *Geophys. Monogr. Ser.*, vol. 120, American Geophysical Union, Washington, D. C., 2000b.

Shaw, B. E., J. M. Carlson, and J. S. Langer, Patterns of seismic activity preceding large earthquakes, *J. Geophys. Res.*, *97*, 479, 1997.

Shebalin, P. N., Automatic duplicate identification in set of earthquake catalogues merged together, *U.S. Geol. Surv. Open File Rep.*, *92-401*, Appendix II, 1992.

Shebalin, P., I. Zaliapin, and V. Keilis-Borok, Premonitory raise of the earthquakes' correlation range: Lesser Antilles, *Phys. Earth Planet. Inter.*, *122*, 241–249, 2000.

Smalley, R. F., D. L. Turcotte, and S. A. Solla, A renormalization group approach to the stick-slip behavior of faults, *J. Geophys. Res.*, *90*, 1894–1900, 1985.

Soloviev, A. A., I. A. Vorobieva, and G. F. Panza, Modeling of block-structure dynamics: Parametric study for Vrancea, *Pure Appl. Geophys.*, *156*, 395–420, 1999.

Sornette, D., and C. G. Sammis, Complex critical exponents from renormalization group theory of earthquakes: Implications for earthquake predictions, *J. Phys. I*, *5*, 607–619, 1995.

Turcotte, D. L., *Fractals and Chaos in Geology and Geophysics*, 2nd ed., Cambridge Univ. Press, New York, 1997.

Turcotte, D. L., W. I. Newman, and A. Gabrielov, A statistical physics approach to earthquakes, in *Geocomplexity and the Physics of Earthquakes*, *Geophys. Monogr. Ser.*, vol. 120, edited by J. B. Rundle, D. L. Turcotte, and W. Klein, pp. 83–96, AGU, Washington, D.C., 2000.

Vere-Jones, D., Earthquake prediction—A statistician view, *J. Phys. Earth*, *26*, 129–146, 1978.

Vilkovich, E. V., and M. G. Shnirman, Epicenter migration waves: Examples and models, *Comput. Seismol.*, *14*, 27–36, 1982.

Vorobieva, I. A., Prediction of a subsequent large earthquake, *Phys. Earth Planet. Inter.*, *111*, 197–206, 1999.

Yamashita, T., Simulation of seismicity due to fluid migration in a fault zone, *Geophys. J. Int.*, *132*, 674–686, 1998.

Yamashita, T., Pore creation due to fault slip in a fluid-permeated fault zone and its effect on seismicity: Generation mechanism of earthquake swarm, *Pure Appl. Geophys.*, *155*, 625–647, 1999.

Zöller, G., S. Hainzl, and J. Kurths, Observation of growing correlation length as an indicator for critical point behavior prior to large earthquakes, *J. Geophys. Res.*, *106*, 2167–2176, 2001.

G. Axen, Department of Earth and Space Sciences, University of California, Los Angeles, 595 Circle Drive East, Los Angeles, CA 90095-1567, USA. (gaxen@ess.ucla.edu)

V. Keilis-Borok and I. Zaliapin, Institute of Geophysics and Planetary Physics, University of California, Los Angeles, 3845 Slichter Hall, Los Angeles, CA 90095-1567, USA. (vkb@ess.ucla.edu; zal@ess.ucla.edu)

Annu. Rev. Earth Planet. Sci. 2002. 30:1–33
DOI: 10.1146/annurev.earth.30.100301.083856

EARTHQUAKE PREDICTION: State-of-the-Art and Emerging Possibilities

Vladimir Keilis-Borok[1,2]

[1]*International Institute of Earthquake Prediction Theory and Mathematical Geophysics, Russian Academy of Sciences, Warshavskoe sh. 79, korp. 2, 113556 Moscow, Russia*
[2]*Institute of Geophysics and Planetary Physics and Department of Earth and Space Sciences, University of California, Los Angeles, California 90095-1567; e-mail: vkb@ess.ucla.edu, vkborok@mitp.ru*

Key Words earthquake preparedness, complexity, critical transitions, premonitory seismicity patterns

FOREWORD

1. Earthquake prediction is pivotal both for reduction of the damage from earthquakes and for fundamental understanding of lithosphere dynamics. That twofold goal, usual for prediction research, brings up the key questions considered here: (*a*) What predictions are already possible? (*b*) How can damage from earthquakes be reduced on the basis of such predictions, given their limited accuracy? (*c*) What fundamental knowledge has been gained in earthquake prediction research? The common underlying question is, what comes next?

2. This problem is of urgent practical importance because earthquakes pose a rapidly growing threat to survival and sustainable development of our civilization. This is due to the well-known interrelated developments: proliferation of radioactive waste disposals, high dams, nuclear power plants, lifelines, and other objects whose damage poses an unacceptable risk; self-destruction of megacities; destabilization of the environment; and growing socio-economic volatility of the global village. For all of these reasons, seismic risk has escalated also in numerous regions of low seismicity. Today, a single earthquake may take up to a million lives, cause material damage up to $\$10^{12}$, raze a megacity, trigger a global economic depression, render a large territory uninhabitable, and destabilize the military balance in a region.

Earthquake prediction is necessary to undertake disaster preparedness measures, reducing the damage from the earthquakes. This requires that the accuracy of prediction be known, but, contrary to common belief, a timely prediction of low accuracy may be very useful.

0084-6597/02/0519-0001$14.00

3. Earthquake prediction is necessary also for fundamental understanding of the dynamics of the lithosphere, particularly in the timescales of 10^2 years and less. So far, this problem is in the same stage as the theory of gravity was between T. Brahe and J. Kepler: the study of heuristic regularities that are necessary to develop a fundamental theory.
4. Here we review the research that extends to formally define prediction algorithms and to their tests by advance prediction. Being a part of much broader efforts in earthquake prediction, this is presently most essential both for damage reduction and for understanding the lithosphere. Methodologically, this research integrates theoretical modeling and analysis of observations.
5. Why was this topic suggested for the prefatory chapter of the *Annual Review of Earth and Planetary Sciences*? First, as the reader will see, the earthquake prediction problem is connected, one way or another, with most of the solid Earth sciences, tying together an immense variety of field and processes in the wide range of time- and space scales. Algorithmic prediction, if successful, provides one of the major hopes for bringing internal order in that diversity of topics and methods. Indeed, since the times of Galileo, if not through the whole history of science, prediction has been a major tool of fundamental research, a source of heuristic constraints and hypotheses, and the fina test of theories.

Second, the earthquake prediction problem happens to be closely relevant to what I believe is a current frontier of the solid Earth sciences: emergence of a fundamental concept that would succeed plate tectonics, provide a fundamental base for prediction and (with luck) control of geological and geotechnical disasters, and establish links with "universal" scenarios of critical transitions in nonlinear (complex) systems.
6. The earthquake prediction realm still exhibits a striking gap in mutual awareness. This gap is amazingly large even for such a huge conglomerate of problems and professions; for earthquake preparedness it bodes ill—the chance to undertake preparedness measures should not be missed.
7. This gap might be partly due to the lack of a common language. To reach a possibly wider audience, I wrote this review in qualitative terms (retaining, I hope, a reasonable precision), although the studies considered here are entirely quantitative, with a substantial (to put it gently) mathematical component.
8. The following topics are covered by this review. (*a*) Structure of the earthquake-prone fault network: hierarchy of blocks and faults, and nucleation of earthquakes in mosaic nodes at the faults' intersections and junctions. (*b*) Fault networks as a stockpile of instability: a multitude of mechanisms that destabilize the strength-stress fiel and turn the network into a complex system with earthquakes for critical phenomena and the predictability of that system. (*c*) Prediction algorithms: premonitory seismicity patterns and the performance of the algorithms in advance prediction worldwide. (*d*) Error diagrams—a tool for validation of prediction methods. (*e*) Four paradigms in earthquake prediction: basic types of precursors,

long-range correlations in the fault network, partial similarity of precursors worldwide, and their dual origin—some precursors are common for many complex systems, others are Earth-specific (*f*) Earthquake prediction and earthquake preparedness. (*g*) Emerging possibilities, yet unexplored.

9. This paper is a preview to the monographic treatise on earthquake prediction, now in preparation, by a team from the International Institute of Earthquake Prediction Theory and Mathematical Geophysics, Russian Academy of Sciences. Most of the specifi finding discussed here were obtained in that Institute as well as in the following institutions: in France, the Institute of the Physics of the Earth (Paris) and Observatory of Nice; in the United States, Cornell and Purdue Universities, University of California, Los Angeles, University of Southern California, Massachussetts Institute of Technology, and U.S. Geological Survey; in Italy, the Abdus Salam International Center for Theoretical Physics, Universities of Rome ("La Sapienza") and Trieste.

INTRODUCTION

Seismicity and Geotectonics

Earthquakes occur in some parts of the outer shell of the solid Earth, called the lithosphere; its thickness ranges from a few kilometers near the mid-ocean ridges to a few hundred kilometers in certain continental regions. Below the lithosphere, down to a depth of 2900 km, lies Earth's mantle, which is partially melted in its upper 10^2 km.

Large-scale convection currents in the mantle, with characteristic velocities of centimeters per year, and internal processes within the lithosphere itself put the lithosphere in a state of permanent motion, highly irregular in space and time. Two major distinctive features of the lithosphere are responsible for the generation of earthquakes: (*a*) It is subject to fracturing. Exceptions are the soft soil at its very top, and the lower depths where a combination of stress, strength, and temperature is unfavorable for fracturing. (*b*) It is hierarchically divided into volumes ("blocks") driven by tectonic forces; they move relative to each other along the faults that separate them. In seismically active regions, earthquakes produce a significan part of the relative motion of the blocks. This makes seismicity an indelible component of the tectonic development of Earth.

An earthquake starts as an episode of rupture and discontinuous displacement in a certain part of a fault system, which becomes an earthquake source. These episodes alternate with slower deformations in the "stick-slip" sequences. Nearly a million earthquakes with magnitude 2 or more are registered each year worldwide. About a hundred of them cause considerable damage and once or twice in a decade a catastrophic earthquake occurs. Worldwide distribution of seismicity is shown in Figure 1. Epicenters of earthquakes are concentrated in seismic belts, encompassing tectonic plates—the major blocks comprising the lithosphere.

Stages of Prediction

Due to the multiscale nature and complexity of the processes expressed in seismicity, the problem of earthquake prediction consists of a consecutive step-by-step narrowing down of the time-space domain where a strong earthquake should be expected. Five major stages of earthquake prediction are commonly distinguished. The background stage provides "territorial" distribution of the recurrence time for destructive earthquakes of different magnitudes, up to the maximal possible one. The subsequent stages, fuzzily divided, include time-prediction; they differ in characteristic duration of alarms—the time intervals, where an earthquake is predicted (these intervals are indicated in brackets below): long-term (tens of years), intermediate-term (years), short-term (months to weeks), and immediate (days and less).

Data

The problem of earthquake prediction is particularly challenging because the bulk of the earthquake-prone lithosphere is inaccessible for the direct measurement of earthquake-related stress and strength fields The lithosphere is virtually a black box in a state of permanent activity. It generates a multitude of observable time-dependent fields which provide potential data sources for earthquake prediction. Prediction algorithms reviewed here diagnose the approach of a strong earthquake by analysis of one of such fields—seismi activity in a wide magnitude range.

FAULT NETWORK: HIERARCHICAL STRUCTURE

Almost all earthquakes occur in the fault network, delineating the major structural elements of the lithosphere (Keilis-Borok 1990a, Turcotte 1997): a hierarchy of blocks separated by boundary zones, with densely fractured nodes at junctions and intersections of these zones.

Blocks

The lithosphere is divided into a hierarchy of volumes, or blocks, that move relative to each other. The largest blocks are ten or so major tectonic plates of continental size. They include smaller blocks such as shields or mountain belts. After 15 to 20 consecutive divisions, we come to nearly 10^{25} grains of rocks

Boundary Zones

The blocks are separated by less rigid boundary zones, whose widths are 10–100 times smaller than the characteristic dimension of the blocks that they separate. The zones between the largest blocks are known as fault zones (10^2–10^1 km), then as faults (km to tens of meters), sliding surfaces (meters to centimeters), and, finall , as interfaces between the grains of rocks. Except at the lowest levels of hierarchy, each boundary zone has a similar hierarchical structure with more dense division: It consists of its own blocks, divided by boundary zones, etc. For brevity, we refer to all boundary zones as faults, unless distinction is necessary.

Nodes

Even more densely fractured mosaic structures, called nodes, are formed around the intersections and junctions of the faults. Their origin is due, roughly, to collision of the corners of the blocks (Gabrielov et al. 1996, King 1983). The formalized definitio of nodes is developed by Alekseevskaya et al. (1977).

Nodes are well known in structural geology and geomorphology and play a prominent textbook role in geological prospecting as attractors of mineral deposits. Their origin and strong impact on seismicity, sometimes overlooked in earthquake studies, are discussed below in the section Geometric Instability. The systems of boundary zones and nodes are called here the fault networks.

FAULT NETWORK: THE STOCKPILE OF INSTABILITY

The boundary layers of different rank, from the Circum Pacifi seismic belt, with the giant triple junctions for the nodes, to an interface between the grains of rocks, with the corners of the grains for the nodes, play a similar role in lithosphere dynamics. Specificall , although tectonic energy is stored in the whole volume of the lithosphere and well beneath, the tectonic energy release (through earthquakes and slow deformations) is to a large extent controlled by the processes in the relatively thin fault networks. This disparity is due to the following reasons.

First, as in any solid body, deformations and fracturing in the lithosphere are controlled by the {stress-strength} field At the same time, the strength of the fault network is weakened by denser fragmentation and higher permeability to fluid (compared to the blocks). For that reason, tectonic deformations, earthquakes included, are concentrated in the fault networks, whereas the blocks move more as a whole, with smaller internal deformations. In the timescale most relevant to the earthquake prediction problem, tens of years or less, a major part of lithosphere dynamics is realized through the deformation of the fault networks and relative movement of the blocks.

Second, the strength of the fault networks is not only smaller but also highly unstable, since it is sensitive to many processes there. Two types of instability coexist in the fault network: "physical," originated by a physical or chemical mechanism at the elementary (micro) level, and "geometric," controlled by the geometry of the fault network on a global (macro) level.

Physical Instability

The simplest source of instability is the abrupt triggering of an earthquake when the stress exceeds the strength on some segment of a fault. Some other sources of physical instability are described below.

RHEBINDER EFFECT, OR STRESS CORROSION

Description Many solid substances lose their strength when they come in contact with certain surface-active liquids. The liquid diminishes the surface tension and, consequently, the strength. When the strength drops, cracks may appear under a small stress, even gravitational stress might suffice This triggers expansion of fatigue: Liquid penetrates the cracks and they grow, with liquid propelling forward until it dissipates. This mechanism, called the Rhebinder effect (Gabrielov & Keilis-Borok 1983), requires very little energy to generate fracturing. It was discovered firs for metals and ceramics. Then, such combinations of solid substances and surface-active liquids were recognized among the common ingredients of the lithosphere, e.g., a basalt and sulphur solution. When they meet, a grid of cracks permeates the basalt, and the efficien strength may instantly drop by a factor 10 or more. Thus, the Rhebinder effect brings into the dynamics of the lithosphere a strong and specifi instability controlled by the stress fiel and by the geochemistry of the fluids

Relevance to prediction The weakened areas, where the cracks concentrate, may have only certain configurations depending on singularities of the stress field Examples of such configuration are shown in Figure 2. The thin lines show trajectories or cracks. Each heavy line is a separatrix, dividing different patterns of the cracks.

How would an observer of the lithosphere interpret Figure 2? Suppose the source of a flui appears in the place shown by the arrows. It will concentrate in the shaded areas and their strength will drop. A slight displacement of the source across a separatrix may divert the fatigue to a different place and completely change its configuration

This effect might explain many phenomena highly relevant to the development of a strong earthquake. For example, the diagrams show formation of weak links (four top panels) and asperities (two bottom panels). However, such explanations have at least two limitations. First, configuration shown in Figure 2 are the local ones. Inhomogeneity of stress and strength field and dissipation of fluid may prevent their formation in the scale of the observed premonitory patterns, which is up to hundreds of kilometers. More likely, these configuration are the elements that compose a nonlocal infrastructure of fatigue. Second, the Rhebinder effect is not a single mechanism of physical instability. Even the fluid alone may generate other equally strong mechanisms, such as the one described below.

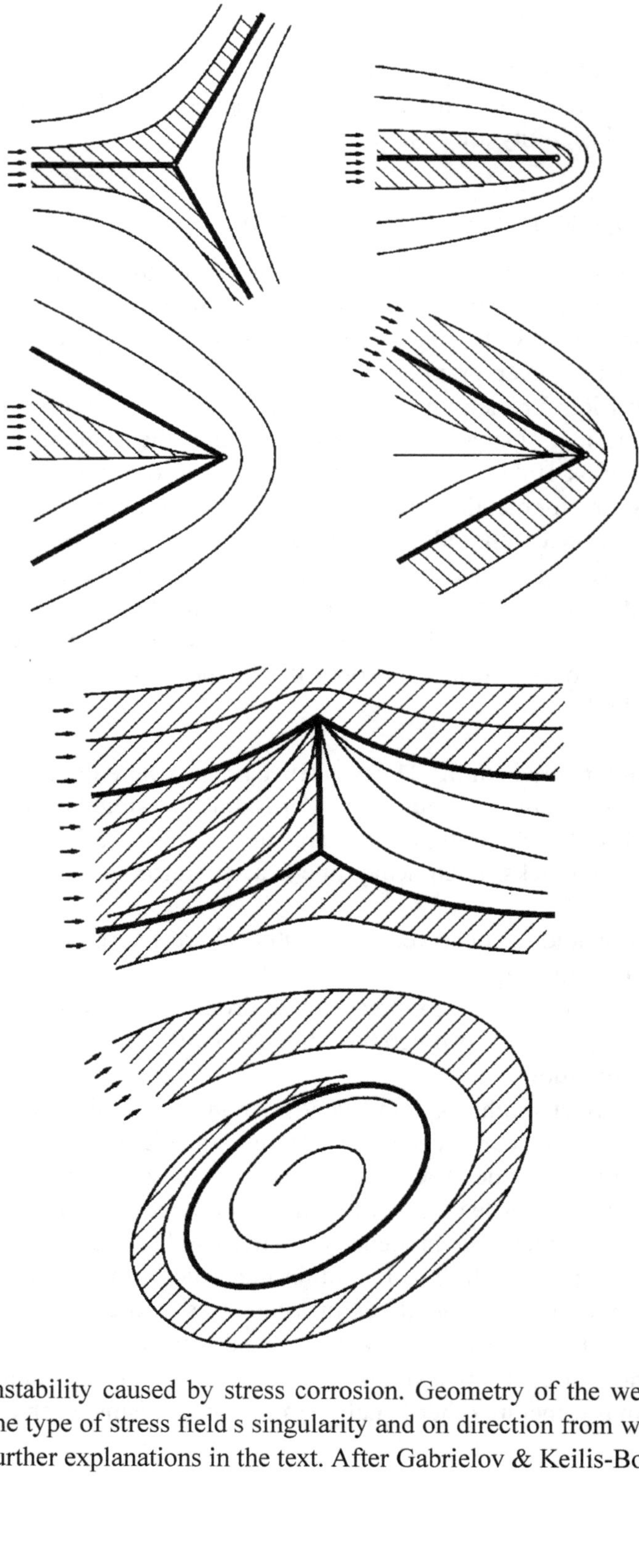

Figure 2 Instability caused by stress corrosion. Geometry of the weakened areas depends on the type of stress field s singularity and on direction from which the flui arrives; see further explanations in the text. After Gabrielov & Keilis-Borok 1983.

NONLINEAR FILTRATION

Description One of the competing mechanisms is filtratio of the fluid through the fault zones. Barenblatt et al. (1983) modeled this process as the relative movement of impermeable blocks separated by a porous layer. The latter is connected with a source of a flui that migrates along the gradient of pressure. The flui acts as the lubricator, decreasing the friction and triggering the episodes of fast movements—the "slips." If porosity is subcritical (below a certain threshold) the slip, once started, will increase the friction and self-decelerate. At most, the flui will trigger vacillating creep or a slow earthquake. However, if porosity exceeds a critical threshold, the slip will decrease the friction and the continuously forming microcracks will self-accelerate, grow, and merge at an escalating rate.

The porosity can be raised above the critical threshold by infiltratio of a flui itself; this will increase the tension and the pores will expand. Critical porosity propagates along the fault zone with finit velocity comparable at an observed velocity of migration of seismicity.

Relevance to prediction Nonlinear filtratio also explains many features of real seismicity, including the premonitory rise of seismic activity and of earthquake clustering. Moreover, it implies certain premonitory changes in the dynamics of flui and electromagnetic field However, the same limitations hold as in the case of stress corrosion. First, such instabilities may rise simultaneously within boundary zones of different rank and interact along the hierarchy, creating more complex patterns of filtration-generate instability. And second, this is not a single major source of physical instability.

OTHER MECHANISMS Further examples illustrate the diversity of mechanisms, causing the physical instability of fault networks.

"Fingers of fluids" These finger spring out at the front of the fluid migrating in a porous medium.

Dissolution of rocks Its impact is magnifie by the Riecke effect—increase of solubility of rocks with pressure. This effect leads to a mass transfer: Solid material is dissolved under high stress and carried out in solution along the stress gradient to areas of smaller stress where it precipitates.

Petrochemical transitions Some petrochemical transitions tie up or release the fluids as in formation or decomposition of serpentines, respectively. Others cause a rapid drop of density, such as in transformation of calcite into aragonite. This would create a vacuum and unlock the fault; the vacuum will be closed at once by hydrostatic pressure, but the rupture might be triggered.

Sensitivity of dynamic friction The sensitivity of dynamic friction to the local physical environment is another mechanism causing instability of the fault networks.

Mechanical processes Several mechanical processes, such as multiple fracturing, buckling, viscous fl w, etc., contribute to the instability of the fault networks.

Impact of pressure and temperature The above mechanisms are also influence by pressure and temperature.

This list is by no means complete. With such a diversity of mechanisms, physical instability of the fault network can hardly be attributed to any single mechanism alone.

Geometric Instability

The geometry of the fault network is often incompatible with tectonic movements. This leads to accumulation of stress, deformation, fracturing, and change of the fault geometry, jointly destabilizing the fault network. Two integral measures of that instability have been found by Gabrielov et al. (1996): (*a*) geometric incompatibility, concentrated in mosaic nodes, and (*b*) kinematic incompatibility spread over the whole lithosphere, including the blocks. Each measure estimates an integrated effect of tectonic movements in a wide range of timescales, from seismicity to neotectonics.

GEOMETRIC INCOMPATIBILITY

Description The phenomenon of geometric incompatibility is illustrated in Figure 3. It shows an intersection of two strike-slip faults separating the moving blocks. If the movements indicated by the arrows on Figure 3*a* could continue, the corners *A* and *C* would penetrate into each other and an intersection point would split into a parallelogram (Figure 3*b*). In reality, such a parallegram cannot be formed; instead, collision of the corners will trigger accumulation of stress and deformations near the intersection, followed by the fracturing and change of the fault geometry. Geometric incompatibility is a quantitative measure of this process. If the corners of the blocks tend to overlap (Figure 3*a*,*b*), the ensuing compression locks up the intersection. If all the corners tend to diverge (Figure 3*c*,*d*), the ensuing tension will unlock the intersection.

Formation of the nodes Accumulation of stress at the corners of tectonic blocks was firs described by McKenzie & Morgan (1969) for a triple junction. They found a condition under which a single junction "can retain its geometry as the plates move" so that the stress will not accumulate. King (1983) suggested that in a general case, when that condition is not satisfied the ensuing fracturing will not dissolve the stress accumulation, but will only redistribute it among newly formed corners. As a result, a hierarchy of progressively smaller and smaller faults is formed around an initial intersection; this is a node, recognizable by its densely mosaic structure, probably having self-similar fractal geometry.

Stokes-type theorem In reality, we encounter not a single intersection, as in Figure 3, but clusters of intersections in a node, and interacting nodes in a fault

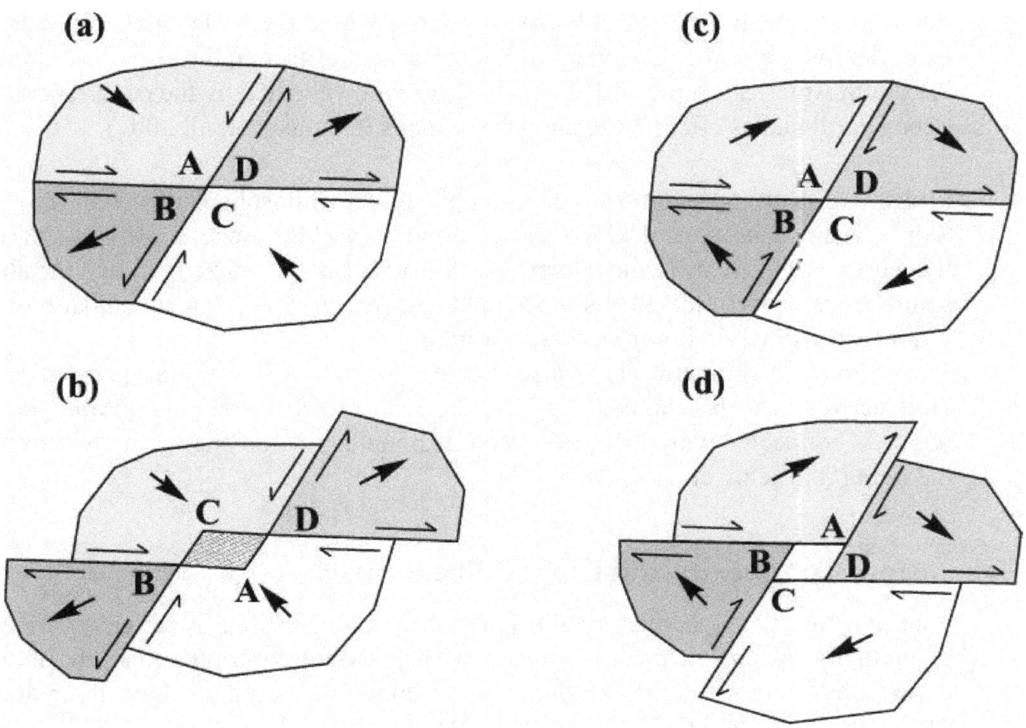

Figure 3 Geometric incompatibility near a single intersection of faults. *a*, *c*—initial position of the blocks; *b*, *d*—physically unrealizable extrapolation of initial movement. *a*, *b*—locked node; *c*, *d*—unlocked node. After Gabrielov et al. 1996.

network. Luckily, geometric incompatibility is additive. Gabrielov et al. (1996) found the analog of the Stokes theorem connecting the total geometric incompatibility of a fault network within a boundary with observations on that boundary. This allows one to estimate from outside geometric incompatibility in a complicated structure, such as an ensemble of nodes. This theorem reflect the fact that geometric incompatibility in different nodes is interdependent because the nodes are connected through the movements of blocks and on the faults. A strong earthquake in a node might change its incompatibility, thus affecting the generation of earthquakes in other nodes.

Such interplay is described by Prozorov & Shreider (1990). They observed that a strong earthquake is followed by "long-range aftershocks," a rise of seismic activity in the area where the next strong earthquake is going to occur within 10 years or so.

Earthquakes nucleation There is compelling evidence that strong earthquakes nucleate within the nodes, or more precisely, within certain specifi nodes that can be recognized by geological and geomorphologic criteria (Gelfand et al. 1976).

Such nodes have been mapped for many regions with different level of seismicity, from the Pyrenees, the Alps, and the Caucasus to California, the Himalayas, and the South American Andes. That analysis was well validated by the occurrence of strong earthquakes after publication of the maps (Gorshkov et al. 2001).

KINEMATIC INCOMPATIBILITY Let us apply to the lithosphere the well-known Saint Venant condition of kinematic compatibility (McKenzie & Parker 1967). It ensures that the relative movements on the faults can be realized through the absolute movements of the blocks separated by the faults. Kinematic incompatibility is a measure of deviation from that condition.

Kinematic incompatibility is also additive, its values in different parts of the fault network can be summed up, and the total kinematic incompatibility in a region is connected with observations on its boundary by an analog of the Stokes theorem (Gabrielov et al. 1996).

Conclusion: Complexity and Critical Phenomena

Summing up, the dynamics of the lithosphere is controlled by a wide variety of mutually dependent mechanisms, to a large extent concentrated in the fault network and interacting across and along its hierarchy. Each mechanism creates strong instability of the {strength-stress} field particularly of the strength. Except under very special circumstances, no single mechanism prevails so that the others can be neglected.

Even the primary element of the lithosphere—a grain of a rock—may act simultaneously as a material point; a visco-elastic body; an aggregate of crystals; and a source or absorber of energy, fluids and volume. Moreover, its body and surface are involved in quite different processes.

To assemble the set of corresponding equations is unrealistic and probably impossible: a maxim in nonlinear dynamics tells "one cannot understand a chaotic system by breaking it apart" (Crutchfiel et al. 1986). Rather, one may hope for a generalized theory (or at least a model) that directly represents the gross integrated behavior of the lithosphere. That brings us to the concept that ". . . in the timescale relevant to earthquake prediction problem, 10^2 years and less, the mechanisms destabilizing the strength of the fault network turn the lithosphere into a nonlinear hierarchical dissipative system" (Keilis-Borok 1990a). Strong earthquakes are critical phenomena in that system (Sornette & Sammis 1995, Turcotte 1997, Sornette 2000, Rundle et al. 2000).

Complex systems are not predictable with absolute precision. However, after a coarse-graining (i.e., averaging), in a not-too-detailed scale, premonitory phenomena emerge and a system becomes predictable, up to the limits (e.g., Gell-Mann 1994, Holland 1995). Accordingly, prediction of complex systems requires a holistic approach, "from the whole to details," in consecutive approximations, starting with the most robust coarse-graining of the processes considered.

GENERAL SCHEME OF PREDICTION

An Early Example

The firs prediction algorithm of the type considered here was introduced by Keilis-Borok & Malinovskaya (1964). Prediction of a strong earthquake was based on the rise of seismic activity in the medium magnitude range. The total area of rupture surfaces in the earthquake sources was chosen as a measure of seismic activity define by the function

$$\Sigma(t) = \sum_i 10^{Bm_i}, \quad m_i < M. \tag{1}$$

Here, m_i is the magnitude of i-th earthquake (a logarithmic measure of energy release) and M is the magnitude of a strong earthquake targeted for prediction. Summation is taken over the earthquakes that occurred within the sliding time window $(t - s, t)$ in the region considered. The value of parameter B was chosen from condition that each term under summation is coarsely proportional to the rupture area in a source.

The premonitory seismicity pattern Σ was diagnosed by the condition $\Sigma(t) \geq C_\Sigma$. The threshold C_Σ is self-adapting to a target magnitude M; it is proportional to the rupture area in the source of a single strong earthquake. The emergence of pattern Σ before 20 strong earthquakes worldwide was demonstrated by Keilis-Borok & Malinovskaya (1964). Figure 4 shows an example for the catastrophic in Assam earthquake in India in 1950, M = 8.6.

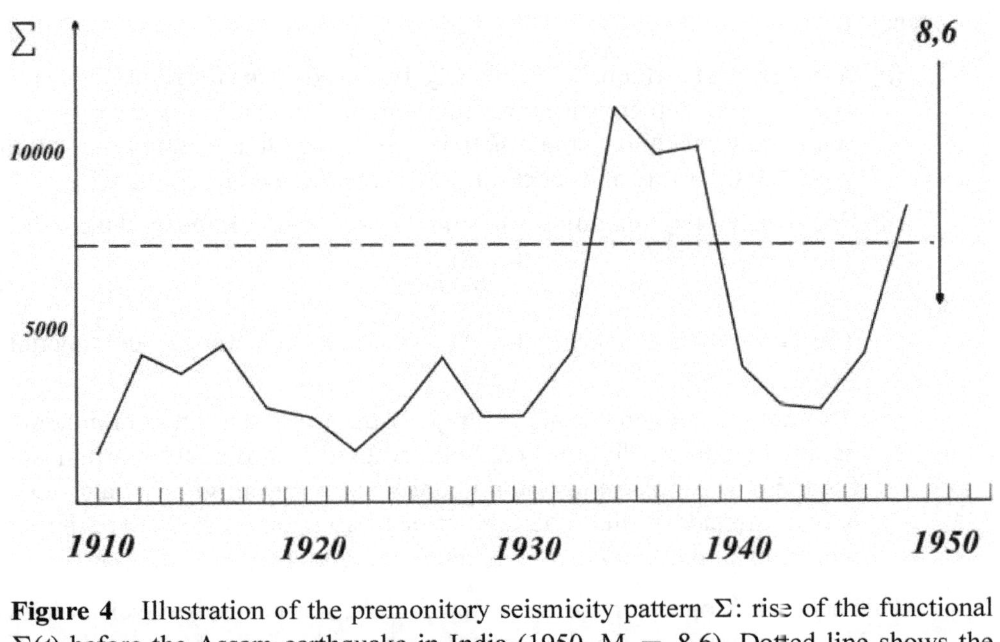

Figure 4 Illustration of the premonitory seismicity pattern Σ: rise of the functional $\Sigma(t)$ before the Assam earthquake in India (1950, M = 8.6). Dotted line shows the threshold C_Σ definin this pattern. After Keilis-Borok & Malinovskaya 1964.

Pattern Σ was the firs premonitory seismicity pattern that demonstrated the major features of the patterns discovered later: long-range correlations and similarity (see Four Paradigms below).

Formulation of the Problem

We consider prediction as a pattern recognition problem: Given is the dynamics of seismicity (and/or other relevant fields in a certain area prior to a moment t; to predict whether a strong earthquake will or will not occur within that area during the subsequent time interval $(t, t+\Delta)$; note that prediction may include reduction of the area.

In terms of pattern recognition, the "object of recognition" is a moment t. The problem is to recognize whether or not it belongs to the time interval Δ preceding a strong earthquake. That interval was named *TIP*, for the time of increased probability of a strong earthquake.

Pattern recognition of infrequent events (Gelfand et al. 1976, Keilis-Borok & Press 1980) happens to be very efficien in such an approach to prediction. That methodology has been developed by the school of I. Gelfand for the study of rare phenomena of highly complex origin, a situation where classical statistical methods are inapplicable.

The probabilistic side of prediction is reflecte in the rates of its errors.

Data Analysis

Under the approach described above, the data analysis comprises the following steps:

(*a*) A sequence of earthquakes is robustly described by the functionals $F_k(t)$, k = 1, 2, . . ., each depicting a certain premonitory seismicity pattern (Figure 5). With a few exceptions, the functions are define in a sliding time window $(t - s, t)$; the value of a function is attributed to the end of the window.

(*b*) Emergence of a premonitory seismicity pattern is define by the condition

$$F_k(t) \geq C_k. \tag{2}$$

The threshold C_k is usually define as a certain percentile of the functional F_k.

(*c*) An alarm is triggered when a single pattern or a certain combination of patterns emerges; different patterns are used in different algorithms (see Prediction). An alarm lasts for a certain time period; in some algorithms, it is terminated if and when the strong earthquake occurs. The possible outcomes of prediction are illustrated in Figure 6.

This scheme is open for the use of other data, not necessarily seismological ones. The key element in development of such an algorithm is obviously to fin the functionals $F_k(t)$ that provide a good prediction.

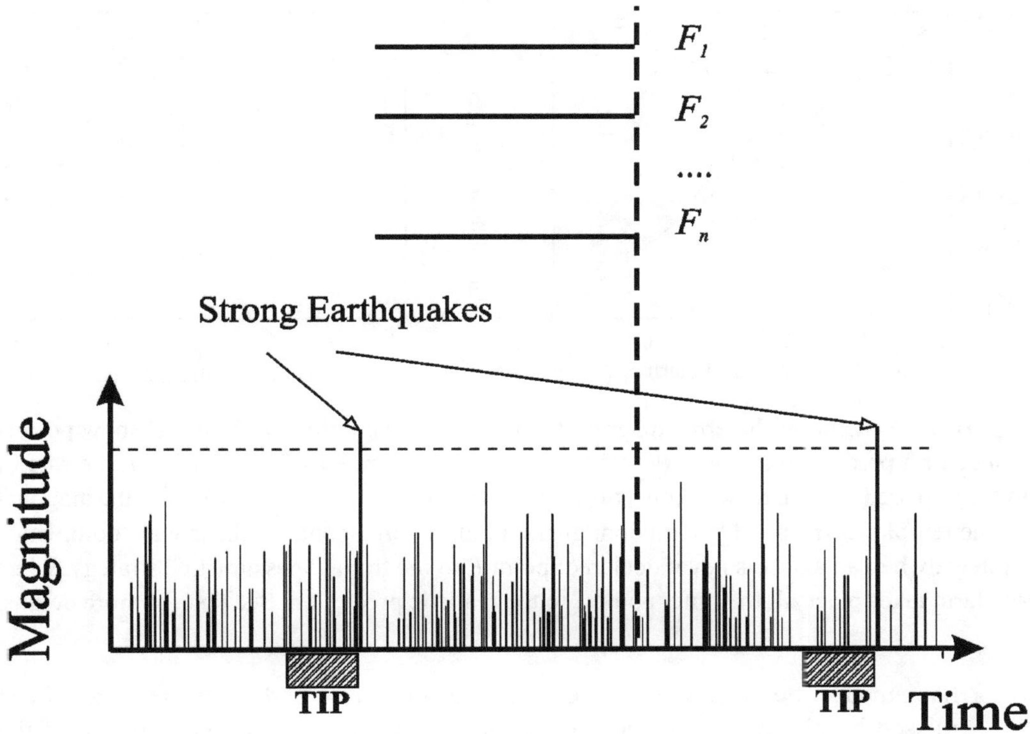

Figure 5 General scheme of prediction. See explanations in the text. After Keilis-Borok 1990a.

Error Diagrams

Error diagrams introduced in seismology by Molchan (1997) show the trade-off between different errors of prediction. They are pivotal in the development and validation of prediction methods, as well as in using predictions for enhancing earthquake preparedness.

DEFINITION Consider a prediction algorithm applied to a certain territory during the time period *T*. A certain number *A* of alarms is declared and A_f of them

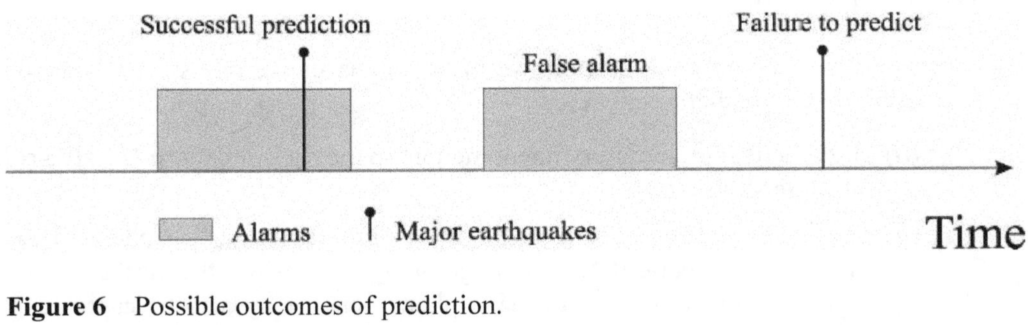

Figure 6 Possible outcomes of prediction.

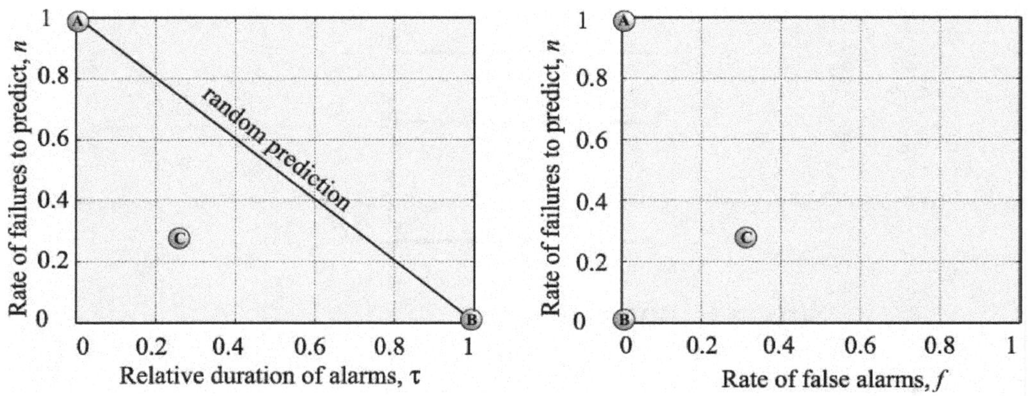

Figure 7 Scheme of the error diagram (Molchan, 1997). Points *A*, *B*, and *C* show performance of a prediction method: the trade-off between the rate of false alarms, *f*; the rate of failures to predict, *n*; and the relative time-space occupied by alarms, τ. Points on the diagonal on the left plot correspond to a random guess. Point *A* corresponds to the trivial "optimistic" strategy, when an alarm is never declared; point *B* to the trivial "pessimistic" strategy, when an alarm takes place all the time; point *C* to the realistic prediction. See text for more details.

happened to be false. *N* strong earthquakes occurred, and N_f of them have been missed by alarms. Altogether, the alarms cover the time *D*. Performance of the algorithm is characterized by three dimensionless parameters: the total relative duration of alarms, $\tau = D/T$; the rate of failures to predict, $n = N_f/N$; and the rate of false alarms, $f = A_f/A$.

The values of τ, *n*, and *f* are summed up on an error diagram, schematically illustrated in Figure 7. Different points correspond to different combinations of adjustable elements.

To validate an algorithm, an exhaustive set of numerical experiments is designed (e.g., Gelfand et al. 1976); they take a lion's share of effort in the development of an algorithm. A prediction algorithm makes sense only if its performance is (*a*) sufficientl better than a random guess, and (*b*) not too sensitive to variation of adjustable elements. An error diagram is a powerful tool for checking these conditions.

FOUR PARADIGMS

Summarizing the previous section, we discern consecutive milestones in the development of a prediction method.

(*a*) Hypothetical premonitory phenomenon; in the case of pattern Σ, it is a rise of the seismic activity.

(*b*) Qualitative definitio of a seismicity pattern capturing that phenomenon. In the case of pattern Σ, it is large area unlocked by the earthquakes in a medium magnitude range. That phenomenon may also be captured by other patterns; for example, by the number of earthquakes, not weighted by magnitude.

(*c*) The formal quantitative definitio of that pattern by a functional define on an earthquake sequence; this is $\Sigma(t)$ in the case of pattern Σ and $F_k(t)$ in the general scheme of prediction (see Figure 5).

(*d*) The prediction algorithm based on such functionals or on their combinations (see Data Analysis above).

The paradigms formulated here provide orientation for research in these directions. They have been found in the quest for premonitory seismicity patterns in the observed and modeled seismicity. There are compelling reasons to also apply them to premonitory phenomena in other relevant fields An overview of seismicity patterns and prediction algorithms is given in the two subsequent sections.

Paradigm I. Basic Types of Premonitory Phenomena

The approach of a strong earthquake is indicated by the following phenomena, reflectin the changes in the basic characteristics of seismicity: (*a*) rise of seismic activity, (*b*) rise of earthquake clustering in space and time, (*c*) rise of the earthquakes' correlation range, (*d*) transformation of frequency—magnitude relation, (*e*) rise of irregularity in space and time, (*f*) reversal of territorial distribution of seismicity, (*g*) rise of correlation between different components (decrease of dimensionality, and (*h*) rise of response to excitation.

The place of premonitory phenomena in the general scheme of earthquake prediction is discussed at the beginning of this section. The essence of these phenomena is illustrated in the next section by specifi seismicity patterns that capture them. Here, we give just a few explanations to make the description of this paradigm self-sustaining. Seismic activity (*a*) is the rate of earthquake occurrence; earthquake clustering (*b*) is the tendency of the earthquakes to occur closely in time and space; earthquakes' correlation range (*c*) is the distance within which the earthquakes do not occur independently; transformation of the frequency magnitude relation (*d*) favors relatively stronger earthquakes (see Figure 10); reversal of territorial distribution of seismicity (*f*) consists of the rise of activity in the areas, where average activity is relatively low, and vice versa; for the last two phenomena, (*g*) and (*h*), we so far have only the common general definition not yet specifie for the problem considered.

HOW THESE PHENOMENA HAVE BEEN FOUND Premonitory seismicity patterns of the firs two types, (*a*) and (*b*), have been found firs in the observations, and then in the models (Keilis-Borok 1990b, Keilis-Borok & Shebalin 1999, Gabrielov et al. 2000); patterns of the next three types, (*c*)–(*e*), have been found in the reverse order, firs on the models (Gabrielov et al. 2000, Newman et al. 1995, Sornette 2000), and then in observations; reversal of territorial distribution of seismicity (*f*) has been found in observations (Shebalin & Keilis-Borok 1999) and is not explored on the models yet; the last two phenomena remain purely hypothetical. Patterns of the firs two types—rise of intensity and clustering—have been validated by statistically significan predictions of real earthquakes (Molchan et al. 1990, Keilis-Borok & Shebalin 1999).

Paradigm II. Long-Range Correlations

The generation of an earthquake is not localized around its future source. A sequence of earthquakes is generated by a fault network, rather than each earthquake being generated by a segment of a single fault. Accordingly, the premonitory signals of an approaching earthquake come not only from a narrow vicinity of the incipient source but from a much wider area.

SIZE OF THE AREAS WHERE PREMONITORY PHENOMENA ARE FORMED Let M be the magnitude of a strong earthquake, and $L(M)$ the characteristic length of its source. At the intermediate-term stage of prediction (in the timescale of years), premonitory phenomena may be formed within distance $10L(M)$ from the source; it might be reduced down to $3L$–L in a second approximation (Kossobokov et al. 1990). At the long-term stage, in the timescale of tens of years, that distance reaches approximately 100 L. For example, according to Press & Allen (1995), the Parkfield CA, earthquake, with M close to 6, $L \approx 10$ km, "... is not likely to occur until activity picks up in the Great Basin or the Gulf of California," nearly 800 km away.

HISTORICAL PERSPECTIVE An early and probably the firs estimation of the area where premonitory patterns are formed was obtained for pattern Σ (See An Early Example). Table 1 shows similar estimations for other intermediate-term patterns. It is noteworthy that Charles Richter, who was generally skeptical about the

TABLE 1 Linear size of the regions, where premonitory phenomena are observed

Size, $d(M)$	Measure/algorithm	Reference[b]
Up to $10L$[a]	Total area of sources	Keilis-Borok & Malinovskaya 1964
$3L$–$5L$	CN	Keilis-Borok & Rotwain 1990
$5L$–$10L$	M8	Keilis-Borok & Kossobokov 1990
L–$3L$	M8 & MSc	Kossobokov et al. 1990
$\sim 5L$[c]	Benioff strain release	Bufe & Varnes 1993, Bowman et al. 1998
$\sim 100L$	Seismic activity	Press & Allen 1995

Notes: (*a*) L is the linear size of rupture in the source of an approaching strong earthquake. (*b*) References are given to the recent (known to the authors) publications rather than to original ones. (*c*) The last two rows are related to timescale of tens of years.

feasibility of earthquake prediction, made exception to the pattern Σ specificall because it was based on long-range correlations. He wrote (Richter 1964): "... It is important that [the authors] confir the necessity of considering a very extensive region including the center of the approaching event. It is very rarely true that the major event is preceded by increasing activity in its immediate vicinity."

MECHANISMS OF LONG-RANGE CORRELATIONS To explain the long-range correlations, several mechanisms (not mutually exclusive) have been suggested. They may be divided into two groups.

1. Some explanations attribute long-range correlations to a large-scale process controlling stress and strength in the lithosphere. Among such processes are microrotation of tectonic plates (Press & Allen 1995); interaction of crustal blocks (Soloviev & Vorobieva 1999); microfluctuatio in the direction of mantle currents (Keilis-Borok & Shebalin 1999); migration of pore fluid in fault systems (Barenblatt et al. 1983); hydrodynamic waves in the upper mantle, triggering strong earthquakes (Romanowicz 1993); perturbations of the ductile layer beneath the seismically active zone (Aki 1996); inelasticity and inhomogeneity of lithosphere, etc. Such mechanisms would act under different circumstances, separately or jointly. Altogether, they make long-range correlations inevitable.
2. In another approach, the lithosphere is regarded as a nonlinear chaotic or complex system; then the long-range correlations are again inevitable, as a general feature of chaotic systems in a near-critical state (Keilis-Borok 1990a, Turcotte 1997, Newman et al. 1994, Bowman et al. 1998, Rundle et al. 2000, Sornette 2000).

Paradigm III. Similarity

Premonitory phenomena are similar (identical after normalization) in extremely diverse environments and in a broad energy range. The similarity is not unlimited, however, and regional variations of premonitory phenomena do emerge.

Premonitory seismicity patterns happen to retain a prediction power through a large variety of environments: microfractures in laboratory samples; induced seismicity; earthquakes in subduction zones, major strike-slip fault zones, rift zones, and finall platforms (Keilis-Borok & Shebalin 1999). The corresponding seismic energy release ranges from ergs to 10^{25} ergs. However, the performance of prediction algorithms does vary from case to case.

An opportunity to explore, albeit qualitatively, the further frontiers of similarity is provided by registration of the 111 flashe of energy radiated from the neutron star with celestial coordinates 1806–20 (Kossobokov et al. 2000). These flashe are attributed to the "starquakes," i.e., fractures in the neutron star crust. Starquakes and earthquakes originate in exceedingly different environments: Earth's crust is composed of grains of rocks whereas the crust of the neutron star is a lattice of heavy nuclei; the radius of Earth and the neutron star is 6731 km and 10 km, respectively; density is $5.5 \cdot 10^6$ g/m^3 and 10^{14} g/m^3; energy release during a quake is up to 10^{26} ergs and 10^{46} ergs.

Figure 8 compares the emergence of premonitory patterns before a major starquake and an earthquake in the Aquaba Gulf, on December 11, 1995, $M = 7.3$. The functionals capturing these patterns are taken from the earthquake prediction algorithms (see Prediction, below): functionals Σ, N, Z, and L capture the intensity of earthquakes fl w; B captures the earthquakes' clustering.

The patterns work in both cases, their fantastic difference notwithstanding. Only the timescale for the starquakes had to be readjusted a posteriori.

Paradigm IV. Dual Nature of Premonitory Phenomena

Some of the premonitory phenomena are "universal," common for hierarchical complex nonlinear systems of different origin; others are specifi to geometry of the fault network, or to a certain physical mechanism controlling the {strength-stress} fiel in the lithosphere.

UNIVERSAL PREMONITORY PHENOMENA The known premonitory seismicity patterns happen to be universal in the sense that they emerge in the lattice models, not specifi to Earth only; accordingly, one may expect them in systems of quite different origin. Such models are developed in statistical physics and nonlinear dynamics. The most complete set of premonitory seismicity patterns was reproduced by the model of colliding cascades (Gabrielov et al. 2000, Zaliapin et al. 2001). A premonitory increase of correlation range was firs found in this model and then, with the same definition in the observed seismicity.

EARTH-SPECIFIC PREMONITORY PHENOMENA Phenomena of this kind have not yet been define in a clear-cut way to be incorporated directly in the prediction algorithms. Here we hypothesize what such phenomena might be.

Reversals of geometric incompatibility might produce asperities, relaxation barriers, weakest links, alternation of seismicity, and creep ("loud" and "silent" earthquakes)—phenomena that are highly relevant to prediction. If this conjecture is correct, such phenomena would migrate from node to node with the velocity typical for migration of seismicity: tens to hundreds km/year.

Kinematic incompatibility is relevant to prediction because it captures the stress accumulation in the fault network, blocks included.

PREDICTION

Here we provide an overview of prediction algorithms based on premonitory seismicity patterns. Specificall , we consider algorithms for intermediate-term prediction, which are currently at different stages of validation. There is a large family of such algorithms, and an even larger one of individual premonitory patterns.

The algorithms considered are based on the same general scheme of data analysis and the same underlying concepts. These common elements are described in previous sections.

Premonitory Seismicity Patterns

Each of the premonitory phenomena listed in Four Paradigms (Paradigm I) is captured by different seismicity patterns described here. These patterns are formed by the earthquakes within a given territory, magnitude range, and time interval; for most of the patterns, aftershocks are eliminated from consideration. Earthquake energy E and size of a seismic source are roughly estimated from the magnitude.

RISE OF SEISMIC ACTIVITY

Total area of seismic sources, or pattern Σ The predictive power of pattern Σ (see An Early Example above) has been demonstrated for many regions of the world (Keilis-Borok & Malinovskaya 1964, Keilis-Borok et al. 1980). The area of a source is assumed to be proportional to $E^{2/3}$.

Active zone size (AZS) The *AZS* captures the size of the area where seismic activity has risen. It was found in a model by Pepke et al. (1994) and then in the observed seismicity of California by Kossobokov & Carlson (1995). That pattern also reflect the premonitory rise of correlation range, discussed below.

Benioff strain release (ε) ε captures the weighted number of earthquakes, with the weights proportional to $E^{1/2}$ (Bowman et al. 1998, Bufe & Varnes 1993). Many recent studies advocate a power growth of the cumulative Benioff strain release prior to a strong earthquake.

Number of earthquakes (N) N captures a simple measure of seismic activity—the number of earthquakes in a certain magnitude range, without weighting them according to magnitude. Knopoff et al. (1996) explored N's individual performance for California. A particular feature of this measure is that it does not depend much on the relatively stronger and less numerous earthquakes.

Concentration of the main shocks in space (Z) Z is estimated as the ratio of the average diameter the earthquake sources to the average distance between them. This pattern was firs found in the classical laboratory experiments of Zhurkov (1968).

Deviation of seismic activity from the long-term trend (L) *L* captures the increasing rate of the growth of seismic activity.

Time interval during which a given number of earthquakes occurred (R) A rise of activity is obviously indicated by a decrease of *R* (Shreider 1999). An unusual feature of that pattern is that count of time is replaced by count of earthquakes, as in the studies of acoustic emission.

Note on seismic quiescence Several studies report premonitory decrease of seismic activity (Shreider 1999, Wyss & Habermann 1987). This is not a contradiction because the rise and fall of activity may take place in different spatio-temporal scales (e.g., see Kossobokov et al. 1990) or succeed each other. For example, switching from low to high activity is one of the patterns used in the composite algorithm CN (Keilis-Borok & Rotwain 1990).

RISE OF EARTHQUAKE'S CLUSTERING

Burst of aftershocks or "Pattern B" Pattern *B* is captured by a main shock in a medium magnitude range with a large number of aftershocks in the firs few days. Molchan et al. (1990) established high statistical significanc of prediction by this pattern for 13 regions worldwide.

Swarm (S) *S* is a space-time cluster of main shocks with nearly the same magnitudes. Predictive power of this pattern is demonstrated for Italy, California and Nevada, Japan, and New Zealand (Keilis-Borok et al. 1980, Caputo et al. 1983).

Pattern BG This pattern is a cluster of earthquakes, aftershocks included, regardless of their magnitudes (Gabrielov et al. 1983). The remarkable feature of this pattern is that the magnitude should be known only for strong earthquakes, targeted for prediction. (Pattern *BG* was originally designed to use a regional catalog, containing only macroseismic intensities but not magnitudes of earthquakes.)

RISE OF EARTHQUAKE CORRELATION RANGE Premonitory seismicity patterns of this type recently attracted wide interest, and we give them a little more attention here.

Long-range aftershocks (LRA) *LRA* capture the earthquakes that occur immediately after a strong main shock at a large distance from the latter. According to Prozorov & Shreider (1990), such earthquakes form a long-term premonitory seismicity pattern: They occur near the epicenter of a future strong earthquake, to be expected within 10 years or so.

The two following patterns were found firs in synthetic seismicity generated by the Colliding Cascades model (Gabrielov et al. 2000) and then in observations.

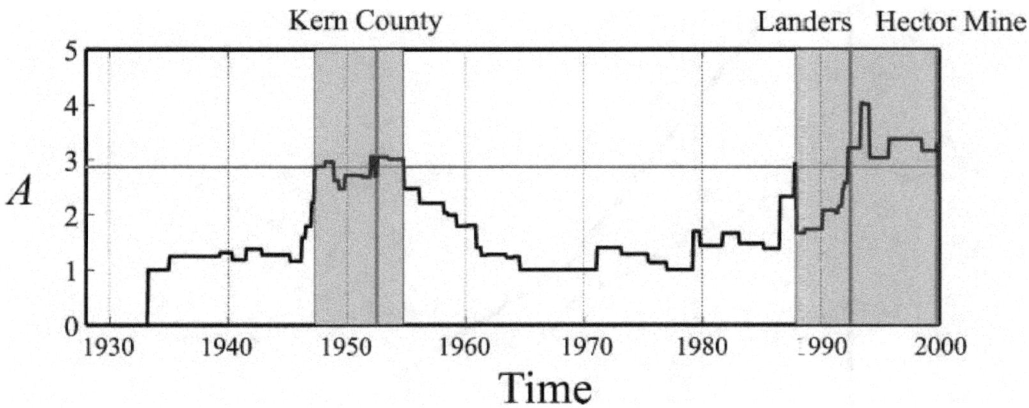

Figure 9 Performance of premonitory pattern Accord in Southern California (retrospective analysis). Function *A*(*t*) depicts emergence of the pattern Accord. The alarms are shown in gray; the horizontal black line represents the threshold for declaring an alarm; vertical black lines show times of the three largest earthquakes. Function *A* obviously increases prior to these earthquakes. This result is stable to variations of adjustable elements of the prediction algorithm.

Pattern "Accord" This pattern is a nearly simultaneous activation of several major fault zones in the region, hence its name. Pattern *Accord* links the geometry of the fault network with universal (not Earth-specific features of seismicity. Figure 9 shows its highly encouraging, albeit retrospective, application to Southern California (Zaliapin et al. 2000). That region comprises seven commonly recognized major fault zones. Seismic activity in each zone was measured by the total area of ruptures in seismic sources (as in pattern Σ). Scaled at predicting the earthquakes with $M > 7.5$, the pattern *Accord* emerges prior to the three largest earthquakes of the region: Kern County, in 1952; Landers, in 1992; and Hector Mine, in 1999, and at no other times. Rescaled at the lower target magnitudes, $6.5 \leq M < 7.5$, the pattern precedes six out of nine strong earthquakes, with the total duration of alarms 28% of the time considered. Predictions yield acceptable success-to-failure scores in a wide range of adjustable elements of the algorithm.

Pattern ROC (radius of correlation) Pattern *ROC* captures the nearly simultaneous occurrence of medium magnitude earthquakes at a large distance. Its successful (also retrospective) application to Lesser Antilles is described by Shebalin et al. (2000). A promising feature of this pattern is a higher accuracy of time-prediction; the duration of alarm is only a few months.

Distribution of links The distribution of links in a single-link cluster that connects epicenters of consecutive earthquakes was explored for seismicity of Southern California by Zoeller et al. (2001).

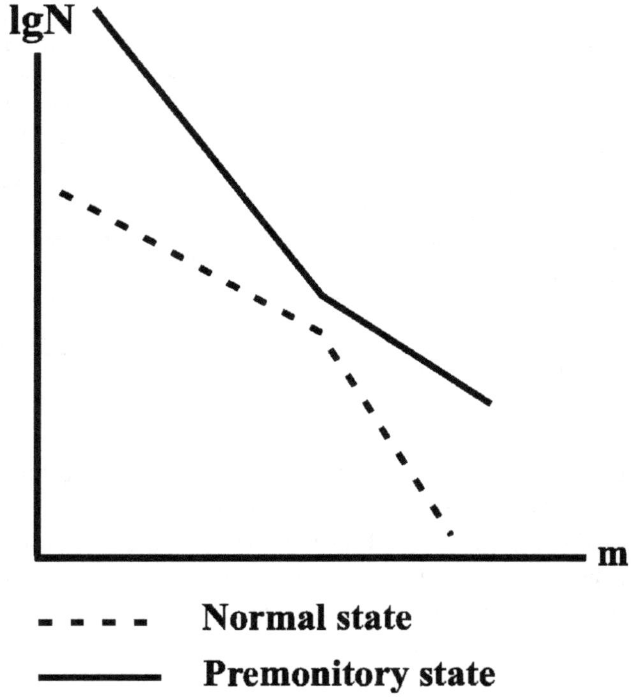

Figure 10 Premonitory changes of the Gutenberg-Richter relation. N is the occurrence rate of earthquakes, m is the magnitude (logarithmic measure of energy). The dashed line represents time far from a strong earthquake; the solid line represents time close to it.

CHANGE OF THE MAGNITUDE DISTRIBUTION The relation between the occurence rate N and the energy E of earthquakes is a fundamental feature of the dynamics of seismicity commonly known as the Gutenberg-Richter relation. It is often regarded as a power-law, which is actually an approximation valid only after a substantial averaging over time and territory, and within a limited energy range. Considering that relation in consecutive time intervals, one might see temporal variations preceding a strong earthquake. Two premonitory patterns of that type are schematically shown in Figure 10.

Upward bend (U) U captures the upward bend of the right-hand end of the $N(m)$ curve. Such a bend shows that relatively strong earthquakes gained a larger share of the whole seismicity. This pattern was firs reported in laboratory experiments. Narkunskaya & Shirman (1994) analyzed the pattern in seismicity generated by a lattice-type hierarchical model of fracture development, and, in less detail, in observations.

Change of convexity (γ) This was firs found in laboratory experiments with steel samples and then in observations for Southern California (Rotwain et al. 1997).

IRREGULARITY OF THE EARTHQUAKE SEQUENCE

Variation of magnitudes This is one of the patterns used jointly in algorithm SSE for predicting the second strong earthquake in a pair (see Composite Prediction Algorithms).

REVERSAL OF TERRITORIAL DISTRIBUTION OF SEISMICITY

Seismic reversal (SR) *SR* is the only pattern of this type so far suggested. It consists of the rise of activity on the faults, where average seismicity is relatively low, and, vice versa, drop of activity on the relatively active faults. This pattern was found in seismicity of Lesser Antilles (Shebalin & Keilis-Borok 1999).

CONCLUSION The above list is by no means complete, but well illustrates the multitude of reproducible premonitory seismicity patterns. Why should we consider such a multitude for earthquake prediction?

First, a set of patterns, even if correlated by definition is more reliable than a single one, due to the complex nature of seismicity and unavoidable noise (Zaliapin et al. 2001). Second, a still unknown scenario of the development of a strong earthquake might include the advent of several independent phenomena. In that case, regardless of errors, an ensemble of phenomena might yield a better prediction. For these reasons, the composite prediction algorithms, using several premonitory patterns, are designed.

Composite Prediction Algorithms

Composite algorithms use a combination of individual precursors. Here we describe four such algorithms, relatively better validated by advance prediction in many regions worldwide.

ALGORITHM M8 This algorithm was designed by retrospective analysis of seismicity preceding the greatest ($M \geq 8$) earthquakes worldwide, hence its name. We describe it using as an example the recent prediction of major earthquake in Southern Sumatera, Indonesia, on June 4, 2000, $M = 8.0$. That prediction was part of the Russian-American experiment in advance prediction of strongest earthquakes worldwide. Figure 11 illustrates how this experiment works. The area targeted for prediction covers 80% of the major seismic belts—the territory for which earthquake catalog is sufficientl complete to apply the algorithm. This territory is scanned by overlapping circles with diameter of $12° \approx 1333$ km. Prediction is made separately for each circle. Figure 11 shows alarms issued in January 2000, to be updated in six months.

The algorithm for issuing alarms is illustrated in Figure 12. It shows a case history of prediction of the Southern Sumatera earthquake. Three circles close to the epicenter of that earthquake are shown in Figure 12. Within each circle several characteristics of seismicity are estimated. Three characteristics, number of earthquakes (N), concentration of sources (Z), and deviation from long-term

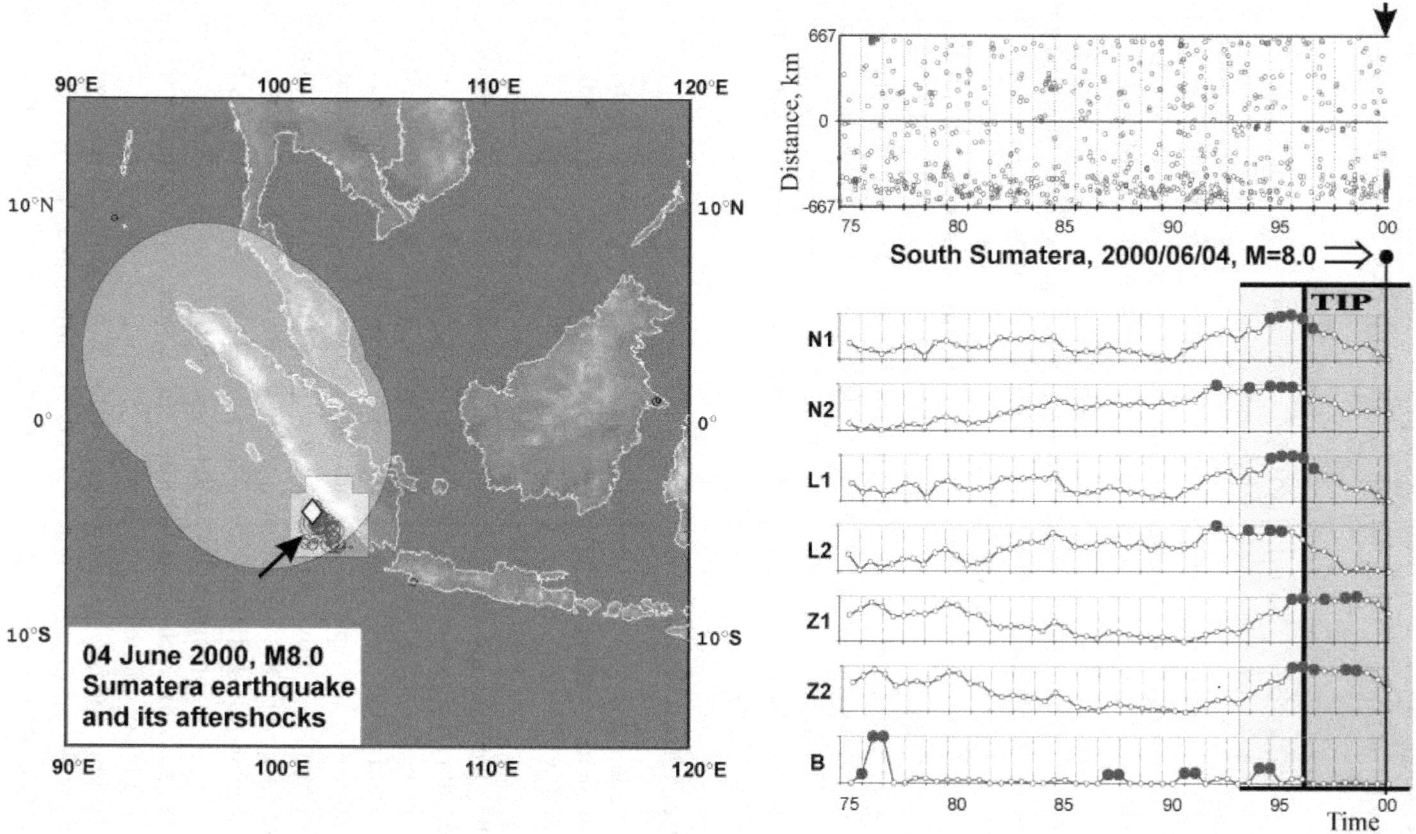

Figure 12 Advance prediction of the Southern Sumatera earthquake, June 4, 2000, M = 8.0. *Large light circles*—M8 alarms; *rectangular area*—MSc alarm; *black circles*—epicenters of the Sumatera earthquake and its aftershocks; *white diamond*—epicenter of a large foreshock, M = 4.7. Seismicity preceding the earthquake and functionals used for prediction are shown at the right; arrows show the time of the major earthquake. After Romashkova & Kossobokov 2001.

trend (*L*), depict seismic activity averaged over a large time window (six years); each is computed in two magnitude intervals to make description of seismicity more complete. One characteristic (*B*) depicts earthquake clustering. An alarm, or TIP, is issued for fi e years when clustering, and at least fi e other characteristics become "very large" at approximately the same time.

The alarm shown in Figure 12 appeared as early as July 1, 1996, with expiration on June 30, 2001. Four years after the onset of that alarm, on June 4, 2000, an earthquake of magnitude 8.0 occurred in S. Sumatera; its epicenter is shown on Figure 12 by a black circle; smaller circles show the aftershocks, roughly outlining the earthquake source. We see that the prediction was correct: This earthquake occurred within the territory and time-period covered by the alarm. The polygon in Figure 12 shows the smaller area of alarm; it is determined in a second approximation by the algorithm MSc (see below).

ALGORITHM MSc ("MENDOCINO SCENARIO") This algorithm was developed by retrospective analysis of seismicity prior to the Eureka earthquake (1980, M = 7.2) near Cape Mendocino in California, hence its name (Kossobokov et al. 1990). It is aimed at the reduction of the territory where a strong earthquake is already predicted by some other algorithm. Seismic activity in the territory of the original alarm is analyzed in more details. The alarm is confine to the area where seismicity is high but irregular, interrupted by short intervals of quiescence.

ALGORITHM CN The algorithm CN was developed (Keilis-Borok & Rotwain 1990) by retrospective analysis of seismicity preceding the earthquakes with M $\geq$ 6.5 in California and the adjacent part of Nevada, hence its name. In that algorithm, seismicity is described by nine functionals, depicting seismic activity, its temporal variation, and bursts of aftershocks. Only large, medium, and small values of each function are distinguished. Qualitatively, an alarm starts when earthquake clustering is high; seismicity is irregular, high, and growing; and the increase of seismicity was preceded by quiescence.

ALGORITHM SSE (SECOND STRONG EARTHQUAKE) This algorithm was developed by retrospective analysis of 21 strong earthquakes in California and Nevada (Vorobieva & Levshina 1994). It is aimed at the following problem: A strong earthquake with magnitude M occurred and the beginning of its aftershock sequence is given; to predict whether a second strong earthquake, with magnitude ($M - 1$) or larger, will occur within a certain distance $R(M)$ from the firs one during T months thereafter (in practice $T = 18$ months).

Prediction is based on several characteristics of the aftershocks' sequence, depicting its intensity, irregularity, and territorial spreading. In addition, the algorithm uses activity preceding the firs earthquake. This algorithm is applied to the nine regions shown in Figure 13. The choice of regions is restricted for the following reasons: First, the algorithm does not succeed in Circum-Pacifi subduction zones, reminding us that the similarity of premonitory phenomena is not unlimited; second, in many regions, the available earthquake catalogs are insufficientl complete.

Case history The firs strong earthquake in that case was the Landers earthquake in Southern California, in 1992, M = 7.6. The map in Figure 14 shows the epicenters of this earthquake and its aftershocks. Algorithm SSE predicted that a second strong earthquake, M ≥ 6.6, would occur within 18 months after the Landers earthquake at the distance 200 km from its epicenter (Levshina & Vorobieva 1992). The corresponding alarm expired on December 28, 1993. The Northridge earthquake with M = 6.8 occurred within the above distance on January 17, 1994,

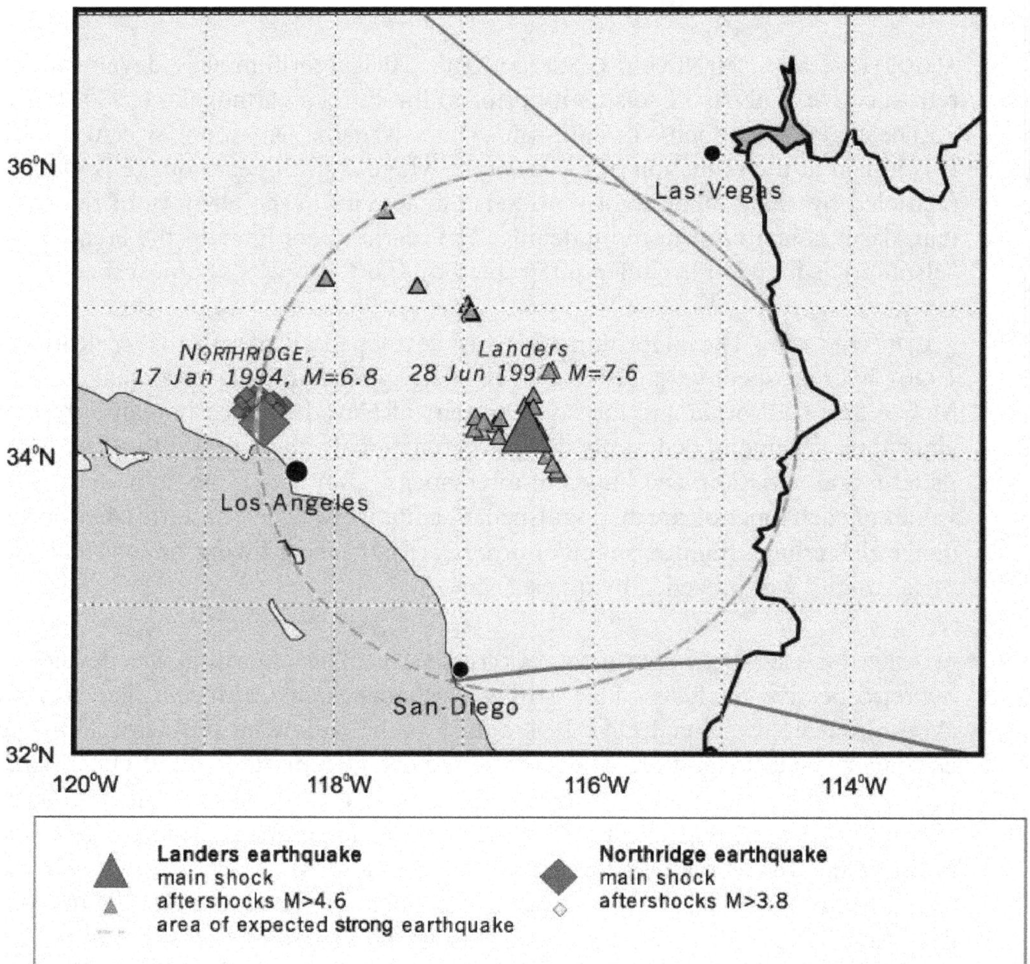

Figure 14 Advance prediction of the Northridge earthquake made by algorithm SSE. See details in the text. After Levshina & Verobieva 1992, Vorobieva 1999.

19 days after the expiration of the alarm; accordingly, this prediction is counted as a false alarm in the test of the algorithm. The algorithm SSE was then applied to the Northridge earthquake and generated no alarm. That prediction happened to be correct: no strong earthquakes occurred during the 18 months following the Northridge earthquake.

Importance Prediction of a second strong earthquake is specially important for several reasons: (*a*) The firs earthquake can destabilize high-risk constructions and natural sources of hazard, like mountain slopes, glaciers, etc. (*b*) Disaster management services are already on the spot and need to know whether to relax or maintain safety measures (e.g., whether to let people return to their homes). (*c*) Public expectations are higher than for prediction in general: Failure to predict the firs earthquake might be met with fatalism, but once it has happened, attention is riveted on the scientifi community.

Advanced Prediction Worldwide

Here we describe the results of testing the composite algorithms, M8, MSc, and SSE, by advance prediction. We consider these algorithms because testing is continuing; recently, the predictions have been file on a web site. The complete track record of the predictions, errors included, are available at http://www.mitp.ru or http://www.phys.ualberta.ca/mirrors/mitp.

ALGORITHMS M8 AND MSc The performance of algorithms M8 and MSc (Kossobokov et al. 1999) in predicting major earthquakes, with magnitude 8.0+ worldwide, is summed up in Table 2. Algorithm M8 successfully predicted all six strong earthquakes that occurred from 1992–2000. Algorithm MSc predicted fi e more accurately and missed one.

ALGORITHM SSE The performance of algorithm SSE (Vorobieva 1999) is summed up in Table 3. Advance prediction started in 1989. Up to the year 1998, 20 strong

TABLE 2 Performance of the prediction algorithms M8 and MSc for earthquakes of magnitude 8.0 and more. After (Kossobokov et al. 1999)*

Test period	Strong earthquakes	Predicted by		Space-time covered by alarms, %	
		M8	MSc	M8	MSc
1985–2000	8	8	7	34.9	18.0
1992–2000	6	6	5	30.2	15.3

*Data for the period 1997–2000 are taken from http://www.mitp.ru (mirror at http://www.phys.ualberta.ca/mirrors/mitp).

TABLE 3 Performance of the prediction algorithm SSE. After (Vorobieva 1999)

PREDICTIONS:		Did a subsequent strong earthquake occur?			
		YES		NO	
Test period	**Strong earthquakes**	**Total**	**Correct**	**Total**	**Correct**
1989–1998	20	6	4	14	13

earthquakes occurred and 6 of them were followed by a second strong earthquake; the algorithm made 17 correct predictions and 3 mistakes.

EARTHQUAKE PREDICTION AND EARTHQUAKE PREPAREDNESS

Currently realistic earthquake predictions have limited accuracy. How can one use them to reduce damage from the earthquakes? The key is to escalate or de-escalate preparedness depending on (*a*) the content of a current alarm (what is predicted and where); (*b*) the probability of a false alarm; and (*c*) the cost/benefi ratio of disaster-preparedness measures. Such is the standard practice in preparedness for all disasters, war included. A costly mistake—that only a precise short-term prediction is practically useful (besides estimation of seismic hazard)—emerges from time to time in the seismological literature. Actually, as in the case of defense, prediction may be useful if its accuracy is known, even if it is not high.

Diversity of Damage

Earthquakes may destroy buildings, lifelines, etc.; trigger fires release toxic, radioactive, and genetically active materials; and trigger floods avalanches, land-slides, tsunamis, etc. Equally dangerous are the socio-economic and political consequences of earthquakes: disruption of vital services—supply, medical, fi nancial, law enforcement, etc.; epidemics; of production and unemployment; military destabilization; profiteering crime, and disruptive anxiety of population. The socio-economic consequences may be inflicte also by the undue release of predictions.

Disaster Preparedness Measures

Such diversity of damage requires a hierarchy of disaster-preparedness measures, from building codes and insurance to mobilization of post-disaster services to a red alert. It takes different times, from decades to seconds, to undertake different measures; having different costs, they can be maintained for different time periods.

Accordingly, no single stage can replace another one for damage reduction, and no single measure is sufficien alone.

Many important measures do not require a particularly precise prediction. An example is the 1994 Northridge earthquake, California, mentioned above, which caused economic damage on the order of $30 billion. Its prediction, published well in advance (Levshina & Vorobieva 1992), was not precise: The alarm covered an area 400 km in diameter and time period of 18 months. However, an inexpensive, low-key response to this prediction (e.g., out-of-turn safety inspection) would be well justifie if even a few percent of the damage is prevented.

Optimization

The framework for the optimal choice of disaster-preparedness measures undertaken in response to an alarm was developed by Molchan (1997); earlier analysis of the problem is described in (Kantorovich & Keilis-Borok 1991). Optimization is based on the error diagrams and the trade-off between cost and benefit of different measures.

WHAT COMES NEXT?

Expansion of Observational Base

Yet unknown precursors may exist in a wealth of observed field potentially relevant to earthquake prediction: stress and strain, slow tectonic movements, satellite topography, flui regimes, geochemistry, electric and magnetic fields etc. It seems promising to explore such precursors with the common scaling and common types of premonitory phenomena—similar to those found for premonitory seismicity patterns.

Precursors Specific to Tectonic Environment

The search for precursors specifi to the tectonic environment includes premonitory changes in geometric and kinematic incompatibilities; different precursors in the nodes, fault zones, and blocks; and, finall , region-specifi precursors. Valuable for the last purpose is a rich collection of errors and successful predictions in more than 20 regions worldwide (Keilis-Borok & Shebalin 1999).

Scenarios of Transition to a Strong Earthquake

There is evidence that premonitory seismicity patterns appear in a certain order depending on the type of the pattern and on its scaling in time, space, and magnitude range. This experience suggests that different precursors—not necessarily only seismological—might be integrated in a unifie scenario of development of a strong earthquake, extending through consecutive stages of prediction.

Conclusion

Altogether, these lines of research are converging on the following goals, which seem marginally feasible: development of the next generation of prediction algorithms, 5- to 10-times more accurate than existing ones; prediction of earthquakes in low seismicity regions with rare devastating earthquakes; and prediction of geotechnical and other geological disasters.

ACKNOWLEDGMENTS

Dr. I. Zaliapin made a large and important contribution to work for this review. My colleagues from the International Institute of Earthquake Prediction Theory and Mathematical Geophysics, Moscow, willingly provided invaluable original materials of their research.

Visit the Annual Reviews home page at www.annualreviews.org

LITERATURE CITED

Aki K. 1996. Scale dependence in earthquake phenomena and its relevance to earthquake prediction. *Proc. Natl. Acad. Sci. USA* 93:3748–55

Alekseevskaya M, Gabrielov A, Gelfand I, Gvishiani A, Rantsman E. 1977. Formal morphostructural zoning of mountain territories. *J. Geophys.* 43:227–33

Barenblatt GI, Keilis-Borok VI, Monin AS. 1983. Filtration model of earthquake sequence. *Dokl. Akad. Nauk SSSR* 269(4):831–34 (In Russian)

Bowman DD, Ouillon G, Sammis CG, Sornette A, Sornette D. 1998. An observational test of the critical earthquake concept. *J. Geophys. Res.* 103:24359–72

Bufe CG, Varnes DJ. 1993. Predictive modeling of the seismic cycle of the greater San Francisco bay region *J. Geophys. Res.* 98:9871–83

Caputo M, Console R, Gabrielov AM, Keilis-Borok VI, Sidorenko TV. 1983. Long-term premonitory seismicity patterns in Italy. *Geophys. J. R. Astron. Soc.* 75:71–75

Crutchfiel JP, Farmer JD, Packard NH, Shaw RS. 1986. Chaos. *Sci. Am.* 255:46–57

Gabrielov A, Keilis-Borok VI, Jackson DD. 1996. Geometric incompatibility in a fault system. *Proc. Natl. Acad. Sci. USA* 93(9): 3838–42

Gabrielov AM, Caputo M, Keilis-Borok VI, Console R, Sidorenko TV. 1983. Long-term premonitory seismicity patterns in Italy. *Geophys. J. R. Astron. Soc.* 75:71–75

Gabrielov AM, Keilis-Borok VI. 1983. Patterns of stress corrosion: geometry of the principal stresses *Pure Appl. Geophys.* 121:477–94

Gabrielov AM, Zaliapin IV, Newman WI, Keilis-Borok VI. 2000. Colliding cascades model for earthquake prediction. *J. Geophys. Int.* 143:427–37

Gelfand IM, Guberman ShA, Keilis-Borok VI, Knopoff L, Press F, et al. 1976. Pattern recognition applied to earthquake epicenters in California. *Phys. Earth Planet. Inter.* 11:227–83

Gell-Mann M. 1994. *The Quark and the Jaguar: Adventures in the Simple and the Complex.* New York: Freeman

Gorshkov AI, Kossobokov VG, Rantsman EYa, Soloviev AA. 2001. Recognition of earthquake prone areas: validity of results obtained from 1972 to 2000. *Vych. Seism.* 32:48–57

Holland JH. 1995. *Hidden Order: How Adaptation Builds Complexity*. Reading, MA: Addison-Wesley

Kantorovich LV, Keilis-Borok VI. 1991. Earthquake prediction and decision-making: social, economic and civil protection aspects. In *Int. Conf. Earthquake Prediction: State-of-the-Art, Strasbourg, Fr., Sci.-Tech. Contrib., CSEM-EMSC*, pp. 586–93

Keilis-Borok VI. 1990a. The lithosphere of the Earth as a non-linear system with implications for earthquake prediction. *Rev. Geophys.* 28(1):19–34 (Transl. into Chinese)

Keilis-Borok VI, ed. 1990b. Intermediate-term earthquake prediction: models, phenomenology, worldwide tests. *Phys. Earth Planet. Inter.* 61:1–144

Keilis-Borok VI, Knopoff L, Rotwain IM. 1980. Bursts of aftershocks, long-term precursors of strong earthquakes. *Nature* 283:258–63

Keilis-Borok VI, Kossobokov VG. 1990. Premonitory activation of earthquake fl w: algorithm M8. *Phys. Earth Planet. Inter.* 61:73–83

Keilis-Borok VI, Malinovskaya LN. 1964. One regularity in the occurrence of strong earthquakes. *J. Geophys. Res.* 69(14):3019–24

Keilis-Borok VI, Press F. 1980. On seismological applications of pattern recognition. In *Source Mechanism and Earthquake Prediction Applications*, ed. CJ Allegre, pp. 51–60. Paris: Ed. Cent. Natl. Rech. Sci.

Keilis-Borok VI, Rotwain IM. 1990. Diagnosis of time of increased probability of strong earthquakes in different regions of the world: algorithm CN. *Phys. Earth Planet. Inter.* 61: 57–72

Keilis-Borok VI, Shebalin PN, eds. 1999. Dynamics of lithosphere and earthquake prediction. *Phys. Earth Planet. Inter.* 111:179–330

King GCP. 1983. The accommodation of large strain in the upper lithosphere of the Earth and other solids by self-similar fault systems: the geometrical origin of b-value. *Pure Appl. Geophys.* 121:761–815

Knopoff L, Levshina T, Keilis-Borok VI, Mattoni C. 1996. Increased long-range intermediate-magnitude earthquake activity prior to strong earthquakes in California. *J. Geophys. Res.* 101:5779–96

Kossobokov VG, Carlson JM. 1995. Active zone size vs. activity. A study of different seismicity patterns in the context of prediction algorithm M8. *J. Geophys. Res.* 100: 6431–41

Kossobokov VG, Keilis-Borok VI, Cheng B. 2000. Similarities of multiple fracturing on a neutron star and on the Earth. *Phys. Rev. E* 61(4):3529–33

Kossobokov VG, Keilis-Borok VI, Smith SW. 1990. Localization of intermediate-term earthquake prediction. *J. Geophys. Res.* 95: 19763–72

Kossobokov VG, Romashkova LL, Keilis-Borok VI, Healy JH. 1999. Testing earthquake prediction algorithms: statistically significan advance prediction of the largest earthquakes in the Circum-Pacific 1992–1997. *Phys. Earth Planet. Inter.* 111:187–96

Levshina T, Vorobieva I. 1992. Application of algorithm for prediction of a strong repeated earthquake to the Joshua Tree and Landers earthquakes. *EOS, Trans. Am. Goephys. Union* 73:382

McKenzie DP, Morgan WJ. 1969. Evolution of triple junctions. *Nature* 224:125–33

McKenzie DP, Parker RL. 1967. The North Pacific an example of tectonics on a sphere. *Nature* 216:1276–80

Molchan GM. 1997. Earthquake prediction as a decision-making problem. *Pure Appl. Geophys.* 149:233–47

Molchan GM, Dmitrieva OE, Rotwain IM, Dewey J. 1990. Statistical analysis of the results of earthquake prediction, based on burst of aftershocks. *Phys. Earth Planet. Inter.* 61:128–39

Narkunskaya GS, Shnirman MG. 1994. An algorithm of earthquake prediction. In *Computational Seismology and Geodynamics*, 1:20–24. Washington DC: Am. Geophys. Union

Newman WI, Gabrielov A, Turcotte DL, eds. 1994. *Nonlinear Dynamics and Predictability of Geophysical Phenomena. Geophys.*

Monogr. Ser. 83. Washington, DC: Am. Geophys. Union

Newman WI, Turcotte DL, Gabrielov A. 1995. Log-periodic behavior of a hierarchical failure model with applications to precursory seismic activation. *Phys. Rev. E* 52:4827–35

Pepke GF, Carlson JM, Shaw BE. 1994. Prediction of large events on a dynamical model of fault. *J. Geophys. Res.* 99:6769–88

Press A, Allen C. 1995. Pattern of seismic release in the southern California region. *J. Geophys. Res.* 100:6421–30

Prozorov AG, Schreider SYu. 1990. Real time test of the long-range aftershock algorithm as a tool for mid-term earthquake prediction in Southern California. *Pure Appl. Geophys.* 133:329–47

Richter CF. 1964. Discussion of paper by VI Keilis-Borok and LN Malinovskaya, "One regularity in the occurrence of strong earthquakes." *J. Geophys. Res.* 69(14):3025

Romanowicz B. 1993. Spatiotemporal patterns in the energy-release of great earthquakes. *Science* 260:1923–26

Romashkova LL, Kossobokov VG. 2001. Seismicity dynamics prior to and after the largest earthquakes worldwide 1985–2000. *Vych. Seism.* 32:162–89

Rotwain I, Keilis-Borok V, Botvina L. 1997. Premonitory transformation of steel fracturing and seismicity. *Phys. Earth Planet. Inter.* 101:61–71

Rundle BJ, Turcotte DL, Klein W, eds. 2000. *Geocomplexity and the Physics of Earthquakes.* Washington, DC: Am. Geophys. Union

Shebalin P, Zaliapin I, Keilis-Borok V. 2000. Premonitory raise of the earthquakes' correlation range: Lesser Antilles. *Phys. Earth Planet. Inter.* 122:241–49

Shebalin PN, Keilis-Borok VI. 1999. Phenomenon of local "seismic reversal" before strong earthquakes. *Phys. Earth Planet Inter.* 111:215–27

Shreider SYu. 1999. Formal definitio of premonitory seismic quiescence. *Phys. Earth Planet. Inter.* 61:113–27

Soloviev A, Vorobieva I. 1999. Long-range interaction between synthetic earthquakes in the model of block structure dynamics. *5th Workshop Non-Linear Dyn. Earthquake Prediction, 4–22 Oct.* Trieste: ICTP, Preprint H4. SMR/1150-4. 18 pp.

Sornette D. 2000. *Critical Phenomena in Natural Sciences. Chaos, Fractals, Self-organization and Disorder: Concepts & Tools.* Springer Ser. Synerg., Heidelberg. New York: Springer-Verlag. 432 pp.

Sornette D, Sammis CG. 1995. Complex critical exponents from renormalization group theory of earthquakes: implications for earthquake predictions *J. Phys. I* 5:607–19

Turcotte DL. 1997. *Fractals and Chaos in Geology and Geophysics.* Cambridge: Cambridge Univ. Press. 2nd ed.

Vorobieva IA. 1999. Prediction of a subsequent strong earthquake. *Phys. Earth Planet. Inter.* 111:197–206

Vorobieva IA, Levshina TA. 1994. Prediction of a second large earthquake based on aftershock sequence. In *Computational Seismology and Geodynamics*, 2:27–36. Washington, DC: Am. Geophys. Union

Wyss M, Habermann RE. 1987. Precursory seismic quiescence. Physical and observational basis for intermediate-term earthquake prediction. *US Geol. Surv. Open-File Rep.* 2:526–36

Zaliapin I, Keilis-Borok VI, Axen G. 2002. Premonitory spreading of seismicity over the fault network in S. California: precursor accord. *J. Geophys. Res.* In press

Zaliapin I, Keilis-Borok VI, Ghil M. 2001. Boolean delay model of colliding cascades. II: Prediction of critical transitions. *J. Stat. Phys.* Submitted

Zoeller G, Hainzl S, Kurths J. 2001. Observation of growing correlation length as an indicator for critical point behavior prior to large earthquakes. *J. Geophys. Res.* 106:2167–76

Zhurkov SN. 1968. Kinetic concept for strength of solids. *Vestn. Akad. Nauk SSSR* 3:46–52

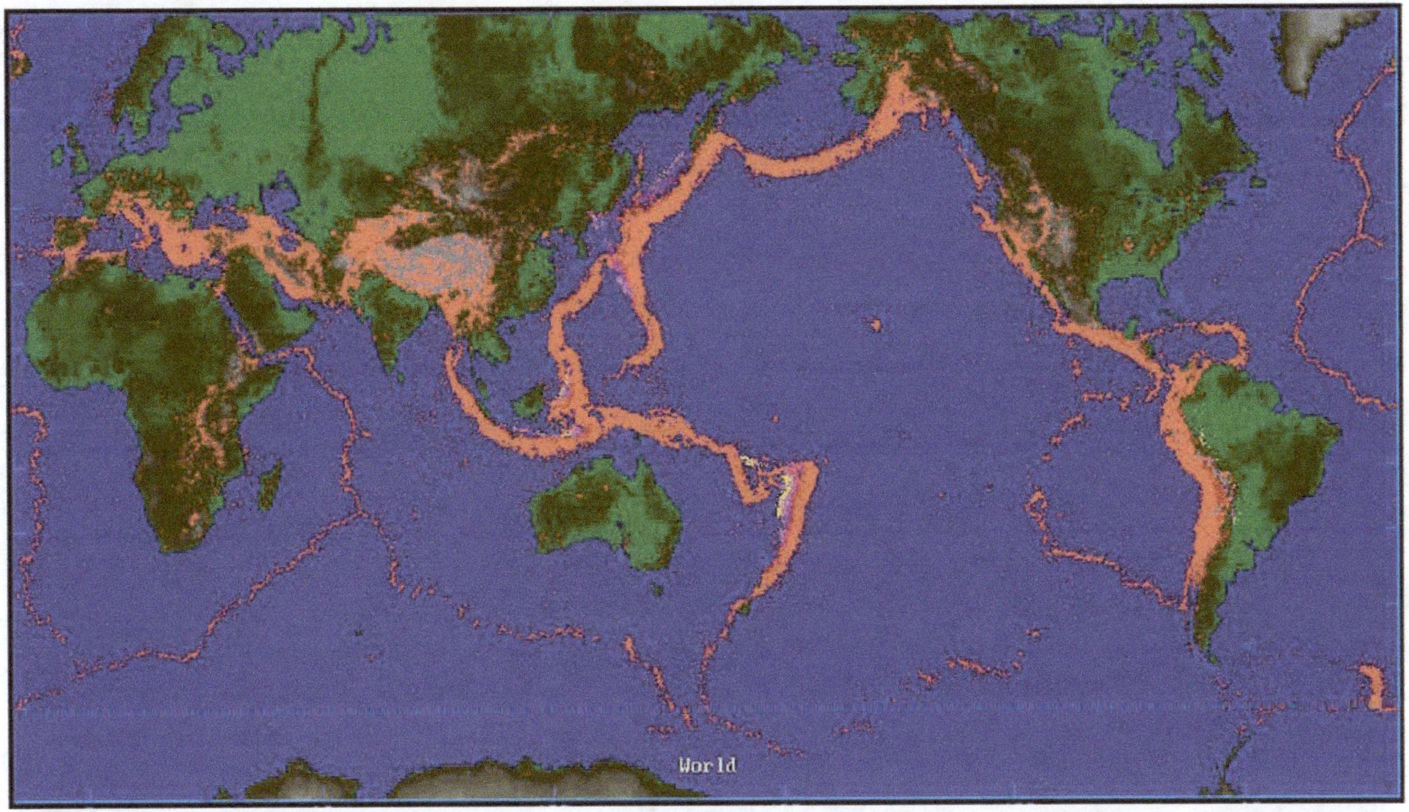

Figure 1 Global distribution of seismicity. *Red* points show earthquakes with magnitude greater than 4.0 for the last century. They clearly outline the global fault network. From http://www.mitp.ru.

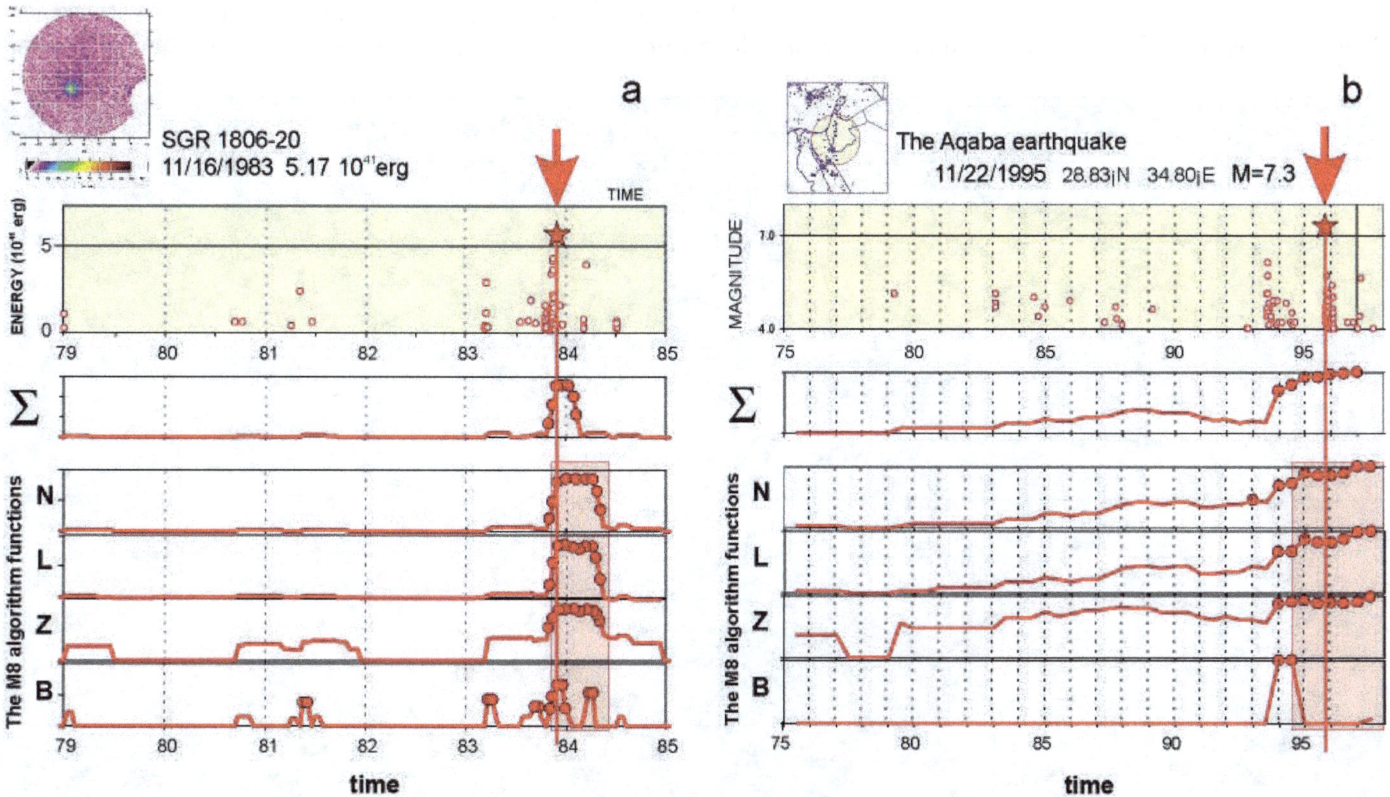

Figure 8 Frontier of similarity of premonitory seismicity patterns. (*a*) Starquake, registered on November 16, 1983; magnitude is approximately 20. (*b*) Aqaba earthquake, November 22, 1995; M = 7.3. *Top panels*: sequence of starquakes (*left*) and earthquakes (*right*). Stars indicate the major events, targeted for prediction. Other panels show the emergence of different premonitory seismicity patterns; note that they have been developed for seismicity of the Earth only. *B* is the measure of clustering, other functionals are different measures of seismic activity (see Prediction). *Solid dots* indicate emergence of a premonitory pattern. After Kossobokov et al. 2000.

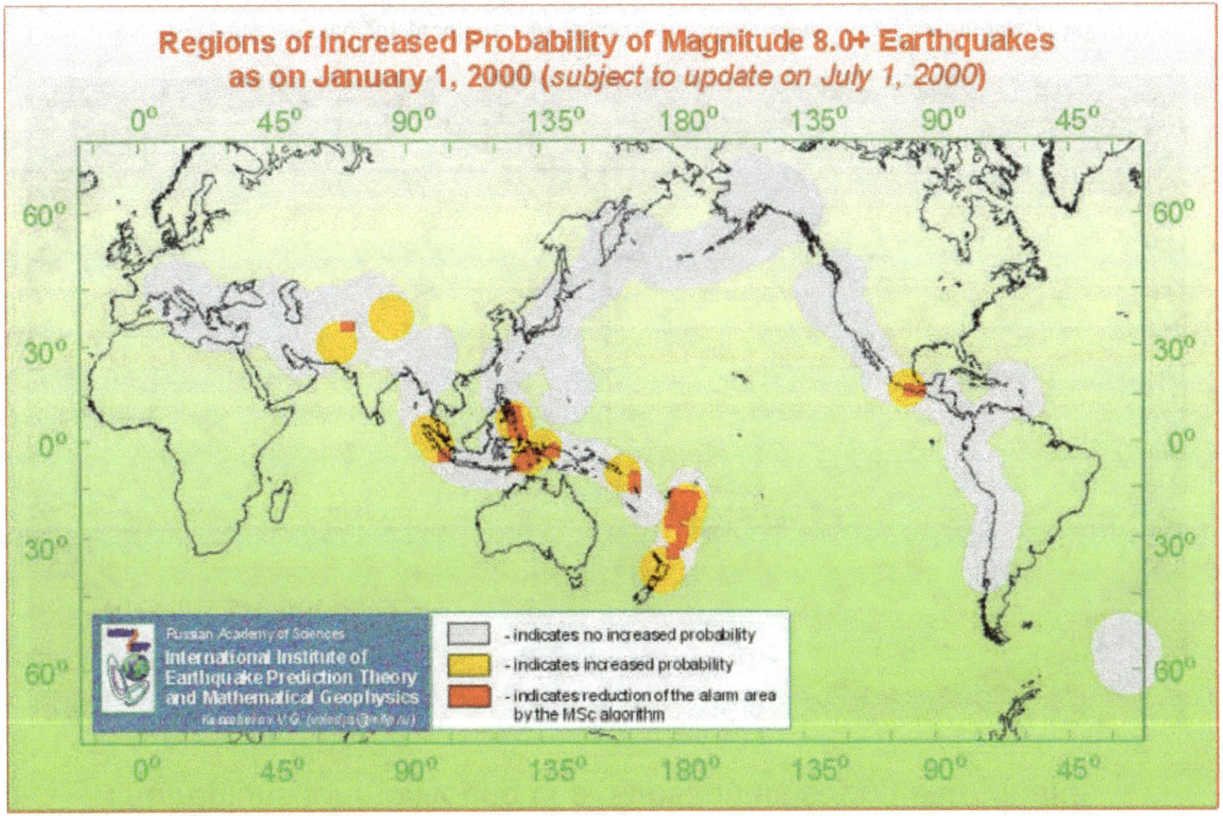

Figure 11 Global prediction issued by algorithm M8 in January, 2000. *Gray*—territory targeted for prediction; *yellow*—alarms by the algorithm M8; *red*—alarms by the algorithm MSc. See details in the text. From http://www.mitp.ru (mirror at http://www.phys.ualberta.ca/mirrors/mitp).

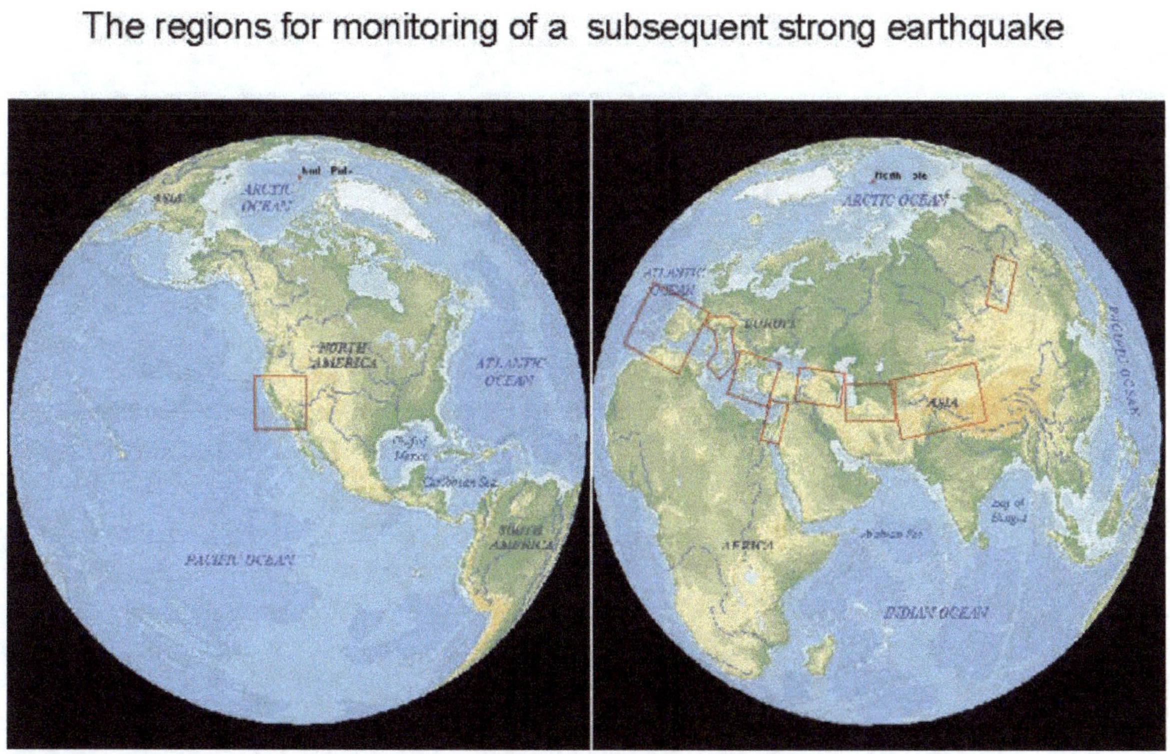

Figure 13 Regions targeted for prediction by algorithm SSE (second strong earthquake). Taken from http://www.mitp.ru (mirror at http://www.phys.ualberta.ca/mirrors/mitp).

ON PREDICTABILITY OF HOMICIDE SURGES IN MEGACITIES

V.I. KEILIS-BOROK[1,2], D.J. GASCON[3], A.A. SOLOVIEV ,
M.D. INTRILIGATOR[4], R. PICHARDO[5], F.E. WINBERG[1]

[1] *International Institute of Earthquake Prediction Theory and Mathematical Geophysics, Russian Academy of Sciences Warshavskoye shosse 79 kor. 2, Moscow 117556, Russia*

[2] *Institute of Geophysics and Planetary Physics, University of California, Los Angeles 405 Hilgard av., Los Angeles, CA 90095- 1567, USA*

[3] *Assistant Chief (ret), Los Angeles Police Department 150 N. Los Angeles Street, Rm. 611, Los Angeles, CA 90012, USA*

[4] *Department of Economics, University of California, Los Angeles Box 951477, Los Angeles, CA 90095-1477, USA*

[5] *Crime Analysis Section, Los Angeles Police Department 6464 Sunset Blvd. #520, Hollywood, CA 90028, USA*

Dynamics of crimes reflects important aspects of sustainability of our society and the risk of its destabilisation – a prelude to a disaster. Here, we consider a prominent feature of crime dynamics – surge of the homicides in a megacity. Our study integrates the professional expertise of the police officers and of the scientists working on pattern recognition of infrequent events. The latter is a type of artificial intelligence methodology that has been successful in predicting infrequently occurring phenomena that result from highly complex processes.

In this paper we analyse statistics of several types of crimes in Los Angeles over the period 1975-2002. Our analysis focuses on how these statistics change before a sharp and lasting rise ("a surge") of the homicide rate. The goal is to find an algorithm for predicting such a surge by monitoring the rates of different crimes.

Our hope for feasibility of that goal comes from two sources. First is the set of available crime statistics, showing that a surge of major crimes is preceded by the rise of less severe crimes. Second is recent research in the prediction of critical phenomena (i.e. abrupt overall changes) in various complex non-linear systems, such as those in theoretical physics, earth sciences, social sciences, etc.

Data. Out of a multitude of relevant data we analyse statistics of robberies, assaults, burglaries, and the homicides themselves.

Results. Our findings may be summarised as follows: Episodes of a rise of burglaries and assaults simultaneously occur 4 to 11 months before a homicide surge, while robberies decline. Later on, closer to the rise in homicides, robberies start to rise. These changes are given unambiguous and quantitative definitions, which are used to formulate a hypothetical algorithm for the prediction of homicide surges.

In retrospective analysis we have found that this algorithm is applicable through all the years considered despite substantial changes both in socio-economic conditions and in the counting of crimes. Moreover, it gives satisfactory results for the prediction of homicide surges in New York City as well. Sensitivity tests show that predictions are stable to variations of the adjustable elements of the algorithm.

What did we learn? The existing qualitative portrayals of crime escalation are complemented here by a quantitatively defined set of precursors to homicide surges. The same set emerges before each surge through the time period under consideration. That implies the existence of a "universal" scenario of crime escalation, independent of the concrete reasons triggering each surge. These findings provide heuristic constraints for the modeling of crime dynamics and indicate promising lines of further research.

Perspective. Decisive validation of our findings requires experimentation in *advance prediction*, for which this study sets up a base. Particularly encouraging for this further research is the wealth of yet untapped possibilities: we have used so far only a small part of the data and mathematical models that are currently available and that are relevant to crime dynamics.

On the practical side, our results enhance our capability to identify a situation that is "ripe" for homicide surges and, accordingly, to escalate the crime prevention measures. In the broader scheme of things, a surge of crime is one of potential ripple effects of natural disasters. Accordingly the risk of a natural disaster is higher in such a situation.

1. Introduction

Understanding and prediction of crime dynamics is one of the problems important for coping with the risks threatening humanity. These risks are to a large extent concentrated in megacities, whose role in the global village is rapidly growing along with their vulnerability to natural and socio-economic disasters. Present study is focused on the crime dynamics in Los Angeles; its experience, we believe, might be useful for studying similar problem in other megacities.

1.1. PREDICTION TARGET

We consider prediction of a specific phenomenon in crime dynamics: a large and lasting increase in the homicide rate. Qualitatively, this phenomenon is illustrated in Figure 1; we call it by the acronym *SHS*, for "Start of the Homicide Surge". Our goal is to find a method to predict an *SHS* by monitoring the relevant indicators. Among a multitude of such indicators, we consider here statistics on assaults, burglaries, robberies and the homicides themselves.

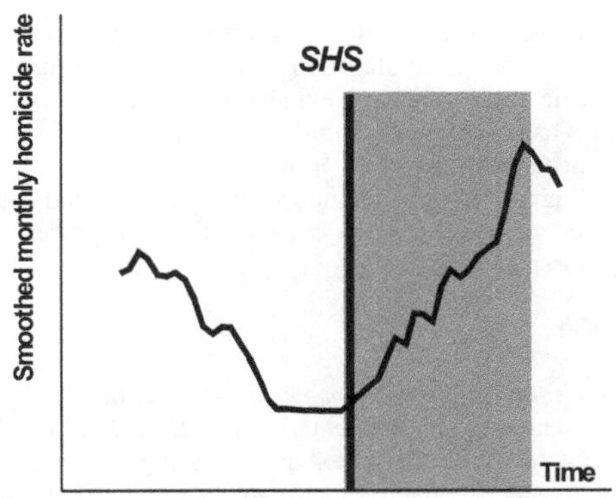

Figure 1. Target of prediction – the Start of the Homicide Surge ("*SHS*"); schematic definition. Gray bar marks the period of homicide surge.

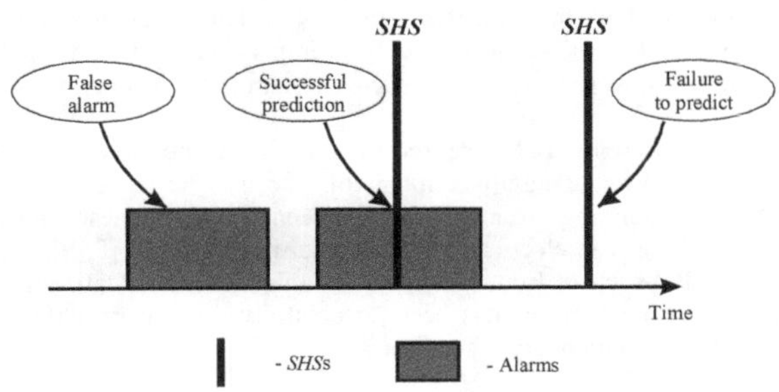

Figure 2. Possible outcomes of prediction.

1.2. THE PROBLEM

Our goal is to develop a method for predicting the surge of homicides by monitoring the relevant observed indicators. We hope to recognise the "premonitory" patterns formed by such indicators when an *SHS* approaches. In terms of pattern recognition we look for an algorithm (a "recognition rule") that solves the following problem:

given the time series of certain crime rates (or of other relevant indicators) prior to a moment of time *t*,

to predict whether an episode of *SHS* will or will not occur during the subsequent time period $(t, t+\tau)$; in other words, whether the lasting surge of homicides will or will not start during that period.

If the prediction is "yes", this period will be the "period of alarm." The possible outcomes of such a prediction are illustrated in Figure 2.

The probabilistic component of this prediction is represented by the estimated probabilities of errors – both false alarms on one side and failures to predict on the other. That probabilistic component is inevitable since we consider a highly complex non-stationary process using imprecise crime statistics. Moreover, the predictability of a chaotic system is, in principle, limited.

Such "yes or no" prediction of specific extraordinary phenomena is different from predictions in a more traditional sense - extrapolation of a process in time, which is better supported by classical theory.

1.3. METHODOLOGY

We use *pattern recognition of infrequent events* – a methodology developed by the artificial intelligence school of the mathematician I.M. Gelfand [1] for the analysis of infrequent phenomena of highly complex origin. Using this methodology, we here conduct a so-called "technical" analysis that involves a heuristic search for phenomena preceding episodes of *SHS*. A distinctive feature of this methodology is the robustness of the analysis, which helps to overcome both the complexity of the process considered and the chronic imperfection of the data; in that aspect it is akin to exploratory data analysis, as developed by the statistics school of J. Tukey [2]. Robust analysis – "a transdisciplinary crude look at the whole" – is imperative in a study of any complex system [3]. The surest way *not* to predict such a system is to consider it in too fine detail [4].

Close to the present study are recent studies of the prediction of economic recessions and surges of unemployment [5, 6]. We use the same pattern recognition algorithm, called "Hamming distance," that has been applied in these studies, as well as in predictions of American elections [7] and in seismology, e.g. [8, 9]. The essence of the algorithm will be clear from the way we analyse crime statistics here. Pattern recognition of infrequent events has been successfully used in geophysics, geological prospecting, medicine, and many other areas.

1.4. CONTENT

Following is a schematic outline of our analysis:

Data comprise the monthly rates of homicides, robberies, assaults, and burglaries for Los Angeles, 1975 – 2002 (Section 2).

Five targets of prediction (*SHS*) are defined during the time period under consideration (Sections 3, 7). Those are the moments when a years-long trend of the homicide rate turns from decline to a long steep rise.

We have found *premonitory changes of crime statistics* as illustrated in Figure 3. Within several months before a homicide surge, burglaries and assaults simultaneously escalate, while robberies decline (Section 4). On the basis of these changes we suggest a *hypothetical prediction algorithm* (Section 5). In retrospect, it provides a robust satisfactory prediction (Sections 6, 7). However it has to be further validated by application to independent data. As always in prediction research, *the final validation of our algorithm requires prediction in advance*, for which this study sets up a base.

Later on, closer to a homicide surge, robberies also escalate (Section 8). These changes will be explored elsewhere.

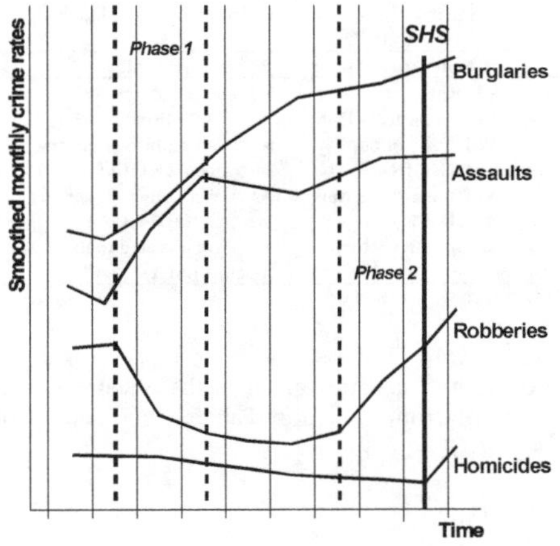

Figure 3. Scheme of premonitory changes in crime statistics.

1.5. COMMON NOTATION

Our analysis focuses on trends in the crime rates. We estimate these trends by linear regression, using the following notations:

$$C(m),\ m = 1, 2, \ldots$$

is the time series of a monthly indicator, and m is the sequence number of a month.

$$W^C(m/q, p) = K^C(q, p)m + B^C(q, p),\ q \le m \le p, \qquad (1)$$

is the local linear least-squares regression of the function $C(m)$ within the sliding time window over the time period (q, p).

2. The Data

We use the following data sources:

(i) The National Archive of Criminal Justice Data (NACJD), placed on the web site (http://www.icpsr.umich.edu/NACJD/index.html). Carlson [10] gives its description. This site contains data for the years 1975-1993.

(ii) Data bank of the Los Angeles Police Department (LAPD Information Technology Division); it contains similar data for the years 1990 – May 2001.

TABLE 1. Types of crimes considered
(after [10]; abbreviations are indicated in brackets)

Homicide	Robberies	Assaults	Burglaries
• All (**H**)	• All (**Rob**)	• All (**A**)*	• Unlawful not forcible entry (**UNFE**)
	• With firearms (**FRob**)	• With firearms (**FA**)	• Attempted forcible entry (**AFE**)*
	• With knife or cutting instrument (**KCIR**)	• With knife or cutting instrument (**KCIA**)	
	• With other dangerous weapon (**ODWR**)	• With other dangerous weapon (**ODWA**)*	
	• Strong-arm robberies (**SAR**)*	• Aggravated injury assaults (**AIA**)*	

* Analysed in sensitivity tests only (Section 6)

Out of numerous crime statistics given in these sources, we analyse the monthly rates of the four types of crimes listed in Table 1, homicides, robberies, assaults, and burglaries.

3. Prediction Targets

Here and in the next two sections we analyse the data for 1975 – 1993 as taken from the National Archive of Criminal Justice Data [10].

Definition. Let $H(m)$, $m = 1,2\ldots$, be the time series of the monthly number of all homicides. Figure 4 shows the plot of $H(m)$ in Los Angeles, per 3,000,000 inhabitants of the city. To identify the episodes of *SHS* (Fig. 1) we smooth out the seasonal variations, which are clearly seen in Figure 4, by replacing $H(m)$ by its linear least square regression (1): $H^*(m) = W^H(m/m\text{-}6,\ m+6)$. Since $H^*(m)$ is defined on the time interval $(m - 6,\ m + 6)$, it depends on the future. Thus, it is admissible to define prediction targets (but not precursors).

The function $H^*(m)$ is shown in Figure 4 by the thick curve. Three time periods of a lasting homicide rise are clearly seen: 1977 – 1980, 1988 - 1992 and a relatively shorter period 1985 - 1986. We choose as prediction targets the starting months of these periods: 04:1977, 03:1985, and 08:1988. They are marked in Figure 4 by vertical lines.

4. Premonitory Trends of Single Types of Crimes

Here we analyse the monthly data on seven types of crimes out of 13 types listed in Table 1. We look for "premonitory" trends of each crime that tend to appear more frequently as an *SHS* approaches. Prediction itself is based on the collective behaviour of these trends, as analysed in the next Section. Orientation on a *set* of precursors has been found to be rather successful in prediction research: an ensemble of "imprecise" precursors usually gives better predictions than a single "precise" precursor [11, 12].

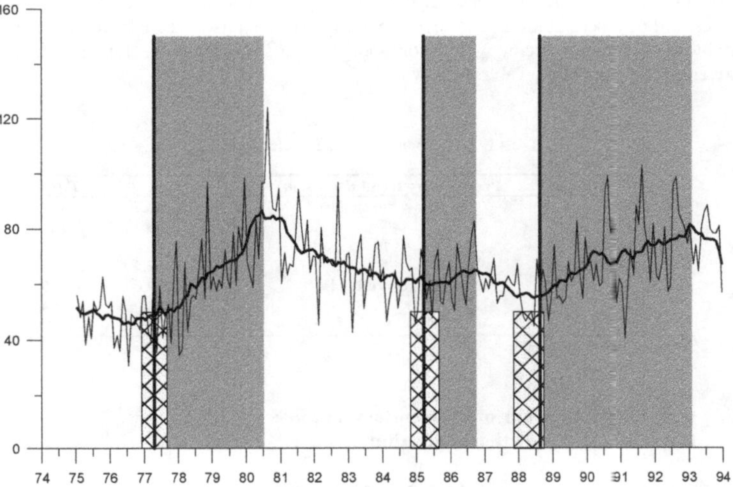

Figure 4. Total monthly number of homicides in Los Angeles city, 1975-1993. Data are taken from the National Archive of Criminal Justice Data [10]. Thin curve – original time series, *H(m)*, per 3,000,000 inhabitants. Thick curve - smoothed series $H^*(m)$, with seasonal variations eliminated as described in Section 1. Vertical lines show the targets of prediction – episodes of *SHS* (Section 3). Gray bars are the periods of homicide surge. Checkered bars are the alarms declared by the hypothetical prediction algorithm (Section 5).

Our hypothetical premonitory pattern (Fig. 3) comprises simultaneous escalation of assaults and burglaries and decline of robberies. We quantify this hypothesis as follows.

First, we *smooth out seasonal variations* replacing the time series $C(m)$ of each type of crime by its regression (1): $C^*(m) = W^C(m/m\text{-}12, m)$. Regression is done over the prior 12 months and does not depend on the future, so that it can be used for prediction.

Next, we approximate the trends of the crimes by the regression coefficients $K^C(m\text{-}12, m)$. The functions K^C also are also independent on information on the future. Fig. 5 shows these functions for 7 types of crime considered.

Finally, following the pattern recognition approach, we give the robust "crude look at the whole" set of time series shown in Figure 5. We discretize them on the lowest level of resolution: a binary one, that distinguishes only the trends above and below a threshold $T^C(Q^C)$. It is defined as a percentile of a level Q^C, that is, by the condition that $K^C(m\text{-}s, m)$ exceeds $T^C(Q^C)$ during Q^C percent of the months considered. We choose the values of Q^C indicated in Table 2.

Observations suggest a hypothesis (Fig. 3) that the trends of assaults and burglaries more frequently and the trends of robberies – less frequently exceed the respective threshlds as an *SHS*, approaches. For convenience, we will give the same code, "1", to the "premonitory" trend of each crime, regardless of whether it is above or below the threshold of discretization. The seven monthly crime statistics considered here are thus reduced to a binary vector with 7 components. The crime history, thus transformed, is given in the Appendix, Table A1.

TABLE 2. Premonitory trends for selected crime types

#	Crime type	Premonitory trend $K^C(m\text{-}s, m)$	s	Q^C, %	$T^C(Q^C)$
1	Rob	Below threshold	12	66.7	-3.69
2	FRob	"	12	66.7	-1.29
3	KCIR	"	12	50.0	1.73
4	ODWR	"	12	87.5	-3.87
5	FA	Above threshold	12	50.0	1.89
6	KCIA	"	12	50.0	1.94
7	UNFE	"	12	50.0	-1.32

See notations in the text.

5. Collective Behaviour of Premonitory Trends: Hypothetical Prediction Algorithm

Here, we consider how the approach of a homicide surge is reflected in the *collective* behaviour of the trends. The simplest description of this behaviour is $\Delta(m)$ - the number of *non*-premonitory trends at a given month m. If our identification of premonitory trends is correct then $\Delta(m)$ should be low in the proximity of an *SHS*. By definition $\Delta(m)$ is the number of zeros in the binary code of the monthly situation. This is the so-called "Hamming distance" between that code and the code of the "pure" premonitory situation, {1,1,1,1,1,1,1} when all seven trends listed in Table 2 are premonitory [5, 13, 14].

The values of $\Delta(m)$ are given in the Appendix, Table A1. Figure 6 shows the change of $\Delta(m)$ with time. The value of $\Delta(m)$ may vary from 0 to 7 but the minimal observed value is 1; the corresponding lines in Table A1 are marked by "+". That value appears within 4 to 11 months before an *SHS* and at no other time. An examination of the temporal change of $\Delta(m)$ in Table A1 suggests the following hypothetical prediction algorithm:

An alarm is declared for 9 months each time when $\Delta(m) \leq D$ for two consecutive months (regardless of whether these two months belong or not to an already declared alarm).

Possible outcomes of such a prediction are illustrated in Figure 2. The condition $\Delta(m) \leq D$ means, by definition, that D or less trends are not premonitory at the month m. A count of $\Delta(m)$ in Table A1 suggests that we take $D = 1$. A waiting period of 9 months is introduced because the premonitory trends do not appear right before an *SHS*. The requirement that this condition holds two months in a row makes prediction more reliable and reduces the total duration of alarms.

The alarms obtained by this algorithm are shown in Figure 4 by the grey bars. The total duration of these alarms is 30 months, representing 14 percent of all months considered. In real prediction that score would be quite satisfactory.

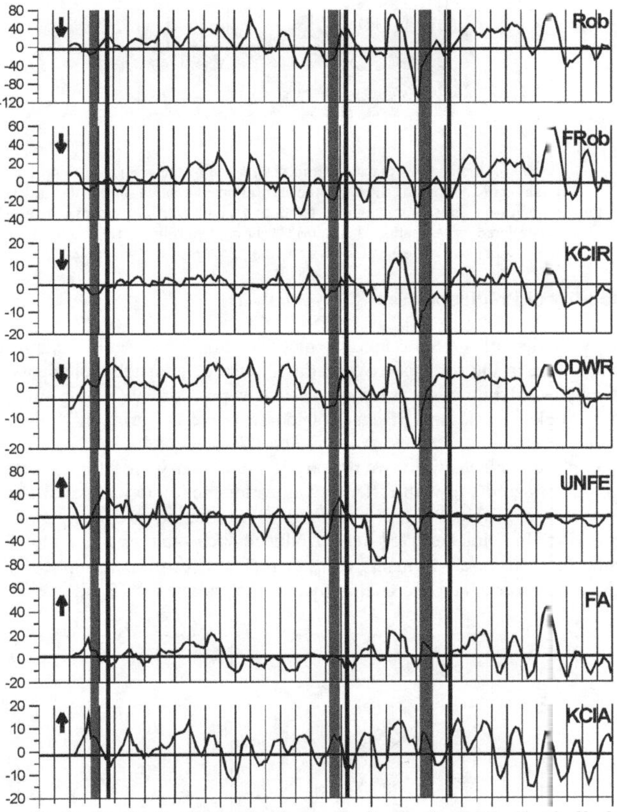

Figure 5. The regression coefficients $K^C(m\text{-}12, m)$ for seven crime types. See the definition in Section 4 and notations in Table. 1. Original data are taken from the National Archive of Criminal Justice Data [10]. Horizontal lines and arrows show respectively discretization thresholds and premonitory trends in accordance with Table 2. Vertical lines show episodes of *SHS*. Gray bars indicate months when $\Delta(m) \leq 1$.

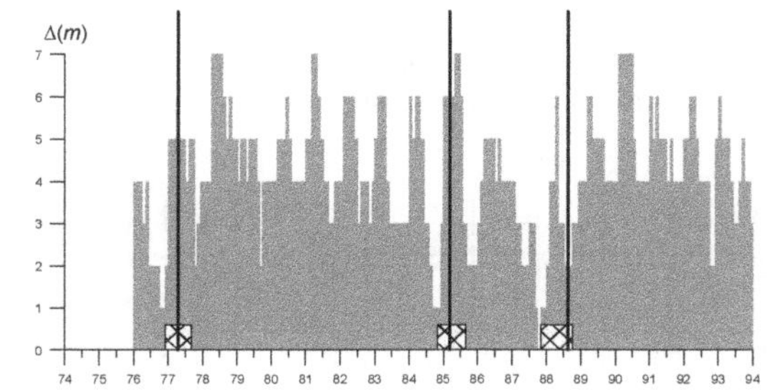

Figure 6. Homicide surges and alarms determined by the prediction algorithm. Start of a homicide surge is shown by the vertical line. Function $\Delta(m)$ is the number of crime statistics *not* showing premonitory trends at a month m. Alarms (shown by checkered bars) are declared for 9 months, when $\Delta(m) \leq 1$ during two consecutive month. Adjustable parameters correspond to version 10 of the algorithm (see Table 3).

6. Stability of Prediction (Sensitivity Analysis)

Inevitably in lieu of a set of fundamental equations for crime dynamics we have a certain freedom in the retrospective ad hoc choice of adjustable elements: the types of crimes considered, numerical parameters, such as percentiles Q^C, etc. An algorithm thus developed makes sense only if it is not too sensitive to variation of these choices; as Enrico Fermi put it, "*with four exponents I can fit an elephant*".

To explore that sensitivity we repeat the prediction with different sets of the kinds of crimes considered and with different values for the numerical parameters. These sets are described in Table 3. The outcomes of prediction are compared on the error diagrams (Fig. 7). Molchan [15] has introduced such diagrams as a tool for evaluating prediction methods and optimising disaster preparedness. Their application to research in prediction of recessions and unemployment is described in [5, 6].

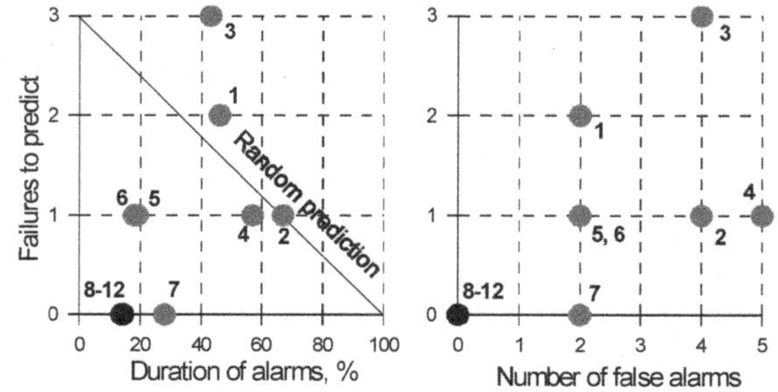

Figure 7. Error diagram.
Numbers near the dots identify the variant of the algorithm in Table 3. Black dots show the variants suggested for advance prediction. See explanations in the text.

The “basic” variant (Section 4) is # 10 in Table 3. We now discuss the variations considered. *Variation of the percentiles* Q^C, defining discretization thresholds (#8, 9, 11, 12). Lowering them, we obviously increase the total duration of alarms, but the results of prediction do not change much and remain acceptable. *Using only two kinds of crimes* (#12) we obtain comparable results. However it would be risky to make advance prediction with only two indicators. *The limits of acceptable variations* are reached in the other variants (#1–7). We tried to find a premonitory *rise* of robberies, simultaneous with rise of other crimes and consider other kinds of crimes; in all variants its performance remains unacceptable.

TABLE 3. Variation of the adjustable elements

<table>
<tr><td colspan="2"></td><td colspan="12">Variants</td></tr>
<tr><td colspan="2"></td><td>1</td><td>2</td><td>3</td><td>4</td><td>5</td><td>6</td><td>7</td><td>8</td><td>9</td><td>10</td><td>11</td><td>12</td></tr>
<tr><td colspan="2">Value of D*</td><td>3</td><td colspan="2">5</td><td>8</td><td colspan="2">4</td><td>3</td><td colspan="4">1</td><td>0</td></tr>
<tr><td colspan="2">Crime type</td><td colspan="12">Trend and percentile</td></tr>
<tr><td rowspan="2">Rob</td><td>Premonitory trend</td><td colspan="3">upward</td><td colspan="8">downward</td><td rowspan="2"></td></tr>
<tr><td>Q^C, %</td><td>33</td><td>25</td><td>20</td><td>80</td><td colspan="6">67</td><td>50</td></tr>
<tr><td rowspan="2">FRob</td><td>Premonitory trend</td><td colspan="3" rowspan="2"></td><td colspan="8">downward</td><td rowspan="2"></td></tr>
<tr><td>Q^C, %</td><td>80</td><td colspan="6">67</td><td>50</td></tr>
<tr><td rowspan="2">KCIR</td><td>Premonitory trend</td><td colspan="3" rowspan="2"></td><td colspan="9">downward</td></tr>
<tr><td>Q^C, %</td><td>80</td><td colspan="5">67</td><td colspan="2">50</td><td>67</td></tr>
<tr><td rowspan="2">ODWR</td><td>Premonitory trend</td><td colspan="3">upward</td><td colspan="8">downward</td><td rowspan="2"></td></tr>
<tr><td>Q^C, %</td><td>33</td><td>25</td><td>20</td><td>80</td><td>67</td><td colspan="6">87.5</td></tr>
<tr><td rowspan="2">SAR</td><td>Premonitory trend</td><td colspan="3">upward</td><td colspan="4">downward</td><td colspan="5" rowspan="2"></td></tr>
<tr><td>Q^C, %</td><td>33</td><td>25</td><td>20</td><td>80</td><td colspan="3">67</td></tr>
<tr><td rowspan="2">A</td><td>Premonitory trend</td><td colspan="7">upward</td><td colspan="5" rowspan="2"></td></tr>
<tr><td>Q^C, %</td><td>33</td><td>25</td><td>20</td><td>20</td><td colspan="3">33</td></tr>
<tr><td rowspan="2">FA</td><td>Premonitory trend</td><td colspan="11">upward</td><td rowspan="2"></td></tr>
<tr><td>Q^C, %</td><td>33</td><td>25</td><td>20</td><td>20</td><td colspan="4">33</td><td colspan="3">50</td></tr>
<tr><td rowspan="2">KCIA</td><td>Premonitory trend</td><td colspan="11">upward</td><td rowspan="2"></td></tr>
<tr><td>Q^C, %</td><td>33</td><td>25</td><td>20</td><td>20</td><td colspan="4">33</td><td colspan="3">50</td></tr>
<tr><td rowspan="2">ODWA</td><td>Premonitory trend</td><td colspan="7">upward</td><td colspan="5" rowspan="2"></td></tr>
<tr><td>Q^C, %</td><td>33</td><td>25</td><td>20</td><td>20</td><td colspan="3">33</td></tr>
<tr><td rowspan="2">AIA</td><td>Premonitory trend</td><td colspan="7">upward</td><td colspan="5" rowspan="2"></td></tr>
<tr><td>Q^C, %</td><td>33</td><td>25</td><td>20</td><td>20</td><td colspan="3">33</td></tr>
<tr><td rowspan="2">UNFE</td><td>Premonitory trend</td><td colspan="3" rowspan="2"></td><td colspan="3">upward</td><td rowspan="2"></td><td colspan="5">Upward</td></tr>
<tr><td>Q^C, %</td><td>20</td><td colspan="2">33</td><td>33</td><td colspan="3">50</td><td>33</td></tr>
<tr><td rowspan="2">AFE</td><td>Premonitory trend</td><td colspan="3" rowspan="2"></td><td colspan="3">upward</td><td colspan="6" rowspan="2"></td></tr>
<tr><td>Q^C, %</td><td>20</td><td colspan="2">33</td></tr>
</table>

*The values that give relatively best performance for that variant.

For advance prediction variants 8–11 might be used in parallel. Such parallel predictions might better suit the needs of a decision-maker, determining possible disaster preparedness measures [12, 16].

7. Applications to Independent Data

Here we test our algorithm by application to "out of sample" data not used in its development. Such tests are always necessary to validate and/or improve a prediction algorithm. Such a test is possible since our algorithm is self-adaptive: the thresholds $T^C(Q^C)$ are not fixed but are adapted to crime statistics, as the percentile of a level Q^C.

7.1. LOS ANGELES, 1994-2002

So far we used the data source [10] covering the years 1975 – 1993. To extend the analysis past 1993, we have the data of the LAPD Information Technology Division, covering the time period from January 1990 to May 2002. Comparing the data for the overlapping three years we find that they are reasonably close, particularly after smoothing.

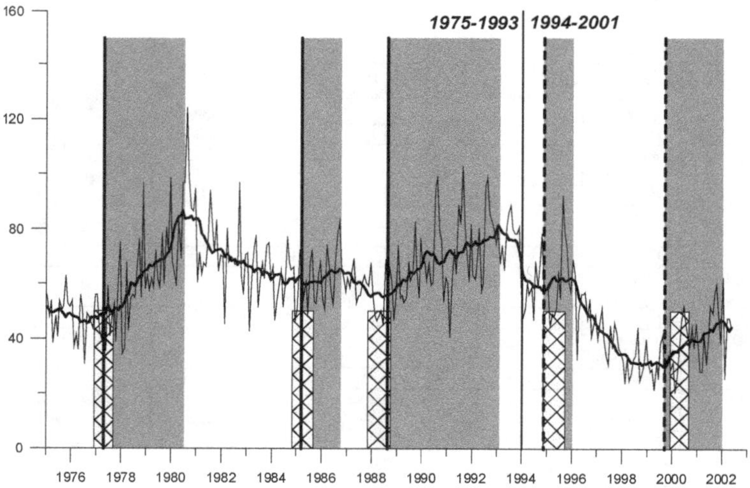

Figure 8. Performance of prediction algorithm through 1975-2002.
Data from the National Archive of Criminal Justice Data [10] for 1975 – 1993 have been used to develop the algorithm. It was than applied to the data from the Data Bank of the Los Angeles Police Department (LAPD Information Technology Division) for subsequent 9 years. Notations are the same as in Figure 4. Dashed vertical lines indicate *SHS* episodes that occurred after 1993.

Figure 8 shows the homicide rates through the whole period from 1975 to May 2002. Two *SHS* episodes are identified in the later period 1994-2001. They are indicated in Figure 8 by dashed vertical lines. The first episode is captured by an alarm, which starts in the month of *SHS* without a lead time. The second episode is missed in that an alarm has started two months after it. That error has to be put on the record; nevertheless the prediction remains informative: during these two months homicide rose by only a few percent, giving no indication that a lasting homicide surge had started.

7.2. NEW YORK CITY

Figure 9 shows the monthly total homicide rates in New York City per 7 million inhabitants of the city. We identified two *SHS* episodes (02:1978 and 02:1985). Our prediction algorithm gives two alarms, as shown in Figure 9 by chequered bars. One of them predicts the second *SHS*, while the first one is missed. We consider another alarm as a false one; this has to be confirmed by processing the data for the period after 1993. Though the failure to predict and a false alarm are disappointing, the results as a whole appear to be useful: one of the two *SHS* is captured by alarms lasting together 21 months, amounting to 10 percent of the time interval considered.

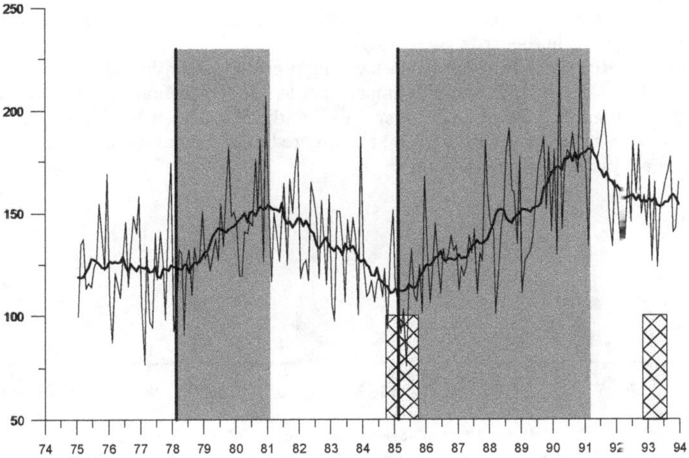

Figure 9. Application of the prediction algorithm to New York City. Notations are the same as in Figure 4. Data are taken from the National Archive of Criminal Justice Data [10]. Homicide statistics is shown per 7,000,000 inhabitants.

8. On a More Precise Prediction

Here, we outline a conjecture, one that we believe is worth exploring in the future. We have observed two consecutive patterns of the crimes considered. The first one precedes an *SHS* with a lead time of 4 to 11 months; it is formally defined and explored in Sections 4-6. We will discuss now in more detail the second pattern. It emerges with a shorter lead time, promising a more accurate prediction of the time of an incipient *SHS*.

A distinctive trait of the second pattern is a steep simultaneous rise of the different types of robberies. Let us replace this pattern by a less specific one that is more broadly defined: the absence of a steep decline. By definition, that pattern will be captured by the zeros in the first four columns of Table A1. Counting them, we find that three or more emerge within 6 months before each *SHS*. This result suggests the following second approximation to the prediction algorithm described above. Consider the period of alarm declared by the algorithm; let us call it "the first phase alarm". Within that period *a "second phase alarm" is declared for 6 months after the first month when* $\Delta_1(m) \leq 1$. Here $\Delta_1(m)$ is the number of ones in the codes of the robberies (the first four columns in Table A1). In the absence of the first-phase alarm the second one is not declared.

Alarms obtained by this rule are shown in Figure 10. The alarms became much shorter; their total duration drops to 18 months, that is, from 14 percent to 8 percent of all the months considered. We will possibly get even better results directly capturing *a rise* of robberies, but that probably requires weekly if not daily crime rates (since the lead time of the rise in robberies is relatively short).

Using the trend of homicides themselves might provide a similar possibility. Values of the function $K^H(m\text{-}12, m)$, which estimates that trend (see Section 4) are given in the Appendix (column K^H). Within each alarm we see the months when $K^H(m\text{-}12, m) > 0$. Starting alarms at these months, we might further reduce the duration of alarms without having an additional failure to predict.

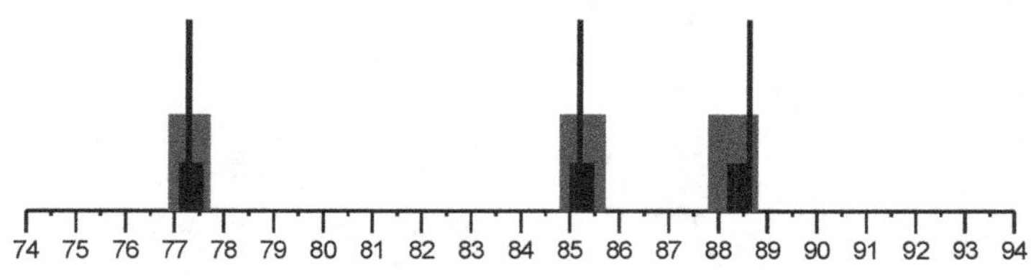

Figure 10. Possible reduction of duration of alarms.
Vertical lines –starting points of a homicide surge (*SHS*). Grey bars - alarms obtained by the suggested algorithm. Black bars – alarms obtained in a hypothetical second approximation.

9. Discussion

1. Our conclusions might be summed up as follows. We analysed crime statistics in the city of Los Angeles for the period 1975 – 2001, exploring the possibility of anticipating a turn of the homicide rate from decline to a surge. We have found that such a turn is

preceded within 4 to 11 months by a specific pattern of the crime statistics: both burglaries and assaults escalate, while robberies decline, along with the homicides themselves. Both changes, escalation and decline, are not monotonic, but occur sporadically, each lasting 2-6 months.

Based on this pattern we formulated a prediction algorithm, giving it a robust and unambiguous definition. It is self-adapting to average crime statistics, so that we could apply it to New York City as well. The major limitation of this study is that, as is inevitable for an initial study, only a small number of homicide surges were available for analysis. The algorithm remains hypothetical until it is validated by advance prediction. It is encouraging, however, that those predictions are stable as to variations in the adjustable elements of the algorithm.

Closer to the surge of homicides, the robberies also turn from decline to rise. This indicates the possibility of a second approximation to the prediction, with twice better the accuracy (that is with a twofold reduction in the duration of alarms).

2. Our analysis captures the consecutive escalation of different crimes: first – of burglaries and assaults only, then of robberies, then of homicides. That sequence, albeit hypothetical so far, seems natural, being in good accord with previous experience in the following areas.

(*i*) The sequence reflects a more general phenomenon, commonly known in law enforcement practice: a consecutive escalation of more and more severe crimes, signalling that a surge of major crimes is approaching. We give a quantitative definition of a specific manifestation of this phenomenon. Similar escalation has been found in French suburban areas [17].

(*ii*) The sequence is also in accord with a well-known "universal" feature of many hierarchical complex systems: the rise of permanent background activity ("static") of the system culminated by a fast major change – a "critical transition". That feature happens to be common for different physical and socio-economic systems. It is reproduced by the "universal" models of hierarchical complex systems, such as those developed in theoretical physics, e.g., [3, 12, 18-26].

That feature was also observed in many very different real world systems. For example, in earthquakes prone regions the "static" includes background seismicity. Premonitory escalation of seismic activity is a well-known precursor to major earthquakes, which is used in many earthquake prediction algorithms [21 27-29]. In an economy the "static" includes various macroeconomic indicators. Their premonitory escalation has been successfully used in the prediction of recessions and surges of unemployment [5, 6].

Our results are also in accord with a distinctive common trait of precursors established in many of these studies: premonitory evolution of background activity is not monotonic, but realised sporadically, in a sequence of relatively short intermittent changes.

The universality of premonitory phenomena is limited and cannot be taken for granted in studying any specific system. Nevertheless, it is worth exploring in crime dynamics other known types of premonitory patterns, e.g. the clustering of background activity and the rise of the correlation range [19, 28, 30].

3. What is the place of our study in the broad field of prediction of crime dynamics? Specific features of our approach might be summed up as follows.

(i) We are trying to predict not the whole dynamics of homicides, but only the relatively rare phenomena - episodes of *SHS*.

(ii) Accordingly, we are looking for a quantitative and precisely defined prediction algorithm of the *"yes or no"* variety: at any moment of time such an algorithm would indicate whether or not such an episode should be expected within a fixed time interval.

(iii) Our analysis is intentionally robust, which makes the prediction algorithm more reliable and applicable in different circumstances. In our case the performance of the algorithm did not change through the period considered even though Los Angeles has witnessed many changes relevant to crime over this period. This stability is achieved at a price, however, in that the time of a homicide surge is predicted with limited accuracy and the duration of a surge even more so.

4. Our approach – a heuristic "technical" analysis - is not competing with but complementary to the cause-and-effect "fundamental" analysis. The cause that triggered a specific homicide surge is usually known, at least in retrospect. This might be, for example, a rise in drug use, a rise in unemployment, a natural disaster etc. However, that does not render predictions considered in this study redundant. On the contrary, our approach might predict an unstable situation when a homicide surge might be triggered, thus enhancing the reliability of cause-and-effect predictions.

5. It is encouraging for further studies in this direction that we used here only a small part of the relevant and available data that can be incorporated in our analysis. Among these are other types of crimes [31], economic and demographic indicators [32] and the territorial distribution of crimes. It seems worthwhile to try the same approach with other targets of prediction – e.g. surges of all violent crimes; and to other areas, e.g. separate Bureaus of the city of Los Angeles, or to other major cities. In a broader scheme of things, our analysis discriminates stable situations from unstable, where the risk of different disasters is higher.

6. At the same time it would be important to set up an experiment in advance prediction of homicide surges in Los Angeles using the algorithm hypothesised here. Successes and errors will both provide for evaluation of this algorithm and for developing a better one.

Acknowledgements. We are grateful to Professor Wellford Wilms and Dr. Robert Mehlman (University of California, Los Angeles) for valuable comments; and to Marina Dmitrenko and Tatiana Prokhorova (International Institute of Earthquake Prediction Theory and Mathematical Geophysics, Russian Academy of Sciences), and Cecile Coronel (Los Angeles Police Department) for the difficult work of data collection and pre-processing. This study was made possible by the 21st Century Collaborative Activity Award for Studying Complex Systems, granted by the James S. McDonnell Foundation (Project "Understanding and Prediction of Critical Transitions in Complex Systems").

10. References

1. Gelfand, I., Keilis-Borok, V., Knopoff, L., Press, F., Rantsman, E., Rotwain, I., and Sadovsky, A. (1976) Pattern recognition applied to earthquake epicenters in California, *Phys. Earth Planet. Inter.* **11**, 227–283.

2. Tukey, J.W. (1977) *Exploratory Data Analysis. Addison-Wesley Series in Behavioral Science: Quantitative Methods*, Addison-Wesley, Reading, Mass.
3. Gell-Mann, M. (1994) *The Quark and the Jaguar: Adventures in the Simple and the Complex*, W.H. Freeman and Company, New York.
4. Crutchfield, J.P., Farmer, J.D., Packard, N.H., and Shaw, R.S. (1986) Chaos, *Sci. Am.* **255**, 46-57.
5. Keilis-Borok, V., Stock, J.H., Soloviev, A., and Mikhalev, P. (2000) Pre-recession pattern of six economic indicators in the USA, *Journal of Forecasting* **19**, 65-80.
6. Keilis-Borok, V.I., Soloviev, A.A., Allègre, C.B., Sobolevskii, A.N., and Intriligator, M.D. (2001) Dynamics of macroeconomic indicators before the rize of unemployment in Western Europe and the USA, Sixth Workshop on Non-Linear Dynamics and Earthquake Prediction, 15 - 27 October 2001, H4.SMR/1330-11, ICTP, Trieste.
7. Keilis-Borok, V.I. and Lichtman, A.J. (1993) The self-organization of American society in Presidential and Senatorial elections, in Yu.A. Kravtsov (ed.), *Limits of Predictability*, Springer-Verlag, Berlin-Heidelberg, pp. 223-238.
8. Kosobokov, V.G. (1983) Recognition of the sites of strong earthquakes in East Central Asia and Anatolia by Hamming's method, in V.I. Keilis-Borok and A.L. Levshin (eds.), *Mathematical models of the structure of the Earth and the earthquake prediction, Comput. Seismol., 14*, Allerton Press, New York, pp. 78-82.
9. Vorobieva, I.A. (1999) Prediction of a subsequent large earthquake, *Phys. Earth Planet. Inter.* **111**, 197-206.
10. Carlson, S.M. (1998) Uniform Crime Reports: Monthly Weapon-specific Crime and Arrest Time Series, 1975-1993 (National, State, and 12-City Data), ICPSR 6792, Inter-university Consortium for Political and Social Research, P.O. Box 1248, Ann Arbor, Michigan 48106.
11. Keilis-Borok, V.I. and Rotwain, I.M. (1990) Diagnosis of time of increased probability of strong earthquakes in different regions of the world: algorithm CN, *Phys. Earth Planet. Inter.* **61**, 57-72.
12. Zaliapin, I., Keilis-Borok, V., and Ghil, M. (2001) A Boolean delay equation model of colliding cascades. Part II: Predictions of critical transitions, Sixth Workshop on Non-Linear Dynamics and Earthquake Prediction, 15 - 27 October 2001, H4.SMR/1330-2, ICTP, Trieste.
13. Gvishiani, A.D. and Kosobokov, V.G. (1981) On foundations of the pattern recognition results applied to earthquake-prone areas, *Proceedings of Ac. Sci. USSR: Physics of the Earth* **2**, 21-36 (in Russian).
14. Lichtman, A. and Keilis-Borok, V.I. (1989) Aggregate-level analysis and prediction of midterm senatorial elections in the United States, 1974-1986, *Proc. Natl. Acad. Sci. USA* **86**, 10176-10180.
15. Molchan, G.M. (1997) Earthquake prediction as a decision-making problem, *Pure Appl. Geophys.* **149**, 233-237.
16. Kossobokov, V.G., Keilis-Borok, V.I., Turcotte, D.L., and Malamud, B.D. (2000) Implications of a statistical physics approach for earthquake hazard assessment and forecasting, *Pure Appl. Geophys.* **157**, 2323-2349.
17. Bui Trong, L. (2003) Risk of collective youth violence in French suburbs. A clinical scale of evaluation, an alert system, *this volume*, 199-221.
18. Allègre, C.J., Shebalin, P., Le Mouel, J.-L., and Narteau, C. (1998) Energetic balance in scaling organization of fracture tectonics, *Phys. Earth Planet. Inter.* **106**, 139-153.
19. Gabrielov, A., Zaliapin, I., Newman, W.I., and Keilis-Borok, V.I. (2000) Colliding cascades model for earthquake prediction, *Geophys. J. Int.* **143**, 427-437.
20. Holland, J.H. (1995) *Hidden Order: How Adaptation Builds Complexity*, Addison-Wesley, Reading, Mass.
21. Newman, W.I., Turcotte, D.L., and Gabrielov, A. (1995) Log-periodic behaviour of a hierarchical failure model with applications to precursory seismic activation, *Phys. Rev. E* **52**, 4827-4835.
22. Rundle, B.J., Turcotte, D.L., and Klein, W., eds. (2000) *Geocomplexity and the Physics of Earthquakes*, Am. Geophys. Union, Washington, DC.
23. Shnirman, M.G. and Blanter, E.M. (1998) Self-organized criticality in a mixed hierarchical system, *Phys. Rev. Letters*, **81**, 5445-5448.
24. Sornette, D. (2000) *Critical Phenomena in Natural Sciences. Chaos, Fractals, Self-organization and Disorder: Concepts & Tools*, Springer-Verlag, Berlin-Heidelberg.
25. Turcotte, D.L. (1997) *Fractals and Chaos in Geology and Geophysics*, 2nd ed., University Press, Cambridge.
26. Yamashita, T. and Knopoff, L. (1992) Model for intermediate-term precursory clustering of earthquakes, *J. Geophys. Res.*, **97**, 19873-19879.

27. Keilis-Borok, V.I. and Kossobokov, V.G. (1990) Premonitory activation of earthquake flow: algorithm M8, *Phys. Earth Planet. Inter.* **61**, 73-83.
28. Keilis-Borok, V.I. (2002) Earthquake prediction: State-of-the-art and emerging possibilities, *Annu. Rev. Earth Planet Sci.* **30**, 1-33.
29. Kossobokov, V.G. and Carlson, J.M. (1995) Active zone size vs. activity: A study of different seismicity patterns in the context of the prediction algorithm M8, *J. Geophys. Res.* **100**, 6431-6441.
30. Shebalin, P., Zaliapin, I., and Keilis-Borok, V. (2000) Premonitory raise of the earthquakes' correlation range: Lesser Antilles, *Phys. Earth Planet. Inter.* **122**, 241-249.
31. Bursik, R.J., Jr., Grasmick, H.G., and Chamlin, M.B. (1990) The effect of longitudinal arrest patterns on the development of robbery trends at the neighborhood level, *Criminology* **28**, 431-450.
32. Messner, S.F. (1983) Regional differences in the economic correlates of the urban homicide rate. *Criminology* **21**, 477-488.

Appendix

TABLE A1. Binary codes of the trends for 7 types of crimes and values of $K^H(m, m\text{-}12)$
See notations in Table 1. Discretization is defined in Table 2

#	Month	Discretized trends							Δ		K^H
		Rob	FRob	KCIR	ODWR	FA	KCIA	UNFE			
1	1976:01	1	0	0	1	0	1	0	4		0.44
2	1976:02	1	0	0	0	1	1	0	4		0.21
3	1976:03	1	0	0	0	1	1	0	4		-0.07
4	1976:04	0	0	0	1	1	1	1	3		-0.28
5	1976:05	0	0	0	1	1	0	1	4		-0.35
6	1976:06	0	1	1	1	1	0	1	2		-0.33
7	1976:07	0	1	1	1	1	0	1	2		-1.18
8	1976:08	0	1	1	1	1	0	1	2		-0.96
9	1976:09	0	1	1	1	1	0	1	2		-0.87
10	1976:10	0	1	1	1	1	1	1	**1**	+	-0.79
11	1976:11	0	1	1	1	1	1	1	**1**	+	-0.37
12	1976:12	0	0	1	1	1	1	1	2		0.16
13	1977:01	0	0	0	1	0	1	0	5		0.68
14	1977:02	0	0	0	1	0	1	0	5		0.84
15	1977:03	0	0	0	1	0	1	0	5		0.38
16	**1977:04**	0	0	0	0	0	1	0	6		0.85
17	1977:05	0	0	0	1	0	1	0	5		0.07
18	1977:06	0	0	0	1	0	1	0	5		0.71
19	1977:07	0	1	0	1	0	1	0	4		0.58
20	1977:08	0	1	0	0	0	1	0	5		0.03
21	1977:09	0	1	0	0	0	1	0	5		0.18
22	1977:10	0	1	1	0	1	1	1	2		-0.17
23	1977:11	0	1	0	0	1	1	1	3		0.24
24	1977:12	0	0	0	0	1	1	1	4		0.81
25	1978:01	0	0	0	0	1	1	1	4		0.38
26	1978:02	0	0	0	0	1	1	1	4		0.16
27	1978:03	0	0	0	0	1	1	1	4		0.53
28	1978:04	0	0	0	0	0	0	0	7		-0.09
29	1978:05	0	0	0	0	0	0	0	7		0.55
30	1978:06	0	0	0	0	0	0	0	7		0.20
31	1978:07	0	0	0	0	0	0	0	7		0.57
32	1978:08	0	0	0	1	0	0	0	6		0.66
33	1978:09	0	0	0	1	0	1	0	5		1.19
34	1978:10	0	0	0	0	0	1	0	6		1.07
35	1978:11	0	0	0	0	1	1	0	5		1.72
36	1978:12	0	0	0	0	1	1	0	5		1.81
37	1979:01	0	0	0	0	1	1	1	4		2.52
38	1979:02	0	0	0	0	1	1	0	5		1.66
39	1979:03	0	0	0	0	1	1	0	5		0.88
40	1979:04	0	0	0	0	1	1	1	4		0.98
41	1979:05	0	0	0	0	1	0	1	5		0.66
42	1979:06	0	0	0	0	1	0	1	5		0.27
43	1979:07	0	0	0	0	1	0	1	5		-0.21
44	1979:08	0	1	0	0	1	0	1	4		-0.14
45	1979:09	0	1	0	1	1	1	1	2		-0.36
46	1979:10	0	0	0	0	1	1	1	4		0.45
47	1979:11	0	0	0	0	1	1	1	4		0.19
48	1979:12	0	0	0	0	1	1	1	4		2.15
49	1980:01	0	0	0	0	1	1	1	4		1.82
50	1980:02	0	0	0	0	1	1	1	4		1.49
51	1980:03	0	0	0	0	1	1	0	5		0.82
52	1980:04	0	0	0	0	1	1	0	5		1.08
53	1980:05	0	0	0	0	1	1	0	5		0.63
54	1980:06	0	0	0	0	1	0	0	6		1.51
55	1980:07	0	0	0	0	1	0	1	5		1.86
56	1980:08	0	0	0	0	1	1	1	4		2.79
57	1980:09	0	0	0	0	1	1	1	4		3.26
58	1980:10	0	0	0	0	1	1	1	4		2.56
59	1980:11	0	0	0	0	1	1	1	4		2.43
60	1980:12	0	0	0	0	1	1	1	4		2.22
61	1981:01	0	0	0	0	1	1	0	5		1.78
62	1981:02	0	0	0	0	1	1	0	5		0.85
63	1981:03	0	0	0	0	0	0	0	7		-0.57
64	1981:04	0	0	0	0	0	0	0	7		-2.02
65	1981:05	0	0	0	1	0	0	0	6		-2.64
66	1981:06	0	1	0	1	0	0	0	5		-3.35
67	1981:07	0	1	1	1	0	0	0	4		-2.51
68	1981:08	0	1	1	1	0	0	0	4		-2.10
69	1981:09	0	1	0	1	0	1	1	3		-0.92
70	1981:10	0	1	0	1	0	1	1	3		-0.02
71	1981:11	0	0	0	1	0	1	1	4		-0.10

72	1981:12	0	0	0	1	0	1	1	4		0.09
73	1982:01	0	0	0	1	0	1	1	4		0.70
74	1982:02	0	0	0	1	0	0	0	6		-0.63
75	1982:03	0	0	0	1	0	0	0	6		-0.27
76	1982:04	0	0	0	1	0	0	0	6		-0.71
77	1982:05	0	0	0	1	0	0	0	6		-0.97
78	1982:06	0	0	1	1	0	0	0	5		-1.31
79	1982:07	1	1	1	1	0	0	0	3		-1.32
80	1982:08	1	1	1	0	0	0	0	4		-0.63
81	1982:09	1	1	1	0	0	0	0	4		0.73
82	1982:10	0	1	0	0	0	1	1	4		0.45
83	1982:11	0	1	1	0	0	1	1	3		0.43
84	1982:12	0	1	0	0	0	1	1	4		0.33
85	1983:01	0	0	0	0	0	1	1	5		0.48
86	1983:02	0	0	0	0	0	1	0	6		-0.35
87	1983:03	0	0	0	1	0	0	0	6		-1.32
88	1983:04	0	0	0	1	0	0	0	6		-0.62
89	1983:05	0	1	0	1	0	0	1	4		-0.22
90	1983:06	0	1	1	1	0	0	1	3		-0.73
91	1983:07	0	1	1	1	0	0	1	3		-0.85
92	1983:08	0	1	1	1	0	0	1	3		-0.78
93	1983:09	0	1	1	1	0	0	1	3		-0.92
94	1983:10	0	1	1	1	0	0	1	3		0.59
95	1983:11	0	1	1	0	0	1	1	3		0.90
96	1983:12	0	1	1	0	1	0	1	3		0.41
97	1984:01	0	0	0	0	1	0	0	6		0.58
98	1984:02	0	1	0	0	1	0	0	5		0.20
99	1984:03	0	0	0	0	1	0	0	6		-0.93
100	1984:04	0	0	1	0	0	0	0	6		-1.16
101	1984:05	0	0	1	1	0	0	0	5		-0.76
102	1984:06	1	1	1	1	0	0	0	3		-0.57
103	1984:07	1	1	1	1	0	0	0	3		-1.04
104	1984:08	1	1	1	1	0	0	1	2		-0.55
105	1984:09	1	1	1	1	1	0	1	**1**	+	-0.21
106	1984:10	1	1	1	1	0	1	1	**1**	+	-0.20
107	1984:11	1	1	1	1	0	1	1	**1**	+	0.33
108	1984:12	0	1	0	1	0	1	1	3		0.25
109	1985:01	0	0	0	0	0	1	0	6		0.01
110	1985:02	0	0	0	0	0	1	0	6		0.73
111	**1985:03**	0	0	0	0	0	1	0	6		0.45
112	1985:04	0	0	0	0	0	1	0	6		-0.08
113	1985:05	0	0	0	0	0	0	0	7		-0.15
114	1985:06	0	0	0	0	0	0	0	7		-0.51
115	1985:07	0	1	0	0	0	0	0	6		-0.61
116	1985:08	0	1	0	1	1	0	1	3		-0.34
117	1985:09	0	1	1	1	1	0	1	2		-0.08
118	1985:10	0	1	1	1	1	0	1	2		-0.11
119	1985:11	0	1	1	1	1	0	1	2		0.19
120	1985:12	0	1	1	1	1	0	1	2		0.70
121	1986:01	0	1	0	1	1	0	1	3		-0.04
122	1986:02	0	0	1	1	1	0	0	4		-0.74
123	1986:03	0	0	1	1	0	0	0	5		0.29
124	1986:04	0	0	1	1	0	0	0	5		0.77
125	1986:05	0	0	1	1	0	0	0	5		0.42
126	1986:06	0	0	1	1	0	0	0	5		0.15
127	1986:07	0	0	1	1	1	0	0	4		-0.08
128	1986:08	0	0	0	0	1	0	1	5		0.76
29	1986:09	0	0	0	0	1	1	1	4		1.90

130	1986:10	0	0	0	0	1	1	1	4		1.56
131	1986:11	0	0	0	0	1	1	1	4		0.98
132	1986:12	0	0	0	0	1	1	1	4		0.59
133	1987:01	0	0	0	0	1	1	1	4		0.66
134	1987:02	1	0	0	0	1	1	1	3		0.11
135	1987:03	1	0	0	0	1	1	1	3		-0.43
136	1987:04	1	0	1	1	1	0	1	2		0.14
137	1987:05	1	1	1	1	0	0	1	2		-0.20
138	1987:06	1	1	1	1	0	0	1	2		-0.76
139	1987:07	1	1	1	1	0	0	0	3		-1.52
140	1987:08	1	1	1	1	0	0	0	3		-1.48
141	1987:09	1	1	1	1	1	0	1	**1**	+	-0.90
142	1987:10	1	1	1	1	1	1	1	**0**	+	-0.13
143	1987:11	0	1	1	1	1	1	1	**1**	+	-0.01
144	1987:12	0	1	1	1	1	1	1	**1**	+	0.29
145	1988:01	0	1	1	1	1	1	0	2		-0.01
146	1988:02	0	0	0	1	1	1	0	4		-0.35
147	1988:03	0	0	0	1	1	1	0	4		-0.84
148	1988:04	0	0	0	1	0	0	0	6		-0.96
149	1988:05	0	1	1	1	0	1	0	3		-0.69
150	1988:06	0	1	1	1	0	1	0	3		-1.21
151	1988:07	0	1	1	1	0	1	1	2		-0.68
152	**1988:08**	0	1	1	1	0	1	1	2		-0.66
153	1988:09	0	1	1	0	1	1	1	2		-0.13
154	1988:10	0	1	0	0	1	1	1	3		0.27
155	1988:11	0	1	0	0	1	1	1	3		0.04
156	1988:12	0	0	0	0	1	1	1	4		0.15
157	1989:01	0	0	0	0	1	1	1	4		1.38
158	1989:02	0	0	0	0	1	1	1	4		1.41
159	1989:03	0	0	0	0	1	0	0	6		1.13
160	1989:04	0	0	0	0	1	0	0	6		0.62
161	1989:05	0	0	0	0	1	0	1	5		0.21
162	1989:06	0	0	0	0	1	0	1	5		0.34
163	1989:07	0	0	0	0	1	0	1	5		-0.15
164	1989:08	0	0	0	0	1	0	1	5		0.39
165	1989:09	0	0	0	0	1	1	1	4		1.19
166	1989:10	0	0	0	0	1	1	1	4		1.47
167	1989:11	0	0	0	0	1	1	1	4		1.20
168	1989:12	0	0	0	0	1	1	1	4		1.18
169	1990:01	0	0	0	0	1	1	1	4		0.97
170	1990:02	0	0	0	0	0	0	0	7		1.60
171	1990:03	0	0	0	0	0	0	0	7		1.15
172	1990:04	0	0	0	0	0	0	0	7		0.92
173	1990:05	0	0	0	0	0	0	0	7		0.21
174	1990:06	0	0	0	0	0	0	0	7		-0.44
175	1990:07	0	0	0	0	1	1	0	5		0.74
176	1990:08	0	0	0	0	1	1	1	4		1.63
177	1990:09	0	0	0	0	1	1	1	4		1.64
178	1990:10	0	0	0	0	1	1	1	4		2.25
179	1990:11	0	0	0	0	1	1	1	4		1.24
180	1990:12	0	0	0	0	1	1	1	4		0.20
181	1991:01	0	0	0	0	0	1	0	6		0.04
182	1991:02	0	0	0	1	0	1	0	5		-1.16
183	1991:03	0	0	0	1	0	0	0	6		-1.38
184	1991:04	0	0	1	1	0	0	0	5		-1.40
185	1991:05	0	0	1	1	0	0	0	5		-1.60
186	1991:06	0	0	1	1	0	0	0	5		-1.14
187	1991:07	0	0	1	1	1	0	0	4		-1.10

188	1991:08	0	0	0	1	1	0	0	5	1.12
189	1991:09	0	0	0	1	1	0	1	4	2.63
190	1991:10	0	0	0	0	1	1	1	4	3.04
191	1991:11	0	0	0	0	1	1	1	4	2.86
192	1991:12	0	0	0	0	1	1	1	4	2.60
193	1992:01	0	0	0	0	1	1	0	5	1.84
194	1992:02	0	0	0	0	1	1	0	5	0.95
195	1992:03	0	0	0	0	1	0	0	6	-0.58
196	1992:04	0	0	0	1	0	0	0	6	-0.88
197	1992:05	0	0	1	1	0	0	0	5	-1.70
198	1992:06	0	1	1	1	0	0	0	4	-2.75
199	1992:07	0	1	1	1	0	0	0	4	-1.21
200	1992:08	0	1	1	1	0	0	0	4	0.10
201	1992:09	0	1	1	1	0	0	0	4	1.72
202	1992:10	0	1	1	1	1	0	1	2	2.35
203	1992:11	0	1	1	1	1	0	1	2	2.47
204	1992:12	0	0	1	1	0	0	0	5	1.76
205	1993:01	0	0	0	1	0	0	0	6	2.04
206	1993:02	1	0	0	1	0	0	0	5	1.04
207	1993:03	1	0	0	1	0	0	0	5	0.21
208	1993:04	1	0	0	1	0	0	0	5	-0.74
209	1993:05	1	0	1	1	0	0	0	4	-0.46
210	1993:06	1	1	1	1	0	0	0	3	-1.05
211	1993:07	1	1	1	1	0	0	0	3	-1.60
212	1993:08	0	1	1	1	0	0	0	4	-1.00
213	1993:09	0	0	0	1	0	0	1	5	-0.20
214	1993:10	0	0	0	1	0	1	1	4	0.14
215	1993:11	0	0	0	1	0	1	1	4	0.52
216	1993:12	0	0	1	1	0	1	1	3	-0.12

PERGAMON

Available at
www.ElsevierComputerScience.com
POWERED BY SCIENCE DIRECT

Pattern Recognition 38 (2005) 423–435

PATTERN RECOGNITION
THE JOURNAL OF THE PATTERN RECOGNITION SOCIETY

www.elsevier.com/locate/patcog

Patterns of macroeconomic indicators preceding the unemployment rise in Western Europe and the USA

V.I. Keilis-Borok[a, b, *], A.A. Soloviev[a], C.B. Allègre[c], A.N. Sobolevskii[a, d], M.D. Intriligator[e]

[a] International Institute of Earthquake Prediction Theory and Mathematical Geophysics, Russian Academy of Sciences, Warshavskoye shosse 79 kor. 2, Moscow 117556, Russia

[b] Department of Earth and Space Sciences, University of California, Los Angeles, Institute of Geophysics and Planetary Physics, 405 Hilgard Ave., IGPP, Los Angeles, CA 90095-1567, USA

[c] Department of Social and Labor Studies, University of Paris I, Pantheon-Sorbonne, Place du Partheon 75005, Paris, France

[d] Department of Physics, Moscow State University, Vorobievy Gory, Moscow 119899, Russia

[e] Department of Economics, University of California, Los Angeles, Box 951477, Los Angeles, CA 90095-1477, USA

Received 1 December 2003; received in revised form 30 July 2004; accepted 18 August 2004

Abstract

This study concerns the fields of economics, labor relations, and the dynamics of complex systems. We consider a specific phenomenon in the dynamics of unemployment—episodes of a sharp increase in the unemployment rate, called here "fast acceleration of unemployment" (FAU). Our study is a "technical" analysis that is a heuristic search of phenomena preceding FAUs. We use the methodology of *pattern recognition of infrequent events* developed by the artificial intelligence school of Gelfand for a study of rare phenomena of highly complex origin, that, by their nature, limit the possibilities of using classical statistical or econometric methods. Our goal is to identify by an analysis of macroeconomic indicators a robust and rigidly defined prediction algorithm of the "*yes or no*" variety indicating at any time moment, whether a FAU should be expected or not within the subsequent months. Considering unemployment in France between 1962 and 1997, we have found a specific "premonitory" pattern of three macroeconomic indicators that may be used for algorithmic prediction of FAUs. Among seven FAUs identified within these years six are preceded within 12 months by this pattern that appears at no other time. The application of this algorithm to Germany, Italy and the USA yields similar results. Such predictability reflects the fact that the economy, like other complex systems, exhibits regular collective behavior patterns. The final test, as in any prediction research, should be advance prediction. The first such prediction, for the USA for early 2000, has been correct.

Keywords: Unemployment; Prediction; Pattern recognition of infrequent events; Macroeconomic indicators; Complex system; Critical phenomenon

1. Introduction

1.1. The problem: prediction of FAUs

We consider here a specific phenomenon in the dynamics of unemployment: a sharp increase in the rate

* Corresponding author: Department of Earth and Space Sciences, University of California, Los Angeles, Institute of Geophysics and Planetary Physics, 405 Hilgard Ave., IGPP, Los Angeles, CA 90095-1567, USA. Tel.: +1 310 206 5667; fax: +1 310 206 3051.
E-mail address: vkb@ess.ucla.edu (V.I. Keilis-Borok).

0031-3203/$30.00 © 2004 Pattern Recognition Society. Published by Elsevier Ltd. All rights reserved.
doi:10.1016/j.patcog.2004.08.005

of unemployment growth. Qualitatively this phenomenon is illustrated in Fig. 1. The thin line is the monthly number of unemployed $u(t)$, including seasonal variations. After smoothing $u(t)$ to eliminate such variations, we obtain the function $U(t)$. The target of our prediction is the starting time of a strong and lasting increase in the unemployment rate $U(t)$. An example is the turning point indicated by the arrow in Fig. 1. We call this target by the acronym FAU, for "fast acceleration of unemployment". Our goal is to design an algorithm for predicting such FAUs by an analysis of macroeconomic indicators.

Such an algorithm, if found and validated, may be useful in two ways: (i) As a quantitative and reproducible description of phenomena that are premonitory to FAUs that would provide empirical constraints for the theoretical modeling of unemployment. (ii) As a practical tool, complementing existing methods for predicting unemployment. These are the usual twofold goals of modeling and prediction in any such research. In Sections 2–5 we develop such an algorithm for France, while in Sections 6 and 7 we test it on independent data for Germany, Italy, and USA. Finally, we use the algorithm to predict unemployment growth in the USA.

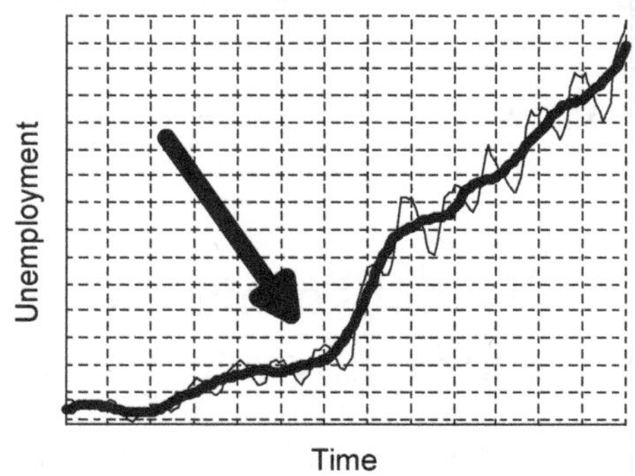

Fig. 1. Fast acceleration of unemployment (FAU): schematic definition. Thin line—monthly unemployment; with seasonal quasiperiodic variations. Thick line—monthly unemployment, with seasonal variations smoothed away. The arrow indicates a FAU—the sharp bend of the smoothed curve. The moment of a FAU is the target of prediction.

1.2. Methodology

This study belongs to the so-called "technical" analysis, consisting of a heuristic search for phenomena preceding FAUs. The alternative would be a "fundamental" analysis, focusing on "cause-and-effect" mechanisms leading to a FAU. The methodology we use for technical analysis is the *pattern recognition of infrequent events*. It was developed by the artificial intelligence school of the Russian mathematician Gelfand [1] for the study of rare phenomena of highly complex origin. This approach differs from but complements classical statistical and econometric methods such as regression analysis and ARIMA [2]; see also [3–5]. For comparison of this approach with the multiple regression analysis see [6] where a preliminary algorithm for the prediction of economic recessions is developed. We do not use multiple regression analysis but we use the linear regression coefficients for single macroeconomic indicators to approximate the trends. The pattern recognition approach has been successfully applied also to the prediction of the outcome of American elections [7,8] as well as in seismology and earthquake prediction [1,9–12], geological prospecting [13] and in many other fields, as given in the references in these papers. We use here the simplest version of this methodology, called the "Hamming distance" [7,12,14, and refs. therein]. It is applied for classification of binary vectors into two classes on the basis of learning samples, which are used to determine a reference binary vector (kernel) with components that are typical for one class. The binary vectors under consideration are classified as belonging to this class if the Hamming distance from them to the kernel is not greater than a certain threshold.

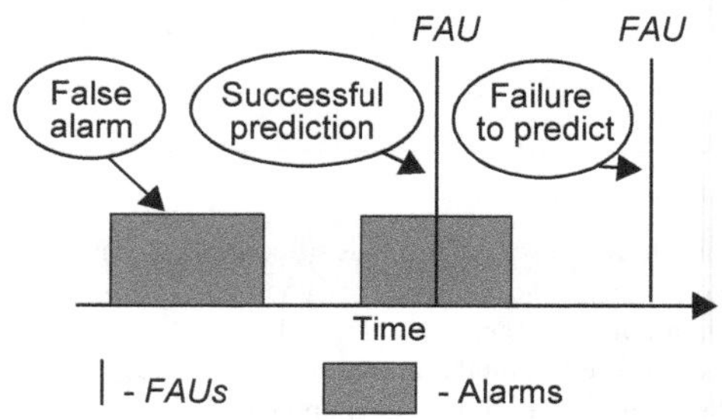

Fig. 2. Possible outcomes of prediction.

1.3. What is the place of our approach in unemployment prediction research?

The specific features of our approach might be summed up as follows.

(i) We are trying to predict not the whole dynamics of unemployment but only relatively rare and extraordinary phenomena—the FAUs.
(ii) We are looking for a quantitative and precisely defined prediction algorithm of the "*yes or no*" variety, where, at any moment of time, the algorithm would indicate whether or not a FAU should be expected within the subsequent r months. The possible outcomes of such a prediction are illustrated in Fig. 2. The probabilistic nature of such prediction is reflected in the estimation of the probabilities of false alarms, failures to predict, and of the relative time occupied by alarms, all of which are discussed below.

(iii) Our analysis is intentionally robust, which makes the prediction algorithm more reliable and applicable to different FAUs despite the diversity of their specific causes. This is achieved at a price, however, in that the exactness of prediction is limited.

(iv) Our analysis does not compete with but rather complements the short-term predictions of the "*cause and effect*" kind. The cause that triggered a specific FAU is usually known, at least in retrospect. It may be, for example, a certain governmental decision, a change in international trade, a major event leading to changes in expectations (such as September 11, 2001) etc. Accordingly, one might predict an imminent FAU (possibly not so rigidly defined) when a triggering event occurs. That does not exclude predictability of FAUs with a longer lead-time, as in the present study. On the contrary, it seems plausible that a FAU may be triggered only if and when the situation becomes "ripe" for a FAU; otherwise the government would not make that decision; unemployment would be less sensitive to international trade etc. If that conjecture is correct our approach can predict such a "ripe" situation. This may be of independent interest for the modeling and containment of unemployment, and it may help in the short-term prediction of FAUs.

(v) The dynamics of the economy, including changes in unemployment, may have some basic features that are common in various complex processes and found in models of statistical physics [12,15–21]. Our results may provide for such modeling of the scaling and parameterization of the processes considered and a formal definition of the relevant critical phenomena.

2. The data

We used databases issued by the Organization for Economic Cooperation and Development [22] and the International Monetary Fund [23]. Past FAUs were identified by an analysis of the monthly statistics of unemployment. To explore the predictability of FAUs we analyzed the monthly indicators listed below. For France our data sources are sufficiently complete for the time period between January 1965 and May 1997, and it is this period that we consider here.

Composite characteristics of the French national economy:

1. *IP*: Industrial production index, composed of weighted production levels in numerous sectors of the economy, in % relative to the index for 1990.
2. *L*: Long-term interest rate on 10-year government bonds, in %.
3. *S*: Short-term interest rate on 3-month bills, in %.

The analogues of these indicators for the USA have been successfully used in research on predicting American economic recessions [6,24].

Characteristics of more narrow areas of the French economy, which are sensitive to its overall state:

4. *NC*: The number of new passenger car registrations, in thousands of units.
5. *EI*: The expected prospects for the national industrial sector.
6. *EP*: The expected prospects for selected manufacturers.
7. *EO*: The estimated current volume of orders.

The last three indicators are subjective; they distinguish the "good" from "bad" situations by polling 2500 manufacturers, with the expectations weighted by the size of their businesses. They are available only for France.

Two indicators related to the American economy:

8. *FF/$*: Value of US dollar in French francs.
9. *AR*: The state of the American economy: is it close to a recession or not? This indicator shows whether a pre-recession alarm for the USA is or is not declared by the algorithm described in Ref. [6]. Note that we consider here not the American recessions themselves but a certain precursor to these recessions, which includes no European indicators.

3. Identification of FAUs

Common notation here and below is $W^f(l/q, p)$—the local linear least-squares regression of a function $f(m)$ within the sliding time window (q, p)

$$W^f(l/q, p) = K^f(q, p)l + B^f(q, p), \quad q \leqslant l \leqslant p. \qquad (1)$$

3.1. Definitions

Here, we formalize the definition of FAU that was qualitatively described in Section 1.1 (Fig. 1). As before, we denote by $u(m)$ the number of unemployed in a month $m (m = 1, 2, \ldots)$. FAUs are defined as follows. First, smoothing out the seasonal variation of u we obtain function $U(m) = W^u(m/m - 6, m + 6)$—a value of the regression (1) over the time interval $(m - 6, m + 6)$. Next, we determine a function $F(m/s) = K^U(m + s, m) - K^U(m, m - s)$—the difference between the linear trends in regression (1) of $U(m)$ within s subsequent months and s preceding months. This function with $s = 24$ months is used as a coarse measure of unemployment acceleration. Finally, the FAUs are defined by the local maxima of $F(m)$ exceeding a certain threshold $\boldsymbol{F}$. The time m^* and the height F^* of such a maximum are, respectively, the time and the magnitude of a FAU.

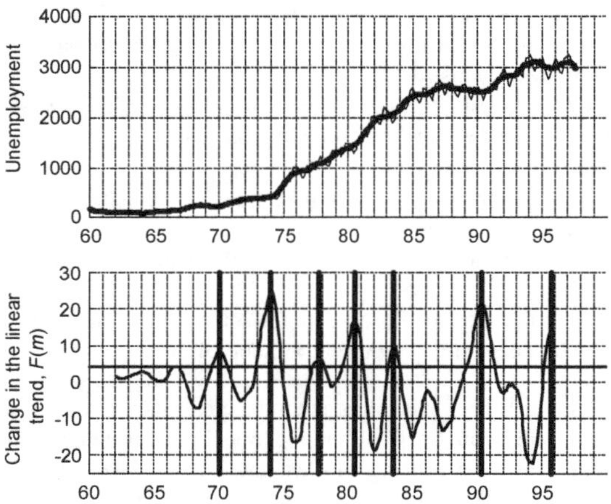

Fig. 3. Unemployment in France. Top: Monthly unemployment, thousands of people. Thin line: $u(m)$, data from the OECD database; note the seasonal variations. Thick line: $U(m)$, data smoothed over one year. Bottom: Determination of FAUs. $F(m)$ shows the change in the linear trend of unemployment $U(m)$. FAUs are attributed to the local maxima of $F(m)$ exceeding threshold $\boldsymbol{F} = 4.0$ shown by solid horizontal line. The thick vertical lines show moments of the FAUs.

Acceleration ends in a month m_e of the subsequent local minimum of $F(m)$.

3.2. FAUs in France

Monthly unemployment in France is shown in Fig. 3. The function $F(m)$ for the time period considered, from January 1965 to May 1997, is also shown in Fig. 3. One may see in Fig. 3 that the threshold 4 identifies obviously outstanding peaks of $F(m)$. Seven FAUs identified by the condition $F^* \geqslant F = 4$ are listed in Table 1. As we see, each such FAU is the beginning of a long unemployment rise, lasting 16–24 months. Since it is determined after a strong smoothing of the unemployment rate, the meaningful accuracy of prediction may hardly be better than about 2 months. Three "major" FAUs, marked in bold in Table 1, are distinctly larger than the others.

Besides the magnitude F^*, the scale of FAU phenomenon may be characterized by the "*glut*" of unemployment G (Fig. 4)

$$G = \sum \{U(l) - W^U(l/m^* - s, m^*)\}, \quad m^* \leqslant l \leqslant m_e. \quad (2)$$

Here m^* and m_e are as defined above the first and the last month of the FAU. The first term is the actual cumulative loss of employment, in person-months, during the FAU. The second term shows what this loss would have been had unemployment not accelerated but rather grew at the same rate as right before the FAU.

The values of G calculated for FAUs by means of (2) are given in the last line of Table 1.

4. Premonitory trends of single indicators

Here, we explore the "premonitory" trends of the indicators, which tend to occur more frequently as a FAU approaches.

Definition: We approximate the trends by the regression coefficients $K^I(m, m - s)$ that are defined as in (1) with f replaced by the symbol of an indicator. The value of K^I may be used for prediction since it is attributed to the *end* of the time window where it is determined; accordingly it does *not* depend on information on the future. (Our definition of the FAUs is applicable only in retrospect, two years after a FAU occurs in order to ensure a reliable identification of past FAUs).

Discretization: An example of further analysis is given in Fig. 5. It shows the indicator $L(m)$ and its trend $K^L(m/s) = K^L(m, m - s)$. The plot of $K^L(m/s)$ suggests that its peaks (i.e. the relatively steep upward trends of L) emerge more frequently before the FAUs and in that sense they are "premonitory" to FAUs. To give a robust quantitative definition of a premonitory trend, we define the values of $K^L(m/s)$ on the lowest level of resolution, distinguishing only the values above and below a threshold $T^L(Q)$. It is defined as a percentile of a level Q, that is, by the condition that $K^L(m/s)$ exceeds $T^L(Q)$ during Q% of the months considered.

Premonitory trends: Similar analyses of other indicators are summarized in Table 2 . Premonitory behavior of these indicators has a transparent qualitative explanation.

(i) A steep rise in macroeconomic indicators (Nos. 1–3 in Table 1) reflects the "overheating" of the economy. For industrial production (No. 1) this may sound counterintuitive, since the rise of production is supposed to create more jobs.

Table 1
FAUs in France, 1965–1997

Time, *year: month*	'70:01	'74:01	'77:09	'80:07	'83:07	'90:05	'95:09
"Magnitude" F^*	7.6	22.6	5.3	15.7	9.2	20.3	9.4
Duration ($m_e - m^*$) (months)	22	24	16	20	19	21	20
G (person × months)	1993	5570	371	2760	1400	3420	2960

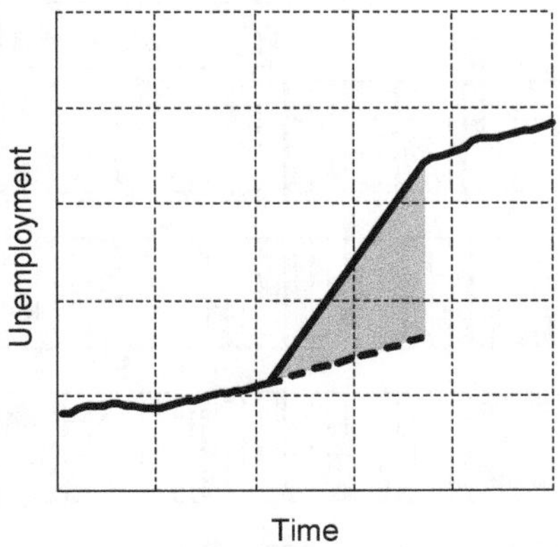

Fig. 4. Glut of unemployment: schematic definition. Its measure is the shaded area, showing the amount of person-months of jobs lost. The dashed line is the extrapolation of the pre-FAU trend.

Table 2
Trends and thresholds

Indicator	Premonitory trend	s	Q (%)
IP: industrial production index	Upward	12	50
L: interest rate, long-term bonds	Upward	12	33
S: interest rate, short-term bills	Upward	12	25
NC: new passenger cars registrations	Downward	6	33
EI: prospects for industrial sector	Downward	6	33
EP: prospects for selected manufacturers	Downward	6	33
EO: orders	Downward	6	33
$FF/\$$: french francs per USD, exchange rate	Downward	6	33
AR: recession alarm in the US	Is current		

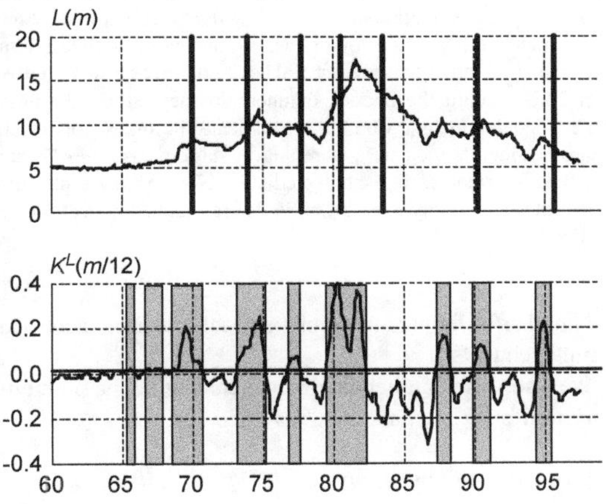

Fig. 5. Robust discretization of an economic indicator. Top: $L(m)$, interest rate on long-term (10 year) governmental bonds (%). Source: OECD. Bottom: $K^L(m/12)$, linear growth of $L(m)$ in the sliding time window $(m-12, m)$. The thin horizontal line is the upper 33% percentile for K^L so that K^L exceeds this threshold 33% of time. The gray bars mark time intervals when the value of $K^L(m/12)$ exceeds the threshold. The thick vertical lines show moments of FAUs. Note that after a quiet period before 1967 "large" values of $K^L(m/12)$ tend to appear more frequently when a FAU is approaching. However, with this threshold they would "predict" the FAUs only in combination with other indexes.

However, premonitory to FAUs is only a steep rise, above 50%, which creates oversupply (see also item 4 in the Discussion below).

(ii) A steep decline of the next four indicators reflects the expectations of the general public (No. 4) and the appraisal of economic trends by the business community (Nos. 5–7). These indicators seem better substantiated, than, say, public opinion polls.

(iii) The last two indicators (Nos. 8 and 9) are related to the American economy. Their premonitory behavior would be more difficult to foresee—an equally plausible explanation could be probably found for the opposite behavior. Accordingly, these indicators yield particularly strong constraints on the modeling of the dynamics of unemployment.

Summing up, we have reduced the description of the unemployment-relevant situation to a monthly time series of a binary vector with 9 components, as is usual in the pattern recognition of infrequent events. For convenience, we give the same code, 1, to the "premonitory" trend of each indicator, regardless of whether it is an upward or a downward one.

5. Ensemble of premonitory trends

Here, we consider how the approach of a FAU is reflected in the *collective behavior* of the indicators. Its simplest description is function $\Delta(m)$—the number of non-premonitory indicators for the month m. If our identification of premonitory trends is correct then the value of $\Delta(m)$ should decrease as a FAU approaches. By definition $\Delta(m)$ is the number of zeros in the binary code of the situation. This is the so-called "Hamming distance" between that code and the code of the "perfect" premonitory situation, when all the components are equal to 1, that is, all the trends are all premonitory. This measure was used in different applications of the pattern recognition—to elections [7], recessions [6], and earthquakes [12,25,26].

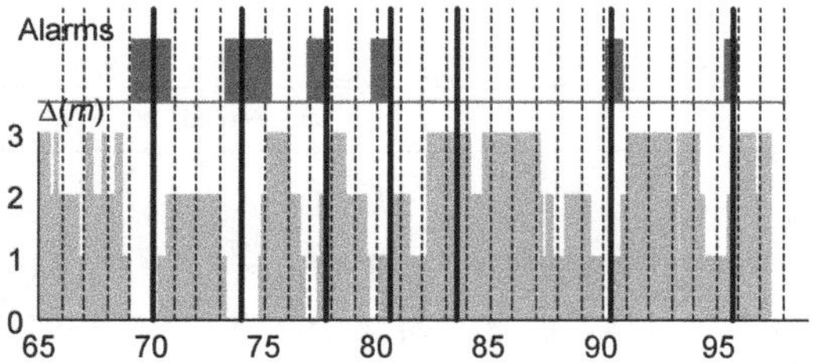

Fig. 6. Collective performance of premonitory trends. Function $\Delta(m)$ is the number of *non-premonitory* trends at month m. Vertical lines show FAUs. Alarms (shown by dark gray bars) are declared for 6 months after each month when $\Delta(m) = 0$.

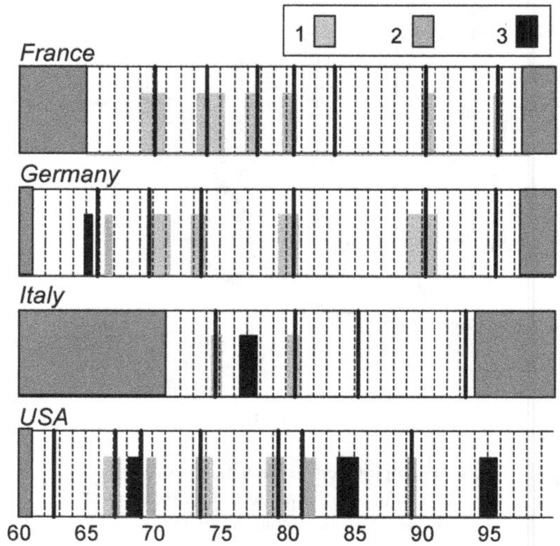

Fig. 7. Retrospective prediction for four countries: FAUs and alarms obtained by the prediction algorithm. The thick vertical lines show the moments of FAUs in a country. Bars—the alarms with different outcome: 1—alarms that predict FAUs, 2—alarms starting shortly after FAUs within the periods of unemployment surge, 3—false alarms. Shaded areas on both sides indicate the times, for which data on economic indicators were unavailable. In case 2 FAUs are counted, for rigor sake, as not predicted. Note however that the alarms starting shortly after these FAUs are practically useful (see Section 7.2).

5.1. Three composite indicators

We consider first only the composite indicators *IP*, *L*, and *S*. The value of $\Delta(m)$ may vary in this case from 0 to 3. Change of $\Delta(m)$ through the time considered is juxtaposed with FAUs in Fig. 6. We see that the minimal value $\Delta(m)=0$ appears within 1–12 months before a FAU and at no other time. Data on Fig. 6 suggest the following hypothetical prediction algorithm: *An alarm is declared for 6 months after each month with* $\Delta(m)=0$ (*regardless of whether this month belongs or not to an already determined alarm*). A waiting period of 6 months is introduced because in three cases (1977, 1980, and 1995) our premonitory pattern does not appear right before a FAU. Results of prediction are shown in the top row of Fig. 7. We see that this algorithm predicts 6 out of 7 FAUs, including all three major ones.

Next, we need a stability test. The above algorithm is developed retrospectively, with a certain freedom in the ad hoc choice of indicators and numerical parameters. The results of such a prediction should be stable to their variations. To estimate that stability we have to repeat the prediction with different groups of indicators and values of parameters.

5.2. Other indicators

We considered five other groups of indicators (Table 3). The results are compared in the error diagram in Fig. 8. Error diagrams as a tool for the evaluation of a prediction method and for the optimal choice of a response to a prediction are described in Refs. [12,27]. Numerical values of τ (the total duration of alarms in per cent of the total time considered), $\boldsymbol{n}$ (the number of failures to predict), and $\boldsymbol{f}$ (the number of false alarms) are given in the last three columns of Table 3.

We see that composite indicators (group 1) give relatively the best predictions, alone or with some other indicators. Only the group 3 gives no failures to predict; however that is achieved at the price of 5 false alarms, which are hard to accept. We tried to replace the indicator *EO* in groups 3–6 by a subjective indicator *EI* or *EP*, but in each case the number of failures to predict and false alarms increased simultaneously.

We also varied numerical parameters that define premonitory trends (Q^I, s), alarms (D, T) and FAUs ($\boldsymbol{F}$, s).

5.3. Stability of prediction (sensitivity analysis)

The prediction described in the paper involves a retrospective analysis, with a certain freedom in the ad hoc choice of indicators and of adjustable numerical parameters. We have explored the stability of these predictions, repeating them with various indicators and parameters and comparing the ensuing alarms. For that purpose, we generalize the rule for declaring the alarms as follows: an alarm covers a continuous sequence of the months with $\Delta(m) \leqslant D$; this sequence is extended to r subsequent months, where the hypothetical prediction algorithm formulated in the paper corresponds to the case $D = 0$, $r = 6$.

Variation of adjustable parameters: We varied both the percentile Q (by 5–10%) and the window s for determination of the regression coefficients (from 9 to 15). The outcomes of several experiments on the variation of the adjustable parameters are shown in Table 4. These experiments were made for the prediction based on the first group of indicators: *IP*, *L*, and *S* (line 1 in Table 3).

Table 3
Variation of the groups of indicators used for prediction

# No.	Indicators	D	r	τ (%)	n	f
1	*IP, L, S*	0	6	23	1	0
2	*IP, L, S, FF/$*	0	6	13	3	1
3	*NC, EO*	0	3	31	0	5
4	*IP, L, S, NC, EO*	1	6	26	1	1
5	*IP, L, S, NC, EO, FF/$*	1	6	19	2	1
6	*IP, L, S, NC, EO, AR*	2	6	28	1	0

The last three columns show the performance of a prediction algorithm.

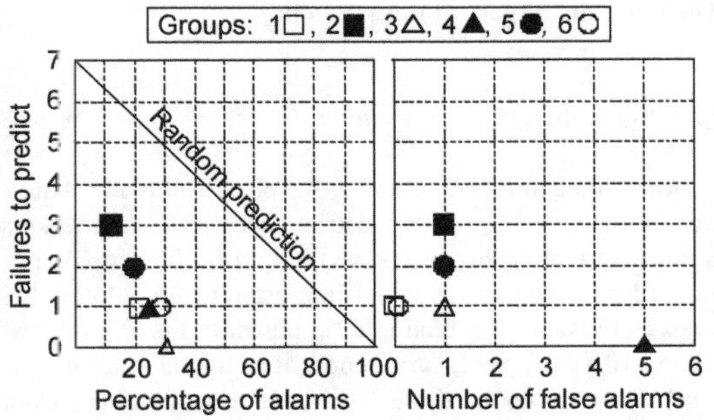

Fig. 8. Error diagram for predictions with various groups of indicators. Group 1 includes *IP, L, S*; 2—*IP, L, S, FF/$*; 3—*NC, EO*; 4—*IP, L, S, NC, EO*; 5—*IP, L, S, NC, EO, FF/$*; 6—*IP, L, S, NC, EO, AR*.

Variation of definition of FAU: Two adjustable parameters are used to define the moment of FAUs: the time window s for evaluation of the trend of unemployment and the threshold $\boldsymbol{F}$ for identification of "fast" acceleration. The analysis given in the paper corresponds to the case $s = 24$ months and $\boldsymbol{F} = 4$. Variation of s from 21 to 27 does not shift the moments of FAUs by more than 1–2 months—quite within the meaningful accuracy of prediction. The variation of $\boldsymbol{F}$ in the range from 3.6 to 5.0 does not lead to the emergence or disappearance of the targets of our prediction—the local maxima of $F(m)$.

Summing up, we conclude that the *most stable results are obtained with group 1 indicators—the three composite economic indicators*. Prediction with this group produced no false alarms; it sustains best the change of numerical parameters; and it does not gain by the addition of other indicators.

Next, we test our algorithm on the independent data not used in its development. In Section 6 we apply it to Germany and Italy and in Section 7 to the USA.

6. Germany and Italy

Here we apply the algorithm suggested in Section 5.1 to the data for Germany and Italy. We use *exactly* the same algorithm, including the definition of the trends $K^I(m, m-s)$; the values of adjustable numerical parameters; and the rule for declaration of alarms. Such transfer from country to country is possible since the algorithm is self-adaptive, where the percentiles Q^I common in each application determine the thresholds for discretization of the trends.

Table 4
Variation of adjustable parameters for indicators *IP*, *L*, and *S*

# No	Indicator(s), for which variation of a parameter is made	Variation of the parameter	τ (%)	n	f
Variation of Q					
1	*IP*	60%	23	1	0
2		40%	22	1	0
3	*L*	40%	28	1	1
4		25%	15	2	0
5	*S*	30%	30	1	2
6		20%	23	1	0
Variation of s					
7	*IP, L, S*	15	20	1	0
8		9	24	2	1

The last three columns summarize the performance of a prediction algorithm.

Table 5
Comparative performance of the prediction algorithm for France, Germany, Italy, and the USA

Country	FAUs			Alarms "not predicting" FAUs	
	Total	"Predicted"	"Missed"	After FAUs	"False"
France, 1965–1997	7	6	1	None	None
Germany, 1961–1997	6	4	2	1	1
Italy, 1971–1993	4	2	2	None	1
USA, 1961–1999	7	4	3	2	3

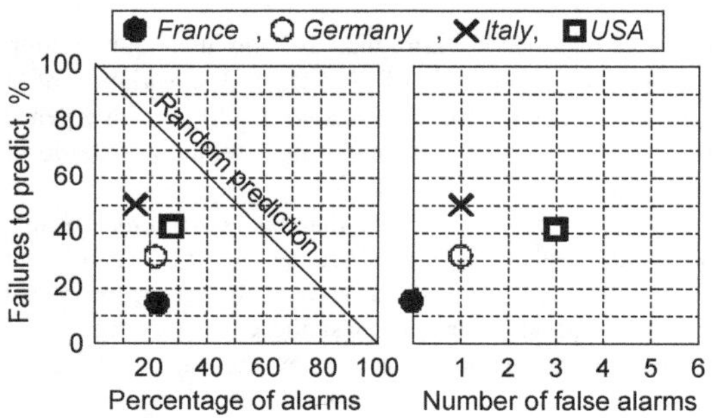

Fig. 9. Error diagram for predictions for different countries.

The results are shown in Fig. 7. Table 5 and Fig. 9 compare the quality of prediction for different countries. Alarms preceding FAUs are usually continuing within the periods of fast unemployment growth (after FAUs), and such continuations should obviously not be regarded as false alarms. Alarms within the periods of fast unemployment growth are also not regarded as false ones. For Germany the quality of prediction is about the same as for France, while for Italy it is lower, though still better than random.

7. The United States

7.1. Identification of FAUs

We use the data on monthly unemployment rates for the US civilian labor force, as given in Ref. [28]. Unlike Europe, unemployment in USA had no general trend during the years considered. One can see this in Fig. 10. The FAUs are the times when unemployment started to rise, that are the local minima of the unemployment rate. They are formally defined as follows. Let $R(m)$ be the smoothed monthly unemployment rate in a month m. Then $R(m)$ has the local minima in a month m^* if for $j=1,2,3,4$ $R(m^*-j) \geqslant R(m^*)$ and $R(m^*+j) > R(m^*)$. Seven such minima are identified within the period 1960–1999 in 1962:08 (9), 1967:03 (3), 1969:02 (28), 1973:07 (24), 1979:05 (19), 1981:03 (21), and 1989:05 (38). The duration of the unemployment rise is given in brackets after the corresponding months m^*, which are the targets of our prediction.

7.2. Application of the algorithm

Data: Indicators *IP*, *S*, and *L* have the following American equivalents. For IP-"industrial production, total". This is index of real (constant dollars) output in the entire economy (dimensionless). For *S*—interest rate on 90-day US treasury bills, at an annual rate (in percent). For *L*—interest rate on 10-year US treasury bonds, at an annual rate (in percent). Time series of these indicators were obtained from the CITIBASE, where *IP*, *FYGM3*, and *FYGT10* are their respective mnemonics.

Alarms and FAUs are juxtaposed in the bottom row of Fig. 7. We see that 4 out of 7 FAUs are captured by alarms; three FAUs, in 1962, 1969, and 1981, are missed; and we have three false alarms, in 1968, 1983, and 1994. The alarms within the periods of unemployment growth are not regarded as false ones. Total count of errors for the *USA* is given in Table 5 and Fig. 9. It is worse than for European countries, though still better than random. Note that this is a rigorous count, giving lower estimate for the algorithm's performance. Such an estimate is necessary for some purposes, e.g. evaluation of statistical significance of predictions, but for other purposes it might be misleading. Next we discuss a more practical estimate.

Information for a potential end user: Let us consider our count of errors from a disaster preparedness point of view. One of the alarms ended in 1968:12, a month before the FAU; we counted it as a false alarm and the subsequent FAU—as missed by prediction. Similarly, we counted as missed the FAU in 1981:03, while it was followed by an alarm a month later.

Since a FAU is a starting point of a long rise in unemployment, lasting about 20 months, a one month difference is not necessary important for a decision-maker, responding to a prediction. Moreover, this difference is within the accuracy of FAUs, since they are determined after considerable smoothing of the unemployment rate. Accordingly, for the end user only the three errors might be worth counting: the failure to predict in 1962 and the false alarms in 1983 and 1994.

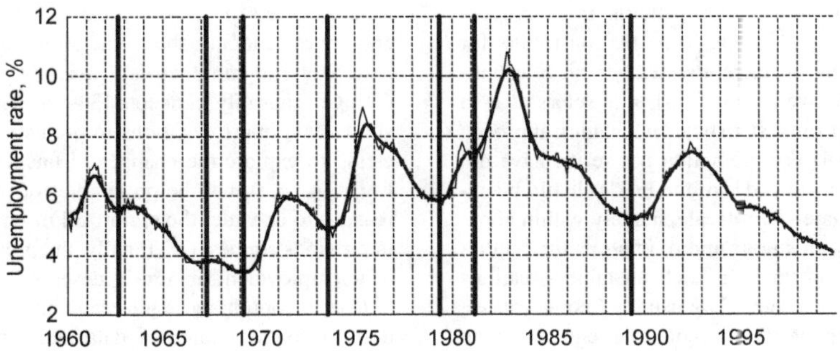

Fig. 10. Unemployment rates in the US. Thin line: $r(m)$, original data. Thick line: $r(m)$, data after smoothing out the seasonal variations. The thick vertical lines show the moments when unemployment started to rise (local minima of smoothed unemployment rate).

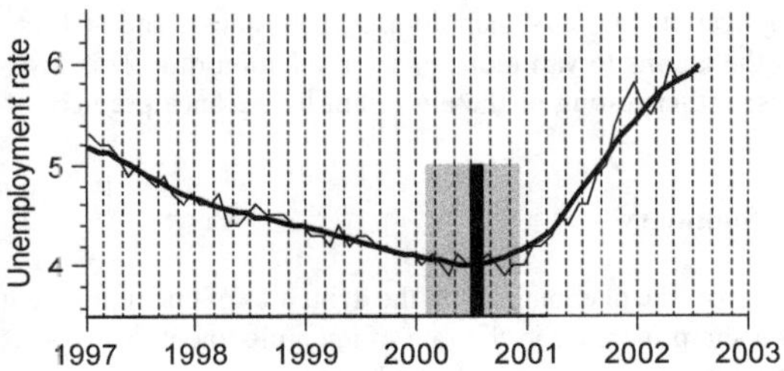

Fig. 11. Advance prediction. Unemployment rate in the US, 1997–2002: thin curve shows original data; thick curve shows the rates with seasonal variations smoothed out. The gray bar shows the alarm period, defined by analysis of macroeconomic indicators. The black vertical line shows the actual start of the unemployment rise, as defined by an analysis of monthly unemployment rates for the US civilian labor force.

7.3. First advance prediction

An advance prediction by the suggested algorithm is shown in Fig. 11. Analyzing the data for USA up to December 2000 we have found that $\Delta(m) = 0$ during the four months, from February to May 2000. Accordingly, the algorithm declared the alarm for the period from February to November 2000 [29]. The data on unemployment rates for the subsequent period, up to July 2002, confirmed that prediction: a FAU materialized in July 2000. Obviously, that confirmation is most encouraging, but much longer experimentation is necessary to validate the algorithm.

Numerous other warnings of a coming rise of unemployment in the USA did appear during the first months of 2000, and even in the popular media. The particular feature of the prediction discussed here, however, is that a formal unambiguous algorithm obtained it and that it indicates a specific time interval when the unemployment will start to rise.

8. Discussion

(1) We have identified a robust pattern of macroeconomic indicators that, since 1962, precedes certain persistently reoccurring structures in the dynamics of the unemployment in France, namely episodes of fast acceleration of unemployment (FAUs). This pattern has emerged within 12 months prior to six out of seven FAUs and at no other time, suggesting a possible prediction algorithm, to be tested by advance predictions. The retrospective alarms defined by this algorithm are encouragingly stable with respect to the choice of indicators and to variations of adjustable numerical parameters.

The pattern recognition methodology used here inherits some of the features of the diffusion indexes of classical business cycle analysis [30–32], reformulating these features in the robust language of pattern recognition. More detailed comments on that connection are given in Ref. [6].

(2) It is known that the economy, like many other complex systems, exhibits regular collective behavior patterns, transcending its immense complexity. Therefore it is not surprising that our premonitory patterns are applicable in such different conditions: (i) through the whole time period considered (the last third of the 20th century) despite extraordinary changes in the economy and the relevant labor market and (ii) to FAUs of different origin. Specifically, the FAUs in Europe reflect the cyclical fluctuations in the economy, with unemployment rising from a moderate or high level to a yet higher one. However the FAUs in the USA reflect the onset of a recession, with unemployment rising from very low levels to high ones. Finally, the prediction algorithm is applicable without readaptation to Germany, Italy, and the USA.

As is usual in the prediction of highly complex processes that applicability is achieved at a price. The magnitude and time of an incipient FAU are indicated with a limited accuracy, and the rate of errors is not negligible.

(3) If confirmed by advance prediction, our results would expand the known limits of predictability of the economy,

and the realm of robust behavior patterns in macroeconomics. (i) We have found strong heuristic constraints for macroeconomic models of unemployment. Besides econometric models, as reviewed in Ref. [2], this refers also to models that are based on a statistical mechanics approach [12,15,17,18,20,21]. (ii) For prediction per se we have defined the situation when episodes of the FAUs should be expected in an intermediate—term scale, namely within about a year. This provides an initial approximation for prediction with a shorter lead-time. In such unstable situations the FAUs may be easier triggered by usually known causes, such as the announcement of new economic regulations, an oil crisis, etc.

(4) Unemployment has been traditionally associated with a decline in the economy. This relation becomes more complex and can even be reversed in a tight labor market. A decrease of unemployment can be regarded as a threat of "inflation by wages", leading to declines in stock market indexes. On the contrary, when a corporation announces massive layoffs of employees, the price of its stock often increases, with the stockowners (especially the pension funds) considering the wage bill a fixed "cost". For those kinds of reasons, Western economic and financial executives no longer consider a low unemployment rate good news and appear to prefer corporate "downsizing", leading to job destruction [33]. Several other factors in the functioning of the economy, by contrast, reduce unemployment. For example, the acceleration of technological progress leads to the fast growth of small and medium-size enterprises in new sectors of the economy and, accordingly, to job creation.

Thus, *a fine balance between the forces leading to job creation and those leading to job destruction determines the level of unemployment. The FAUs reflect a change in that balance.*

(5) The economic integration of the European Union is increasing, along with the general globalization of the economy. This will not necessarily render our analysis irrelevant, since our algorithm is very robust. In any case, economic integration would make it easier to develop new algorithms of this kind, with an even higher level of averaging of the processes considered.

Similarly, our results may not become irrelevant due to some drastic change of the mechanisms controlling the quickly accelerating modern economy. The premonitory patterns considered here probably reflect some type of scenario of transition to critical phenomena (one that is common for many mechanisms in the case of non-linear dynamics). An indirect confirmation is the uniform performance of the prediction algorithm over the last 35 years.

(6) It is encouraging also, that we have used so far only a part of the relevant data and methods. In that sense this study is preliminary and exploratory. Our results suggest the following directions for further research: (i) We have identified here the indicators that are *sufficient* for prediction, but not *all* indicators potentially useful for that purpose. (ii) With less robust predictions, premonitory patterns may be considered in the context of the accelerator hypothesis and, more generally, the cointegration of each indicator separately and in combinations, using Dickey–Fuller tests and Granger causality concepts [34–36]. (iii) In the search for subsequent approximations to prediction it would be interesting to explore the changes of unemployment *during* the FAUs. Also, it may be preferable for purely computational reasons to consider the reciprocal of unemployment as used in previous studies. (iv) Finally, the prediction of a *decrease* of unemployment obviously deserves a parallel study.

(7) Our optimism is guarded, since we had certain freedom as to the choice of data and formulation of the algorithm. Such a retrospectively designed prediction algorithm could possibly depict random coincidences, which eliminate the errors by chance and may not reoccur. We have encouraging evidence supporting the validity of the algorithm described here, however, including its successful application to out-of-sample data; robustness; and stability of the alarms to variations of the prediction rule. This justifies further testing of the algorithm by advance prediction.

9. Summary

A specific phenomenon in the dynamics of unemployment —a sharp increase in the rate of unemployment growth—is considered the target of prediction. It is called by the acronym FAU, for "Fast Acceleration of Unemployment". The goal of the study is to design an algorithm for predicting such FAUs by an analysis of macroeconomic indicators. The prediction is of the yes-or-no type: at each moment the algorithm indicates whether FAU should or should not be expected within the subsequent τ months, τ being the "duration of alarm". This study concerns the fields of economics, labor relations, and the dynamics of complex systems, and it is a "technical" analysis, that is, a heuristic search of phenomena preceding FAUs. The methodology of *pattern recognition of infrequent events* is used. In terms of pattern recognition, time moments are classified on the basis of a series of the macroeconomic indicators into two classes: moments after that FAU should be expected and moments after that FAU should not be expected. A specific "premonitory" pattern of three macroeconomic indicators—(i) the industrial production index, composed of weighted production levels in numerous sectors of the economy; (ii) the long-term interest rate on 10-year government bonds; (iii) the short-term interest rate on 3-month bills—has been found in France between 1962 and 1997. This pattern may be used for the algorithmic prediction of FAUs. Among seven FAUs identified within these years, six are preceded within 12 months by this pattern that appears at no other time. The application of this algorithm to Germany, Italy and the USA yields similar results. The final test, as in any prediction research, should be advance prediction. The first such prediction, for the USA for early 2000, has been successful.

The approach used here involves trying to predict not the whole dynamics of unemployment but only relatively rare and extraordinary phenomena—the FAUs, where, at any moment of time, the algorithm would indicate whether or not a FAU should be expected. The probabilistic nature of such prediction is reflected in the estimation of the probabilities of false alarms, failures to predict, and of the relative time occupied by alarms. This analysis does not compete with but rather complements short-term predictions of the "*cause and effect*" kind, where the cause that triggered a specific FAU is usually known, at least in retrospect. In this approach the dynamics of the economy, including changes in unemployment, may have some basic features that are common in various complex processes that are found in models of statistical physics. Thus, the economy, like many other complex systems, exhibits regular collective behavior patterns, transcending its immense complexity.

Acknowledgements

This work was supported by the James S. McDonnell Foundation within the framework of the 21st Century Collaborative Activity Award for Studying Complex Systems (project "*Understanding and Prediction of Critical Transitions in Complex Systems*"). We are grateful to Professor Claude J. Allègre, Institut de Physique du Globe de Paris, who initiated this study, turning our attention from recessions to unemployment; to Professor Gaston Schaber, Centre d'Etudes de Populations de Pauvrete et de Politiques Socio-Economiques/International Networks for Studies in Technology, Environment, Alternatives, Development (CEPS/INSTEAD), for inspiring discussions and constructive criticism; and to Professor Giora Hanoch of the Hebrew University for most helpful comments and suggestions.

References

[1] I. Gelfand, V. Keilis-Borok, L. Knopoff, F. Press, E. Rantsman, I. Rotwain, A. Sadovsky, Pattern recognition applied to earthquake epicenters in California, Phys. Earth Planet. Inter. 11 (1976) 227–283.

[2] R.F. Engle, D.L. McFadden (Eds.), Handbook of Econometrics, vol. 4, North-Holland Publishing Co., Amsterdam, 1994.

[3] J.H. Stock, M.W. Watson, in: O.J. Blanchard, S.F. Fischer (Eds.), NBER Macroeconomics Annual, 1989, pp. 351–394.

[4] Ph.A. Klein, M.P. Niemira, Forecasting Financial and Economic Cycles, Wiley, New York, 1994.

[5] M. Mostaghimi, F. Rezayat, Probability forecast of a downturn in U.S. economy using classical statistical theory, Empirical Econ. 21 (1996) 255–279.

[6] V. Keilis-Borok, J.H. Stock, A. Soloviev, P. Mikhalev, Pre-recession pattern of six economic indicators in the USA, J. Forecast. 19 (2000) 65–80.

[7] A.J. Lichtman, V.I. Keilis-Borok, Aggregate-level analysis and prediction of midterm senatorial elections in the United States. 1974–1986, Proc. Natl. Acad. Sci. USA 86 (24) (1989) 10176–10180.

[8] V.I. Keilis-Borok, A.J. Lichtman, in: Yu.A. Kravtsov (Ed.), Limits of Predictability, Springer, Berlin, 1993, pp. 223–238.

[9] F. Press, P. Briggs, Earthquakes, Chandler wobble, rotation and geomagnetic changes a pattern recognition approach, Nature 256 (1975) 270–273.

[10] V.I. Keilis-Borok, F. Press, in: C. Allègre (Ed.), Source Mechanisms and Earthquake Prediction, Paris, 1980, pp. 51–60.

[11] F. Press, C. Allen, Patterns of seismic release in the southern California region, J. Geophys. Res. 100 (B4) (1995) 6421–6430.

[12] V.I. Keilis-Borok, A.A. Soloviev (Eds.), Nonlinear Dynamics of the Lithosphere and Earthquake Prediction, Springer, Berlin, Heidelberg, 2003.

[13] F. Press, P. Briggs, Pattern recognition applied to uranium prospecting, Nature 268 (1977) 125–127.

[14] A.D. Gvishiani, V.G. Kossobokov, Proc. Ac. Sci. USSR Phys. Earth 2 (1981) 21–36.

[15] R. Burridge, L. Knopoff, Model and theoretical seismicity, Bull. Seism. Soc. Am. 57 (1967) 341–371.

[16] M. Gell-Mann, The Quark and the Jaguar: Adventures in the Simple and the Complex, W.H. Freeman and Company, New York, 1994.

[17] W.I. Newman, A.M. Gabrielov, D.L. Turcotte (Eds.), Non-linear Dynamics and Predictability of Geophysical Phenomena, Geophysical Monograph Series, IUGG-Am. Geophys. Union, Washington, DC, 1994.

[18] C.J. Allègre, J.-L. Le Mouel, C. Ha Duyen, C. Narteau, Scaling organization of fracture tectonics (SOFT) and earthquake mechanism, Phys. Earth Planet. Inter. 92 (1995) 215–233.

[19] J.H. Holland, Hidden Order: How Adaptation Builds Complexity, Addison-Wesley, Reading, MA, 1995.

[20] E.M. Blanter, M.G. Shnirman, J.-L. Le Mouel, C.J. Allègre, Scaling laws in blocks dynamics and dynamic self-organized criticality, Phys. Earth Planet. Inter. 99 (1997) 295–307.

[21] A. Gabrielov, V. Keilis-Borok, I. Zaliapin, W.I. Newman, Critical transitions in colliding cascades, Phys. Rev. E 62 (2000) 237–249.

[22] OECD, Main Economic Indicators: Historical Statistics 1960–1996, Paris, CD-ROM, 1997.

[23] International Monetary Fund, International Financial Statistics, CD-ROM, 1997.

[24] J.H. Stock, M.W. Watson, in: J.H. Stock, M.W. Watson (Eds.), Business Cycles, Indicators, and Forecasting, NBER Studies in Business Cycles, vol. 28, 1993, pp. 95–156.

[25] V.I. Keilis-Borok, V.G. Kossobokov, Premonitory activation of earthquake flow: algorithm M8, Phys. Earth Planet. Inter. 61 (1990) 73–83.

[26] I.A. Vorobieva, T.A. Levshina, in: D.K. Chowdhury (Ed.), Computational Seismology and Geodynamics, vol 2, Am. Geophys. Union, Washington, DC, 1994, pp. 27–36.

[27] G.M. Molchan, Earthquake prediction as a decision-making problem, Pure Appl. Geophys. 149 (1997) 233–237.

[28] U.S. Department of Labor, Bureau of Labor Statistics, Labor Force Statistics from the Current Population Survey. Web site: <http://stats.bls.gov/webapps/legacy/cpsatab5.htm>.

[29] V.I. Keilis-Borok, A.A. Soloviev, C.B. Allègre, A.N. Sobolevskii, M.D. Intriligator, Sixth Workshop on Non-Linear Dynamics and Earthquake Prediction, 15–27 October 2001, H4.SMR/1330-11. ICTP, Trieste, 2001.

[30] W.C. Mitchell, A.F. Burns, Measuring Business Cycles, NBER, New York, 1946.

[31] W.C. Mitchell, What Happens During Business Cycles: A Progress Report, NBER, New York, 1951.

[32] G.H. Moore (Ed.), Business Cycle Indicators, NBER, New York, 1961.

[33] D.L. Brito, M.D. Intriligator, E.R. Worth, in: G. Eliasson, C. Green (Eds.), Microfoundations of Economic Growth: A Schumpeterian Perspective, University of Michigan Press, Ann Arbor, 1998.

[34] N.R. Ericson, Empirical modeling of money demand, Empirical Econ. 23 (1998) 295–315.

[35] N.R. Ericson, J.G. MacKinnon, International Finance Discussion Papers 655, December, Board of Governors of the Federal Reserve System, 1999.

[36] M.W. Watson, in: R.F. Engle, D.L. McFadden (Eds.), Handbook of Econometrics, vol. IV, North-Holland Publishing Co., Amsterdam, 1994.

About the Author—VLADIMIR I. KEILIS-BOROK received his M.Sc. degree in Geophysics from the Moscow Institute (educational) of Exploration Geophysics in 1943, his Ph.D. (1948), and D.Sc. (1953) degrees in Mathematical Geophysics from the Geophysical Institute of the USSR Academy of Science. He was founding Director of the International Institute of Earthquake Prediction Theory and Mathematical Geophysics, Russian Academy of Science. Now he is Research Group Leader at this Institute and Professor in Residence at the Institute of Geophysics and Planetary Physics and the Department of Earth and Space Studies of the University of California, Los Angeles. He has been elected to the American Academy of Arts and Sciences, the US National Academy of Sciences, the Russian Academy of Sciences, the Royal Astronomical Society, the Austrian Academy of Sciences, the Pontifical Academy of Sciences, and the Academia Europaea. He published more than 250 papers in the areas of instability of the lithosphere, rocks–fluids interaction, earthquake prediction, earthquake preparedness, seismic risk, reconstruction of fault networks, seismological inversion, identification of underground nuclear explosions, theory of seismic waves, critical phenomena in chaotic systems, socio-economic predictions, and pattern recognition. He is coeditor, with Alexandre A. Soloviev, of the recent monograph "Nonlinear Dynamics of the Lithosphere and Earthquake Prediction" (Springer, 2003).

About the Author—ALEXANDRE A. SOLOVIEV received his M.Sc. degree in Mathematics from M.V. Lomonosov Moscow State University in 1971, his Ph.D. degree in Theoretical Mechanics from Institute of Applied Mathematics, Academy of Science USSR in 1975, and his D.Sc. degree in Physics of the Solid Earth from the O.J. Shmidt Institute of the Physics of the Earth, Academy of Science USSR in 1990. He is Director of International Institute of Earthquake Prediction Theory and Mathematical Geophysics, Russian Academy of Science and a corresponding member of the Russian Academy of Science. He is the author of more than 150 journal articles and other publications in the areas of modeling of lithosphere faults-and-blocks dynamics and seismicity, application of the pattern recognition methods for determination of unstable structures in the lithosphere, particularly—earthquake prone areas, predictability of socio-economic instabilities, pattern recognition of infrequent events, magnetic dynamo, earthquake prediction research, computation of movement of artificial satellites. He is coeditor, with Vladimir I. Keilis-Borok, of the monograph "Nonlinear Dynamics of the Lithosphere and Earthquake Prediction" (Springer, 2003).

About the Author—CLAUDE ALLÈGRE (NEE SIMON) received her degrees at the University of Sorbonne: Advanced B.Sc. degree (Diplome d'Etudes Superieures Specialisees) in Labor, Administration and Management (1963); M.Sc. degree (Diplome d'Etudes Approfondies) in Social and Political Sciences (1966); and Ph.D. degree in Economy (1971). She has worked in Sorbonne since her graduation; her current position is Associate Professor of the Department of Labor and Social Studies, University of Paris 1, Pantheon-Sorbonne. Her fields of research include economic strategy and the structure of industry, large scale movements of the labor market, labor productivity and endogenous factors of profitability, the structure and dynamics of the cost of labor, the instability of employment, and multinational corporations and labor markets in developing countries.

About the Author—ANDREI N. SOBOLEVSKII received his M.Sc. degree in Physics in 1996 and his Ph.D. degree in Theoretical Physics in 1999, both from M.V. Lomonosov Moscow State University. He is Assistant Professor of the Physics Department of M.V. Lomonosov Moscow State University and Research Fellow of International Institute of Earthquake Prediction Theory and Mathematical Geophysics, Russian Ac. Sci. He is the author of 13 journal articles and other publications in the areas of dynamical systems, partial differential equations (viscosity solutions), ballistic aggregation, combinatorial and convex optimization, and nonstationary statistical physics (critical phenomena).

About the Author—MICHAEL D. INTRILIGATOR received his S.B. (Bachelor of Science) degree in Economics from the Massachusetts Institute of Technology in 1959, his M.A. in Economics from Yale University in 1960, and his Ph.D. degree from the Massachusetts Institute of Technology in 1963. He is Professor of Economics at the University of California, Los Angeles (UCLA), where he is also Professor of Political Science, Professor of Policy Studies in the School of Public Policy and Social Research, and Co-Director of the Jacob Marschak Interdisciplinary Colloquium on Mathematics in the Behavioral Sciences. He is also a Senior Fellow of the Milken Institute in Santa Monica, California and the Gorbachev Foundation of North America in Boston, Massachusetts. He is a Fellow of the Econometric Society, a foreign member of the Russian Academy of Science, an AAAS Fellow of the American

Association for the Advancement of Science, and an elected member of the Council on Foreign Relations (New York) and the International Institute for Strategic Studies (London). He is the author of more than 200 journal articles and other publications in the areas of economic theory and mathematical economics, econometrics, health economics, reform of the Russian economy, and strategy and arms control. He is the author of Mathematical Optimization and Economic Theory (Prentice-Hall, 1971, also translated into Spanish and Russian) and of Econometric Models, Techniques, and Applications (Prentice-Hall, 1978, also translated into Greek and Spanish; second edition, with Ronald G. Bodkin and Cheng Hsiao, 1996). He is coeditor, with Kenneth J. Arrow, of the Handbook of Mathematical Economics (North-Holland, 1981, 1982, 1985); coeditor, with Zvi Griliches, of the Handbook of Econometrics (North-Holland, 1982, 1983, 1986), and coeditor with Kenneth J. Arrow of the book series, "Handbooks in Economics" published by Elsevier.

Available online at www.sciencedirect.com
SCIENCE @ DIRECT®
ELSEVIER
Tectonophysics 413 (2006) 63–75
TECTONOPHYSICS
www.elsevier.com/locate/tecto

Short-term earthquake prediction by reverse analysis of lithosphere dynamics

P. Shebalin [a,d], V. Keilis-Borok [a,b,c,*], A. Gabrielov [e], I. Zaliapin [a,b], D. Turcotte [f]

[a] *International Institute for Earthquake Prediction Theory and Mathematical Geophysics, Russian Ac. Sci., Warshavskoe sh., 79, korp. 2, Moscow, 113556, Russia*
[b] *Institute of Geophysics and Planetary Physics, University of California, Los Angeles, CA 90095-1567, USA*
[c] *Department of Earth and Space Sciences, University of California, Los Angeles, CA 90095-1567, USA*
[d] *Institut de Physique du Globe de Paris, 4 Place Jussieu, 75252, Paris Cedex 05, France*
[e] *Departments of Mathematics and Earth and Atmospheric Sciences, Purdue University, West Lafayette, IN 47907-1395, USA*
[f] *Department of Geology, University of California, Davis, CA 95616, USA*

Received 24 December 2004; received in revised form 14 June 2005; accepted 4 October 2005
Available online 13 December 2005

Abstract

Short-term earthquake prediction, months in advance, is an elusive goal of earth sciences, of great importance for fundamental science and for disaster preparedness. Here, we describe a methodology for short-term prediction named RTP (Reverse Tracing of Precursors). Using this methodology the San Simeon earthquake in Central California (magnitude 6.5, Dec. 22, 2003) and the Tokachi-Oki earthquake in Northern Japan (magnitude 8.1, Sept. 25, 2003) were predicted 6 and 7 months in advance, respectively. The physical basis of RTP can be summed up as follows: An earthquake is generated by two interacting processes in a fault network: an accumulation of energy that the earthquake will release and a rise of instability triggering this release. Energy is carried by the stress field, instability is carried by the difference between the stress and strength fields. Both processes can be detected and characterized by "premonitory" patterns of seismicity or other relevant fields. Here, we consider an ensemble of premonitory seismicity patterns. RTP methodology is able to reconstruct these patterns by tracing their sequence backwards in time. The principles of RTP are not specific to earthquakes and may be applicable to critical transitions in a wide class of hierarchical non-linear systems.

Keywords: Reverse tracing of precursors; Short-term earthquake prediction

1. Introduction

There is increasing evidence that variations in regional seismicity occur prior to intermediate and large earthquakes (Keilis-Borok et al., 1980; Mogi, 1981; Caputo et al., 1983; Sykes, 1983; Keilis-Borok, 1990; Ma et al., 1990; Molchan et al., 1990; Knopoff et al., 1996; Bowman et al., 1998). Using pattern recognition techniques a series of algorithms have been developed which provide intermediate-term and long-term predictions with lead times of years to decades respectively (Keilis-Borok and Shebalin, 1999; Keilis-Borok, 2002; Keilis-Borok and Soloviev, 2003; Rundle et al., 2003). In this paper an algorithm is introduced that has successfully made short-term earthquake predictions, i.e.

* Corresponding author. Institute of Geophysics and Planetary Physics, University of California, Los Angeles, CA 90095-1567, USA.
E-mail address: vkb@ess.ucla.edu (V. Keilis-Borok).

0040-1951/$ - see front matter © 2005 Elsevier B.V. All rights reserved.
doi:10.1016/j.tecto.2005.10.033

months in advance (Shebalin et al., 2004; Keilis-Borok et al., 2004). This algorithm is currently tested by advance prediction in several seismically active regions and its performance is yet to be validated. However, the first successes along with the novelty of the methodology used in this algorithm and a multitude of its possible applications motivate us to describe here its underlying ideas and techniques.

It should be emphasized that there are two quite different approaches to earthquake prediction. The first is to make continuous predictions; in terms of earthquakes, this requires the specification of earthquake risk at all spatial points at each time instant. This approach is very useful when predicting a large number of small to intermediate earthquake since one can directly compare the observed and predicted seismic rates (probabilities, intensities, etc.) using the log-likelihood paradigm (Daley and Vere-Jones, 2004) or the least-square discrepancy (Whittle, 1963). The second approach that is used here is binary: an earthquake is forecast (predicted) for a specified area and time window, called *alarm region* or *alarm*. This approach is better justified when predicting extremely rare large events, so the direct comparison of the predicted continuous rate with a couple of observed earthquakes is rather problematic (Molchan, 1990, 2003). Our goal is to narrow down the area and time duration of alarms, within which a target earthquake is expected. Prediction is targeted at the large and therefore rare earthquakes; in a typical alarm area they occur on average once in 10–20 years. Our prediction should capture the target within an interval 20–30 times smaller, since a short-term alarm lasts months. Thus our alarms should be equally rare and each correct alarm would typically capture only one target.

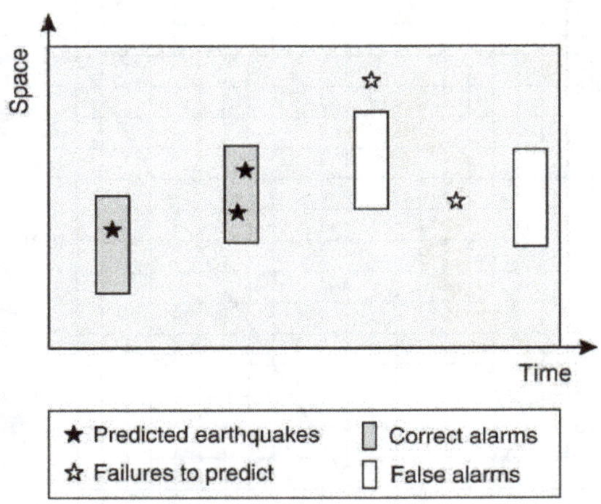

Fig. 1. Possible outcomes of prediction. For simplicity the territory where the prediction is made is represented by a 1D 'Space' axis. Rectangles—space–time areas covered by correct (gray) and false (white) alarms respectively.

An early theoretical discussion of the necessity of a binary approach instead of a continuous one in predicting rare point events is given in (Lindgren, 1975, 1985; De Mare, 1980). The difference between continuous and binary predictions has been widely recognized in weather forecasting (Jolliffe and Stephenson, 2003). An example of binary forecast is a tornado warning issued for a specified area and time window. Tornado warnings are analogous to the earthquake alarms considered in this paper. Possible outcomes of such predictions are illustrated in Fig. 1. In this scheme, we have two types of errors: failures to predict (target earthquake outside alarm region) and false alarms (no target earthquakes within an alarm); a prediction algorithm is also characterized by the total time–space covered by the alarms. Probability of errors of different types is estimated using a sequence of predictions and is visually represented in the error diagrams (Sect. 3 below; Molchan, 1990, 2003). An analogous approach in weather forecasting is the *relative operating characteristic* diagram (Jolliffe and Stephenson, 2003).

2. Reverse Tracing of Precursors (RTP)

We will now outline the RTP (Reverse Tracing of Precursors) approach to short-term earthquake forecasting. Technical details are given in the Appendix. Three aspects of RTP are important:

(i) *Precursory chains* that reflect the premonitory increase of the earthquakes' correlation range; qualitatively, these chains are the dense, long, and rapidly formed sequences of small and medium sized earthquakes. Their definition generalizes premonitory seismicity patterns ROC and ACCORD. Heuristically, the pattern ROC ensures the ongoing increase of earthquake correlation-range, expressed via the pair-wise correlation function; while ACCORD reflects simultaneous activation of several major parts of the regional fault network. They represent complimentary approaches to detecting the earthquake correlation. Formal definitions of these patterns as well as their performance in synthetic and observed seismicity can be found in (Gabrielov et al., 2000; Shebalin et al., 2000; Zaliapin et al., 2002, 2003; Keilis-Borok et al., 2002). An alternative approach to measuring the earthquake correlation was introduced in (Zöller and Hainzl, 2001; Zöller et al., 2001).

(ii) *Intermediate-term patterns*, originally found in the modeled and observed seismicity (Prozorov and Schreider, 1990; Keilis-Borok and Shebalin, 1999; Gabrielov et al., 2000; Keilis-Borok, 2002; Keilis-Borok and Soloviev, 2003; Zaliapin et al., 2003). They reflect four major types of premonitory phe-

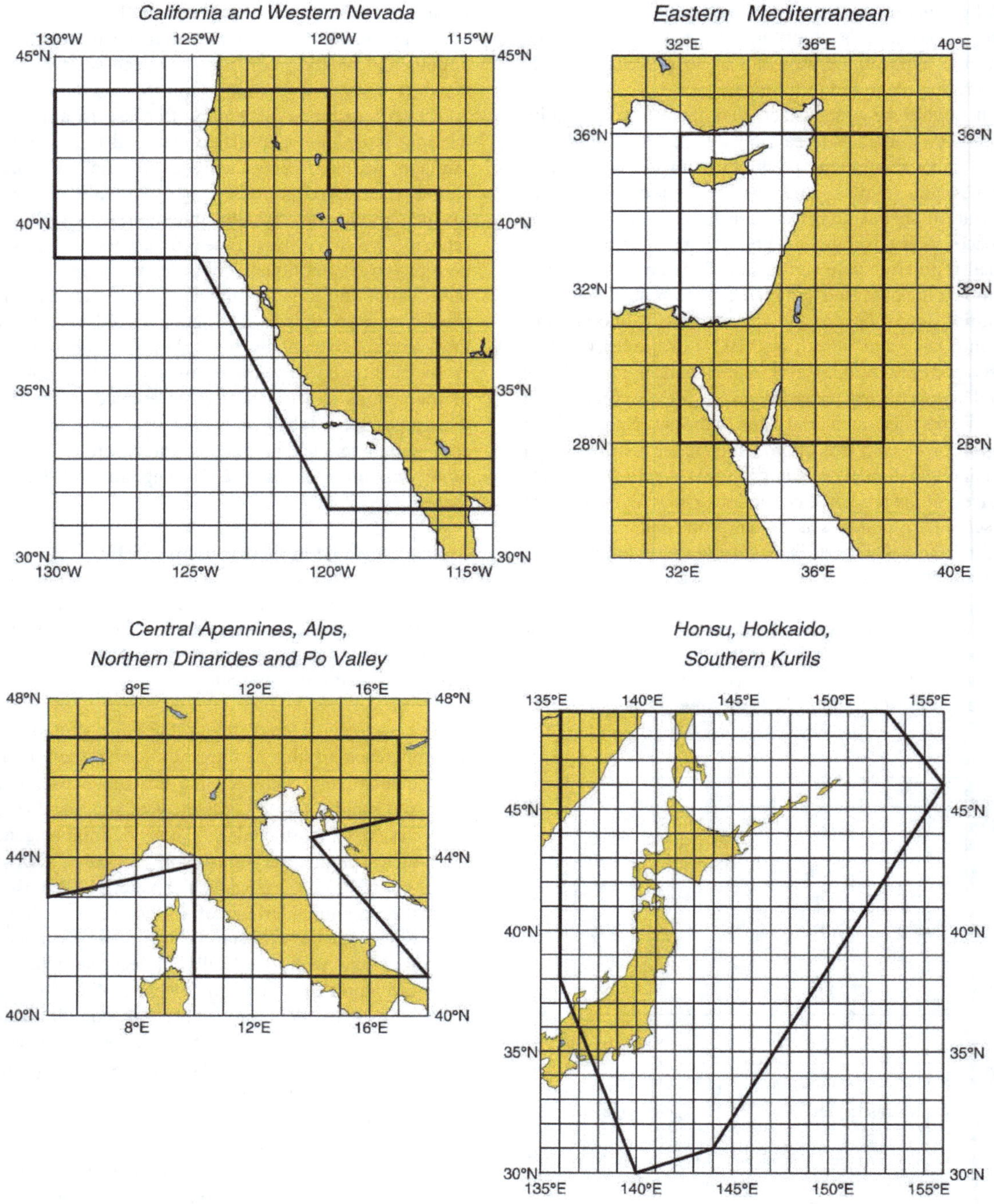

Fig. 2. Regions where the proposed algorithm was tested by advance prediction. See text for details.

nomena: rise of seismic activity, rise of earthquakes' clustering, rise of earthquakes correlation range, and a transformation of the magnitude–frequency (Gutenberg–Richter) relation towards an increasing share of relatively large magnitudes.

(iii) *Pattern recognition of infrequent events* is used to define the precursory combination of the patterns. Specifically, we used the Hamming algorithm which in our case is analogous to voting (Keilis-Borok and Soloviev, 2003); this algorithm is a standard tool in making a decision by considering several "opinions". Formally, the Hamming distance between two Boolean vectors of the same length is defined as the number of their non-coincident symbols. In our problem, the Hamming distance is the number of emergent intermediate-term premonitory patterns.

RTP analysis consists of the following stages. First, we detect chains—the "candidates" for the short-term precursors. We have found that precursory chains emerge within months before most of the target earthquakes. However, up to 90% of the chains are not followed so closely by strong earthquakes and in prediction they would cause false alarms. To eliminate false alarms, we next determine which intermediate-term precursors have occurred in the vicinity of each candidate within few years preceding it. Finally, we apply pattern recognition: knowing for each candidate what intermediate-term patterns have preceded it, we recognize which chains are precursory and which are false alarms. Specifically, we discriminate precursory and non-precursory chains using a set of M individual intermediate-term premonitory patterns. Some of them give premonitory signal (emerge) while other do not. The current state of the patterns is represented by a $M \times 1$ Boolean vector indicating which pattern emerge (1) and which is not (0). A zero vector would indicate that none of the patterns emerge and the chain is most probably not precursory, while a vector consisting of all ones that all the patterns emerge and the chain is most probably precursory. Hamming distance, defined as the number of ones in our Boolean vector, shows how far the vector is from a zero one; in other words, how many patterns "voted" for making the chain precursory. If sufficient number of votes is accumulated (the threshold is established during the learning) the chain is considered precursory. This brings us to the prediction proper. The emergence of each precursory chain starts an alarm: a target earthquake is expected during τ months after the chain was formed and in its formally defined vicinity.

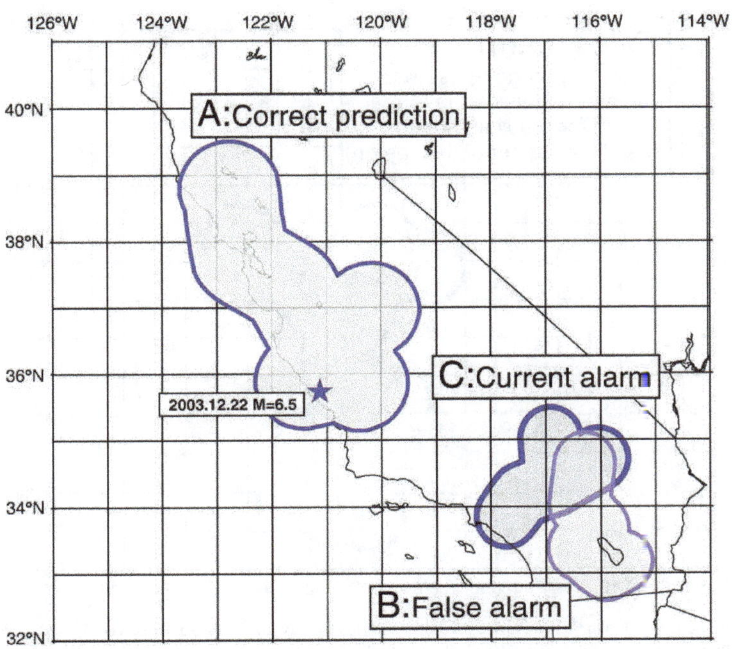

Fig. 3. Results of advance prediction in California. The advance prediction started in July 2003. Three alarms were issued: one correct (marked A), one false (B), one current (C).

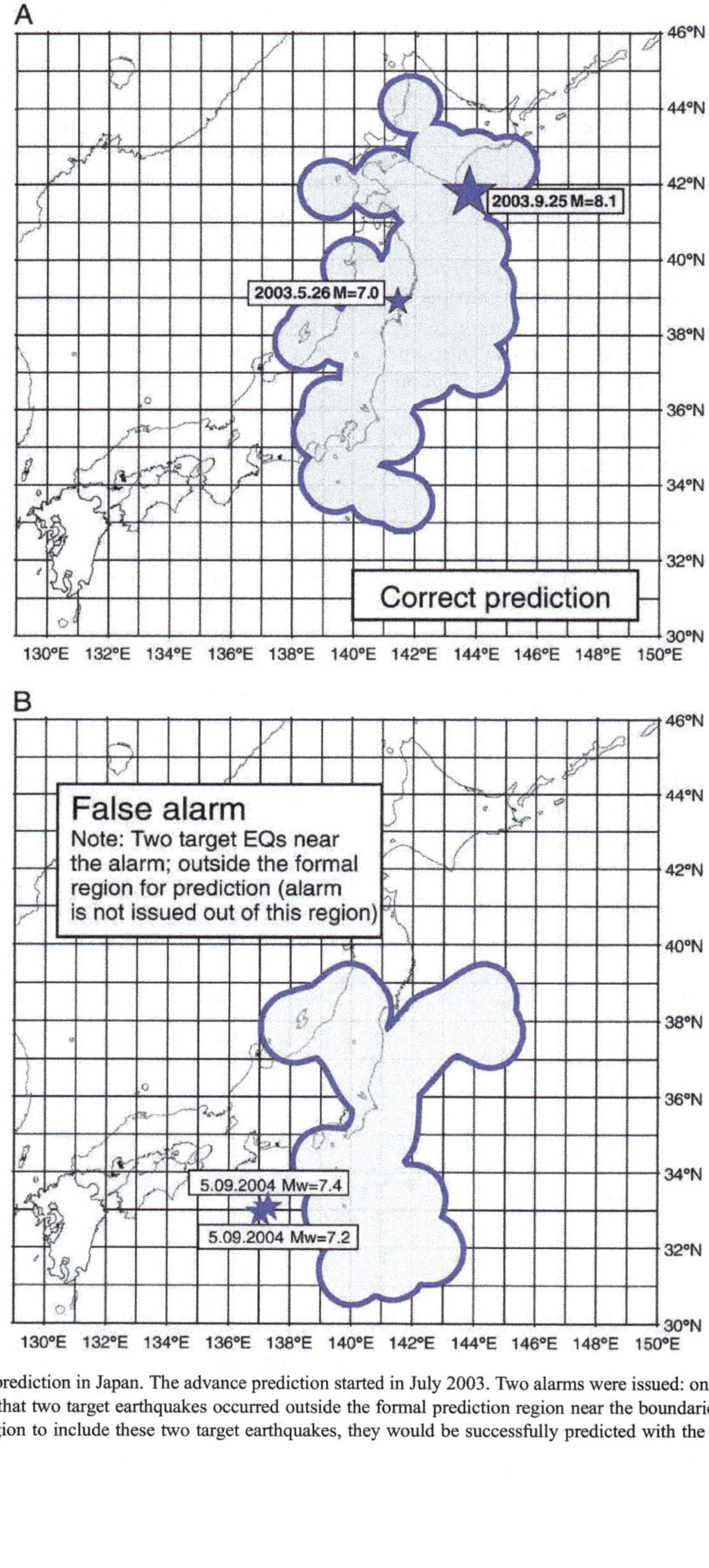

Fig. 4. Results of advance prediction in Japan. The advance prediction started in July 2003. Two alarms were issued: one correct (panel A), and one false (panel B). We notice that two target earthquakes occurred outside the formal prediction region near the boundaries of our false alarm; if one extended the prediction region to include these two target earthquakes, they would be successfully predicted with the current values of algorithm parameters.

Thus, the precursory chain indicates the narrow area of a possibly complex shape (the chain vicinity) where intermediate-term precursors should be looked for. Their presence in turn validates the chain, as a short-term precursor. A chain is considered first although it emerges later—hence our analysis is called *reverse*.

3. Performance

We have tested our algorithm by advance prediction in Southern and Central California using the earthquake catalog ANSS/CNSS starting from January 1965. First, the data for 1965–1994 have been used for "learning", i.e. self-adaptation of some of the parameters (see Appendix A3). Then, the resulting rule was tested on independent data (i.e. the data not used for learning) for the period from January 1995 to May 2003. In June 2003 we have launched advance prediction. The only target earthquake that happened in the prediction region during the advance phase of the experiment (the San Simeon earthquake, December 22, 2003, $M=6.5$) was successfully predicted. The alarm capturing this earthquake started on May 5, 2003—the date when the precursory chain triggering the alarm was completed. This alarm was reported on June 21 of the same year (Aki et al., 2003).

The algorithm has also been applied to the territories of Japan; Central Apennines, Alps, Northern Dinarides and Po valley; and Eastern Mediterranean (Figs. 2–5) with magnitude of target earthquakes $M \geq 7$, $M \geq 5.5$ and $M \geq 6.5$ respectively. In Japan, the learning was performed during 1975–2003, and advance prediction started on July 1, 2003 (Shebalin et al., 2003). In Central Apennines, Alps, Northern Dinarides and Po valley the learning was performed during 1970–1990, and advance prediction started on May 12, 2004. In Eastern Mediterranean the learning was performed during 1983–2003, and advance prediction started on May 12, 2004. The intermediate-term patterns (see Appendix A3) showed amazing self-adjustment: they were applicable within all three regions, and to all chains within each region, with the same values of their four numerical parameters.

The Tokachi-Oki, Japan, earthquake, 25 September 2003, $M=8.1$, has been also predicted in advance: the alarm started on 27 March, 2003 and was reported on 2 July 2003 (Shebalin et al., 2004).

During the time period covered by our advance prediction experiment, two target earthquakes have occurred; both of them have been predicted. Three false alarms were issued; one alarm is current. Figs. 3–5 and Table 1 summarize the results of the experiment. It is worth noticing for further research that a large earthquake ($M_L=5.7$, $M_W=5.3$) occurred within the alarm issued in Northern Dinarides; and that two target earthquakes ($M_W=7.4$, $M_W=7.2$) occurred near one of the alarms issued in Japan, but outside the formal prediction region. A retrospective analysis for an extended

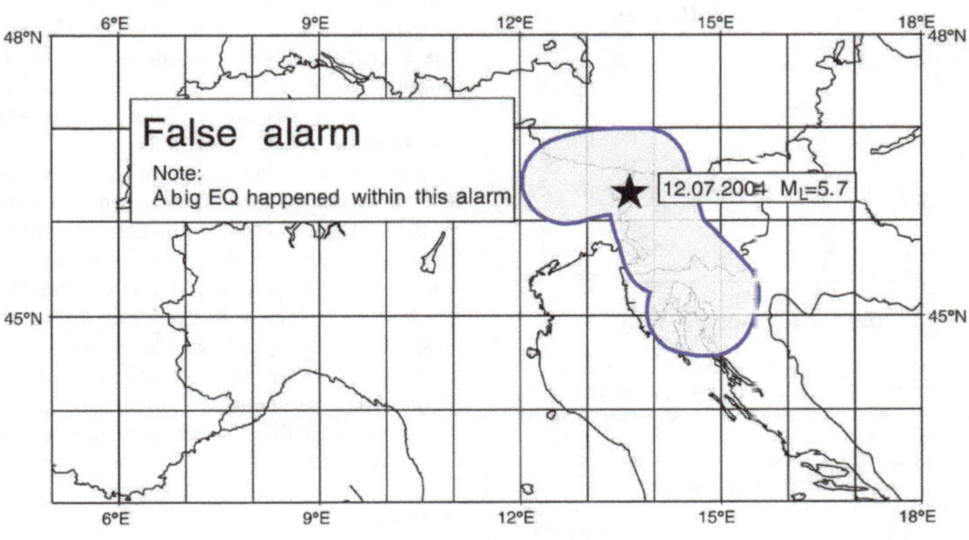

Fig. 5. Results of advance prediction in Northern Dinarides. The advance prediction started in May 2004. One false alarm was issued. We notice that a big earthquake happened within our false alarm; this earthquake formally does not fit the prediction as its magnitude $M_W=5.3$ (prediction was done for $M_W>=5.5$).

Table 1
Results of the advance prediction experiment

Region/target earthquakes	Alarm duration	Probability that a target earthquake will occur at random in the time–area of alarm[a]
Japan $M_{JMA} \geq 7.0$	27 Mar 2003–27 Jan 2004 Tokachi-Oki, 25 Sep 2003, $M_W = 8.3$	0.25
Central and Southern California $M_{ANSS} \geq 6.4$	5 May 2003–27 Feb 2004 San Simeon, 22 Dec 2004, $M = 6.5$	0.05
Southern California $M_{ANSS} \geq 6.4$	29 Oct 2003–05 Sep 2004 False alarm	0.08
Honsu, Japan $M_W \geq 7.2$	8 Feb 2004–8 Nov 2004 False alarm (note: two quakes, $M_W = 7.4$ and $M_W = 7.2$ outside formal region boundaries)	0.07
Northern Dinarides $M_W \geq 5.5$	29 Feb 2004–29 Nov 2004 False alarm (note: $M_W = 5.3$, $M_L = 5.7$ in the area of alarm)	0.07
Southern California $M_{ANSS} \geq 6.4$	14 Nov 2004–14 Aug 2005 Current alarm	0.05

[a] See Appendix A7 for the explanations of how this estimation was obtained.

region gives a successful prediction for those two earthquakes.

4. Prediction quality

As we mentioned in the Introduction, the problem of evaluating a binary prediction requires special tools.

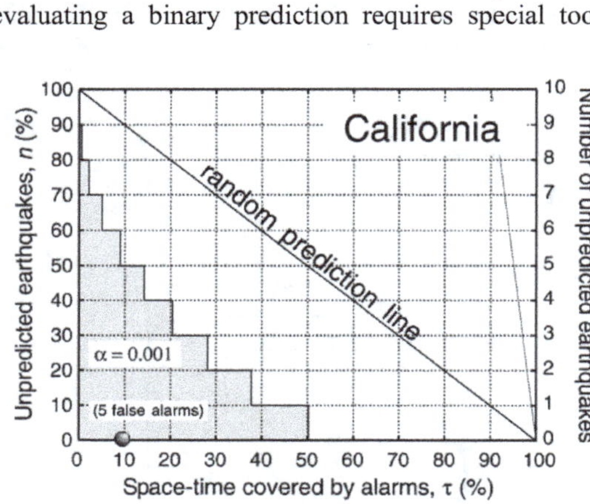

Fig. 6. Significance of prediction in California: an illustration (see Appendix A6 for details). Shaded ball shows performance of the prediction algorithm during the time interval considered. A perfect prediction would lie in the origin. Random binomial predictions (alarm is declared for each elementary spatio-temporal unit with a fixed probability τ) asymptotically occupy the diagonal, but might deviate from it with finite number of target earthquakes. Random predictions with fixed τ fall in the grey area with probability $\alpha = 0.001$. Note that the shape of the grey area depends on the number of the target earthquakes that actually happened within the prediction region.

The main difference from evaluating a continuous prediction is that we can no longer use a single measure of discrepancy between prediction and observations (one faces the same situation in classical hypothesis testing where errors of two types are introduced). We use three interdependent measures of prediction quality, defined in Appendix A5: fraction of unpredicted earthquakes, n; fraction of false alarms, f; and the space–time τ covered by all alarms together, normalized by the whole space–time considered. The space is measured not in km^2 but in long-term average of seismicity. Specifically, we used the average number of mainshocks with $m \geq 4$. The optimal tradeoff between different characteristics depends on a loss function $L(n,f,\tau)$ for preparedness measures (Molchan, 1990, 2003).

The *error diagram* juxtaposes the prediction errors; each particular prediction corresponds to a single point in (n, τ, f) space. The error diagram is used to evaluate the predictive power of our prediction algorithm and its stability. For illustration, the error diagram for our prediction experiment in California during 1964–2005 is shown in Fig. 6; it shows the relative alarm coverage τ (10%) vs. the number of failures to predict (0); the number of false alarms (5) is indicated in parentheses. A more detailed discussion of error diagram approach is given in Appendix A6.

5. Discussion

1. *A possible physical mechanism* underlying the RTP methodology is based on models of dynamical

systems (Gabrielov et al., 2000; Zaliapin et al., 2003) and geodynamical models (Rundquist and Soloviev, 1999). Precursory chains outline the areas where instability is accumulated months before a target earthquake. This instability reveals itself through an increase of the earthquake correlation range. Intermediate-term premonitory seismicity patterns considered reflect the accumulation of energy and instability necessary and sufficient to trigger an earthquake, in the area outlined by a precursory chain, but years before the chain. In more general terms, RTP identifies a small-scale perturbation that carries a memory of the larger scale history of a complex system (in our case, the fault network). Increases of the correlation range are a known symptom of critical transitions in statistical physics and of bifurcation in nonlinear dynamics (Kadanoff, 2000). Typically for premonitory patterns of this kind precursors considered are sporadic short-lived phenomena not necessarily reflecting the steady trends of seismicity. This suggests that both patterns are symptoms but not the causes of a target earthquake: they signal its approach but do not trigger it. Such sporadic precursors to critical phenomena have been found also in socio-economic complex systems (Keilis-Borok et al., 2000).

2. It seems promising to apply RTP analysis to the detection of earthquake precursors in the other relevant and available data such as electromagnetic fields (Uyeda and Park, 2002), fluid regime (Ma et al., 1990), InSAR and GPS (Simons et al., 2002). The first positive result has been obtained with precursors gauging interaction between the ductile and brittle layers of the Earth crust; this opens a highly promising link of geodynamics and nonlinear dynamics approaches to prediction (Jin et al., 2004).

3. *The methodological advantage* of RTP over a direct analysis is in the drastic reduction in dimensionality of the parameter space where premonitory patterns are looked for. We have found here the patterns formed in narrow areas different from case to case, whose shape might be complicated, and with diverse size. To find these areas by a trial-and-error procedure would require trying different shapes, sizes, and locations, which is hardly realistic. Reverse analysis resolves this impasse, determining from the start a limited number of the areas to consider. Thus, RTP analysis provides a common methodological approach to the prediction of avalanches in a wide class of the complex systems, formed separately or jointly by nature and society.

4. The only decisive test of any prediction theory is an experiment in advance prediction. Such an experiment for the methodology described above was launched in June 2003 and is currently maintained by University of California Los Angeles (USA), Russian Academy of Sciences, and Institut de Physique du Globe de Paris (France). The complete results will be published elsewhere. The goal of this paper is to present the essential underlying concepts and report its first successes to a broad range of multidisciplinary experts, attracting their attention to the possibility of exploring premonitory patterns in diverse physical fields using the RTP methodology.

Acknowledgements

We are grateful to K. Aki, M. Ghil, A. Jin, L. Knopoff, J. McWilliams, and A. Soloviev for tough constructive criticism; to A. Kashina for inspired editing; and to D. Shatto for help in preparing the manuscript. This study was supported by the 21st Century Collaborative Activity Award for Studying Complex Systems from the James S. McDonnell Foundation, the National Science Foundation under grant ATM 0327558, and ISTC project 1538. The earthquake catalogs are compiled by Advanced National Seismic System (ANSS), Geophysical Institute of Israel (GII), and Japanese Meteorological Agency (JMA). The JMA catalog was received through the Japan Meteorological Business Support Center.

Appendix A

A.1. Earthquake catalogs

The data used in analysis are provided by the routinely compiled earthquake catalogs, which present at the moment the most accurate and complete information about the dynamics of seismicity. The earthquake catalog is taken from ANSS/CNSS (ANSS/CNSS Worldwide Earthquake Catalog, 1965–2003) and NEIC. We use a common representation of the earthquake catalog $\{t_j, \varphi_j, \lambda_j, M_j, b_j\}$, $j=1, 2, \ldots$ Here t_j is the time of an earthquake, $t_j \geq t_{j-1}$; φ_j and λ_j—latitude and longitude of its epicenter; and M_j magnitude. We consider the earthquakes with magnitude $M \geq M_{\min}$. As in most premonitory patterns of that family (Keilis-Borok, 2002) aftershocks are eliminated from the catalog; however, an integral measure of aftershocks activity b_j is retained for each remaining earthquake (main shocks and foreshocks); b_j is the number of aftershocks occurring immediately after an earthquake (e.g. within 2 days).

A.2. Chains

A chain captures a rise of earthquakes' correlation range in its vicinity. Let us call two earthquakes "neighbors" if their epicenters are closer than r and their times are closer than τ_0. A chain is a sequence of earthquakes where each earthquake has at least one neighbor belonging to that sequence and, therefore, no neighbors outside the sequence. he average density of epicenters decreases with increasing magnitudes. Accordingly, r is normalized as $r=r_0 10^{c(m-2.5)}$, where m is the smallest magnitude in the pair. The R-vicinity of a chain is outlined by the smoothed envelope of the circles of a radius R drawn around each epicenter in the chain. We consider only the chains with two sufficiently large characteristics: number of earthquakes $k \geq k_0$, maximal distance between epicenters $l \geq l_0$. Two parameters of the chains are common for all the regions: $r_0=50$ km, $c=0.35$. Other parameters are common for all chains within a region, but differ between regions as follows: Southern California, $\tau_0=20$ days, $k_0=6$, $M_{min}=2.9$, $l_0=175$ km; Central California, $\tau_0=30$ days, $k_0=10$, $M_{min}=2.9$, $l_0=250$ km; in Japan, $\tau_0=20$ days, $k_0=10$, $M_{min}=3.6$, $l_0=350$ km, $\gamma_0=0.4$; Eastern Mediterranean, $\tau_0=40$ days, $k_0=6$, $M_{min}=3.0$, $l_0=200$ km.

A.3. Intermediate-term patterns

We look for intermediate-term patterns in the R-vicinity of each chain within T years preceding it. To detect a pattern P we compute a function $F_P(t_j)$ defined in the "event window" (Keilis-Borok and Soloviev, 2003), i.e. on the sequence of N consecutive earthquakes with indexes $j-N+1$, $j-N+2, \ldots, j$. In R-vicinity of each chain we normalize seismicity by the lower magnitude cutoff M^*. The latter is derived from magnitude–frequency relation, by the condition $n(M^*)=n^*$; here $n(M^*)$ is the annual number of earthquakes with magnitude $M \geq M^*$.

Four functions represent *a rise of activity*. Namely

$$\text{"Activity"}\ F_U(t_j) = \frac{N}{t_j - t_{j-N+1}} \tag{A1}$$

is inversely proportional to the time it took to accumulate the most recent N earthquakes;

$$\text{"Sigma"}\ F_\Sigma(t_j) = \sum_{k=j-N+1}^{j} 10^{M_k - M^*} \tag{A2}$$

is shown to be a crude measure of total area of fault-breaks during the most recent N earthquakes (Keilis-Borok, 2002);

"Rise of magnitudes", $F_M(t_j)$

$$= \frac{2}{[N/2]}\left(\sum_{k=j-N+1}^{j-N+[N/2]} M_k - \sum_{k=j-[N/2]+1}^{j} M_k\right) \tag{A3}$$

is the difference between the average magnitude of the last $[N/2]$ earthquakes and that of the first $[N/2]$ earthquakes within a series of N;

"Acceleration" $F_C(t_j)$

$$= \frac{1}{[N/2]}\left(\sum_{k=j-N+1}^{j-N+[N/2]} \frac{1}{t_k t_{k-1}} - \sum_{k=j-[N/2]+1}^{j} \frac{1}{t_k - t_{k-1}}\right) \tag{A4}$$

is connected to the function "Activity". "Acceleration" increases if intercurrence time between earthquakes decreases with time.

Here $[x]$ denotes integer part of x.

Two functions depict *a rise of clustering*. The first one, "Swarm", reflects clustering of mainshocks:

$$F_W(t_j) = 1 - \frac{A_r(t_j)}{\pi r^2 N}. \tag{A5}$$

Here A_r is the area of the union of circles of radius r centered at N epicenters in the sequence. The second one, "b-micro", reflects clustering of aftershocks:

$$F_{b_\mu}(t_j) = \sum_{k=j-N+1}^{j} \sum_{l} 10^{M_{kl} - M^*}. \tag{A6}$$

Here M_{kl}, $l=1, 2, \ldots$ are the magnitudes of the aftershocks of the kth main shock within the first 2 days after the main shock.

The *rise of earthquakes correlation range* is depicted by function

$$\text{"Accord"}\ F_A(t_j) = \frac{A_r(t_j)}{\pi r^2}, \tag{A7}$$

which increases if earthquakes are widely distributed in space and their r-neighbourhoods are barely overlapping. Finally, *the transformation of Gutenberg–Richter relation* is reflected by function

$$\text{"Gamma"}\ F_\gamma(t_j) = \frac{1}{N_{M_k \geq M_{1/2}}} \sum_{M_k \geq M_{1/2}} (M_k - M^*), \tag{A8}$$

which increases if the magnitude distribution is shifted to the larger magnitudes (e.g. if the GR slope is decreasing). Here $M_{1/2}$ is the median of magnitudes of N earthquakes in our sequence.

Altogether the eight functions are determined by five parameters. In each region we used the same eight combinations of these parameters: $n^*=10$ and $R=50$ km or $n^*=20$ and $R=100$ km, $N=10$ or 50, $T=6$ or 24 months, $r=50$ km. Emergence of a pattern at the moment t is captured by the condition $F_P(t) \geq C_P$. Each threshold C_P is determined automatically at the learning stage. It minimises the sum $n+f$; here n is the rate of failures to predict and f is the rate of false alarms in prediction with a single pattern P.

In predicting the San Simeon and Tokachi-Oki earthquakes we used $R=75$ km to define the R-vicinity of a chain. This choice corresponds to the average of $R=50$ km and $R=100$ km used in our definition of the intermediate-term patterns. In the ongoing experiment we use $R=100$ km in Japan (region with the highest magnitude of a target earthquake, $M_W=7.2$) and $R=50$ km in all other regions.

A.4. Prediction

Final stage is recognition of precursory chain and issuing an alarm: A chain is recognised as precursory if it was preceded by C or more intermediate-term patterns out of the ensemble considered. The threshold C controls the trade-off between the rates of false alarms and failures to predict. Emergence of precursory chain triggers an alarm in its R-vicinity for the Δ months; statistics of past alarms suggests $\Delta=9$ months. A precursory chain may keep growing accumulating subsequent earthquakes. In that case the alarm is extended. If a target earthquake occurs in the R-vicinity of a chain, then the chain no longer grows, but the alarm (if it has been diagnosed for that chain) is not called off. After a target earthquake all other chains containing its epicenter within the R-vicinity are disregarded during the period Δ.

A.5. Quality of prediction

Suppose that the prediction was performed during the time interval of length T (year) within the region Ω with the area S (km^2); N large earthquakes occurred within this period; A alarms were declared and A_f of them were false; all the alarms together covered the spatio-temporal volume V_A (year × km^2); N_f target earthquakes were unpredicted. Prediction is described by the following dimensionless errors: the fraction of unpredicted earthquakes, $n=N_f/N$; the relative alarm coverage, $\tau=V_A/(T\times S)$; the fraction of false alarms, $f=A_f/A$.

When calculating the alarm coverage, it might be advantageous to take into account the observed inhomogeneities of the earthquake spatial distribution. In our prediction experiment, the relative alarm coverage for an alarm that spans the time T_A and space S_A is calculated as

$$\tau_A = \frac{T_A}{T}\frac{\int_{S_A} dN_4(r)}{\int_S dN_4(r)} = \frac{T_A}{T}\frac{\#\{\text{EQ with } m\geq 4 \text{ within } S_A\}}{\#\{\text{EQ with } m\geq 4 \text{ within } S\}}. \tag{A9}$$

Here by $N_4(r)$ we denote the 2D point process of earthquakes with magnitude $m\geq 4$. The total alarm coverage is the sum of that for all individual alarms.

A.6. Significance level: random binomial prediction

To evaluate significance of a prediction one typically evaluates the chances of getting the same or better result (same or smaller values of errors) when there is no dependence between alarms and the occurrence of target earthquakes. An extremely simple but easily tractable model of prediction which produces alarms independent of the target earthquakes is *random binomial prediction* (Molchan, 2003): One divides the space–time considered for prediction into M small equal bins and declares alarm in each of them with fixed probability p. Indeed, this approach is highly unrealistic. Nevertheless, considered as a null (random) prediction model, it provides a good coarse estimation of the algorithm predictive power. Significance with respect to a random binomial prediction may serve as a necessary, but not sufficient, condition for validating an algorithm.

It is readily checked that expected values of alarm coverage τ and fraction f of failures to predict in the binomial prediction are given by:

$$E(\tau)=p, \qquad E(f)=1-p$$

so the point corresponding to this prediction is on the diagonal $f=(1-\tau)$ in the 2D (τ,f)-section of the error diagram. The probability to predict exactly $N-N_f$ out of N target earthquakes, assuming that no more than one target earthquake may occur within a single bin, is given by binomial distribution

$$\Pr\{\text{predict } N-N_f \text{ out of } N\} = \binom{N}{N_f} p^{N-N_f}(1-p)^{N_f}. \tag{A10}$$

The probability to predict $N-N_f$ out of N target earthquakes issuing alarm within k bins out of M is given by hypergeometric distribution

$$\Pr\{\text{predict } N-N_f \text{ out of } N \text{ declaring alarm in } k \text{ bins out of } M\} = \frac{\binom{k}{N-N_f}\binom{M-k}{N_f}}{\binom{M}{N}}. \quad \text{(A11)}$$

The number of false alarms can also be obtained, but because of the simplistic binomial rules, the number of binomial alarms (and false alarms) will be significantly larger than that in any realistic prediction (where alarm is typically declared for considerable spatio-temporal area, not for a small bin). Thus here we do not make any inference about false alarms using the binomial prediction model.

Using the above probabilities (A10, A11) one can construct different significance measures for a given prediction with errors (τ^*, n^*). One approach is to use the 2D (τ, n) distribution under the binomial model using (A10) with $p=\tau^*$, and evaluate probability of obtaining a prediction of the same or better quality, say

$$\Pr\{(\tau,n):\tau+n\le\tau^*+n^*\} \text{ or } Pr\{(\tau,n);\tau\le\tau^*\&n\le n^*\}.$$

Another approach is to use (A11) to find the probability to predict the same or larger number of earthquakes with the same total duration of alarms. The difference between using (A10) and (A11) is that in the first case we assume fixed *probability* of declaring an alarm, while in the second—fixed *duration* of alarm. Indeed, in generic cases both approaches give very similar evaluation of prediction performance.

To illustrate the above approach, Fig. 6 shows the error diagram for the results of our prediction experiment in California. Shaded ball represents the errors of our prediction experiment during 1964–2005. The probability for a random binomial prediction with given value of τ to fall within the shaded area (i.e. to predict more than $N(1-n)$ target earthquakes with given τ) is less or equal than 0.001 (0.1%). The point that corresponds to our experiment is well within this area, thus indicating very high predictive power. It should be emphasized that the results presented in this figure combine the information from the learning period, independent data, and advance prediction (we have too few alarms and target earthquakes during the advance phase to use them alone). Thus, this analysis is not equivalent to evaluating the real predictive power of the algorithm, where only advance results must be used. Nevertheless, the grey shadowed area that corresponds to the binomial model gives a good orientation for the expected significance of the results.

A.7. Significance level: empirical estimation

An alternative approach to testing significance of a prediction algorithm involves empirical estimations of occurrence rate for target earthquakes. Thus, the approach is unavoidably approximate due to the small number of target earthquakes; yet it is much more realistic comparing to the random binomial prediction.

Specifically, we assume that target earthquakes form a Poisson process $N(t,r)$ stationary in time but nonhomogeneous in space. The expected number of earthquake within the interval of length t and spatial region R is given by

$$E(N(t,R)) = t\mu(R) \quad \text{(A12)}$$

where $\mu(R)$ is some non-negative measure over the space. In practice, a first-order approximation to this measure can be obtained by considering the number N_R of target earthquakes within the region R per unit of time using observations over S years:

$$\mu(R)\approx N_R/S.$$

With our assumptions, the probability of having exactly k target earthquakes within the region R during time interval of length t is given by Poisson distribution

$$\Pr\{k \text{ target earthquakes within } R\} = e^{-\mu(R)t}\frac{(\mu(R)t)^k}{k!} \quad \text{(A13)}$$

and the probability p to have at least one target earthquake is

$$p := \Pr\{\text{at least one target earthquake within } R\} = 1-e^{-\mu(R)t}. \quad \text{(A14)}$$

When the rate of target earthquakes is small (which is indeed the case in our experiment), we can approximate p as

$$p := \Pr\{\text{at least one target earthquake within} R\} \approx \mu(R)t \approx \frac{N_R t}{S} \quad \text{(A15)}$$

Our final goal is to calculate the probability of predicting $N-N_f$ target earthquakes out of N by a set of alarms $A_i=(t_i, R_i)$ that were declared for regions R_i

and time intervals t_i. We denote by p_i the probability to have at least one target earthquake within A_i.

The probability for a given target earthquake to be predicted is calculated as the probability that it will be predicted by at least one of the alarms:

$$\Pr\{\text{given target EQ is predicted}\}$$

$$= \sum_i \Pr\{\text{given target EQ is within} A_i\}$$

$$= \sum_i \Pr\{\text{given target EQ is within} R_i \text{during} t_i\}$$

$$= \sum_i \Pr\{\text{given target EQ is within} R_i\} \frac{t_i}{T}$$

$$= \sum_i q_i \frac{t_i}{T}$$

Here we used the fact that alarms are not overlapping (by definition); factorization property (A12) of the target event process; and the fact that conditional distribution of the occurrence time of an event from Poisson process is uniform, given that this event occurred within the given time interval.

The probability for a given target earthquake to happen within the spatial region R_i can be estimated as

$$q_i = \Pr\{\text{given target EQ happened within } R_i\} = \frac{n_i}{N_\Omega}.$$

where n_i is the number of target earthquakes within R_i during some period S and N_Ω is the total number of target earthquakes within the region Ω considered for prediction during the same time. Finally

$$Q: = \Pr\{\text{given target EQ is predicted}\} \approx \sum_i \frac{n_i}{N_\Omega} \times \frac{t_i}{T} \approx \sum_i p_i \frac{S}{N_\Omega T},$$

and the distribution of the number of predicted target earthquakes out of N is given by the binomial formula:

$$\Pr\{N - N_f \text{ out of } N \text{ target EQ are predicted}\}$$

$$= \binom{N}{N_f} Q^{N-N_f}(1-Q)^{N_f} \approx \binom{N}{N_f}\left(\sum_i \frac{n_i t_i}{N_\Omega T}\right)^{N-N_f} \left(1 - \sum_i \frac{n_i t_i}{N_\Omega T}\right)^{N_f} \approx \binom{N}{N_f}\left(\sum_i p_i \frac{S}{N_\Omega T}\right)^{N-N_f} \left(1 - \sum_i p_i \frac{S}{N_\Omega T}\right)^{N_f}$$

We apply the above approach to California. Specifically, we consider the region Ω shown in Fig. 2 during the period 1965–2004 ($S=40$ years); there were $N_\Omega=10$ target earthquakes. The advance prediction was performed within the same region during July 2003–June 2005 ($T=2$ years), and resulted in three alarms; $N=1$ target earthquake occurred during this period. The probabilities p_i of having at least one target earthquake within each of the alarms are 5%, 8%, and 5% (see Table 1 and Eq. (A14)). The probability to predict the only target event by chance is estimated as 36%. Notice that this is the conditional probability given the actual number of target earthquakes and alarms. If one does not want to be conditioned by the number of actual target earthquakes, then we need to modify our results using (A13). In the case of California, where we had only one target earthquake, this will give:

$$\Pr\{\text{predict 1 target with our three alarms}\}$$

$$= \Pr\{\text{there is exactly one target}\}$$

$$\times \Pr\{\text{it was predicted}\}.$$

The first probability is estimated using (A13):

$$\Pr\{\text{there is exactly one target}\}$$

$$= \mu(R)T\mathrm{e}^{-\mu(R)T} \approx \frac{N_\Omega}{S} T\mathrm{e}^{-\frac{N_\Omega}{S}T} = \frac{10}{40} 2\mathrm{e}^{-\frac{10}{40}2} \approx 0.3.$$

Thus, the probability to have only one target event and predict it by chance is approximately $0.36 \times 0.3 = 0.12$ (or 12%).

References

Aki, K., Keilis-Borok, V., Gabrielov, A., Jin, A., Liu, Z., Shebalin, P., Zaliapin, I., On the current state of the lithosphere in Central California. Letter with San Simeon prediction sent to a group of leading experts on June 21, 2003. Available at http://www.math.purdue.edu/~agabriel/Workshop/SanSimeon_let.pdf.

ANSS/CNSS Worldwide Earthquake Catalog. (http://quake.geo.berkeley.edu/cnss) (Produced by Advanced National Seismic System (ANSS) and hosted by the Northern California Data Center (NCEDC), 1965–2003).

Bowman, D.D., Ouillon, G., Sammis, C.G., Sornette, A., Sornette, D., 1998. An observational test of the critical earthquake concept. J. Geophys. Res. 103, 24359–24372.

Caputo, M., Console, R., Gabrielov, A.M., Keilis-Borok, V.I., Sidorenko, T.V., 1983. Long-term premonitory seismicity patterns in Italy. Geophys. J. R. Astron. Soc. 75, 71–75.

Daley, D.J., Vere-Jones, D., 2004. Scoring probability forecasts for point processes: the entropy score and information gain. J. Appl. Probab. 41A, 297–312.

De Mare, J., 1980. Optimal prediction of catastrophes with applications to Gaussian processes. Ann. Probab. 8, 841–850.

Gabrielov, A., Keilis-Borok, V., Zaliapin, I., Newman, W.I., 2000. Critical transitions in colliding cascades. Phys. Rev., E 62, 237–249.

Jin, A., Aki, K., Liu, Z., Keilis-Borok, V., 2004. Seismological evidence for the brittle–ductile interaction hypothesis on earthquake loading. Earth Planets Space 56, 823–830.

Jolliffe, I.T., Stephenson, D.B., 2003. Forecast Verification. Wiley, Chichester.

Kadanoff, L.P., 2000. Statistical Physics: Statics, Dynamics, and Renormalization. World Scientific Publishing, Singapore.

Keilis-Borok, V.I., 1990. The lithosphere of the Earth as a non-linear system with implications for earthquake prediction. Rev. Geophys. 28 (1), 19–34.

Keilis-Borok, V., 2002. Earthquake prediction: state-of-the-art and emerging possibilities. Annu. Rev. Earth Planet. Sci. 30, 1–33.

Keilis-Borok, V.I., Shebalin, P.N. (Eds.), 1999. Dynamics of the Lithosphere and Earthquake Prediction, Phys. Earth Planet. Inter. (special issue) vol. 111. , pp. 179–330.

Keilis-Borok, V.I., Soloviev, A.A. (Eds.), 2003. Nonlinear Dynamics of the Lithosphere and Earthquake Prediction. Springer, Heidelberg.

Keilis-Borok, V.I., Knopofff, L., Rotwain, I.M., 1980. Bursts of aftershocks, long-term precursors of strong earthquakes. Nature 283, 259–263.

Keilis-Borok, V., Stock, J.H., Soloviev, A., Mikhalev, P., 2000. Pre-recession pattern of six economic indicators in the USA. J. Forecast. 19, 65–80.

Keilis-Borok, V.I., Shebalin, P.N., Zaliapin, I.V., 2002. Premonitory patterns of seismicity months before a large earthquake: five case histories in Southern California. Proc. Natl. Acad. Sci. 99, 16562–16567.

Keilis-Borok, V., Shebalin, P., Gabrielov, A., Turcotte, D., 2004. Reverse tracing of short-term earthquake precursors. Phys. Earth Planet. Inter. 145, 75–85.

Knopoff, L., Levshina, T., Keilis-Borok, V.I., Mattoni, C., 1996. Increased long-range intermediate-magnitude earthquake activity prior to strong earthquakes in California. J. Geophys. Res. 101, 5779–5796.

Lindgren, G., 1975. Prediction from a random time point. Ann. Probab. 3, 412–423.

Lindgren, G., 1985. Optimal prediction of level crossings in Gaussian processes and sequences. Ann. Probab. 13, 804–824.

Ma, Z., Fu, Z., Zhang, Y., Wang, C., Zhang, G., Liu, D., 1990. Earthquake Prediction: Nine Major Earthquakes in China. Springer-Verlag, New York.

Mogi, K., 1981. Seismicity in Western Japan and long-term forecasting. Earthquake Prediction: an International Review, Maurice Ewing Series, vol. 4. American Geophysical Union, Washington, DC, pp. 43–51.

Molchan, G.M., 1990. Strategies in strong earthquake prediction. Phys. Earth Planet. Inter. 61 (1–2), 84–98.

Molchan, G.M., 2003. Earthquake prediction strategies: a theoretical analysis. In: Keilis-Borok, V.I., Soloviev, A.A. (Eds.), Nonlinear Dynamics of the Lithosphere and Earthquake Prediction. Springer, Heidelberg, pp. 209–237.

Molchan, G.M., Dmitrieva, O.E., Rotwain, I.M., Dewey, J., 1990. Statistical analysis of the results of earthquake prediction, based on burst of aftershocks. Phys. Earth Planet. Inter. 61, 128–139.

Prozorov, A.G., Schreider, S.Yu., 1990. Real time test of the long-range aftershock algorithm as a tool for mid-term earthquake prediction in Southern California. Pure Appl. Geophys. 133, 329–347.

Rundle, J.B., Turcotte, D.L., Shcherbakov, R., Klein, W., Sammis, C., 2003. Statistical physics approach to understanding the multiscale dynamics of earthquake fault systems. Rev. Geophys. 41, 1019.

Rundquist, D.V., Soloviev, A.A., 1999. Numerical modeling of block structure dynamics: an arc subduction zone. Phys. Earth Planet. Inter. 111, 241–252.

Shebalin, P., Zaliapin, I., Keilis-Borok, V.I., 2000. Premonitory rise of the earthquakes' correlation range: lesser Antilles. Phys. Earth Planet. Inter. 122, 241–249.

Shebalin, P., Keilis-Borok, V.I., Zaliapin, I., Uyeda, S., Nagao, T., Tsybin, N., 2003. Short-term Premonitory Rise of the Earthquake Correlation Range//IUGG2003, June 30–July 11, 2003, Sapporo, Japan. Abstracts. A184.

Shebalin, P., Keilis-Borok, V., Zaliapin, I., Uyeda, S., Nagao, T., Tsybin, N., 2004. Advance short-term prediction of the large Tokachi-Oki earthquake, September 25, 2003, $M=8.1$: a case history. Earth Planets Space 56, 715–724.

Simons, M., Fialko, Y., Rivera, L., 2002. Coseismic deformation from the 1999 M_W 7.1 Hector Mine, California, earthquake as inferred from InSAR and GPS observations. Bull. Seismol. Soc. Am. 92, 1390–1402.

Sykes, L.R., 1983. Predicting great earthquakes. In: Kanamori, H., Boschi, E. (Eds.), Earthquakes, Observation, Theory, and Interpretation. North Holland, New York, pp. 398–435.

Uyeda, S., Park, S. (Eds.), 2002. Proceedings of the International Symposium on The Recent Aspects of Electromagnetic Variations Related with Earthquakes, 20 and 21 December 1999, J. Geodyn., vol. 33, pp. 4–5. Special issue.

Whittle, P., 1963. Prediction and Regulation by Linear Least-Squares Methods. English University Press, London.

Zaliapin, I., Keilis-Borok, V.I., Axen, G., 2002. Premonitory spreading of seismicity over the faults' network in southern California: precursor accord. J. Geophys. Res., B 107, 2221.

Zaliapin, I., Keilis-Borok, V.I., Ghil, M., 2003. A Boolean delay equation model of colliding cascades. Part II. Prediction of critical transitions. J. Stat. Phys. 111, 839–861.

Zöller, G., Hainzl, S., 2001. Detecting premonitory seismicity patterns based on critical point dynamics. Nat. Hazards Earth Syst. Sci. 1, 93–98.

Zöller, G., Hainzl, S., Kurths, J., 2001. Observation of growing correlation length as an indicator for critical point behavior prior to large earthquakes. J. Geophys. Res. 106, 2167–2176.

Statistical properties of the cluster dynamics of the systems of statistical mechanics

A. Gabrielov, V. Keilis-Borok, Ya. Sinai, and I. Zaliapin

Introduction

L. Boltzmann was the firs who tried to explain the laws of thermodynamics and kinetics as corollaries of dynamics of large systems of interacting particles (see his classical book [1]). In particular, Boltzmann derived the classical Boltzmann kinetic equation based on his Stoßzahlansatz which gave the firs "dynamical" explanation of irreversibility. The Stoßanzahlsatz can be applied to any system of statistical mechanics which can be considered as a small perturbation of an ideal gas and undergoes the so-called Boltzmann-Grad limit transition (see [L1]) when the length of the free path is of the order of the size of the whole system. O. Lanford (see [3]) gave the firs mathematical proof of the local existence theorem of the solutions of the Boltzmann equation where he showed the convergence of correlation functions to solution of the Boltzmann equation.

There were many attempts to extend the conditions under which the Boltzmann kinetic equation or its generalization works. In this connection we would like to mention the book by Bogolyubov [2] where Bogolyubov stressed several times the idea that in the gaseous phase when the interaction between the particles is short-ranged the system can be decomposed onto finit clusters so that during some random interval of time each cluster moves independently on other clusters as a finite-dimensiona dynamical system. After such random time the system can be decomposed again on other dynamically independent clusters and so on. It is natural to call this type of dynamics as cluster dynamics.

The cluster dynamics was shown to exist in the one-dimensional systems of statistical mechanics in [4]. Consider the infinit system of one-dimensional particles with pair-wise interaction having a hard core and short range. Assume that the distribution of particles is given by the Gibbs canonical distribution. In the one-dimensional case in the thermodynamical limit there is no difference between canonical and grand canonical distributions. According to these distributions in a typical situation on any interval $[-R, R]$ there are empty intervals of the length $O(\ln R)$ having at $t = 0$ no particles. On the other hand, the velocities of the particles have Gaussian distributions Therefore for any T the velocities of the particles on $[-R, R]$ grow typically as $O(\sqrt{\ln R})$. It shows that for sufficientl large R particles cannot go from one end-point of an empty interval to the other one. In other words for any T in the finite-dimensiona dynamics particles in the domains bounded by empty intervals do not interact with external particles.

This gives cluster dynamics. The accurate proof requires some probabilistic estimates which show that the velocities do not become large during the finite-dimensiona dynamics (see [4]). All these ideas were generalized to the multi-dimensional case and low density, *i.e.* to the gaseous phase in [5].

The infinit system of equations of motion of particles (x_i, v_i) of mass $m = 1$ has the form

$$\frac{dx_i}{dt} = v_i, \qquad \frac{dv_i}{dt} = -\sum_j \frac{\partial U(|q_i(t) - q_j(t)|)}{\partial q_i} \tag{1}$$

In the paper by E. Presutti, M. Pulvirenti and B. Tirozzi [6] the authors proved the general existence theorem for solutions of (1) for any limit Gibbs distribution with arbitrary density and inverse temperature. Their method is based upon the reduction of (1) to the corresponding integral equation and the proof of existence of solutions of the integral equation.

The most general results were obtained by R. Dobrushin and J. Fritz (see [7]). They described a large subset in the phase space of (1) for which they proved 1) the existence theorem of solutions of (1) and 2) the fact that the probability of this subset with respect to any natural Gibbs distribution is 1. This gives the possibility to describe the dynamics in the space of Gibbs distributions.

Recently, it turned out that the concept of cluster dynamics has a wider domain of applications in such field as plasma physics, geophysics and others (see [20], [21], [23], [16]). In this connection it is interesting to study the distributions of various characteristics of cluster dynamics, such as the distribution of the size of the clusters, the statistics of particles in a cluster and others. The whole set of problems resembles problems of percolation theory but seems to be more difficul than the percolation theory because it involves dynamical characteristics. There is no hope that something can be done analytically. On the other hand, the use of computers opens wide possibilities in getting numerical results. The corresponding experiments were performed recently and the results are presented below in this paper. We want to stress the appearance of new critical indices which are special for cluster dynamics and presumably are different from similar ones in percolation theory.

In this paper we analyze numerically cluster dynamics in a simplest system – a frictionless elastic billiard. The model and cluster rules are described in Section 1. We report a phase transition in the cluster formation process in Section 2 and describe how it depends on the model parameters in Section 3. Also, we study how the size of the maximal cluster in a finit system scales with the system size (Section 4). Section 5 describes possible applications of cluster dynamics. Details of our numerical simulations are given in Section 6.

1 Model

The results in this paper refer to an elastic frictionless billiard on a square table. Namely, we consider N balls of mass $m = 1$ and radius R placed within the region $\mathcal{T} = \{(x, y) : |x| \leq L/2, |y| \leq L/2\}$. We will call N the size of the billiard. Each ball moves without friction with constant velocity $\boldsymbol{v}_i = (v_x^i, v_y^i)$, $i = 1, \ldots, N$ until it collides with a wall or another ball. All collisions are elastic, which means that the total energy $E = \sum_{i=1}^{N} m\,|\boldsymbol{v}_i|^2/2$, $|\boldsymbol{v}_i|^2 = (v_x^i)^2 + (v_y^i)^2$, of the system remains constant, the total momentum $\boldsymbol{p} = \sum_{i=1}^{N} m\, \boldsymbol{v}_i$ is not affected by ball collisions, and in wall collisions the ball's reflectio and incidence angles are the same. A useful characteristic of the billiard is the density of balls:

$$\rho = \frac{N\,\pi\,R^2}{L^2}. \tag{2}$$

We will need the notion of Δ-cluster [4], which is a group of balls that have affected each others dynamics during the time interval of duration Δ. Formally, we call two balls Δ-*neighbors* at epoch t if they collided during the time interval $[t-\Delta, t]$. Any connected component of this neighbour relation is called a Δ-cluster at epoch t. This definitio ensures that each ball has collided with at least one ball from its Δ-cluster within the time interval $[t-\Delta, t]$. The mass of a cluster is the total mass of its balls. We denote by $N_\Delta(t)$ the total number of Δ-clusters at instant t, and by $M_\Delta^i(t)$ the mass of the i-th largest cluster. Thus, the mass of the maximal Δ-cluster is $M_\Delta^1(t)$, of the second largest cluster $M_\Delta^2(t)$, of the smallest cluster $M_\Delta^{N_\Delta}(t)$, *etc.*

Obviously, at $\Delta = 0$ there exist N clusters of mass m, each corresponding to an individual ball. As Δ increases, the balls start to collide so the number of clusters becomes smaller while their masses increase. In this note we only consider a situation when $\Delta = t$, that is we deal with all the clusters that have been formed during the time interval $[0, \Delta]$. This allows us to drop the dependence on t and work with the number of clusters N_Δ, maximal cluster of mass M_Δ^1, *etc.*

In this paper we focus on the time-dependent distribution (density) $f_t(M)$ of cluster sizes (masses). In a numerical study, to obtain a sample of n clusters at epoch t one runs the model n times, each time choosing a single cluster at epoch t. A standard practice in stochastic geometry suggests to choose the cluster nearest to the origin (that is the cluster that contains the ball nearest to the origin). It happens that the computational load associated with this procedure is prohibitive for our purposes. To make the computations workable, we approximate the cluster size distribution $f_t(M)$ using the entire cluster population from a single realization of the model, which has density $g_t(M)$. For large N, $f_t(M)$ becomes an *area-biased* version of $g_t(M)$, since large clusters have a better chance to lie closer to the origin than smaller ones; and the two distributions are connected via [11]:

$$f_t(M) = \frac{M\,g_t(M)}{\int M\,g_t(M)\,dM}. \tag{3}$$

The results in this paper refer to $g_t(M)$.

2 Phase transition in dynamical clustering

To study the dynamics of clusters we have considered models with $10^{-6} \le \rho \le 10^{-1}$, and $10 \le N \le 10^4$. We run each model until the instant when a cluster of mass $0.95 \times N$ is formed.

Figure 1 shows fractional masses $\mu_\Delta^i = M_\Delta^i/N$ of clusters as a function of time when they have been formed; this figur refers to $\rho = 10^{-3}$, $N = 10^4$ Here we

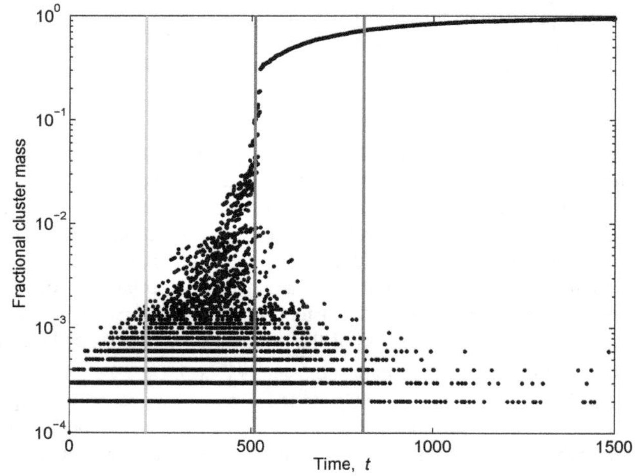

Figure 1. Fractional cluster masses $\mu^i_\Delta = M^i_\Delta/N$ as a function of time when they have been formed in a model with $\rho = 10^{-3}$, $N = 10^4$. Each point corresponds to a single cluster. The horizontal lines at the lower part of the figur are formed by a multitude of clusters of mass $M = 2$ (lowest line), $M = 3$ (second line), *etc.* The point $(0, 10^{-4})$ refers to $N = 10^4$ balls, each of which forms a cluster of unit mass at $t = 0$. Notice the dramatic change in the cluster mass distribution at the moment $t_c \approx 510$ depicted by blue vertical line. Three vertical lines correspond to the three cluster size distributions in Figure 2.

see the following generic qualitative picture that has been observed for *all* the models considered. At $\Delta = 0$, we start with N clusters of mass m. As the model evolves, clusters of increasingly larger masses are formed and the total number of clusters decreases. For some period of time, $0 \le t \le t_c$ (for this model, $t_c \approx 510$) the cluster mass distribution has no notable gaps, in particular the size of the largest cluster does not exceed significantl the size of the second largest cluster. At $t = t_c$ one observes a sharp qualitative change in the cluster formation process. There appears a distinctive largest cluster, which creates a gap in the mass distribution between the largest cluster and the rest of the clusters. Notably, this change happens when the largest cluster is still relatively small: less than 10% of the total system mass.

It will be convenient to defin formally the time instant t_c that corresponds to the change in cluster formation. We defin here t_c as the instant after which there are no clusters larger than $M^1_{t_c}$:

$$t_c = \inf\{t : M^1_t > M^i_s,\ s > t,\ i > 1\}. \tag{4}$$

We will refer to t_c as *critical* time. Note that this definitio uses the information from the times $t > t_c$ (*i.e.*, t_c is not a stopping time), so it cannot be used for operational detection of t_c. It would be useful to have an alternative definitio which wi l use only the information available prior t_c.

Figure 2 shows the empirical cluster mass density for the model with $\rho = 10^{-3}$ and $N = 10^4$ at three time instants depicted by vertical lines in Figure 1.

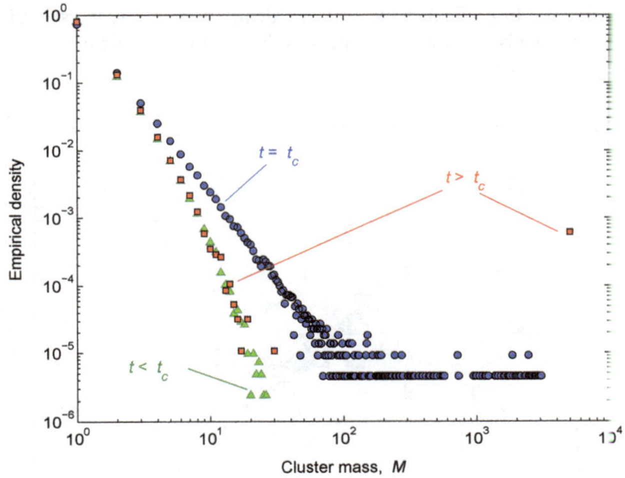

Figure 2. Cluster size distribution at three instants depicted by vertical lines in Figure 1. At $t < t_c$ (green triangles) distribution can be approximated by a power law with exponential taper at the tail; at $t \approx t_c$ (blue balls) it is a pure power law; at $t > t_c$ (red squares) it is a tapered power law plus a δ function at the largest cluster. To produce this figur we used 50 ndependent realizations of the model with $\rho = 10^{-3}$, $N = 10^4$.

For $t \leq t_c$ the cluster mass distribution $g_t(M)$ is well described by a power law with an exponential taper:

$$g_t(M) = M^{-\beta} \exp\left(-\frac{M}{\gamma(t)}\right). \tag{5}$$

Here $\beta \approx 5/2$, and the taper strength is determined by $\gamma(t)$: $\gamma(t) \ll M$ leads to almost an exponential distribution, while $\gamma(t) \gg M$ to a pure power law. Our simulations suggest that $\gamma(t)$ is monotone increasing with $\gamma(t_c) = \infty$, so that $g_t(M)$ transforms from exponential at small times $t \ll t_c$ to a pure power law at the critical epoch t_c. One can readily see the scenario that characterizes systems with phase transition of second kind. In statistical mechanics, equation (5) describes the correlation function $\Gamma(r)$ of a system (*e.g.*, Potts model) near the critical point. A similar power law with exponential taper was suggested for the mass distribution of clusters in a percolation model [12], [13] as the percolation instant approaches; for rupture sizes in a colliding cascades model of earthquakes [14], [15], [17]; for real ruptures in steel samples [16], *etc.* For $t > t_c$ the distribution is decomposed into two components: A δ-function at the mass of the largest cluster, and the tapered power law (5) for the rest of the clusters with $\gamma(t)$ monotone decreasing for $t > t_c$.

This picture is observed for all the models considered independently of the billiard density ρ and size N. In particular, the critical index $\beta \approx 5/2$ is universal: Figure 3 show the cluster mass distribution at t_c for seven different models with $N = 5 \times 10^3$ and $\rho = 10^{-1}, 10^{-2}, \ldots, 10^{-7}$.

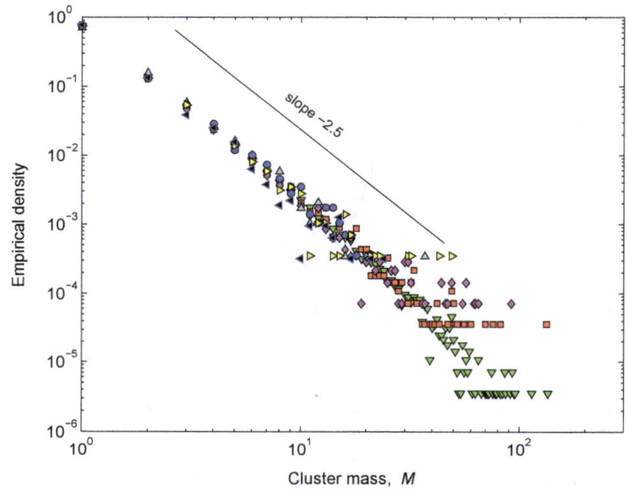

Figure 3. Power-law cluster size distribution at critical instant t_c for seven models with fi ed $N = 5 \times 10^3$ and $\rho = 10^{-1}, 10^{-2}, \ldots, 10^{-7}$. The distribution at critical instant t_c is characterized by a universal index $\beta = 5/2$.

3 Scaling of critical time and mass

Here we study how the critical instant t_c and the fractional mass $\mu_c := M^1_{t_c}/N$ of the largest cluster at t_c depend on the billiard density ρ. For that we run multiple (from 10 to 1000) realizations of the billiard for $\rho = 10^{-1}, 10^{-2}, \ldots, 10^{-7}$ and $N = 5 \times 10^3$. The averaged values of t_c and μ_c are shown in Figures 4, 5 as functions of ρ. These results support the following asymptotic relations, which give very good approximation for $\rho < 10^{-2}$:

$$t_c(\rho \,|\, N = 5 \times 10^3) \approx 0.4\,\rho^{-1}, \quad \mu_c(\rho \,|\, N = 5 \times 10^3) \approx 0.07, \quad \text{as } \rho \to 0. \tag{6}$$

A more detailed analysis (not shown) indicates that the large density correction for critical time is given by:

$$t_c(\rho \mid N = 5 \times 10^3) = 0.4\,\rho^{-1}\left(1 + \rho^{0.4} + o(\rho^{0.4})\right),\ \rho \to 0. \tag{7}$$

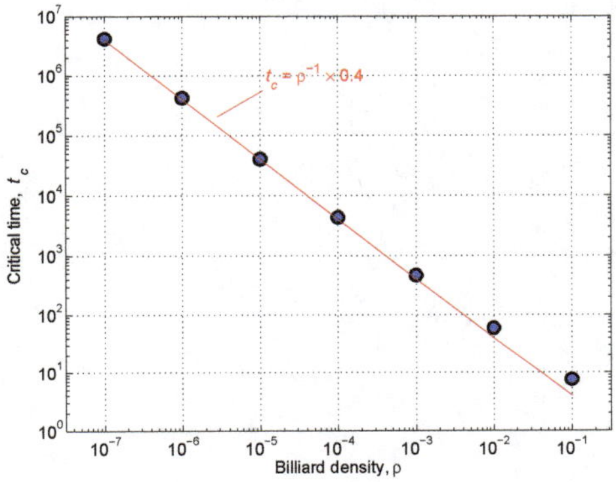

Figure 4. Scaling of critical time t_c with the billiard density ρ at fi ed $N = 5 \times 10^3$.

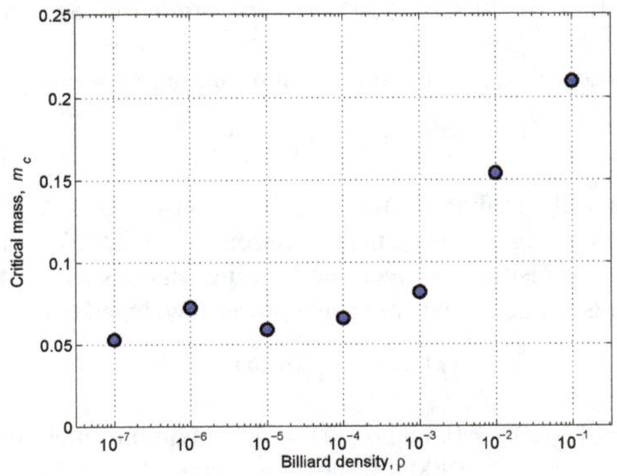

Figure 5. Critical mass μ_c as a function of the billiard density ρ at fi ed $N = 5 \times 10^3$.

4 Size of the maximal cluster

Here we consider the dynamics and scaling of the maximal cluster. Figure 6 shows the dynamics of the fractional mass $\mu_\Delta^1 = M_\Delta^1/N$ of the largest cluster averaged over 10 to 1000 realizations of a billiard; it refers to ten models with $10 \le N \le 10^4$ and fi ed

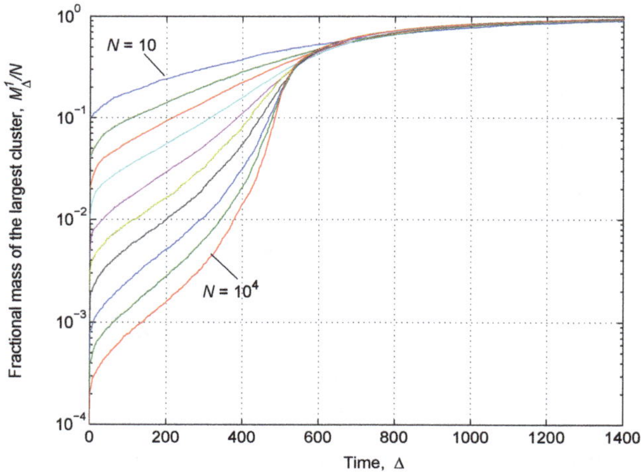

Figure 6. Fractional mass $\mu_\Delta^1 = M_\Delta^1/N$ of the maximal cluster as a function of Δ for ten models with fi ed $\rho = 10^{-3}$ and $10 \le N \le 10^4$ (each line corresponds to a distinct value of N).

$\rho = 10^{-3}$. By construction, we have the following limiting relations:

$$M_0^1 = 1, \quad M_\infty^1 = N. \tag{8}$$

Next we study in detail the scaling of M_Δ^1 as a function of N and Δ. Figure 7 shows M_Δ^1 as a function of N at eight time instants between $\Delta = 0$ and $\Delta = 1400 \approx 3 \times t_c$. One can see a gradual transition between the limiting scalings (8). Approximately linear form of the plots suggests that we have a power law dependence

$$M_\Delta^1 = c(\Delta)\, N^{\alpha(\Delta)}, \tag{9}$$

with $c(0) = c(\infty) = 1$, $\alpha(0) = 0$, and $\alpha(\infty) = 1$. Maximum likelihood estimations of $c(\Delta)$ and $\alpha(\Delta)$ for $0 \le \Delta \le 1400$ are shown in Figure 8. At critical time we have $\alpha(t_c) \approx 2/3$.

A more detailed analysis (not shown) indicates that at initial times ($0 < t < t_c/3$) the maximal cluster size in a finit system of size N is proportional to $\log_{10} N$:

$$M_\Delta^1 = \alpha'(\Delta) \log_{10}(N) + c'(\Delta). \tag{10}$$

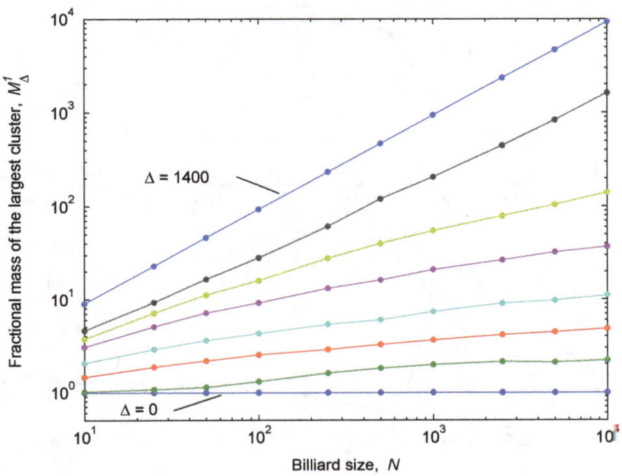

Figure 7. Mass M^1_Δ of the maximal cluster as a function of billiard size N – each line corresponds to a distinct value of Δ. Experiment is done under fixed $\rho = 10^{-3}$.

There exists a natural connection between this transition from logarithmic to power scaling of the maximal cluster size and from exponential to power tail of the cluster size distribution (see equation (5) and Figure 2). Indeed, suppose that cluster sizes are independent and identically distributed with distribution function $F(x)$ and density $f(x)$. Then the size of the maximal among N clusters has distribution $F_{\max}(M) = F^N(M)$ and density $f_{\max}(M) = N\, f\, F^{N-1}$; and the most probable size M^* of the maximal cluster is given by the finite solution of $\frac{d}{dM} f_{\max}(M^*) = 0$. It is readily verified that

$$F(M) = 1 - e^{-\lambda M} \Rightarrow M^* = \frac{\ln(N)}{\lambda} \propto \ln(N), \tag{11}$$

$$F(M) = 1 - M^{-b} \Rightarrow M^* = \left(\frac{N\,b+1}{b+1}\right)^{1/b} \propto N^{1/b} + o(N^{1/b}). \tag{12}$$

The exponential tail of the cluster size distribution at $t < t_c$ thus justifies the logarithmic dependence (10) of the maximal cluster size on N; while the power law cluster size distribution around t_c explains the power law scaling (9). Equations (9) and (12) suggest $b = 1/\alpha$. We notice that the index $b = \beta - 1 \approx 3/2$ of the cluster size distribution observed at t_c (see Figure 3) is indeed equal to $1/\alpha \approx 1/0.66 \approx 3/2$ observed at $t_c \approx 510$ in the top panel of Figure 8.

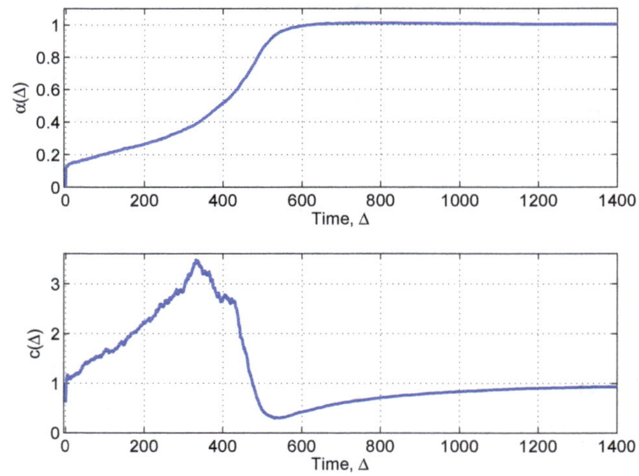

Figure 8. Parameters α and c of scaling law (9) as functions of time Δ. Each line in Figure 7 corresponds to one point in each panel of this figure To produce this figur we used models with fi ed $\rho = 10^{-3}$, and $10 \leq N \leq 10^4$.

5 Possible applications of cluster dynamics

We have demonstrated in Section 2 (Figures 1, 2) above that the distribution of the cluster size in the elastic frictionless billiard changes as the phase transition approaches. A similar phenomenon has been observed in seismicity and economics, where it is related to the approach of an appropriately define critical event. Below we briefl discuss cluster dynamics in seismicity. More detailed discussion and further examples, suggesting universality of that phenomenon, are given in [19], [20], [21].

Seismicity. An earthquake is an episode of rupture and discontinuous displacement in the outer shell of solid Earth, called the lithosphere. Generally speaking, nucleation of a single earthquake is controlled by the clash between stress- and strength-field in its vicinity. Both these fields particularly the strength, are in turn controlled by a multitude of processes generating strong instability. Among them are interactions with other earthquakes, mechanical and chemical interactions of rocks and fluids phase transitions of minerals, heat fl w, non-linear deformations and fracturing, interactions between geospheres, incompatibility between the structure and kinematics, *etc.* These processes evolve in multiple scales, from global to microscopic ones. Altogether they turn the lithosphere into a hierarchical dissipative non-linear ("complex") system. Fundamental equations connecting seismicity with this set of intertwined mechanisms are not yet known.

Scale invariance. Size of an earthquake is usually defined as its magnitude m – a logarithmic measure of the energy E released by the earthquake [8] ($m = 2/3 \log_{10} E + C$). About 10^6 earthquakes with $m > 2$ are recorded annually worldwide, and once in a few years the largest earthquakes with $m > 8$ occur. The magnitude distribution of earthquakes is known in seismology as the Gutenberg–Richter law [9], [10]

$$\log_{10}(N(m)) = a - bm. \tag{13}$$

Here $N(m)$ is the average annual number of earthquakes with magnitude m or more, $b \approx 1$. The Gutenberg–Richter law emerges only after considerable averaging of seismicity over time and territory and gives a good description for small to medium magnitudes. At the relatively large magnitudes, the size distribution bends downwards. The linear relation (13) is equivalent to the power law distribution of earthquake energy:

$$N(E) \propto E^{-\beta}, \quad \beta \approx 2/3. \tag{14}$$

To accommodate the downward slope of the energy distribution at larger magnitudes, it can be approximated by an exponentially tapered distribution:

$$N(E) \propto E^{-\beta} \exp(E/E_0). \tag{15}$$

Strong earthquakes are usually defined as outliers in observed distributions – "extreme events". Such earthquakes are associated with abrupt overall changes of seismicity, and, qualitatively, with phase transitions.

Clustering and phase transitions. Earthquakes come in a hierarchy of clusters formed in a broad range of time-, space-, and magnitude scales, from microscopic to global ones. Clusters emerge, coalesce, split, migrate, and alternate with seismic quiescence ("anti-clusters"). As a strong earthquake approaches, distribution of cluster sizes tends to change in favor of larger clusters. Similarly, the magnitude distribution of individual earthquakes changes in favor of larger magnitudes. In particular the distribution (15) changes similarly to the cluster size distribution (5) described in Section 2. These phenomena have been found in real seismicity of numerous regions worldwide as well as in numerical models of seismicity. They were used in several earthquake prediction algorithms, self-adjusting to statistical properties of seismicity in different regions [21]. Statistical significance of predictions based on that kind of clusters is demonstrated in [23].

Similar changes in scaling relations have been observed before American economic recessions and some socio-economic extreme events [20].

6 Parameters and simulation details

We have simulated the billiard for $1 \leq N \leq 10^4$ and $10^{-8} \leq \rho \leq 1/2$ The table size L was kept constant, while appropriate values of R for any given ρ and N were calculated via (2). We put $m = 1$ and $T = 1/k_B$.

Recall that the Maxwellian velocity distribution of an ideal gas in D-dimensional space is given by

$$f(\mathbf{v}) = \left(\frac{m}{2\pi k_B T}\right)^{D/2} \exp^{-m|\mathbf{v}|^2/(2k_B T)}, \tag{16}$$

where $\mathbf{v} = (v_1, \dots, v_D)$, $|\mathbf{v}|^2 = v_1^2 + \dots + v_D^2$, k_B is Boltzmann constant, T temperature, and m particle mass. Accordingly, the speed $s := |\mathbf{v}|$ of particles in a $2D$ system is given by Chi-distribution with two degrees of freedom:

$$p(s) = s\sqrt{\frac{m}{k_B T}} \exp^{-m s^2/(2k_B T)}. \tag{17}$$

We have verifie that any (non-degenerate) initial velocity distribution eventually transforms to the Maxwellian distribution (16). The rate of convergence is rather fast; say, the isotropic uniform initial velocity distribution for $N = 500$ balls transforms into the Maxwellian distribution after 2 000 ball collisions, which is 4 collisions per ball on average.

We start all our models with the Maxwellian velocity distribution (16) and uniform placement of the balls. Specificall , the balls are placed on the table one-by-one. The center of the firs ball is uniformly distributed within the region

$$\mathcal{T}' = \{(x, y) : |x| < L/2 - R,\ |y| < L/2 - R\},$$

which ensures that it does not intersects with the table walls. The center of $(k+1)$-th ball, $k \geq 1$, is uniformly distributed within the region $\mathcal{T}' \setminus B_k$, where B_k is the union of already placed balls; this ensures the new ball does not intersects with existing balls and the walls.

Acknowledgment. This study was partly supported by NSF grants ATM 0327558 and ATM 0327598.

References

[1] L. Boltzmann, *Vorlesungen über Gastheorie*, 2 vol., J. A. Barth, Leipzig, 1895–1898; English transl.: *Lectures on gas theory*, translated by S. G. Brush, University of California Press, 1964.

[2] N. Bogolyubov, *Problems of Dynamical Theory in Statistical Phyics*, GITTL, Moscow, Leningrad 1946.

[3] O. E. Lanford III, Time evolution of large classical systems, in *Dynamical systems, theory and applications* (E. J. Moser, ed.), Lecture Notes in Phys. 38, Springer-Verlag, Berlin 1975, 1–111.

changes in [4], please check

[4] Ya. G. Sinai, Construction of Cluster Dynamics for Dynamical Systems of Statistical Mechanics, *Vestnik Moskov. Univ. Ser. I Mat. Meh.* **29** (1974), 152–158; English transl. *Moscow Univ. Math. Bull.* **29** (1974), 124–129

[5] Ya. G. Sinai, Constuction of Dynmaics in Infinit Systems of Particles, *Theoret. Math. Phys.* **12** (1973), 487.

[6] E. Presutti, M. Pulvirenti, B. Tirozzi, Time Evolution of Infinit Classical Systems with Singular, Long Range, Two-Body Interactions, *Comm. Math. Phys.* **47** (1976), 81–95.

[7] R. Dobrushin, J. Fritz, Non-equilibrium Dynamics of One-dimensional Infin t Systems with a Hard-Core Interaction, *Comm. Math. Phys.* **55** (1977), 275–292.

[8] H. Kanamori, The energy release in great earthquakes, *J. Geophys. Res.* **82** (1977), 2981–2987.

[9] B. Gutenberg and C. F. Richter, Seismicity of the Earth, *Geol. Soc. Amer.* **34** (Special papers) (1941), 1–131.

[10] B. Gutenberg and C. F. Richter, Frequency of earthquakes in California, *Bull. Seism. Soc. Amer.* **34** (1944), 185–188.

[11] D. Stoyan, W. S. Kendall and J. Mecke, *Stochastic Geometry and Its Applications*, 2nd ed., Wiley, New York 1987.

[12] D. Stauffer and A. Aharony, *Introduction to percolation theory*, 2nd ed., Taylor & Francis, 1994.

[13] A. Margolina, H. Nakanishi, D. Stauffer, and H. E. Stanley, *Monte Carlo and series study of corrections to scaling in two-dimensional percolation. J. Phys. A* **17** (1984), 1683–1701.

[14] A. M. Gabrielov, V. I. Keilis-Borok, I. V. Zaliapin, and W. I. Newman, Critical transitions in colliding cascades, *Phys. Rev. E* **62** (2000), 237–249.

[15] A. M. Gabrielov, I. V. Zaliapin, V. I. Keilis-Borok and W. I. Newman, Colliding Cascades as a Model for Earthquake Prediction, *Geophys. J. Int.* **143** (2000), 427–437.

[16] I. Rotwain, V. Keilis-Borok and L. Botvina, Premonitory transformation of steel fracturing and seismicity, *Phys. Earth Planet. Inter.* **101** (1997), 61–71.

[17] I. Zaliapin, V. Keilis-Borok and M. Ghil, A Boolean Delay Model of Colliding Cascades. II: Prediction of Critical Transitions, *J. Stat. Phys.* **111** (3–4) (2003), 839–861

[18] I. Zaliapin, H. Wong and A. Gabrielov, Inverse cascade in percolation model: Hierarchical description of time-dependent scaling, *Phys. Rev. E* **71** (2005), 066118.

[19] V. Keilis-Borok, J. H. Stock, A. Soloviev and P. Mikhalev, Pre-recession pattern of six economic indicators in the USA, *J. Forecasting* **19** (2000), 65–80, .

[20] V. Keilis-Borok and A. Soloviev, On universal precursors to extreme socio-economic events, *Geophysical Research Abstracts* **7** (2005), Proc. of the EGU General Assembly 2006, Vienna, Austria, 2–7 April 2006, (CD-ROM): EGU06-A-01498, 2005.

[21] V. I. Keilis-Borok and A. A. Soloviev (eds.), *Nonlinear Dynamics of the Lithosphere and Earthquake Prediction*, Springer-Verlag, Heidelberg 2003.

[22] P., Shebalin, V. Keilis-Borok, A. Gabrielov, I. Zaliapin and D. Turcotte, Short-term earthquake prediction by reverse analysis of lithosphere dynamics, *Tectonophysics* **413** (1–2) (2006), 63–75.

[23] G. M. Molchan, O. E. Dmitrieva, I. M. Rotwain and J. Dewey, *Statistical analysis of the results of earthquake prediction, based on bursts of aftershocks. Phys. Earth Planet. Inter.* **61** (1990), 128–139.

Name Index

Subject Index

29. Murray JD (2002) Mathematical Biology, An introduction. Springer, New York
30. Nielsen PE (1993) Peptide nucleic acid (PNA): A model structure for the primordial genetic material. Orig Life Evol Biosph 23:323–327
31. Plasson R, Bersini H, Commeyras A (2004) Recycling Frank: spontaneous emergence of homochirality in noncatalytic systems. Phys Rev 101:16733–16738
32. Plasson R, Kondepudi DK, Bersini H, Commeyras A, Asakura K (2007) Emergence of homochirality in far-from-equilibrium systems: Mechanisms and role in prebiotic chemistry. Chirality 19:589–600
33. Prigogine I, Lefever R (1968) Symmetry breaking instabilities in dissipative systems II. J Chem Phys 48:1695–1700
34. Prigogine I, Nicolis G (1967) On symmetry-breaking instabilities in dissipative systems. Chem J Phys 46:3542–3550
35. Rasmussen S, Chen L, Nilsson M, Abe S (2003) Bridging nonliving and living matter. Artif Life 9:269–316
36. Rosing MT, Bird DK, Sleep NH, Glassley W, Albarede F (2006) The rise of continents–An essay on the geologic consequences of photosynthesis. Palaeogeogr Palaeoclimatol Palaeoecol 232:99–113
37. Russell M (2006) First life. Am Sci 94:32–39
38. Saghatelian A, Yokobayashi Y, Soltani K, Ghadiri MR (2001) A chiroselective peptide replicator. Nature 409:797–801
39. Sandars PGH (2003) A toy model for the generation of homochirality during polymerization. Orig Life Biosph 33:575–587
40. Saunders PT (1994) Evolution without natural selection: further implications of the Daisyworld parable. J Theor Biol 166:365–373
41. Shibata R, Saito Y, Hyuga H (2006) Diffusion accelerates and enhances chirality selection. Phys Rev 74:026117
42. Soai K, Shibata T, Morioka H, Choji K (1995) Asymmetric autocatalysis and amplification of enantiomeric excess of a chiral molecule. Nature 378:767–768
43. Taylor FW (1991) The greenhouse effect and climate change. Rep Prog Phys 54:881–918
44. Turing AM (1952) The chemical basis of morphogenesis. Phil Trans Roy Soc B237:37–72
45. von Bloh W, Block A, Parade M, Schellnhuber HJ (1999) Tutorial Modelling of geosphere-biosphere interactions: the effect of percolation-type habitat fragmentation. Rep Prog Phys A 266:186–106
46. Walker JCG, Hays PB, Kasting JF (1981) A negative feedback mechanism for the long-term stabilization of the Earth's surface temperature. J Geophys Res 86:9776–9782
47. Watson AJ, Lovelock JE (1983) Biological homeostasis of the global environment: the parable of Daisyworld. Tellus B 35:284–289
48. Zhang XZ, Liao HM, Zhou LQ, Quyang Q (2004) Pattern selection in the Belousov–Zhabotinsky reaction with the addition of an activating reactant. J Phys Chem B 108:16990–16994

Books and Reviews

Barbieri M (2003) The organic codes: an introduction to semantic biology. CUP, Cambridge
Brack A (1998) The molecular origins of life. Cambridge University Press, Cambridge
Darwin C (1859) The origin of species by means of natural section. Reprinted by: Penguin books, London, 1985
Haken H (1983) Synergetics – An Introduction. Springer, Berlin
Lovelock JE (1995) Gaia. A new look at life on Earth. Oxford University Press, Oxford
Lunine J (2003) Astrobiology: a multi-disciplinary approach. Pearson Higher Education, Addison-Wesley, San Francisco
Prigogine I (1980) From being to becoming. Time and complexity in the physical sciences. Freeman, New York
Rauchfuß H (2005) Chemische Evolution und der Ursprung des Lebens. Springer, Berlin
Ward PD, Brownlee D (2000) Rare Earth: why complex life is uncommon in the Universe? Copernicus, New York

Extreme Events in Socio-economic and Political Complex Systems, Predictability of

Vladimir Keilis-Borok[1,2], Alexandre Soloviev[2,3], Allan Lichtman[4]
[1] Institute of Geophysics and Planetary Physics and Department of Earth and Space Sciences, University of California, Los Angeles, USA
[2] International Institute of Earthquake Prediction Theory and Mathematical Geophysics, Russian Academy of Science, Moscow, Russia
[3] Abdus Salam International Centre for Theoretical Physics, Trieste, Italy
[4] American University, Washington D.C., USA

Article Outline

Glossary

Complexity A definitive feature of nonlinear systems of interacting elements. It comprises high instability with respect to initial and boundary conditions, and complex but non-random behavior patterns ("order in chaos").

Extreme events Rare events having a large impact. Such events are also known as critical phenomena, disasters, catastrophes, and crises. They persistently reoccur in hierarchical complex systems created, separately or jointly, by nature and society.

Fast acceleration of unemployment (FAU) The start of a strong and lasting increase of the unemployment rate.

Pattern recognition of rare events The methodology of artificial intelligence' kind aimed at studying distinctive features of complex phenomena, in particular – at formulating and testing hypotheses on these features.

Premonitory patterns Patterns of a complex system's behavior that emerge most frequently as an extreme event approaches.

Recession The American National Bureau of Economic Research defines recession as "a significant decline in economic activity spread across the economy, lasting more than a few months". A recession may involve simultaneous decline in coincident measures of overall economic activity such as industrial production, employment, investment, and corporate profits.

Start of the homicide surge (SHS) The start of a strong and lasting increase in the smoothed homicide rate.

Definition of the Subject

At stake in the development of accurate and reliable methods of prediction for social systems is the capacity of scientific reason to improve the human condition. Today's civilization is highly vulnerable to crises arising from extreme events generated by complex and poorly understood systems. Examples include external and civil wars, terrorist attacks, crime waves, economic downturns, and famines, to name just a few. Yet more subtle effects threaten modern society, such as the inability of democratic systems to produce policies responsive to challenges like climate change, global poverty, and resource depletion.

Our capacity to predict the course of events in complex social systems is inherently limited. However, there is a new and promising approach to predicting and understanding complex systems that has emerged through the integration of studies in the social sciences and the mathematics of prediction. This entry describes and analyzes that approach and its real-world applications. These include algorithmic prediction of electoral fortunes of incumbent parties, economic recessions, surges of unemployment, and outbursts of crimes. This leads to important inferences for averting and responding to impending crises and for improving the functioning of modern democratic societies.

That approach was successfully applied also to natural disasters such as earthquakes. Ultimately, improved prediction methods enhance our capacity for understanding the world and for protecting and sustaining our civilization.

Extreme events. Hierarchical complex systems persistently generate extreme events – the rare fast changes that have a strong impact on the system. Depending on connotation they are also known as critical phenomena, disasters, catastrophes, and crises. This article examines the development and application of the algorithmic prediction of extreme socio-economic and political events.

The prediction problem is formulated as follows:

given are time series that describe dynamics of the system up to the current moment of time t and contain potential precursors of an extreme event;

to predict whether an extreme event will or will not occur during the subsequent time period $(t, t + \tau)$; if the answer is "yes", this will be the "*period of alarm*".

As the time goes by, predictions form a discrete sequence of alarms. The possible outcomes of such a prediction are shown in Fig. 1. The actual outcome is determined unambiguously, since the extreme events are identified independently of the prediction either by the actual happening (e. g. by an election result) or by a separate algorithm (e. g. homicide surge) after they occur.

Such "yes or no" prediction is aimed not at analyzing the whole dynamics of the system, but only at identifying the occurrence of rare extreme events. In a broad field of prediction studies this prediction is different from and complementary to the classical Kolmogoroff–Wiener prediction of continuous functions, and to traditional cause-and-effect analysis.

The problem includes estimating the predictions' accuracy: the rates of false alarms and failures to predict, and the total duration of alarms in relation to the total time considered. These characteristics represent the inevitable

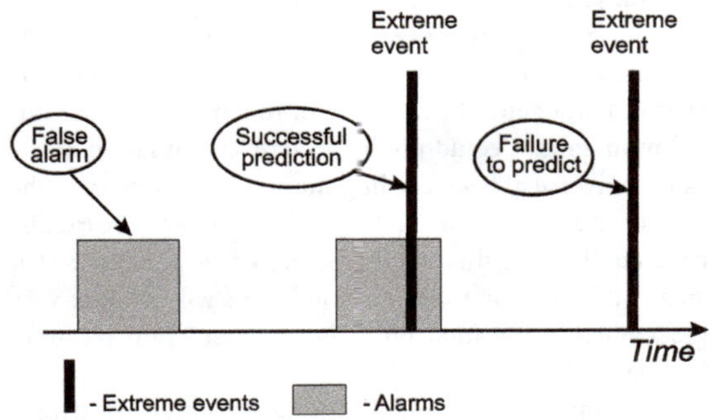

Extreme Events in Socio-economic and Political Complex Systems, Predictability of, Figure 1
Possible outcomes of prediction

probabilistic component of prediction; they provide for statistical validation of a prediction algorithm and for optimizing preparedness to predicted events (e. g. recessions or crime surges).

Twofold importance. The prediction problem is pivotal in two areas:

- *Fundamental understanding of complex systems*. Prediction algorithms quantitatively define phenomena that anticipate extreme events. Such quantitative definition is pivotal for fundamental understanding of a complex system where these events occur, including the intertwined mechanisms of system's development and its basic features, e. g. multiple scaling, correlation range, clustering, fragmentation etc. (see Sects. "Common Elements of Data Analyzes", "Elections", "US Economic Recessions", "Unemployment"). The understanding of complex systems remains a major unsolved problem of modern science, tantamount to transforming our understanding of the natural and human world.
- *Disaster preparedness*. On the practical side prediction is pivotal for coping with a variety of disasters, commonly recognized as major threats to the survival and sustainability of our civilization (e. g. [22]; see also materials of G8-UNESCO World Forum on "Education, Innovation and Research: New Partnership for Sustainable Development", http://g8forum.ictp.it). The reliable advance prediction of extreme events can save lives, contribute to social and economic stability, and to improving the governing of modern societies.

Introduction

Predictability vs. Complexity: The Need for Holistic Approach [7,12,13,15,17,27,32]

Natural science had for many centuries regarded the Universe as a completely predictable machine. As Pierre Simon de Laplace wrote in 1776, "... if we knew exactly the laws of nature and the situation of the universe at the initial moment, we could predict exactly the situation of the same universe at a succeeding moment." However, at the turn of the 20th century (1905) Jules Henry Poincare discovered, that "... this is not always so. It may happen that small differences in the initial conditions will produce very great ones in the final phenomena. Prediction becomes impossible".

This instability to initial conditions is indeed a definitive attribute of complex systems. Nonetheless, through the robust integral description of such systems, it is possible to discover regular behavior patterns that transcend the inherent complexity. For that reason studying complexity requires the holistic approach that proceeds from the whole to details, as opposed to the reductionism approach that proceeds from details to the whole. It is in principle not possible "to understand a complex system by breaking it apart" [13].

Among the regular behavior patterns of complex systems are "premonitory" ones that emerge more frequently as an extreme event approaches. These premonitory patterns make complex systems predictable. The accuracy of predictions, however, is inevitably limited due to the systems' complexity and observational errors.

Premonitory patterns and extreme events are consecutive manifestations of a system's dynamics. These patterns may not trigger extreme events but merely signal the growth of instability, making the system ripe for the emergence of extreme events.

Methodology

The prediction algorithms described here are based on discovering premonitory patterns. The development of the algorithms requires the integration of complementary methods:

- Theoretical and numerical modeling of complex systems; this includes "universal"models considered in statistical physics and non-linear dynamics (e. g. [1,3,5,8,12,15,20,42]), and system-specific models, if available.
- Exploratory data analysis.
- Statistical analysis of limited samples, which is relevant since the prediction targets are by definition rare.
- Practical expertise, even if it is intuitive.
- Risk analysis and theory of optimal control for optimizing prediction strategy along with disaster preparedness.

Pattern Recognition of Rare Events This methodology provides an efficient framework for integrating diverse information into prediction algorithms [4,11,19]. This methodology has been developed by the artificial intelligence school of I. Gelfand for the study of rare phenomena of a highly complex origin. In terminology of pattern recognition, the "object of recognition" is the time moment t. The problem is to recognize whether it belongs to the period of alarm, i. e. to a time interval Δ preceding an extreme event. An alarm starts when certain combinations of premonitory patterns emerges.

Several features of that methodology are important for predicting extreme events in the absence of a complete closed theory that would unambiguously define a prediction algorithm. First, this kind of pattern recognition relies on simple, robust parameters that overcome the bane of complexity analysis – incomplete knowledge of the system's causal mechanisms and chronic imperfections in the available data. In its efficient robustness, pattern recognition of rare events is akin to exploratory data analysis as developed by J. Tukey [50]. Second, unlike other statistical methods, e. g. regression analysis, that methodology can be used for small samples such as presidential elections or economic recessions. Also, it integrates quantitative and judgmental parameters and thereby more fully captures the full dimensions of the prediction problem than procedures that rely strictly on quantitative variables.

Summing up, the methodology described here can help in prediction when there are (1) many causal variables, (2) qualitative knowledge about which variables are important, and (3) limited amounts of data [2].

Besides societal predictions, pattern recognition of rare events has been successfully applied in seismology and earthquake prediction (e. g. [11,19,20,44,46]), geological prospecting (e. g. [45]) and in many other fields. Review can be found in [21,47]. Tutorial materials are available at the web site of the Abdus Salam International Centre for Theoretical Physics (http://cdsagenda5.ictp.it/full_display.php?da=a06219).

Validation of Prediction Algorithms The algorithms include many adjustable elements, from selecting the data and defining the prediction targets, to specifying numerical parameters involved. In lieu of theory that would unambiguously determine these elements they have to be developed retrospectively, by "predicting" past extreme events. The application of the methodology to known events creates the danger of self-deceptive data-fitting: As J. von Neumann put it "*with four exponents I can fit an elephant*". The proper validation of the prediction algorithms requires three consecutive tests.

- *Sensitivity analysis*: testing whether predictions are sensitive to variations of adjustable elements.
- *Out of sample analysis*: application of an algorithm to past data that has not been used in the algorithm's development. The test is considered successful if algorithm retains its accuracy.
- *Predicting future events* – the only decisive test of a prediction algorithm (see for example Sect. "Elections" below).

A highly efficient tool for such tests is the error Diagram, showing major characteristics of prediction accuracy [33,34,35,36,37,38,39]. Its example is given in Fig. 10. Exhaustive sets of these tests are described in [10,11,24,52].

Common Elements of Data Analyzes

The methodology discussed above was used for predicting various kinds of extreme events, as illustrated in the next four Sections. Naturally, from case to case this methodology was used in different ways, according to specifics of phenomena considered. However in all cases data analysis has essential common elements described below.

Sequence of analysis comprises four stages: (i) Defining prediction targets. (ii) Choosing the data (time series), where premonitory patterns will be looked for and summing up a priori constrains on these patterns. (iii) Formulating hypothetical definition of these patterns and developing prediction algorithm; determining the error diagram. (iv) Validating and optimizing that algorithm.

Preliminary transformation of raw data. In predicting recessions (Sect. "US Economic Recessions"), fast acceleration of unemployment (Sect. "Unemployment") and crime surges (Sect. "Homicide Surges") raw data were time series of relevant monthly indicators, hypothetically containing premonitory patterns Let $f(m)$ be such an indicator, with integer m showing time in months. Premonitory behavior of some indicators is better captured by their linear trends.

Let $W^f(l/q, p)$ be the local linear least-squares regression of a function $f(m)$ within the sliding time window (q, p):

$$W^f(l/q, p) = K^f(q, p)l + B^f(q, p), \quad q \le l \le p\,, \quad (1)$$

where integers l, q, and p stand for time in months.

Premonitory behavior of most indicators was captured by the following two functions:

- The trend of $f(m)$ in the s months long window, $(m - s, m)$. For brevity we denote

$$K^f(m/s) = K^f(m - s, m) \quad (2)$$

- The deviation of $f(m)$ from extrapolation of its long-term regression (i. e. regression on a long time window $(q, m - 1)$:

$$R^f(m/q) = f(m) - W^f(m/q, m - 1)\,. \quad (3)$$

Both functions can be used for prediction since their values do not depend on the information about the future (after the month m) which would be anathema in prediction.

Discretization. The prediction algorithms use one or several premonitory patterns. Each pattern is defined at the lowest – binary – level of resolution, as 0 or 1, distinguishing only the presence of absence of a pattern at each moment of time. Then the objects of recognition are described by binary vectors of the same length. This ensures the robustness of the prediction algorithms.

Simple algorithm called Hamming distance is used for classification of binary vectors in applications considered here, [14,20,28]. Each vector is either premonitory or not. Analyzing the samples of vectors of each class ("the learning material"), the algorithm determines a reference binary vector ("kernel") with components typical for the premonitory vector. Let D be the Hamming distance of a vector from the kernel (the number of non-coinciding binary components). The given vector is recognized as premonitory class, if D is below a certain threshold D^*. This criterion takes advantage of the clustering of precursors in time.

Summing up, these elements of the pattern recognition approach are common for its numerous applications, their diversity notwithstanding. Experience in the specific applications is described in Sects. "Elections", "US Economic Recessions", "Unemployment", "Homicide Surges". The conceptual summary of the accumulated experience is given in the final Sect. "Summary: Findings and Emerging Possibilities".

Elections

This Section describes algorithms for predicting the outcome of the US Presidential and mid-term Senatorial elections [28,29,30,31]. Elections' time is set by the law as follows.

- National elections are held every even-numbered year, on the first Tuesday after the first Monday in November (i. e., between November 2 and November 8, inclusively).
- Presidential elections are held once every 4 years, i. e. on every other election day. People in each of the 50 states and District of Columbia are voting separately for "electors" pledged to one or another of the Presidential candidates. These electors make up the "Electoral College" which directly elects the President. Since 1860, when the present two-party system was basically established, the Electoral College reversed the decision of the popular vote only three times, in 1888, 1912, and 2000. Algorithmic prediction of such reversals is not developed so far.
- A third of Senators are elected for a 6-year term every election day; "mid-term" elections held in the middle of a Presidential term are considered here.

Methodology

The prediction target is an electoral defeat of an "incumbent" party, i. e. the party holding the contested seat. Accordingly, the prediction problem is formulated as whether the incumbent party will retain this seat or lose it to the challenging party (*and not whether Republican or Democrat will win*). As is shown below, that formulation is crucial for predicting the outcomes of elections considered.

Data. The pre-election situation is described by robust common sense parameters defined at the lowest (binary) level of resolution, as the *yes* or *no* answers to the questionnaires given below (Tables 1, 2). The questions are formulated in such a way that the answer *no* favors the victory of the challenging party. According to the Hamming distance analysis (Sect. "Common Elements of Data Analyzes") the victory of the challenging party is predicted when the number of answers *no* exceeds a threshold D^*.

Mid-term Senatorial Elections

The prediction algorithm was developed by a retrospective analysis of the data on three elections, 1974, 1978, and 1982. The questionnaire is shown in Table 1. Victory of the challenger is predicted if the number of answers *no* is 5 or more [28,29,30].

The meaning of these questions may be broader than their literal interpretation. For example, financial contributions (key 5 in Table 2) not only provide the resources required for an effective campaign, but may also constitute a poll in which the preferences are weighed by the money attached.

Predicting future elections. This algorithm (without any changes from year to year and from state to state) was applied in advance to the five subsequent elections, 1986–2002. Predictions are shown in Fig. 2. Altogether, 150 seats were put up for election. For each seat a separate prediction was made, 128 predictions were correct, and 22 – wrong.

Statistical significance of this score is 99.9%. In other words the probability to get such a score by chance is below 0.1% [28,29,30]. For some elections these predictions might be considered as trivial, since they coincide with prevailing expectation of experts. Such elections are identified by *Congressional Review*. Eliminating them from the score still results in 99% significance.

Presidential Elections

The prediction algorithm was developed by a retrospective analysis of the data on the past 31 elections, 1860–1980;

Extreme Events in Socio-economic and Political Complex Systems, Predictability of, Table 1
Questionnaire for mid-term Senatorial Elections [28]

1.	(Incumbency): The incumbent -party candidate is the sitting senator.
2.	(Stature): The incumbent -party candidate is a major national figure.
3.	(Contest): There was no serious contest for the incumbent -party nomination.
4.	(Party mandate): The incumbent party won the seat with 60% or more of the vote in the previous election.
5.	(Support): The incumbent -party candidate outspends the challenger by 10% or more.
6.	(Obscurity): The challenging -party candidate is not a major national figure or a past or present governor or member of Congress.
7.	(Opposition): The incumbent party is not the party of the President.
8.	(Contest): There is no serious contest for the challenging -party nomination (the nominee gains a majority of the votes cast in the first primary and beats the second-place finisher at least two to one).

Extreme Events in Socio-economic and Political Complex Systems, Predictability of, Table 2
Questionnaire for Presidential elections [29,30]

KEY 1	(Party Mandate): After the midterm elections, the incumbent party holds more seats in the US House of Representatives than it did after the previous midterm elections.
KEY 2	(Contest): There is no serious contest for the incumbent -party nomination.
KEY 3	(Incumbency): The incumbent -party candidate is the sitting president.
KEY 4	(Third party): There is significant third-party or independent campaign.
KEY 5	(Short-term economy): The economy is not in recession during the election campaign.
KEY 6	(Long-term economy): Real per -capita economic growth during the term equals or exceeds mean growth during the previous two terms.
KEY 7	(Policy change): The incumbent administration effects major changes in national policy.
KEY 8	(Social unrest): There is no sustained social unrest during the term.
KEY 9	(Scandal): The incumbent administration is unattained by a major scandal.
KEY 10	(Foreign/military failure): The incumbent administration suffers no major failure in foreign or military affairs.
KEY 11	(Foreign/military success): The incumbent administration achieves a major success in foreign or military affairs.
KEY 12	(Incumbent charisma): The incumbent -party candidate is charismatic or a national hero.
KEY 13	(Challenger charisma): The challenging -party candidate is not charismatic or a national hero.

that covers the period between victories of A. Lincoln and R. Reagan inclusively. The questionnaire is shown in Table 2. Victory for the challenger is predicted if the number of answers *no* is 6 or more [29,30].

Predicting of future elections. This algorithm (without any changes from year to year state) was applied in advance to the six subsequent elections, 1984–2004. Predictions are shown in Fig. 3. All of them happened to be correct. In 2000 the decision of popular majority was reversed by the Electoral College; such reversals are not targeted by this algorithm [29,30].

Understanding Elections

Collective behavior. The finding that aggregate-level parameters can reliably anticipate the outcome of both presidential and senatorial elections points to an electoral behavior highly integrated not only for the nation as a whole but also within the diverse American states.

- A presidential election is determined by the collective, integrated estimation of performance of incumbent administration during the previous four years.
- In case of senatorial elections the electorate has more diffused expectations of performance but puts more importance on political experience and status than in the case of presidential elections. Senate incumbents, unlike presidential ones, do not suffer from a bad economy or benefit from a good one. (This suggests that rather than punishing the party holding a Senate seat for hard times, the voters may instead regard the incumbent party as a safe port in a storm).

Similarity. For each election year in all states the outcomes of elections follow the same pattern that transcends the diversities of the situations in each of the individual elections.

The same pattern of the choice of the US President prevails since 1860, i. e. since election of A Lincoln, despite

0	1	2	3	4	5	6	7
			OK98				
			CO98				
			FL98				
			GA98				
			HA98	TN02			
			ID98	SC02			
			MA98	NC02			
			ND98	NE02			
			PN98	KY02			
			SD98	IA02			
			UT98	CO02			
			FL94	AL02			
			HA94	AK98			
			IN94	CA98			
			MT94	CT98			
			NB94	NE98			
			NJ94	OR98			
			TX94	SC98			
			WA94	VT98			
		AS98	WV94	WA98			
		KA98	WI94	CT94			
		LA98	AK90	MD94			
		MI98	IN90	NV94			
		NH98	KN90	WY94			
		MS94	ME90	CO90			
	AL98	NM94	MA90	HA90			
	AZ98	ND94	MT90	KY90			
	IO98	RI94	NB90	MI90			
	DL94	VT94	NC90	AZ86			
	MA94	AS90	TX90	CO86			
	NY94	IO90	WY90	ID86			
	AL90	MS90	AR86	LA86			
	DE90	NM90	CA86	NY86			
	IL90	OR90	IL86	OK86	WI98	MN94	
	LA90	RI90	IN86	WI86	CA94	MO94	
	OK90	SD90	IA86	NC86	ID90	VA94	
	SC90	VA90	NH86	WA86	PA86	NH90	
	TN90	WV90	OR86	MN90	IL98	IN98	
	HI86	AK86	VT86	OK94	ME94	OH98	
	OH86	CT86	TN94	PA94	AL86	MI94	
UT94	SC86	KS86	TX02	TN294	FL86	MD86	KY98
GA90	UT86	KY86	OK02	NC98	GA86	NV86	AZ94
NJ90	NH02	ND86	NJ02	NY98	MO86	SD86	OH94
0	1	2	3	4	5	6	7

OK98 – incumbent won, KY98 – challenger won, errors are highlighted.

Extreme Events in Socio-economic and Political Complex Systems, Predictability of, Figure 2

Made-in-advance predictions of the mid-term senatorial elections (1986–2002). Each election is represented by the two-letter state abbreviation with the election year shown by two last digits. Each column shows elections with certain number *D* of answers *"no"* to the questionnaire given in Table 1 (such answers are favorable to challenging party). Value of *D*, indicated at the top, is the Hamming distance from the kernel

D (number of answers NO)	0	1	2	3	4	5	6	7	8	9
Predictions published months in advance			1984	1988	2004	2000* 1996	1992			
Learning	1904	1956 1936	1944 1940 1868	1964 1928 1916 1908 1900 1872 1864	1972 1924 1880	1948 1888*	1912* 1892	1884 1860	1980 1976 1968 1952 1932 1920 1896	1960 1876*

1904 years when incumbent won popular vote

1892 years when challenger won popular vote

* years when popular vote was reversed by electoral vote

Extreme Events in Socio-economic and Political Complex Systems, Predictability of, Figure 3
Division of presidential elections (1860–2004) by the number *D* of answers *"no"* to the questionnaire given in Table 2 (such answers are favorable to challenging party). *D* is the Hamming distance from the kernel

all the overwhelming changes in the electorate, the economy, the social order and the technology of politics during these 130 years. (For example, the electorate of 1860 did not include the groups, which constitute 3/4 of present electorate, such as women, African Americans, or most of the citizens of the Latin American, South European, Eastern European, and Jewish descent [30].

An alternative (and more traditional) concept of American elections focuses on the division of voters into interest and attitudinal groups. By this concept the goal of the contestants is to attract the maximum number of voting blocks with minimal antagonism from other blocks. Electoral choice depends strongly on the factors irrelevant to the essence of the electoral dilemma (e. g. on the campaign tactics). The drawbacks of this concept are discussed in [18,30]. In sum, the work on presidential and senatorial elections described above suggests the following new ways of understanding American politics and perhaps the politics of other societies as well.

1. Fundamental shifts in the composition of the electorate, the technology of campaigning, the prevailing economic and social conditions, and the key issues of campaigns do not necessarily change the pragmatic basis on which voters choose their leaders.
2. It is governing not campaigning that counts in the outcomes of presidential elections.
3. Different factors may decide the outcome of executive as compared to legislative elections.
4. Conventional campaigning will not improve the prospects for candidates faced with an unfavorable combination of fundamental historical factors. Disadvantaged candidates have an incentive to adopt innovative campaigns that break the pattern of conventional politics.
5. All candidates would benefit from using campaigns to build a foundation for governing in the future.

US Economic Recessions

US National Bureau of Economic Research (NBER) has identified the seven recessions that occurred in the US since 1960 (Table 3). The starting points of a recession and of the recovery from it follow the months marked by a peak and a trough of economic activity, respectively.

A peak indicates the last month before a recession, and a trough – the last month of a recession.

Prediction targets considered are the first month after the peak and after the trough ("the turns to the worst and to the best", respectively). The start of the first recession, in 1960, is not among the targets, since the data do not cover a sufficient period of time preceding the recession.

The data used for prediction comprise the following six monthly leading economic indicators obtained from the CITIBASE data base, Jan 1960–June 2000 (abbreviations are the same, as in [49])

G10FF = FYGT10 – FEDFUN Difference between the annual interest rate on 10 year US Treasury bonds, and federal fund annual interest rate.

IP Industrial Production, total: index of real (constant dollars, dimensionless) output in the entire economy.

Extreme Events in Socio-economic and Political Complex Systems, Predictability of, Table 3
American Economic Recessions since 1960

#	Peaks	Troughs
1	1960:04	1961:02
2	1969:12	1970:11
3	1973:11	1975:03
4	1980:01	1980:07
5	1981:07	1982:11
6	1990:07	1991:03
7	2001:03	2001:11

This represents mainly the manufacturing industry, because of the difficulties in measuring the quantity of the output in services (such as travel agents, banking, etc.).

LHELL Index of "help wanted" advertising. This is put together by a private publishing company that measures the amount of job advertising (column-inches) in a number of major newspapers.

LUINC Average weekly number of people claiming unemployment insurance.

INVMTQ Total inventories in manufacturing and trade, in real dollars. Includes intermediate inventories (for example held by manufacturers, ready to be sent to retailers) and final goods inventories (goods on the shelves in stores).

FYGM3 Interest rate on 90 day US treasury bills at an annual rate (in percent).

These indicators were already known [48,49], as those that correlate with a recession's approach.

Prediction of a Recession Start

Single indicators exhibit the following premonitory patterns:

G10FF: small value
IP and INVMTQ: small deviation from the long-term trend R^f (3)
FYGM3: large deviation from the long-term trend R^f
LHELL: small trend K^f (2)
LUINC: large trend K^f

The prediction algorithm triggers an alarm after a month when most of the patterns emerge simultaneously. It lasts Δ months and can be extended by the same rule, if premonitory patterns keep emerging. Formal quantitative definition of the algorithm can be found in [23] along with its validation by sensitivity and out-of-sample analyzes.

Alarms and recessions are juxtaposed in Fig. 4. We see that five recessions occurring between 1961 and 2000 were predicted by an alarm. The sixth recession started in April 2001, one month before the corresponding alarm. (Recession of 1960 was not considered for prediction, since data analyzed start just before it.)

Only the first six recessions listed in Table 1 were considered in the developing of the algorithm [23]. Duration of each alarm was between 1 and 14 months. Total duration of all alarms was 38 months, or 13.6% of the time interval considered. There were no false alarms. No alarms were yielded so far by subsequent prediction in advance and no recession was identified during that time.

Prediction of a Recession End

Prediction targets are the starting points of recovery from recessions; these points are indicated in the last column of Table 3.

The data comprise the same six indicators that indicate the approach of a recession (see Subsect. "Prediction of a Recession Start"); they are analyzed only within the recessions' periods.

Data analysis shows intriguing regularity illustrated in Fig. 5:

- Financial indicators change in opposite directions before the recession and before the recovery.
- Economic indicators change in the same direction before the recession and the recovery; but the change is stronger before the recovery, i. e., the economic situation worsens.

Prediction algorithm is formulated in the same terms as in the previous case but an alarm is triggered after *three* consecutive months when most of the patterns emerge simultaneously. The alarms predict when the recovery will start. Alarms and prediction targets are juxtaposed in Fig. 6. Du-

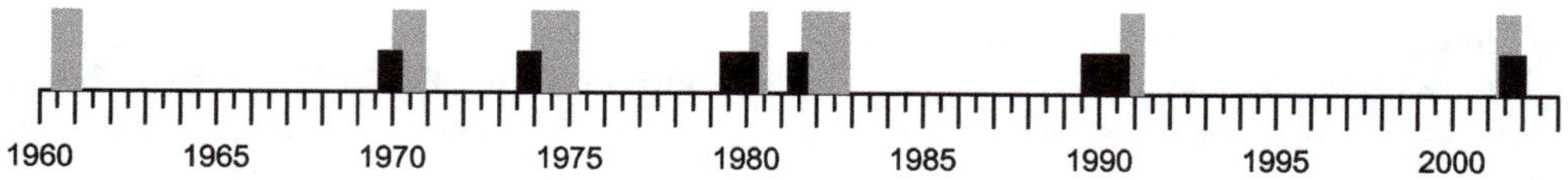

Extreme Events in Socio-economic and Political Complex Systems, Predictability of, Figure 4
Alarms (*black bars*) and recessions (*gray bars*)

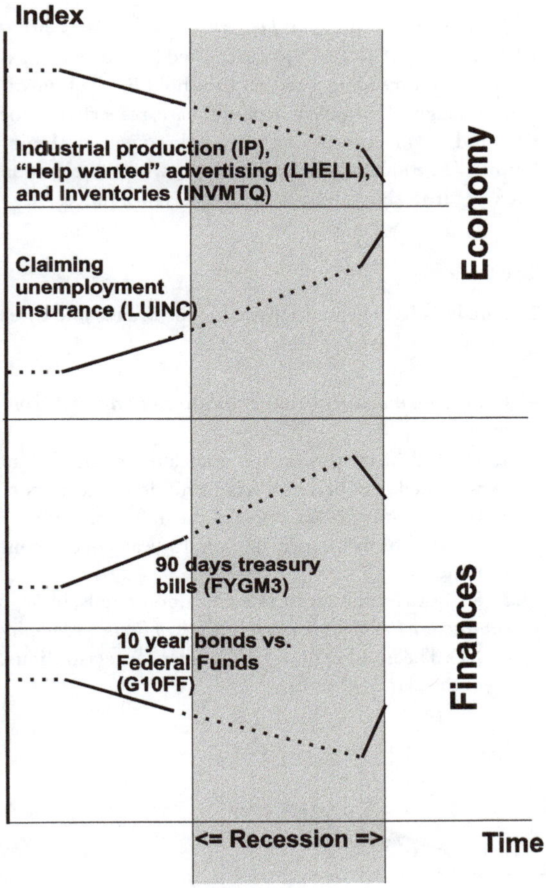

Extreme Events in Socio-economic and Political Complex Systems, Predictability of, Figure 5
Premonitory changes of indicators before the start of a recession and before its end. See explanations in the text

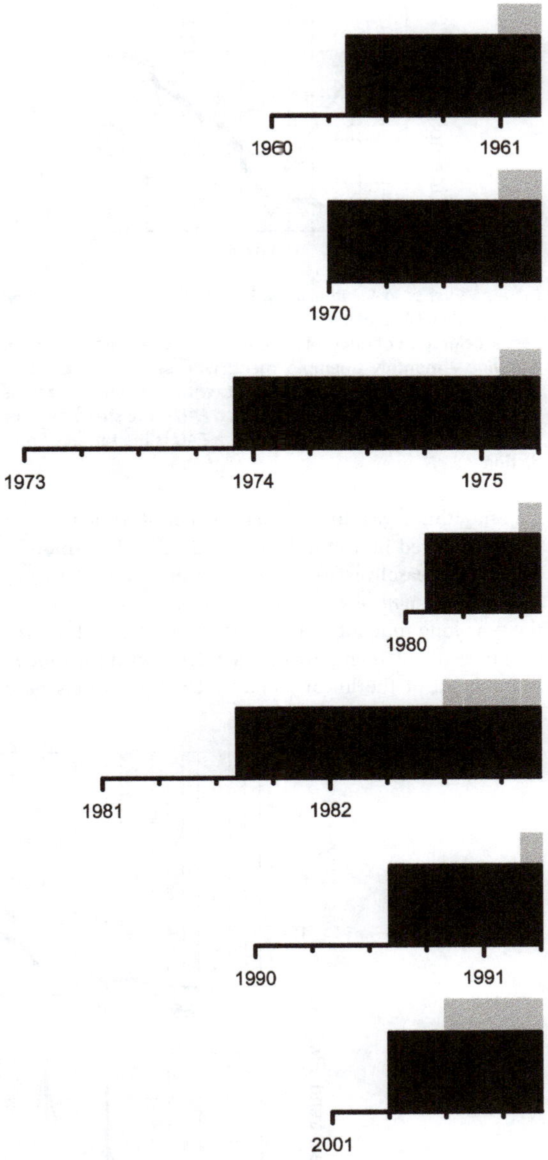

Extreme Events in Socio-economic and Political Complex Systems, Predictability of, Figure 6
Prediction of recovery from a recession. *Black bars* – periods of recessions. *Gray bars* – alarms preceding the end of a recession

ration of a single alarm is one to five months. Total duration of alarms is 16 months, which is 22% of time covered by all recessions. There are neither false alarms nor failures to predict.

Unemployment

Here we describe uniform prediction of the sharp and lasting unemployment surge in France, Germany, Italy, and the USA [25].

Prediction Target

A prediction target is schematically illustrated in Fig. 7. Thin curve shows monthly unemployment with seasonal variations. On the thick curve seasonal variations are smoothed away. The arrow indicates a sharp upward bend of the smoothed curve. The moment of that bend is the prediction target. It is called by the acronym *FAU*, for "Fast Acceleration of Unemployment".

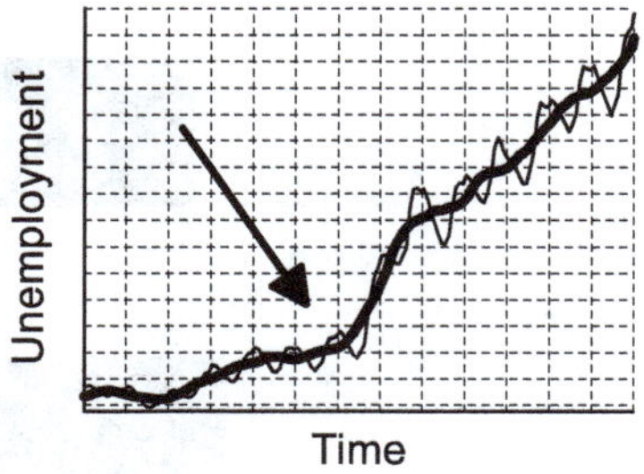

Extreme Events in Socio-economic and Political Complex Systems, Predictability of, Figure 7

Fast acceleration of unemployment (*FAU*): schematic definition. *Thin line* – monthly unemployment; with seasonal variations. *Thick line* – monthly unemployment, with seasonal variations smoothed away. The *arrow* indicates a *FAU* – the sharp bend of the smoothed curve. The moment of a *FAU* is the target of prediction

Smoothing was done as follows: Let $u(m)$ be number of unemployed in a month $m = 1, 2, \ldots$. After smoothing out the seasol variation we obtain time series $U(m) = W^u(m/m-6, m+6)$; this is the linear regression over the year-long time interval $(m-6, m+6)$. A natural robust measure of unemployment acceleration at the time m is the bend of the linear trend of U; in notations used in (1) this is the function $F(m/s) = K^U(m+s, m) - K^U(m, m-s)$. The *FAU*s are identified by the local maxima of $F(m)$ exceeding a certain threshold F. The time m^* and the height F^* of such a maximum are, respectively, the time and the magnitude of a *FAU*. Subsequent local minimum of $F(m)$ identifies the month m_e when acceleration ends. Figure 8 shows thus defined *FAU*s for France.

The Data

The analysis has been initially made for France and three groups of data have been analyzed.

- *Composite macroeconomic indicators of national economy*
  1. **IP**: Industrial production indicator, composed of weighted production levels in numerous sectors of the economy, in % relative to the index for 1990.
  2. **L**: Long-term interest rate on 10-year government bonds, in %.
  3. **S**: Short-term interest rate on 3-month bills, in %.
- *Characteristics of more narrow areas of French economy*
  4. **NC**: The number of new passenger car registrations, in thousands of units.

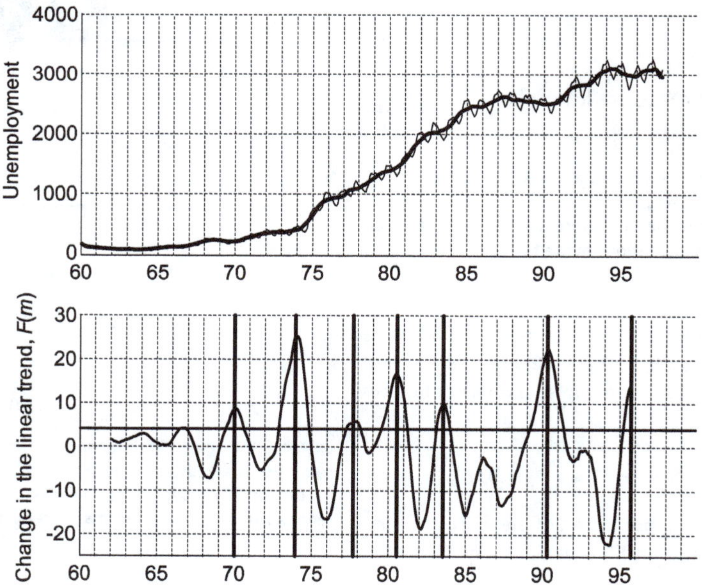

Extreme Events in Socio-economic and Political Complex Systems, Predictability of, Figure 8

Unemployment in France. *Top:* Monthly unemployment, thousands of people. *Thin line*: $u(m)$, data from the OECD database; note the seasonal variations. *Thick line*: $U(m)$, data smoothed over one year. *Bottom:* Determination of *FAUs*. $F(m)$ shows the change in the linear trend of unemployment $U(m)$. *FAUs* are attributed to the local maxima of $F(m)$ exceeding threshold $F = 4.0$ shown by *horizontal line*. The *thick vertical lines* show moments of the *FAUs*

5. **EI**: Expected prospects for the national industrial sector.
6. **EP**: Expected prospects for manufacturers.
7. **EO**: Estimated volume of current orders.

Indicators 5–7 distinguish only "good" and "bad" expectations determined polling 2,500 manufacturers, whose answers are by the size of their businesses.

- *Indicators related to US economy.*

8. **FF/$**: Value of US dollar in French francs.
9. **AR**: The state of the American economy: is it close to a recession or not? This indicator shows the presence or absence of a current pre-recession alarm (see Subsect. "Prediction of a Recession Start").

The data bases with above indicators for Europe are issued by the Organization for Economic Cooperation and Development [43] and the International Monetary Fund [16].

American analogues of indicators **IP**, **L**, and **S** are provided by CITIBASE; they are described in Sect. "US Economic Recessions>" under abbreviations **IP**, **FYGM3** and **FIGT10** respectively.

Prediction

Single indicators exhibit the following premonitory behavior.

- Steep upward trends of composite indicators (#1–#3). This behavior reflects "overheating" of the economy and may sound counterintuitive for industrial production (#1), since the rise of production is supposed to create more jobs. However, a particularly steep rise may create oversupply.
- Steep downward trends of economic expectations by general public (#4) and business community (#5–#8).
- Proximity of an American recession (#9). Before analysis was made such and opposite precursors might be expected for equally plausible reasons, so that this finding, if further confirmed, does provide a constraint on understanding unemployment's dynamics.

Among different combinations of indicators the macroeconomic ones (#1–#3) jointly give relatively better predictions, with smallest rates of errors and highest stability in sensitivity tests.

Retrospective prediction. Macroeconomic indicators were used jointly in the Hamming distance prediction algorithm (Sect. "Common Elements of Data Analyzes"). Being robust and self-adjusting to regional conditions, this algorithm was applied without any changes to the four countries considered here.

Alarms and *FAU*s are juxtaposed in Fig. 9. Error diagram in Fig. 10 shows quality of prediction for different

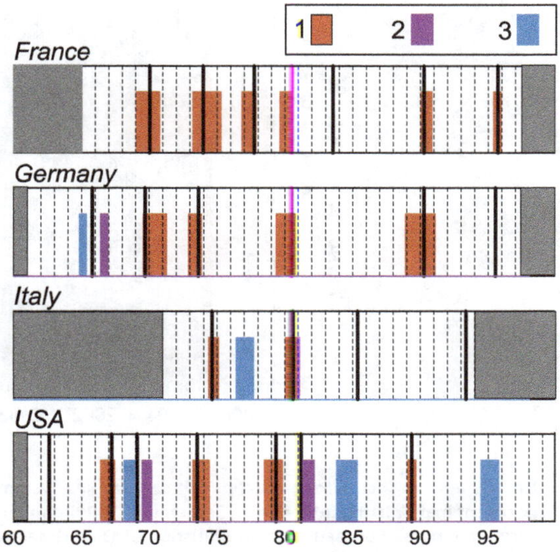

Extreme Events in Socio-economic and Political Complex Systems, Predictability of, Figure 9

Retrospective predictions for four countries: *FAUs* and alarms obtained by the prediction algorithm. The *thick vertical lines* show the moments of *FAUs* in a country. *Bars* – the alarms with different outcome: 1 – alarms that predict *FAUs*, 2 – alarms starting shortly after *FAUs* within the periods of unemployment surge, 3 – false alarms. *Shaded areas* on both sides indicate the times, for which data on economic indicators were unavailable

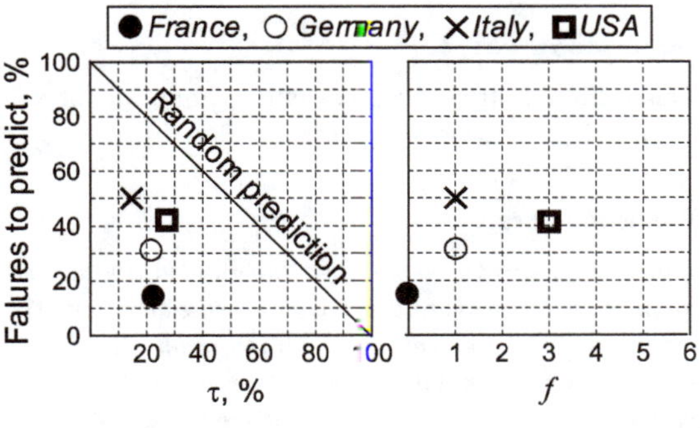

Extreme Events in Socio-economic and Political Complex Systems, Predictability of, Figure 10

Error diagram for prediction of *FAUs* in different countries; τ is total duration of alarms in % to the time interval considered, f – total number of false alarms

countries. For US the quality is lower than for European countries, though still higher than in random predictions.

Prediction of the future FAUs was launched for USA. The results are shown in Fig. 11. It shows that by January 2008 two correct predictions have been made, without ether false alarms or failures to predict. In November

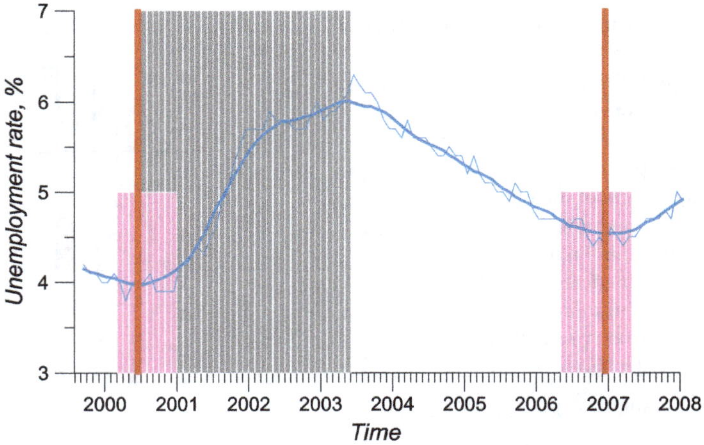

Extreme Events in Socio-economic and Political Complex Systems, Predictability of, Figure 11

Experiment in predicting future FAUs, September (1999)–January (2008). ***Thin blue curve*** **shows monthly unemployment rate in USA, according to data of Bureau of Labor Statistics, US Department of Labor (http://www.data.bls.gov).** ***Thick curve*** **shows this rate with seasonal variation smoothed away.** ***Vertical red lines*** **show prediction targets – the moments of** ***FAU, gray bar*** **– the period of unemployment's growth;** ***pink bars*** **– periods of alarms**

2006 the second prediction was filed on the web site of the Anderson School of Management, University of California, Los Angeles (http://www.uclaforecast.com/). This started the documented experiment in testing the algorithm by predicting future *FAU*s on that website.

Homicide Surges

This section analyzes the prediction of homicide rates in an American megacity – Los Angeles, CA [24].

Prediction Target

A prediction target is the start of a sharp and lasting acceleration of the homicide rate; it is called by the acronym SHS, for "Start of the Homicide Surge." It is formally determined by the analysis of monthly homicides rates, with seasonal variations smoothed out, as described in Subsect. "Prediction Target". Prediction targets thus identified are shown by vertical lines in Figs. 12 and 14 below.

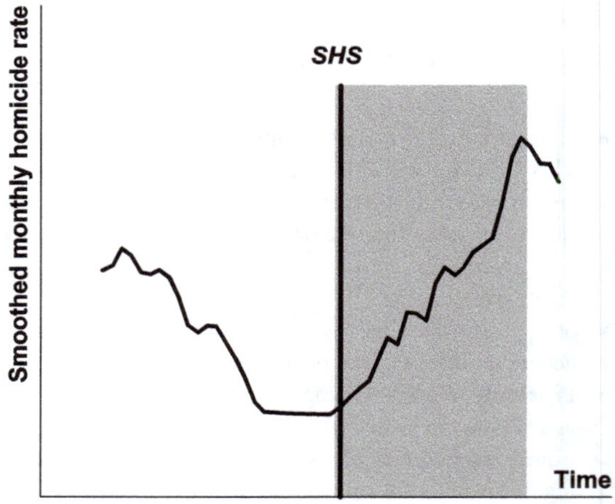

Extreme Events in Socio-economic and Political Complex Systems, Predictability of, Figure 12

Target of prediction – the Start of the Homicide Surge ("*SHS*"); schematic definition. ***Gray bar*** **marks the period of homicide surge**

The Data

The analyzed data include monthly rates of the homicides and 11 types of lesser crimes, listed in Table 2. Definitions of these crimes are given in [6].

The data are taken from two sources:

- The National Archive of Criminal Justice Data, placed on the web site (NACJD), 1975–1993.
- Data bank of the Los Angeles Police Department (LAPD) Information Technology Division), 1990–2003.

The algorithm does not use socio-economic determinants of crime, or other data that might be also useful. The objective was to develop a simple, efficient prediction model; development of comprehensive causal model would be a complementary objective.

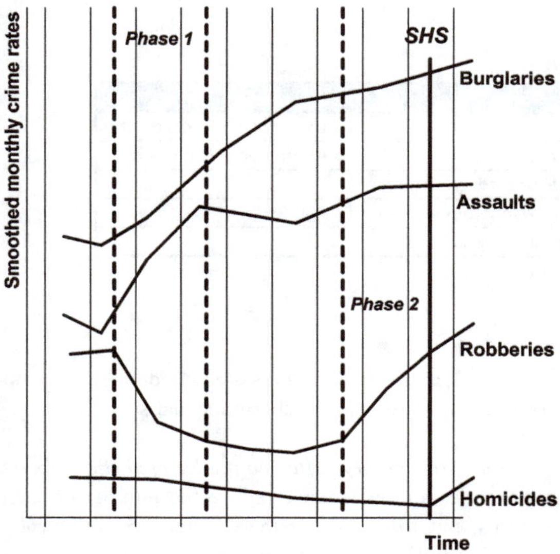

Extreme Events in Socio-economic and Political Complex Systems, Predictability of, Figure 13
Scheme of premonitory changes in crime statistics

Prediction

Premonitory behavior of indicators is illustrated in Fig. 13. The first phase is characterized by an escalation of burglaries and assaults, but not of robberies. Later on, closer to a homicide surge, robberies also increase.

The Prediction algorithm based on Hamming distance (see Sect. "Common Elements of Data Analyzes") uses seven indicators listed in Table 4. Other five indicators marked by * are used in sensitivity tests; and the homicide rate is used for identification of targets *SHS*.

Alarms and homicide surges are juxtaposed in Fig. 14. The *SHS* episode in November 1994 has occurred simultaneously with the corresponding alarm. It is captured by an alarm, which starts in the month of *SHS* without a lead time. Prediction missed the October 1999 episode: it occurred two months *before* the start of the corresponding alarm. Such delays should be taken into account for validating the algorithm. Note, however, that the last prediction did remain informative.

Altogether alarms occupy 15% of the time considered. During phase 2 (as defined in Fig. 13) this rate might be reduced [24].

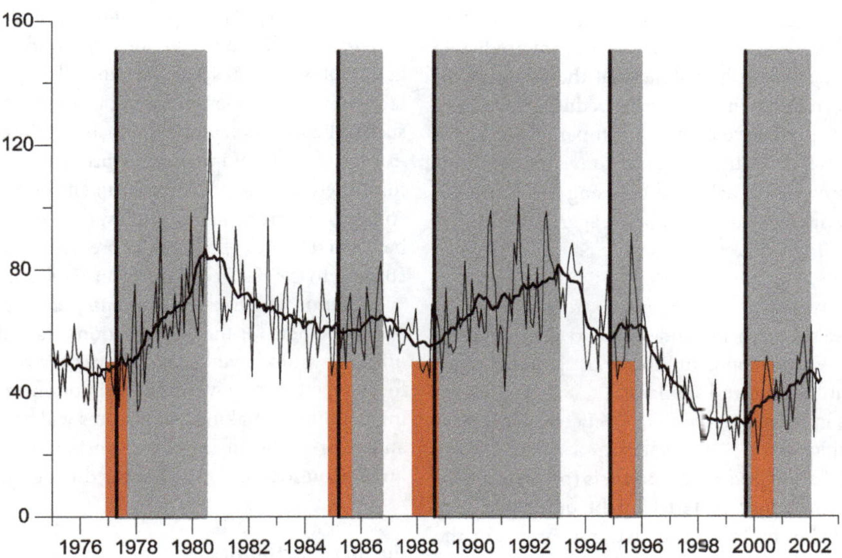

Extreme Events in Socio-economic and Political Complex Systems, Predictability of, Figure 14
Performance of prediction algorithm through 1975–2002. ***Thin curve*** **– original time series, total monthly number of homicides in Los Angeles city, per 3,000,000 inhabitants. Data from NACJD [6] have been used for 1975–1993 and from the Data Bank of the Los Angeles Police Department (LAPD Information Technology Division) for subsequent 9 years.** ***Thick curve*** **– smoothed series, with seasonal variations eliminated.** ***Vertical lines*** **show the targets of prediction – episodes of** ***SHS*** **(Subsect. "Prediction Target").** ***Gray bars*** **show the periods of homicide surge.** ***Red bars*** **show the alarms declared by the prediction algorithm [24]**

Extreme Events in Socio-economic and Political Complex Systems, Predictability of, Table 4
Types of crimes considered (after [6])

Homicide	Robberies	Assaults	Burglaries
• All	• All	• All*	• Unlawful not forcible entry
	• With firearms	• With firearms	• Attempted forcible entry*
	• With knife or cutting instrument	• With knife or cutting instrument	
	• With other dangerous weapon	• With other dangerous weapon*	
	• Strong -arm robberies*	• Aggravated injury assaults*	

*Analyzed in sensitivity tests only

Summary: Findings and Emerging Possibilities

The findings described above enhance predictive understanding of extreme events and indicate yet untapped possibilities for further R&D in that field.

Pattern Recognition Approach

Information extracted from the already available data is indeed increased by this approach. To each problem considered here one may apply the following conclusion of J. Stock, a leading expert in the field: "Prediction/of recessions/requires fitting non-linear, high-dimensional models to a handful of observations generated by a possibly non-stationary economic environment.... The evidence presented here suggests that these simple binary transformations of economic indicators have significant predictive content for recessions. It is striking that these models, in which the information in the data is reduced to binary indicators, have predictive contents comparable to or, in many cases, better than that of more conventional models." Importantly, this is achieved by using not more detailed data and models, but more robust aggregation (Subsect. "Predictability vs. Complexity: The Need for Holistic Approach").

Partial "universality" of premonitory patterns is established by broad research in modeling and data analysis. This includes the common definition of the patterns, their self-adjustment, scaling, and similarity [9,10,20,26,42]; see also references in Sects. "Elections", "US Economic Recessions", "Unemployment", "Homicide Surges").

Relation to "cause and effect" analysis (perpetrators or witnesses?). Premonitory patterns might be either "perpetrators" contributing to causing extreme events, or the "witnesses" – parallel manifestations of the system's development. The cause that triggered a specific extreme event is usually identified, at least in retrospect. It may be, for example, a certain governmental decision, a change in the international situation, a natural disaster, the depletion of natural resources etc. However an actual extreme event might materialize only if the system is destabilized and "ripe" for it. Patterns of each kind signal such a ripe situation.

What premonitory patters to use for prediction? Existing theories and experience reduce the number of such patterns, but too many of them remain hypothetically promising and have to be chosen by a trial and error procedure. Inevitably a prediction algorithm begins with a limited number of promising patterns. They should be sufficient for prediction, but other patterns may be equally or more useful and should be considered in further development of the algorithm. Most relevant "perpetrators" might not be included in the most useful patterns (e. g. due to their sensitivity to too many factors).

Relation to policy-making: prediction and disaster preparedness. Reliable predictions of future extreme events in complex societal systems would allow policy-makers to take remedial action before rather than after the onset of such afflictions as economic disasters, crime surges, etc. As in case of military intelligence predictions would be useful if their accuracy is known, albeit not necessarily high. Analysis of error diagrams allows to regulate the tradeoff between the rates of failures to predict and false alarms according to the needs of a decision-maker.

Relation to governing and campaigning. The findings presented here for the USA elections show that top elected officials would have better chances for reelection, if they focus on effective governing, and not on rhetoric, packaging and image-making. Candidates will benefit themselfes and their parties if they run substantive campaigns that build a foundation for governing during the next term.

Further Possibilities

A wealth of *yet untapped data and models* is readily available for the continuation of the kinds of studies described and analyzed in this article. Following are some immediate possibilities; specific examples can be found in the given references.

- *Continuing experiments in advance prediction*, for which the above findings set up a base (Sect. "Elections"). Successes and errors are equally important [37,38].
- *Incorporating other available data into the analysis* (Sects. "US Economic Recessions","Unemployment")
- *Predicting the same kind of extreme events in different contexts* (Sect. "Unemployment")
- *Predicting the end of a crisis* (Sect. "US Economic Recessions").
- *Multistage prediction with several lead times* (Sect. "Homicide Surges")

Less imminent, but within reach are:

- *"Universal" scenarios of extreme development and low-parametric definition of an ensemble of premonitory patterns* [9,51,52].
- *Validation of an algorithm and joint optimization of prediction and preparedness strategy* [38].
- *Developing prediction algorithms for other types of extreme events.*

The authors would be glad to provide specific information upon request.

Generalizations

The problems considered here have the following common features:

- *The absence of a closed theory* that would unambiguously determine prediction methodology. This leads to the need for intense intertwining of mathematics, statistical physics and non-linear dynamics, a range of societal sciences, and practical experience (Subsect. "Methodology"). In reality this requires long-term collaboration of respective experts. As can be seen from the references to Sects. "Elections", "US Economic Recessions", "Unemployment", "Homicide Surges" previous applications inevitably involved the teams of such experts.
- *Predictions in advance* is the only final validation of the results obtained.
- *The need for holistic analysis* driven to extreme robustness.
- *Considerable, albeit limited, universality* of the premonitory phenomena.

Two classical quotations shed the light on these features:

A. N. Kolmogoroff. "It became clear for me that it is unrealistic to have a hope for the creation of a pure theory [of the turbulent flows of fluids and gases] closed in itself. Due to the absence of such a theory we have to rely upon the hypotheses obtained by processing of the experimental data."

M. Gell-Mann: "... if the parts of a complex system or the various aspects of a complex situation, all defined in advance, are studied carefully by experts on those parts or aspects, and the results of their work are pooled, an adequate description of the whole system or situation does not usually emerge.... The reason, of course, is that these parts or aspects are typically entangled with one another. ... We have to supplement the partial studies with a transdisciplinary crude look at the whole."

In the general scheme of things the problem considered belongs to a much wider field – the quest for a universal theory of complex systems extended to predicting extreme events – the Holy Grail of complexity studies. This quest encompasses the natural and human-made complex systems that comprise what some analysts have called "the global village". It requires entirely new applications of modern science, such as algebraic geometry, combinatorics, and thermodynamics. As a means for anticipating, preventing and responding to natural and manmade disasters and for improving the outcomes of economic and political systems, the methods described here may hold one key for the survival and sustainability of our civilization.

Bibliography

Primary Literature

1. Allègre CJ, Le Mouël J-L, Ha Duyen C, Narteau C (1995) Scaling organization of fracture tectonics (SOFT) and earthquake mechanism. Phys Earth Planet Inter 92:215–233
2. Armstrong JS, Cuzan AG (2005) Index methods for forecasting: An application to american presidential elections. Foresight Int J Appl Forecast 3:10–13
3. Blanter EM, Shnirman MG, Le Mouël JL, Allègre CJ (1997) Scaling laws in blocks dynamics and dynamic self-organized criticality. Phys Earth Planet Inter 99:295–307
4. Bongard MM, Vaintsveig MI, Guberman SA, Izvekova ML, Smirnov MS (1966) The use of self-learning prog in the detection of oil containing layers. Geol Geofiz 6:96–105
5. Burridge R, Knopoff L (1967) Model and theoretical seismicity. Bull Seismol Soc Am 57:341–360
6. Carlson SM (1998) Uniform crime reports: Monthly weapon-specific crime and arrest time series 1975–1993 (National, State, 12-City Data), ICPSR 6792 Inter-university Consortium for Political and Social Research. Ann Arbor
7. Farmer JD, Sidorowich J (1987) Predicting chaotic time series. Phys Rev Lett 59:845
8. Gabrielov A, Keilis-Borok V, Zaliapin I, Newman WI (2000) Critical transitions in colliding cascades. Phys Rev E 62:237–249
9. Gabrielov A, Keilis-Borok V, Zaliapin I (2007) Predictability of extreme events in a branching diffusion model. arXiv:0708.1542

10. Gabrielov AM, Zaliapin IV, Newman WI, Keilis-Borok VI (2000) Colliding cascade model for earthquake prediction. Geophys J Int 143(2):427–437
11. Gelfand IM, Guberman SA, Keilis-Borok VI, Knopoff L, Press F, Ranzman IY, Rotwain IM, Sadovsky AM (1976) Pattern recognition applied to earthquake epicenters in California. Phys Earth Planet Inter 11:227–283
12. Gell-Mann M (1994) The quark and the jaguar: Adventures in the simple and the complex. Freeman, New York
13. Crutchfield JP, Farmer JD, Packard NH, Shaw RS (1986) Chaos Sci Am 255:46–57
14. Gvishiani AD, Kosobokov VG (1981) On found of the pattern recognition results applied to earthquake-prone areas. Izvestiya Acad Sci USSR. Phys Earth 2:21–36
15. Holland JH (1995) Hidden order: How adaptation builds complexity. Addison, Reading
16. IMF (1997) International monetary fund, international financial statistics. CD-ROM
17. Kadanoff LP (1976) Scaling, universality and operator algebras. In: Domb C, Green MS (eds) Phase transitions and critical phenomena, vol 5a. Academic, London, pp 1–34
18. Keilis-Borok VI, Lichtman AJ (1993) The self-organization of American society in presidential and senatorial elections. In: Kravtsov YA (ed) Limits of predictability. Springer, Berlin, pp 223–237
19. Keilis-Borok VI, Press F (1980) On seismological applications of pattern recognition. In: Allègre CJ (ed) Source mechanism and earthquake prediction applications. Editions du centre national du la recherché scientifique, Paris, pp 51–60
20. Keilis-Borok VI, Soloviev AA (eds) (2003) Nonlinear dynamics of the lithosphere and earthquake prediction. Springer, Berlin
21. Keilis-Borok V, Soloviev A (2007) Pattern recognition methods and algorithms. Ninth workshop on non-linear dynamics and earthquake prediction, Trieste ICTP 1864-11
22. Keilis-Borok VI, Sorondo MS (2000) (eds) Science for survival and sustainable development. The proceedings of the study-week of the Pontifical Academy of Sciences, 12–16 March 1999. Pontificiae Academiae Scientiarvm Scripta Varia, Vatican City
23. Keilis-Borok V, Stock JH, Soloviev A, Mikhalev P (2000) Pre-recession pattern of six economic indicators in the USA. J Forecast 19:65–80
24. Keilis-Borok VI, Gascon DJ, Soloviev AA, Intriligator MD, Pichardo R, Winberg FE (2003) On predictability of homicide surges in megacities. In: Beer T, Ismail-Zadeh A (eds) Risk science and sustainability. Kluwer, Dordrecht (NATO Sci Ser II Math, Phys Chem 112), pp 91–110
25. Keilis-Borok VI, Soloviev AA, Allègre CB, Sobolevskii AN, Intriligator MD (2005) Patterns of macroeconomic indicators preceding the unemployment rise in Western Europe and the USA. Pattern Recogn 38(3):423–435
26. Keilis-Borok V, Soloviev A, Gabrielov A, Zaliapin I (2007) Change of scaling before extreme events in complex systems. In: Proceedings of the plenary session on "predictability in science: Accuracy and limitations", Pontificiae Academiae Scientiarvm Scripta Varia, Vatican City
27. Kravtsov YA (ed) (1993) Limits of predictability. Springer, Berlin
28. Lichtman AJ, Keilis-Borok VI (1989) Aggregate-level analysis and prediction of midterm senatorial elections in the United States, 1974–1986. Proc Natl Acad Sci USA 86(24):10176–10180
29. Lichtman AJ (1996) The keys to the White House. Madison Books, Lanham
31. Lichtman AJ (2005) The keys to the White House: Forecast for 2008. Foresight Int J Appl Forecast 3:5–9
30. Lichtman AJ (2008) The keys to the White House, 2008 edn. Rowman/Littlefield, Lanham
32. Ma Z, Fu Z, Zhang Y, Wang C, Zhang G, Liu D (1990) Earthquake prediction: Nine major earthquakes in china. Springer, New York
33. Mason IB (2003) Binary events. In: Jolliffe IT, Stephenson DB (eds) Forecast verification. A practitioner's guide in atmospheric science. Wiley, Chichester, pp 37–76
34. Molchan GM (1990) Strategies in strong earthquake prediction. Phys Earth Planet Inter 61:84–98
35. Molchan GM (1991) Structure of optimal strategies of earthquake prediction. Tectonophysics 193:267–276
36. Molchan GM (1994) Models for optimization of earthquake prediction. In: Chowdhury DK (ed) Computational seismology and geodynamics, vol 1. Am Geophys Un, Washington, pp 1–10
37. Molchan GM (1997) Earthquake prediction as a decision-making problem. Pure Appl Geophys 149:233–237
38. Molchan GM (2003) Earthquake prediction strategies: A theoretical analysis. In: Keilis-Borok VI, Soloviev AA (eds) Nonlinear dynamics of the lithosphere and earthquake prediction. Springer, Berlin, pp 209–237
39. Molchan G, Keilis-Borok V (2008) Earthquake prediction: Probabilistic aspect. Geophys J Int 173(3):1012–1017
40. NACJD: http://www.icpsr.umich.edu/NACJD/index.html
41. NBER: http://www.nber.org/cycles/cyclesmain.html
42. Newman W, Gabrielov A, Turcotte DL (eds) (1994) Nonlinear dynamics and predictability of geophysical phenomena. Am Geophys Un, Int Un Geodesy Geophys, Washington
43. OECD (1997) Main economic indicators: Historical statistics 1960–1996. Paris, CD-ROM
44. Press F, Briggs P (1975) Chandler wobble, earthquakes, rotation and geomagnetic changes. Nature 256:270–273, London
45. Press F, Briggs P (1977) Pattern recognition applied to uranium prospecting. Nature 268:125–127
46. Press F, Allen C (1995) Patterns of seismic release in the southern California region. J Geophys Res 100(B4):6421–6430
47. Soloviev A (2007) Application of the pattern recognition techniques to earthquake-prone areas determination. Ninth workshop on non-linear dynamics and earthquake prediction, Trieste ICTP 1864-9
48. Stock JH, Watson MW (1989) New indexes of leading and coincident economic indicators. NBER Macroecon Ann 4:351–394
49. Stock JH, Watson MW (1993) A procedure for predicting recessions with leading indicators. In: Stock JH, Watson MW (eds) Business cycles, indicators, and forecasting (NBER Studies in Business Cycles, vol 28), pp 95–156
50. Tukey JW (1977) Exploratory data analysis. Addison-wesley series in behavioral science: Quantitative methods. Addison, Reading
51. Turcotte DL, Newman WI, Gabrielov A (2000) A statistical physics approach to earthquakes. In: Geocomplexity and the physics of earthquakes. Am Geophys Un, Washington
52. Zaliapin I, Keilis-Borok V, Ghil M (2003) A Boolean delay model of colliding cascades, II: Prediction of critical transitions. J Stat Phys 111(3–4):839–861

Books and Reviews

Bongard MM (1967) The problem of recognition. Nauka, Moscow

Brito DL, Intriligator MD, Worth ER (1998) In: Eliassson G, Green C (eds) Microfoundations of economic growth: A Schumpeterian perspective. University of Michigan Press, Ann Arbor

Bui Trong L (2003) Risk of collective youth violence in french suburbs. A clinical scale of evaluation, an alert system. In: Beer T, Ismail-Zadeh A (eds) Risk science and sustainability. Kluwer, Dordrecht (NATO Sci Ser II Math Phys Chem 112)

Engle RF, McFadden DL (1994) (eds) Handbook of econometrics, vol 4. North-Holland, Amsterdam

Klein PA, Niemira MP (1994) Forecasting financial and economic cycles. Wiley, New York

Messner SF (1983) Regional differences in the economic correlates of the urban homicide rate. Criminology 21:477–488

Mitchell WC (1951) What happens during business cycles: A progress report. NBER, New York

Mitchell WC, Burns AF (1946) Measuring business cycles. NBER, New York

Moore GH (ed) (1961) Business cycle indicators. NBER, New York

Mostaghimi M, Rezayat F (1996) Probability forecast of a downturn in US economy using classical statistical theory. Empir Econ 21:255–279

Watson MW (1994) In: Engle RF, McFadden DL (eds) Handbook of econometrics, vol IV. North-Holland, Amsterdam

Extreme Value Statistics

MARIO NICODEMI
Department of Physics, University of Warwick,
Coventry, UK

Article Outline

Glossary

Random variable When a coin is tossed two random outcomes are permitted: head or tail. These outcomes can be mapped to numbers in a process which defines a 'random variable': for instance, 'head' and 'tail' could be mapped respectively to +1 and −1. More generally, any function mapping the outcomes of a random process to real numbers is defined a random variable [15]. More technically, a random variable is any function from a probability space to some measurable space, i. e., the space of admitted values of the variable, e. g., real numbers with the Borel σ-algebra. The amount of rainfall in a day or the daily price variation of a stock are two more examples. It's worth to stress that, formally, the outcome of a given random experiment is not a random variable: the random variable is the function describing all the possible outcomes as numbers. Finally, two random variables are said independent when the outcome of either of them has no influence on the other.

Probability distribution The probability of either outcomes, 'head' and 'tail', in tossing a coin is 50%. Similarly, a discrete random variable, X, with values $\{x_1, x_2, \dots\}$ has an associate discrete probability distribution of occurrence $\{p_1, p_2, \dots\}$. More generally, for a random variable on real numbers, X, the corresponding probability distribution [15] is the function returning the probability to find a value of X within a given interval $[x_1, x_2]$ (where x_1 and x_2 are real numbers): $\Pr[x_1 \le X \le x_2]$. In particular, the random variable, X, is fully characterized by its cumulative distribution function, $F(x)$, which is: $F(x) = \Pr[X < x]$ for any x in $\mathcal{R}$. The probability distribution density, $f(x)$, can be often defined as the derivative of $F(x)$: $f(x) = \mathrm{d}F(x)/\mathrm{d}x$.
The probability distribution of two independent random variables, X and Y, is the product of the distributions, F_X and F_Y, of X and Y: $F(x, y) \equiv \Pr[X < x; Y < y] = \Pr[X < x] \cdot \Pr[Y < y] = F_X(x) \cdot F_Y(y)$.

Expected value The expected value [15] of a random variable is its average outcome over many independent experiments. Consider, for instance, a discrete random variable, X, with values in the set $\{x_1, x_2, \dots\}$ and the corresponding probability for each of these values $\{p_1, p_2, \dots\}$. In probability theory, the expected, or average, value of X (denoted $E(X)$) is just the sum: $E(X) = \sum x_i p_i$. For instance, if you have an asset which can give two returns $\{x_1, x_2\}$ with probability $\{p_1, p_2\}$, its expected return is $x_1 p_1 + x_2 p_2$.
In case we have a random variable defined on real numbers and $F(x)$ is its probability distribution function, the expected value of X is: $E(X) = \int X \mathrm{d}F$. As for some $F(x)$ the above integral may not exist, the 'expected value' of a random variable is not always defined.

Variance and moments The variance [15] of a probability distribution is a measure of the average deviations from the mean of the related random variable. In probability theory, the variance is usually defined as the mean squared deviation $E((X - E(X))^2)$, i. e., the expected value of $(X - E(X))^2$. The square root of the

The press release on this paper is available at
http://www.physorg.com/news203674634.html

Variations of trends of indicators describing complex systems: Change of scaling precursory to extreme events

V. I. Keilis-Borok[1,2], A. A. Soloviev[2,3,a)]

[1] *Institute of Geophysics and Planetary Physics and Department of Earth and Space Sciences, University of California, Los Angeles, 405 Hilgard av., IGPP, Los Angeles, CA 90095- 1567, USA*

[2] *International Institute of Earthquake Prediction Theory and Mathematical Geophysics, Russian Academy of Sciences, 84/32 Profsouznaya st., Moscow 117997, Russia*

[3] *The Abdus Salam International Centre for Theoretical Physics, Trieste, Italy*

Socio-economic and natural complex systems persistently generate extreme events also known as disasters, crises, or critical transitions. Here we analyze patterns of background activity preceding extreme events in four complex systems: economic recessions, surges in homicides in a megacity, magnetic storms and strong earthquakes. We use as a starting point the indicators describing the system's behavior and identify changes in an indicator's trend. Those changes constitute our background events (*BE*). We demonstrate a premonitory pattern common to all four systems considered: relatively large magnitude *BE*s become more frequent before extreme event. A premonitory change of scaling has been found in various models and observations. Here we demonstrate this change in scaling of uniformly defined *BE*s in four real complex systems, their enormous differences notwithstanding.

Dynamics of a complex system is controlled by a wide variety of mutually dependent mechanisms that interact across and along the system hierarchy. Each mechanism creates strong instability and except for the very special circumstances, none of these mechanisms alone prevails in the sense that the others can be neglected. Thus creating a theory that could describe the system behavior by means of assembling the set of governing equations is unrealistic and may be misleading as well. A most important feature of complex systems is persistent reoccurrence of extreme events – rare events of low probability but high impact on a system where they occur. In different connotations they are also known as critical transitions, disasters, catastrophes, and crises. Among the regular behavior patterns of complex systems are "premonitory" ones, that emerge more frequently as an extreme event approaches. Premonitory patterns and extreme events are consecutive manifestations of a system's dynamics. These patterns may not trigger extreme events but merely signal the growth of instability, making the system ripe for extreme events. Finding premonitory patterns creates a basis for development of algorithms for prediction of extreme events and is pivotal for fundamental understanding of relevant complex systems. Specifically, the following types of premonitory patterns have been identified formerly: intensity; clustering; range of correlation in space; and change of scaling. In this study the change in scaling of uniformly defined background events has been found before the following extreme events in four complex systems: economic recessions, surges in homicides in a megacity, magnetic storms, and strong earthquakes. The background events are constituted by changes in trends of the observed indicators describing the system behavior. A premonitory pattern common to all four systems considered is formulated as follows: background events with relatively large magnitudes become more frequent before an extreme event.

I. INTRODUCTION

Natural and socio-economic complex systems persistently generate extreme events – also known as critical transitions, disasters, crises etc. Such events are commonly recognized as the major threat to the sustainability, or even survival, of our civilization and predictive understanding of them is commonly regarded as the major frontier problem of basic science.

Here we explore a specific observable phenomenon that precedes extreme events in different complex systems. This phenomenon can be detected by the change of the scaling in the background activity in the system. Scaling itself is a staple of studying complexity. However premonitory transformation of scaling has thus far only been found in seismicity and other forms of multiple fracturing, with major failures (e.g. strong earthquakes) being the extreme event (see Refs. 1-12, for instance).

One can consider the scaling and its changing if the background activity of the system for a certain period is reflected by "events" of different sizes (magnitudes) occurred during this period. In this study we describe the background activity of the system by discrete episodes of trend variations for a certain indicator reflecting the system behavior. These episodes are called background events (*BE*s) and their magnitudes are determined by sizes of relevant variations of the indicator trend. Our hope to find transformation of the *BE* scaling that precedes extreme events is based on the hypothetical assumption that the indicator volatility increases before extreme events and this is expressed in larger variations of its trend. Sequences of *BE*s are analyzed uniformly to find the same precursory phenomenon preceding extreme events of different nature: economic recessions, homicides surges, magnetic storms, and strong earthquakes. In all these cases the similar transformation of scaling has been detected for *BE*s before an extreme event that seems to open the promising line of research in predicting extreme events.

Note that our analysis concerns local variations in the behavior of time series and in this sense it differs from methods that detect long-range dependence in time series, e.g. the rescaled range analysis that is a powerful statistical tool, provides also a method to estimate the Hurst exponent (Refs. 13, 14, for instance), and has been widely applied to economic and financial time series.

II. DEFINITIONS

We consider discrete series of indexes describing complex systems. Let $f(k)$ is one of such series where $k = k_0, k_0{+}1, \dots k_T$ is an integer that stands for consecutive numbers of time points separated by a constant step during a time period $[k_0, k_T]$ under consideration. The indicator trend is estimated by an "average derivative" of function $f(k)$ calculated using its values in several consecutive time points. The change of the indicator trend is determined as a difference between the average derivatives of function $f(k)$ calculated for two sets of time points. A formal definition is formulated as follows.

Let k_c is a current time point. We use as the average derivative calculated for s time points k_c-s, k_c-s+1, …, k_c-1 a coefficient $K^f(k_c-s,k_c-1)$ of the local linear regression $W^f(k/k_c-s,k_c)$ of a function $f(m)$ in the time segment $[k_c-s, k_c-1]$:

$$W^f(k/k_c-s, k_c-1) = K^f(k_c-s,k_c-1)k + B^f(k_c-s,k_c-1),\ k_c-s \le k \le k_c-1, \qquad (1)$$

Coefficients $K^f(k_c-s,k_c-1)$ and $B^f(k_c-s,k_c-1)$ are calculated by means of the least-squares method, i.e. minimizing the sum of squares of differences between values of the regression (1) and $f(k)$ for $k_c-s \le k \le k_c-1$. The change of the indicator trend is characterized by difference $S^f(k_c/s,u) = K^f(k_c-s,k_c-1) - K^f(k_c-s-u,k_c-s-1)$ between average derivatives calculated for s time points preceding k_c and for u time points preceding k_c-s. Here $K^f(k_c-s-u,k_c-s-1)$ is the average derivative for u time points k_c-s-u, $k_c-s-u+1$, …, k_c-s-1 that is calculated by the same way as $K^f(k_c-s,k_c-1)$. In these studies we specify values of u that are much greater than values of s, thus the local change of the indicator trend is considered relatively its long-term trend. Note that the value of $S^f(k_c/s,u)$ is calculated without using the information from "future".

If extreme events in the complex system under consideration are connected with decline in the trend of function $f(k)$ then we consider that a *BE* occurs in time k if $S^f(k/s,u) < 0$ [Fig. 1(a)]. In the case when the extreme events are connected with increase of the trend of function $f(k)$ the *BE* occurs if $S^f(k/s,u) > 0$ [Fig. 1(b)]. In the both cases $\mu^f(k/s,u) = |S^f(k/s,u)|$ is considered as the *BE* magnitude.

Applying the procedure described above we transform series $f(k)$ into sequence of *BE*s occurred in times k_1, k_2, … and having magnitudes $\mu^f(k_1/s,u)$, $\mu^f(k_2/s,u)$, …The result of the transformation is determined by the source series $f(k)$ and the values of the parameters s and u. Note that if the source series is given for a period from time k_0 to time k_T then the *BE*s can be determined only if $k_0 + s + u \le k_T$ and times of their occurrence belong to segment $[k_0 + s + u, k_T]$.

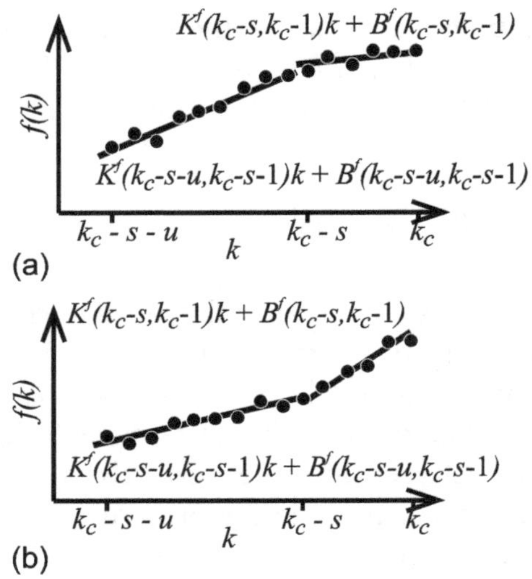

FIG. 1. Definition of *BE*s for index f when the extreme events are connected with decline in its trend, $\mu^f(k_c/s,u) = K^f(k_c{-}s{-}u,k_c{-}s{-}1) - K^f(k_c{-}s,k_c{-}1)$, (a) and when – with increase, $\mu^f(k_c/s,u) = K^f(k_c{-}s,k_c{-}1) - K^f(k_c{-}s{-}u, k_c{-}s{-}1)$, (b). Dots show values of index f, straight lines are the linear least-square regressions on segments $[k_c{-}s{-}u, k_c{-}s{-}1]$ and $[k_c{-}s, k_c{-}1]$.

III. DATA ANALYSIS

We analyze the data relevant to the following extreme events generated by different complex systems indicated in brackets: advents of economic recessions (the U.S. economy); homicide surges [a megacity (Los Angeles)], magnetic storms (the Earth magnetosphere); strong earthquakes (the Earth lithosphere).

3.1. The U.S. economy. The industrial production, total (**IP**) index is selected to define the *BE*s for the U.S. economy. **IP** is a monthly index of real (constant dollars, dimensionless) output in the entire economy. We consider the time period from January 1959 to March 2009. The data on **IP** have been taken from the Economic Research Section of Federal Reserve Bank of St. Louis web site.[15] Fig. 2 shows the plot of this index and the periods of the economic recessions in the U.S. according to the data of the National Bureau of Economic Research.[16]

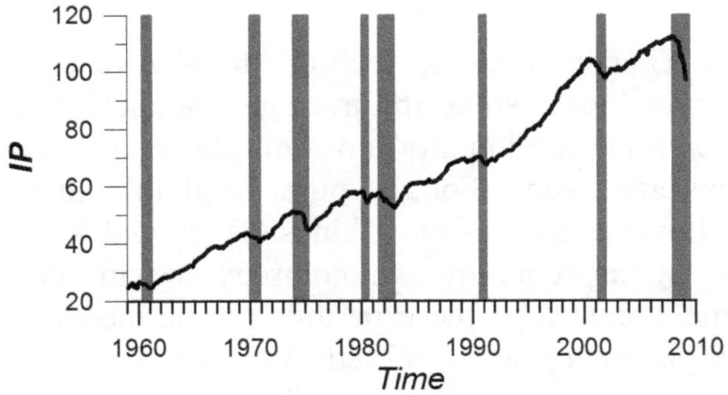

FIG. 2. Industrial production (**IP**) monthly index (black curve) and periods of the economic recessions in the U.S. (gray bars).

We consider a start of an economic recession as an extreme event for the U.S. economy; it is connected with decline in the trend of **IP**. Therefore in accordance with the definition given above a *BE* occurs in month k if $S^{IP}(k/s,u) < 0$ [Fig. 1(a)]. The series of *BE*s have been obtained for different values of s and u. Several corresponding cumulative frequency-magnitude (*FM*) plots are shown in Fig. 3. These plots present the dependence of $P(\mu) = N(\mu)/N(0)$ on μ where $N(\mu)$ is the number of *BE*s with magnitude $\mu^{IP}(k/s,u) \geq \mu$. One can see from Fig. 3 that the dependence of $\log P(\mu)$ on μ is close to a linear one

$$\log P(\mu) = \alpha - \beta \mu \qquad (2)$$

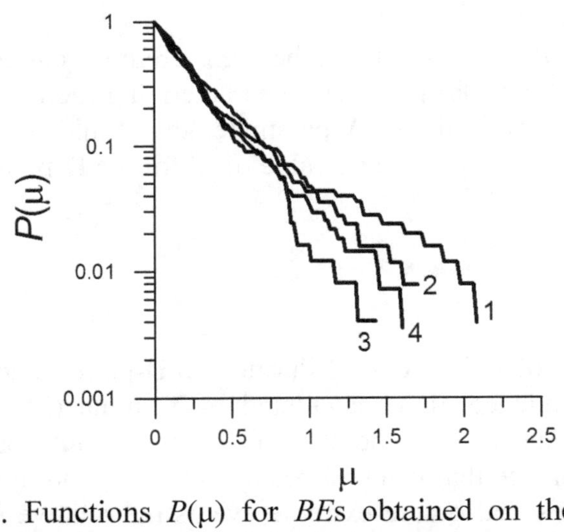

FIG. 3. Functions $P(\mu)$ for *BE*s obtained on the basis of **IP** with the following values (in months) of parameters: $s = 3$ and $u = 48$ (1), $s = 6$ and $u = 48$ (2), $s = 12$ and $u = 48$ (3), $s = 6$ and $u = 24$ (4).

Function $P(\mu)$ characterizes the scaling (the size distribution) of *BE*s. *FM* plots given in Fig. 3 show the average scaling for the whole period under consideration. We inquire into a question: does the scaling change before an extreme event? For the purpose of this analysis time considered is divided into periods of three kinds (**D**, **N**, and **X**), as shown in Fig 4. In the case of the economy as a complex system and starts of recessions as extreme events in it these periods are defined formally as follows. Two integer parameters are specified: $\Delta_D > 0$ and $\Delta_X \geq 0$. Months belonging to periods of recessions and Δ_X months after each recession are included in **X**–period. Among other months Δ_D months before each recession are included in **D**-period. Months that have not been included in **X**- or **D**-periods are included in **N**-period. Denote as $P_D(\mu)$ function $P(\mu)$ obtained in the **D**-period and as $P_N(\mu)$ – in the **N**-period. Difference between these two functions reflects changing of the scaling before extreme events.

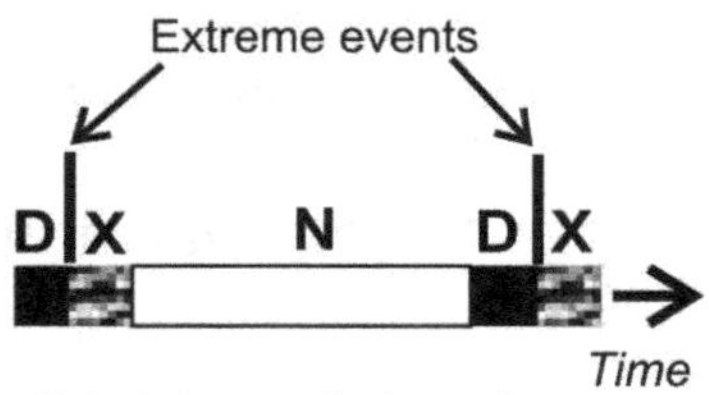

FIG. 4. Division of time into periods of three kinds by their relation to extreme events.

The changing of the scaling of *BE*s before extreme event could be shown by dependence of a scalar parameter characterizing the *BE* scaling for the **D**-period on the value of Δ_D. A parameter that could be used to characterize the *BE* scaling is the average value of μ^2 for the **D**-period:

$$M = \frac{1}{n_D} \sum_{i=1}^{n_D} \mu_i^2, \tag{3}$$

where n_D is the number of *BE*s occurred during the **D**-period and μ_i are their magnitudes. Dependence of M calculated by formula (3) for *BE*s determined with s = 6 and u = 48 months and Δ_X = 1 month on Δ_D is shown in Fig. 5. One can see that lager values of M correspond to smaller values of Δ_D. This means that before extreme events the relative number of *BE*s with large magnitudes is larger. The plots given in Fig. 6 show functions $P_D(\mu)$ and $P_N(\mu)$ obtained for **D**- and **N**-periods corresponding to different values of Δ_D and $\Delta_X = 1$ month. Considered values of Δ_D cover an interval where there are changes in dependence of M on Δ_D (Fig. 5). One can see the obvious difference between $P_D(\mu)$ and $P_N(\mu)$ that is stable to varying the value of Δ_D. This difference is that the value of β in Eq. (2) becomes less before extreme events (advents of economic recessions).

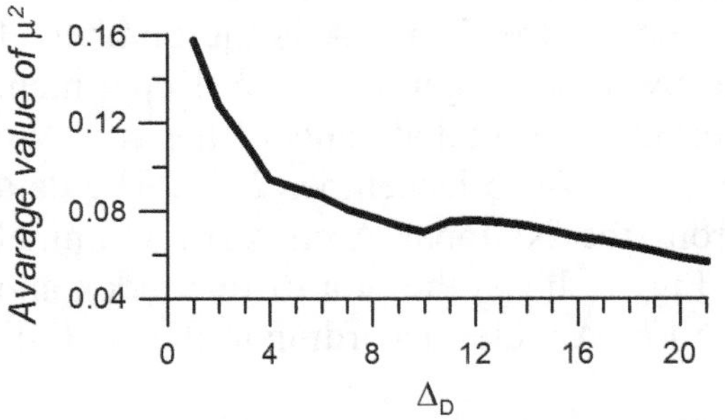

FIG. 5. Dependence of M calculated by formula (3) on Δ_D obtained for BEs determined on the basis of **IP** with $s = 6$ and $u = 48$ months and $\Delta_X = 1$.

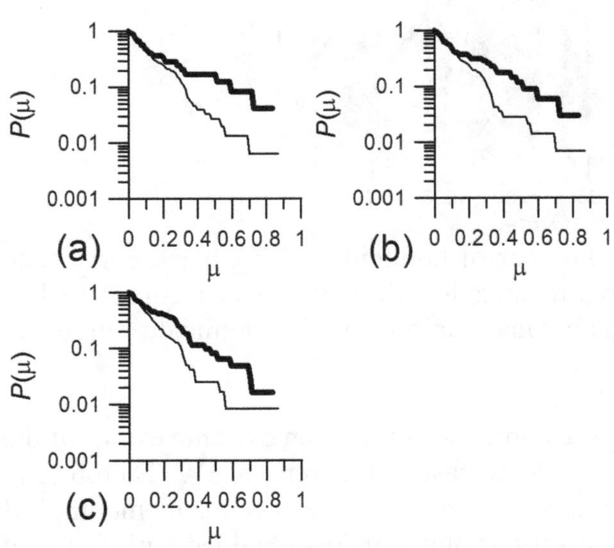

FIG. 6. Functions $P_D(\mu)$ (thick curves) and $P_N(\mu)$ (thin curves) obtained for BEs determined on the basis of **IP** with $s = 6$ and $u = 48$ months and **D**- and **N**-periods corresponding to the following values of parameters: $\Delta_X = 1$ and $\Delta_D = 4$ (a), $\Delta_D = 6$ (b), and $\Delta_D = 12$ (c).

The plots of function $P(M)$ for the whole period (Fig. 3) show existence of BEs with larger magnitudes than magnitudes of BEs occurred during **D**- and **N**-periods (Fig. 6). This is explained by the fact that the whole period includes **X**-periods that contain extreme events and increasing of the system activity induced by them. The maximal magnitude of such *BE*s obtained with the values of s and u corresponding to Fig. 6 and curve 2 in Fig. 3 is 1.70.

3.2. A megacity (Los Angeles). The monthly number of firearm assaults (**FA**) in Los Angeles is selected to define the *BE*s. **FA** is the number of assaults with firearms registered by Los Angeles Police Department during a month in Los Angeles, per 3,000,000 inhabitants of the city. We consider the time period from January 1974 to December 2003. The data on **FA** index have been taken from the National Archive of Criminal Justice Data (NACJD) web site.[17] Fig. 7 shows the plot of this index and the periods of the homicide surge in Los Angeles according to the NACJD data.[17]

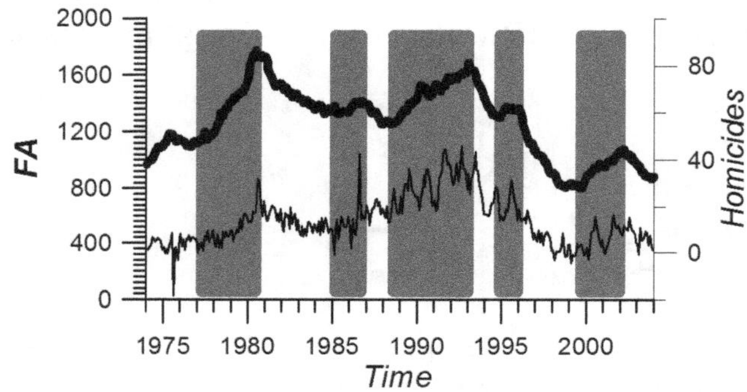

FIG. 7. Smoothed monthly number of homicides (thick black curve) and the monthly number of firearm assaults (**FA**) (thin black curve) in Los Angeles per 3,000,000 inhabitants. Periods of the homicide surge are shown by gray bars.

We consider a start of a homicide surge as an extreme event for the megacity and it is connected with increase in the trend of **FA**. Therefore in accordance with the definition given above a *BE* occurs in month k if $S^{FA}(k/s,u) > 0$ [Fig. 1(b)]. Several *FM* plots for *BE*s obtained with different values of s and u are shown in Fig. 8. Here the dependence of $\log P(\mu)$ on μ does not demonstrate such closeness to a linear one [Eq. (2)] as in the case of *BE*s obtained on the basis of **IP** (Fig. 3).

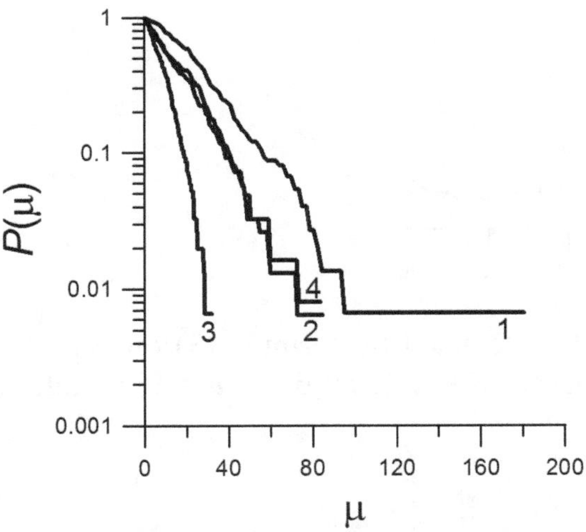

FIG. 8. Functions $P(\mu)$ for *BE*s obtained on the basis of **FA** with the following values (in months) of parameters: $s = 3$ and $u = 48$ (1), $s = 6$ and $u = 48$ (2), $s = 12$ and $u = 48$ (3), $s = 6$ and $u = 96$ (4).

As in the case of recessions we define **D-**, **N-**, and **X**-periods by means of two parameters $\Delta_D > 0$ and $\Delta_X \geq 0$: months belonging to periods of the homicide surge and Δ_X months after these periods are included in **X**–period, among other months Δ_D months before each start of the homicide surge are included in **D**-period, and the months that have not been included in **X**- or **D**-periods are included in **N**-period.

Fig. 9 shows dependence of M calculated by formula (3) for *BE*s determined with $s = 6$ and $u = 48$ months and $\Delta_X = 1$ month on Δ_D. One can see that the maximum value of M corresponds to $\Delta_D = 4$ and M decreases when Δ_D becomes larger. Plots of functions $P_D(\mu)$ and $P_N(\mu)$ obtained for **D**- and **N**-periods corresponding to different values of Δ_D and $\Delta_X = 1$ are given in Fig. 10. One can see that when $\Delta_D = 4$ [Fig. 10(a)] and $\Delta_D = 6$ [Fig. 10(b)] the changing of the *BE* scaling before starts of the homicide surge is like that before recession starts (Fig. 6). At the same time as our analysis shows functions $P_N(\mu)$ shown in Fig. 10 reflects *BE*s that occurred shortly after the ends of periods of homicide surge. This is illustrated by Fig. 11 where $\Delta_X = 3$ and other parameters are the same as in Fig. 10. Here there is the similar difference between functions $P_D(\mu)$ and $P_N(\mu)$ for all three considered values of Δ_D. The maximal magnitude of *BE*s obtained with the values of s and u corresponding to Figs. 10 and 11 and curve 2 in Fig. 8 and occurred during extreme events is 84.64.

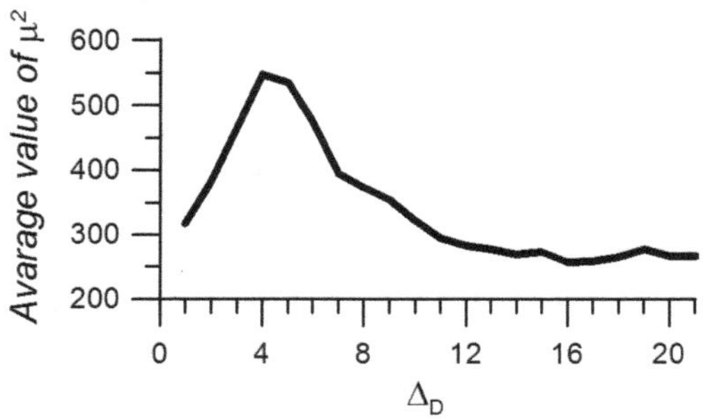

FIG. 9. Dependence of M calculated by formula (3) on Δ_D obtained for *BE*s determined on the basis of **FA** with $s = 6$ and $u = 48$ months and $\Delta_X = 1$.

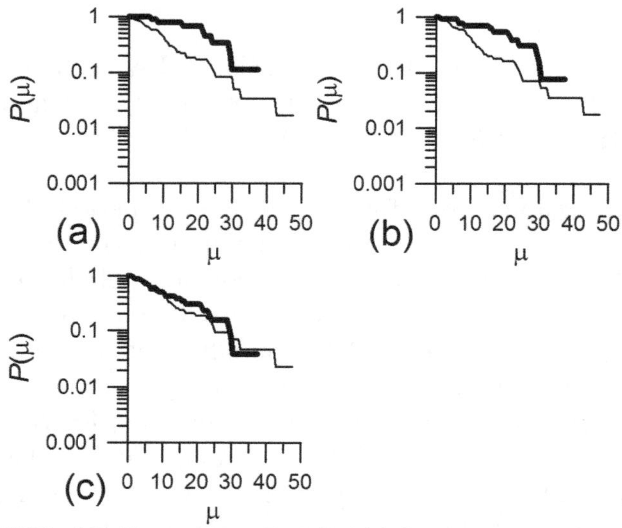

FIG. 10. Functions $P_D(\mu)$ (thick curves) and $P_N(\mu)$ (thin curves) obtained for *BE*s determined on the basis of **FA** with $s = 6$ and $u = 48$ months and **D**- and **N**-periods corresponding to the following values of parameters: $\Delta_X = 1$ and $\Delta_D = 4$ (a), $\Delta_D = 6$ (b), and $\Delta_D = 12$ (c).

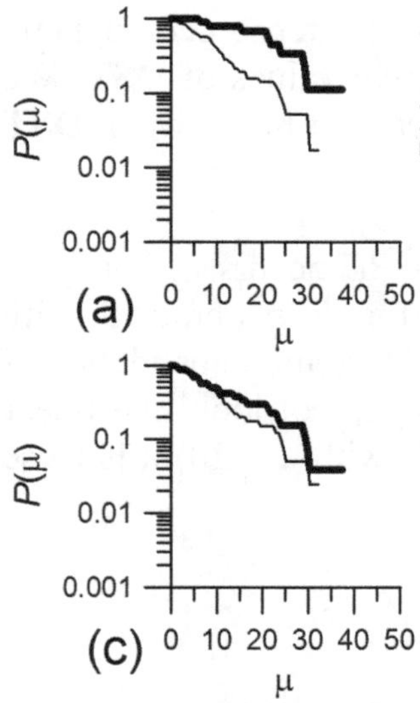

FIG. 11. Functions $P_D(\mu)$ (thick curves) and $P_N(\mu)$ (thin curves) obtained for *BE*s determined on the basis of **FA** with $s = 6$ and $u = 48$ months and **D**- and **N**-periods corresponding to the following values of parameters: $\Delta_X = 3$ and $\Delta_D = 4$ (a), $\Delta_D = 6$ (b), and $\Delta_D = 12$ (c).

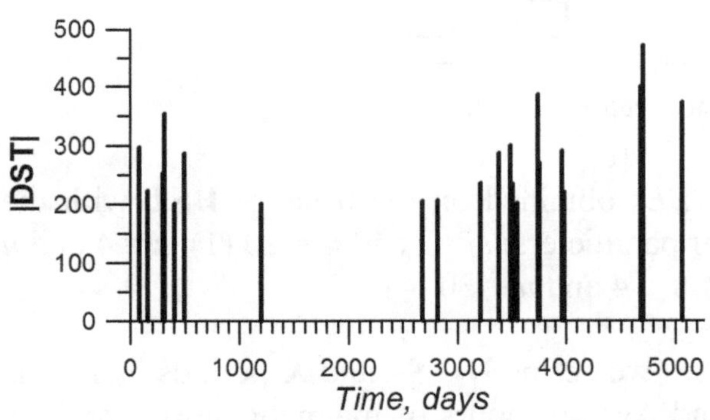

FIG. 12. Magnetic storms (hours when **DST** < -200) shown by vertical lines with heights proportional to |**DST**|. Time is measured in days with 01/01/1991 1h as an initial moment.

3.3. The Earth magnetosphere. The data under consideration are hourly values of Disturbance Storm Time (**DST**) for the period from 01/01/1991 1h to 04/08/2005 13h. **DST** is calculated from an array of magnetometers on the Earth's surface and is a measure of the strength of the ring current, a persistent feature in the Earth's magnetosphere. It is also a measure of geomagnetic activity. The data on DST have been taken from NASA web site.[18] Extreme events are associated with periods of intense geomagnetic activity and characterized by very large and negative values of **DST**. We consider that extreme events (magnetic storms) occur at hours when **DST** < -200. These hours are shown in Fig. 12.

DST is also used to determine *BE*s and a *BE* occurs at hour *k* if $S^{DST}(k/s,u) < 0$ [Fig. 1(a)]. Several *FM* plots for *BE*s obtained with different values of *s* and *u* are shown in Fig. 13. Here the dependence of log $P(\mu)$ on μ in the whole range of values of μ does not look like linear [Eq. (2)]. But if we consider only rather large *BE*s (with $\mu > 20$) it is close to a linear one.

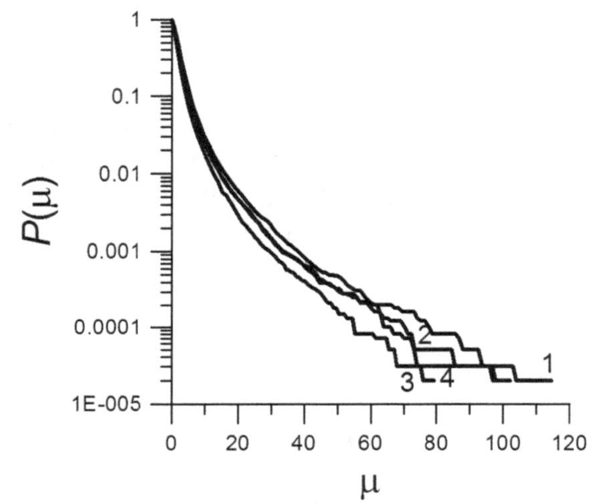

FIG. 13. Functions $P(\mu)$ for *BE*s obtained on the basis of **DST** with the following values (in hours) of parameters: $s = 3$ and $u = 20$ (1), $s = 4$ and $u = 20$ (2), $s = 6$ and $u = 20$ (3), $s = 4$ and $u = 30$ (4).

As in the previous cases we define **D**-, **N**-, and **X**-periods by means of two parameters $\Delta_D > 0$ and $\Delta_X \geq 0$: hours of magnetic storms and Δ_X hours after them are included in **X**–period, among other hours Δ_D hours before each magnetic storm are included in **D**-period, and the hours that have not been included in **X**- or **D**-periods are included in **N**-period. One can see from Fig. 12 that in several cases the moments of magnetic storms form dense groups. To exclude time periods inside such groups and therefore to consider a group as one extreme event we specify $\Delta_X = 96$ hours.

Fig. 14 shows dependence of *M* calculated by formula (3) for *BE*s determined with $s = 4$ and $u = 20$ hours and $\Delta_X = 96$ hours on Δ_D. One can see that the value of *M* decreases monotonously when Δ_D increases. Plots of functions $P_D(\mu)$ and $P_N(\mu)$ that correspond to different values of Δ_D are given in Fig. 15. One can see that the changing of the *BE* scaling before magnetic storms is like that before starts of recessions and periods of the homicide surge (Figs. 6 and 11). The maximal magnitude of *BE*s obtained with the values of *s* and *u* corresponding to Fig. 15 and curve 2 in Fig. 13 and connected with extreme events is 102.01.

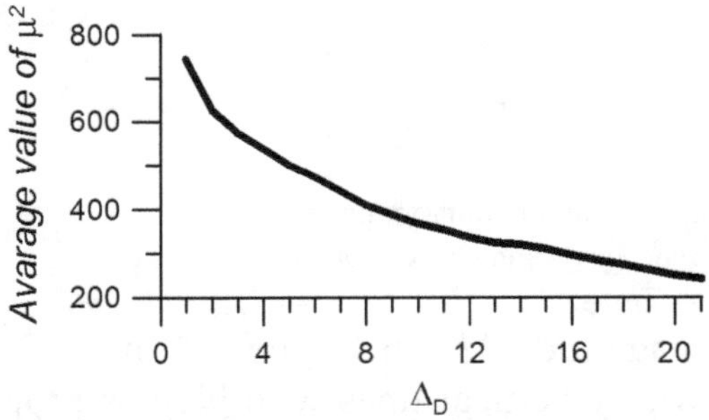

FIG. 14. Dependence of *M* calculated by formula (3) on Δ_D obtained for *BE*s determined on the basis of **DST** with $s = 4$ and $u = 20$ hours and $\Delta_X = 96$.

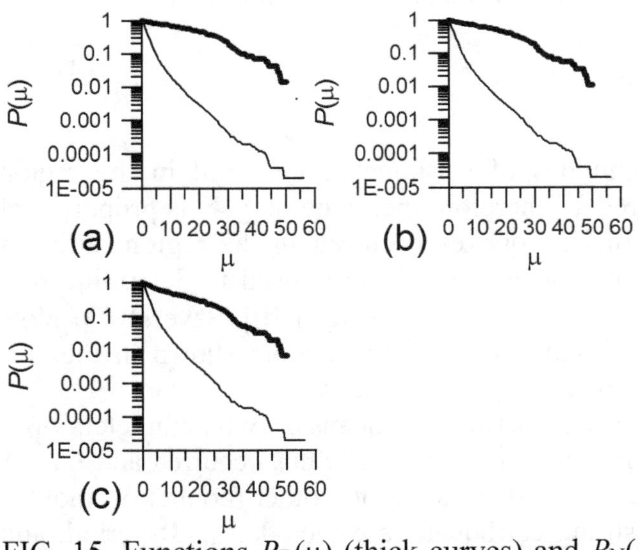

FIG. 15. Functions $P_D(\mu)$ (thick curves) and $P_N(\mu)$ (thin curves) obtained for *BE*s determined on the basis of **DST** with $s = 4$ and $u = 20$ hours and **D**- and **N**-periods corresponding to the following values of parameters: $\Delta_X = 96$ and $\Delta_D = 4$ (a), $\Delta_D = 6$ (b), and $\Delta_D = 12$ (c).

3.4. The Earth lithosphere. We consider the Southern California active region and strong earthquakes (with magnitude $M \geq 6.4$) occurred in 1932-

2005 as extreme events (Fig. 16). The data on seismicity of the Southern California have been taken from The Southern California Earthquake Data Centre (SCEDC) catalogue.[19]

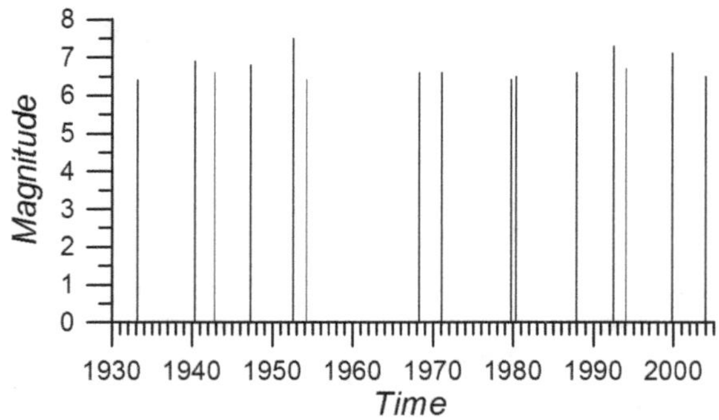

FIG. 16. Earthquakes with magnitude $M \geq 6.4$ occurred in 1932-2005 in the Southern California shown by vertical lines with heights proportional to M.

The seismicity of the Southern California is characterized by monthly values of function $\sum$ that is calculated by formula

$$\Sigma(k) = \sum_i 10^{M_i} . \tag{4}$$

Here M_i are magnitudes of earthquakes occurred in the region during the month numbered k. Therefore the value of $\sum(k)$ is proportional to the total source area of earthquakes occurred in the region during a month under consideration. The function $\sum(k)$ is used to determine *BE*s and a *BE* occurs in month k if $S^{\Sigma}(k/s,u) > 0$ [Fig. 1(b)]. Several *FM* plots for *BE*s obtained with different values of s and u are shown in Fig. 17. Here the dependence of $\log P(\mu)$ on μ is not linear.

We define **D**-, **N**-, and **X**-periods by means of two parameters $\Delta_D > 0$ and $\Delta_X \geq 0$: a month when a strong earthquake has occurred and $\Delta_X - 1$ months after it are included in **X**–period, among other months Δ_D months before each month of a strong earthquake are included in **D**-period, and the months that have not been included in **X**- or **D**-periods are included in **N**-period.

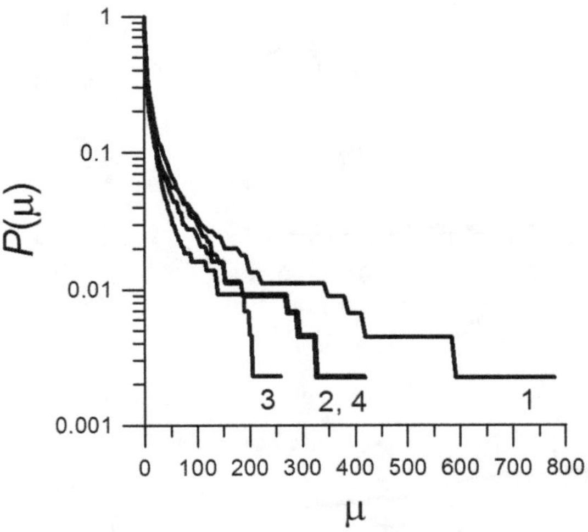

FIG. 17. Functions $P(\mu)$ for *BE*s obtained on the basis of $\sum(m)$ with the following values (in months) of parameters: $s = 4$ and $u = 24$ (1), $s = 6$ and $u = 24$ (2), $s = 8$ and $u = 24$ (3), $s = 6$ and $u = 12$ (4).

Fig. 18 shows dependence of M calculated by formula (3) for *BE*s determined with $s = 6$ and $u = 24$ months and on Δ_D. If $\Delta_X = 1$ month [Fig. 18(a)] then there is an extremely large surge of M when $\Delta_D > 16$. This can be explained by the fact that active periods after the strong earthquakes occurred in 1952 and 1992 (Fig. 16) are included in **D**-period. To avoid this we increase Δ_X and Fig. 18(b) shows dependence of M on Δ_D when $\Delta_X = 24$ months. One can see that the value of M decreases monotonously when $\Delta_D < 10$ and $\Delta_D > 25$. Plots of functions $P_D(\mu)$ and $P_N(\mu)$ obtained with $\Delta_X = 24$ months and different values of Δ_D are given in Fig. 19. One can see that the changing of the *BE* scaling before strong earthquakes is like that before extreme events in other complex systems considered (Figs. 6, 11, and 15). The maximal magnitude of *BE*s obtained with the values of s and u corresponding to Fig. 19 and curve 2 in Fig. 17 and connected with extreme events is 416.43.

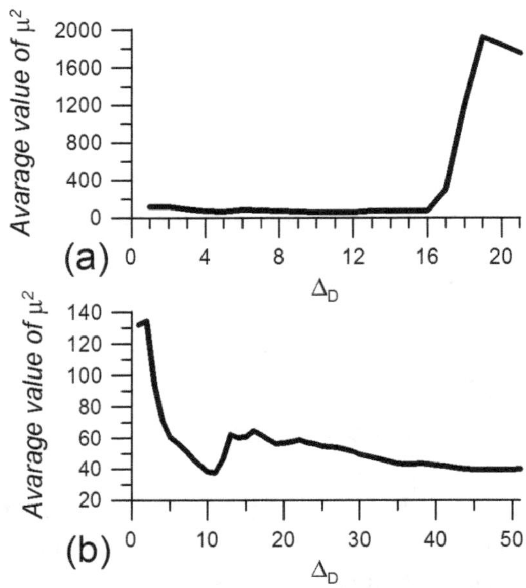

FIG. 18. Dependence of M calculated by formula (3) on Δ_D obtained for BEs determined on the basis of $\sum(m)$ with $s = 6$ and $u = 24$ months and $\Delta_X = 1$ (a) and 24 (b).

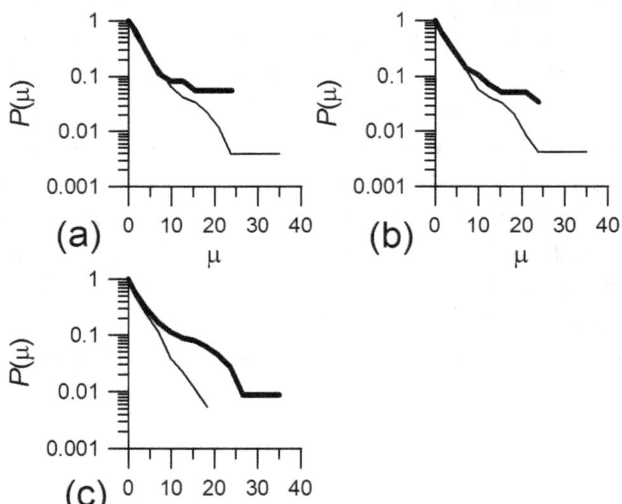

FIG. 19. Functions $P_D(\mu)$ (thick curves) and $P_N(\mu)$ (thin curves) obtained for BEs determined on the basis of $\sum(m)$ with $s = 6$ and $u = 24$ months and **D**- and **N**-periods corresponding to the following values of parameters: $\Delta_X = 24$ and $\Delta_D = 8$ (a), $\Delta_D = 12$ (b), and $\Delta_D = 24$ (c).

IV. CONCLUSION

We have considered four complex systems of rather different nature. In all these systems the phenomenon of changing of scaling before extreme events has been found for the background activity determined by the universal way. The appearance of the phenomenon is stable to variations of parameters determining *BE*s and periods where the phenomenon is searched for. In the case of the lithosphere and earthquakes there is rather small difference between functions $P_D(\mu)$ and $P_N(\mu)$ when parameter Δ_D is small [Fig. 19(a),(b)]. This can be explained by and corresponds to the fact that when Δ_D increases from 10 to 18 the value of M calculated by formula (3) increases as well [Fig. 18(b)].

One may consider our analysis as relevant to the problem of prediction of extreme events in complex systems and the following considerations should be taken into account in this sense.

-- We followed here "premonitory patterns" approach in which prediction is targeted at the rare extreme point events. This approach is complementary to classical Kolmogoroff-Wiener prediction, targeted at extrapolation of a whole process, i.e. mainly at the medium or small events.

-- That approach is complementary also to cause-and-effect analysis. Extreme events and premonitory patterns often are the parallel manifestations of evolution of the complex system.

-- Furthermore premonitory patterns might predict not an extreme event *per se*, but the system's destabilization making it ripe for an extreme event; its triggering then becomes close to inevitable, not requiring a strong impact.

-- Transition to predicting individual extreme events comprises at least the following problems:

parametrization of premonitory change of scaling, sensitivity analysis, and optimization (see Ref. 20);

similar analysis for clustering (e.g., Ref. 21) and correlation range (e.g., Ref. 22).

A hypothetical prediction algorithm could be based on calculation of function M [formula (3)] in a sliding time window, i.e. summing squares of magnitudes of *BE*s occurred in this time window and dividing this sum by the number of these events. The value of M should increase when the sliding time window approaches to an extreme event. Another possibility is to calculate the number of *BE*s with magnitudes exceeding a certain threshold occurred in the sliding time window. This number should also increase when the time window approaches to an extreme event.

ACKNOWLEDGMENTS

The authors are deeply grateful to George Molchan and Cleo Paskal for their help, comments and fruitful discussions, to Mark Moldwin and Mike Hartinger for the data on geomagnetic activity and to the reviewers of the paper for valuable comments and remarks.

[a)]Author to whom correspondence should be addressed. Electronic mail: soloviev@mitp.ru.

References

[1]W. D. Smith, Geophys. J. Int. **86**, 815 (1986).
[2]I. G. Main, P. G. Meredith, and C. Jones, Geophys. J. Int. **96**, 131 (1989).
[3]I. G. Main, P. G. Meredith, and P. R. Sammonds, Tectonophys. **211**, 233 (1992).
[4]J. Henderson, I. Main, P. Meredith, and P. Sammonds, J. Struct. Geol. **14**, 905 (1992).
[5]J. Henderson, I. G. Main, R. G. Pearce, and M. Takeya, Geophys. J. Int. **116**, 217 (1994).
[6]G. S. Narkunskaya and M. G. Shnirman, in *Computational Seismology and Geodynamics Vol. 1*, edited by D. K. Chowdhury, (Am. Geophys. Union, Washington, DC, 1994), pp. 20-24.
[7]I. Rotwain, V. Keilis-Borok, and L. Botvina, Phys. Earth Planet. Inter. **101**, 61 (1997).
[8]S. Wiemer and M. Wyss, J. Geophys. Res. **102**, 15115 (1997).
[9]M. Wyss and S. Wiemer, Science **290**, 1334 (2000).
[10]S. M. Burroughs and S. F. Tebbens, Bull. Seismol. Soc. Am. **92**, 2983 (2002).
[11]D. Amitrano, J. Geophys. Res. **108**, 2044 (2003).
[12]I. Zaliapin, V. Keilis-Borok, and M. Ghil, J. Stat. Phys. **111**, 839 (2003).
[13]H. Hurst, Trans. Am. Soc. Civil Engineers **116**, 770 (1951).
[14]B. Mandelbrot and R. L. Hudson, The (Mis)Behavior of Markets: A Fractal View of Risk, Ruin and Reward. Basic Books, New York, 328 p. (2004).
[15]Economic Research Section of Federal Reserve Bank of St. Louis Web site http://research.stlouisfed.org/fred2/series/INDPRO/
[16]National Bureau of Economic Research Web site http://www.nber.org/cycles/cyclesmain.html

[17]National Archive of Criminal Justice Data Web site http://www.icpsr.umich.edu/NACJD/
[18]NASA Web site http://omniweb.gsfc.nasa.gov/form/dx1.html
[19]SCEDC web site http://www.data.scec.org/about/data_avail.html.
[20]G. M. Molchan, in *Nonlinear Dynamics of the Lithosphere and Earthquake Prediction*, edited by V. I. Keilis-Borok and A. A. Soloviev, (Springer-Verlag, Berlin-Heidelberg, 2003), pp. 209-237.
[21]V. I. Keilis-Borok, L. Knopoff, and I. M. Rotwain, Nature **283**, 258 (1980).
[22]P. Shebalin, Tectonophys. **424**, 335 (2006).

22

On Predictive Understanding of Extreme Events: Pattern Recognition Approach; Prediction Algorithms; Applications to Disaster Preparedness

Vladimir Keilis-Borok,[1,2] Alexandre Soloviev,[2] and Andrei Gabrielov[3]

ABSTRACT

We describe a uniform approach in predicting different extreme events, also known as critical phenomena, disasters, or crises. The following types of extreme events are considered in this chapter: strong earthquakes, economic recessions (their onset and termination), surges of unemployment, surges of crime, and electoral changes of governing party. A uniform approach is possible due to the common feature of these events: each of them is generated by a certain hierarchical dissipative complex system. After a coarse-graining, such systems exhibit regular behavior patterns; among them we look for the *premonitory patterns* that signal approach of an extreme event. We introduce methodology, based on the optimal control theory, assisting disaster management in choosing optimal set of disaster preparedness measures undertaken in response to a prediction. Predictions with their currently realistic (limited) accuracy do allow preventing a considerable part of the damage by a hierarchy of preparedness measures. Accuracy of prediction should be known, but not necessarily high.

22.1. INTRODUCTION

22.1.1. Prediction Problem

Targets of prediction are individual extreme events that are rare but make big impact. Prediction is formulated as a discrete sequence of alarms, each indicating the time window and space where an extreme event is expected (Figure 22.1). An alarm is correct if an extreme event occurs within the predicted time and space; otherwise the alarm is false. Failure to predict is the case where an extreme event occurs outside an alarm.

This approach is complementary to the classical Kolmogoroff-Wiener problem that is concerned with the prediction of a random time series $x(t)$ based on observations available with some time delay τ by the time $t-\tau$.

At the heart of our problem is the absence of a complete theory that would unambiguously define a prediction algorithm. Overcoming that difficulty did require an intense collaboration of experts in mathematics, statistics, exploratory data analysis, and also in the specific extreme events considered. Previous applications have inevitably involved teams of such experts, as can be seen from the list of references. For example, prediction of homicide surges was developed jointly with police officers

[1]*Institute of Geophysics & Planetary Physics and Department of Earth & Space Sciences, University of California, Los Angeles, CA, USA*

[2]*Institute of Earthquake Prediction Theory & Mathematical Geophysics, Russian Academy of Sciences, Moscow, Russia*

[3]*Departments of Mathematics & Earth and Atmospheric Sciences, Purdue University, West Lafayette, IN, USA*

Extreme Events: Observations, Modeling, and Economics, Geophysical Monograph 214, First Edition.
Edited by Mario Chavez, Michael Ghil, and Jaime Urrutia-Fucugauchi.

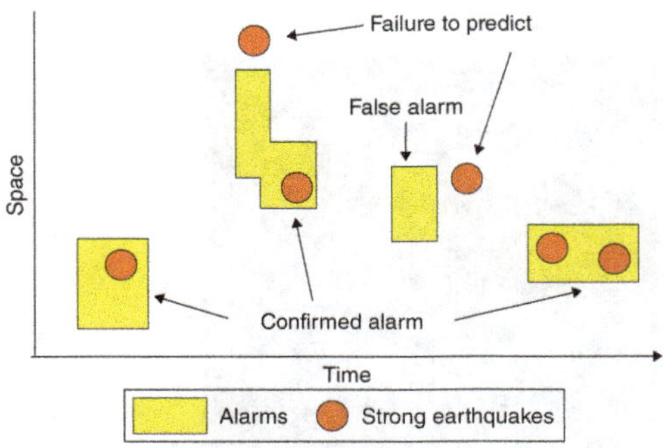

Figure 22.1 Possible outcomes of prediction.

in active service; prediction of unemployment: with experts in labor relations, etc.

Developing a prediction algorithm is naturally divided into the following interconnected stages:

1. *Choosing prediction targets.* These might be either given a priori (e.g., outcome of elections or the start of a recession as it is established by the National Bureau of Economic Research, NBER) or defined independently by data analysis (e.g., strong earthquakes or starting points of homicide surge).

2. *Choosing the background fields where we hope to detect precursors.* For example, prediction of strong earthquakes was based on seismicity patterns in lower magnitude range; prediction of recessions was based on six leading economic indicators. Any potentially relevant field can be considered.

3. *Formulation of a hypothetical prediction algorithm.* That is done by the *pattern recognition of rare events:* the methodology developed by the school of *I. Gelfand* for studying rare events of highly complex origin [*Bongard*, 1970; *Gelfand et al.*, 1976; *Keilis-Borok and Lichtman*, 1993; *Press and Allen*, 1995; *Keilis-Borok et al.*, 2000, 2003, 2005].

4. *Validation of prediction algorithm by prediction in advance.*

Prediction quality is characterized by three scores: rate of false alarms, rate of failures-to-predict, and total space-time occupied by alarms (as percentage of total space-time considered). These characteristics are important to decision makers choosing what, if any, preparedness measures to undertake in response to an alarm.

22.1.2. Predictability

Extreme events targeted by our predictions have a consequential common feature: they are generated by complex (chaotic) systems such as seismically active lithosphere, society, or economy. Complex systems are often regarded as unpredictable in principle. Actually, after a coarse-graining, on a not-too-detailed scale, such systems do exhibit certain regular behavior patterns. Among them are *premonitory patterns* that emerge more frequently as an extreme event approaches. Thus extreme events become predictable up to a limit.

Premonitory patterns might be either "perpetrators" contributing to the triggering of the extreme event, or "witnesses" merely signaling that the system has become unstable, ripe for such an event. An example of a witness is proverbial straws in the wind preceding a storm.

The need for coarse-graining is illustrated in Figure 22.2: a crack is visible only on a less detailed scale. "It is not possible to understand chaotic system by breaking it apart" [*Crutchfield et al.*, 1986].

Taking a holistic approach, "from the whole to details" circumvents the actual complexity and the chronic imperfection of the data. Moreover, it allows to take advantage of the considerable universality of precursors. Quoting *M. Gell-Mann* [1994], "*... if the parts of a complex system or the various aspects of a complex situation, all defined in advance, are studied carefully by experts on those parts or aspects, and the results of their work are pooled, an adequate description of the whole system or situation does not usually emerge. ... The reason, of course, is that these parts or aspects are typically entangled with one another. ... We have to supplement the partial studies with a transdisciplinary crude look at the whole.*"

The general scheme of prediction is illustrated in Figure 22.3. Bold vertical lines mark times of extreme events targeted for prediction. Fine vertical lines show a time series where premonitory patterns are looked for. It is robustly described by the functions $F_k(t)$, $k = 1, 2, \ldots$ usually defined on a sliding time window (t-s, t). Each function captures emergence of a certain pattern. An alarm is triggered when a certain combination of patterns emerge.

In the language of pattern recognition, the "object of recognition" is the time t. The problem is to recognize whether it belongs to the time interval Δ preceding a strong earthquake. That interval is often called the *TIP* (an acronym for the *time of increased probability* of a strong earthquake). Such prediction is aimed not at the whole dynamics of seismicity, but only at the rare extraordinary phenomena, strong earthquakes.

Development of a prediction algorithm by that approach starts with the learning stage where the "learning material," sample of past critical events and the time series hypothetically containing premonitory patterns, is analyzed. This analysis comprises the following steps:

Each time series considered is robustly described by the functions $F_k(t)$, $k = 1, 2, \ldots$, capturing hypothetical

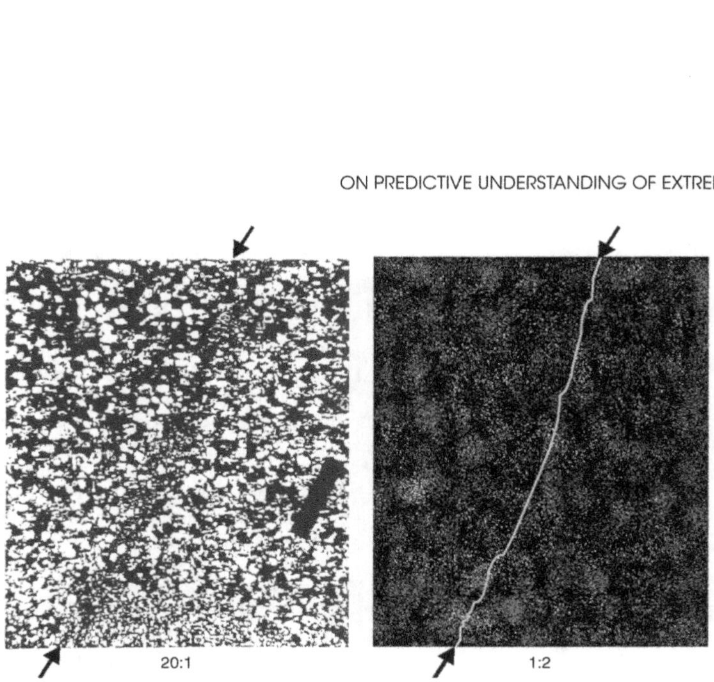

Figure 22.2 The need for coarse-graining (courtesy of *A. Johnson*). A sample with a crack is shown on different scales: more detailed (left) and less detailed (right). The crack is explicitly visible on a less detailed scale.

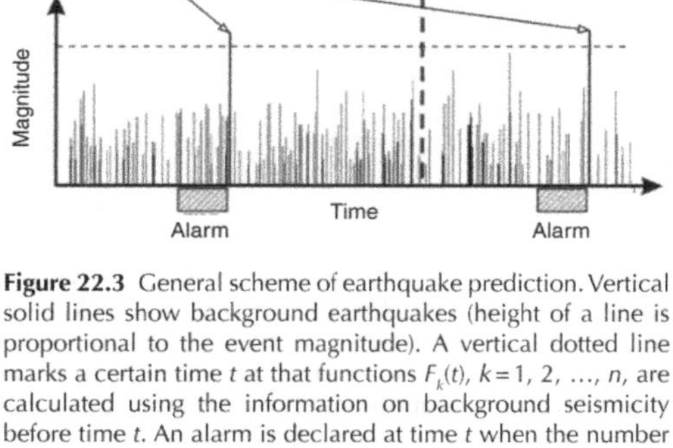

Figure 22.3 General scheme of earthquake prediction. Vertical solid lines show background earthquakes (height of a line is proportional to the event magnitude). A vertical dotted line marks a certain time t at that functions $F_k(t)$, $k = 1, 2, \ldots, n$, are calculated using the information on background seismicity before time t. An alarm is declared at time t when the number of capturing hypothetical patterns functions for that $F_k(t) \geq C_k$. is greater than a specified threshold.

patterns (Figure 22.3). These patterns are selected based on the universal models of complex systems, models of specific systems considered, exploratory data analysis, and practical experience, even if it is intuitive. Pattern recognition of rare events provides an efficient common framework for formulating and testing such hypotheses, their diversity notwithstanding.

Emergence of a premonitory pattern is defined by the condition $F_k(t) \geq C_k$. The threshold C_k is usually defined as a certain percentile of the functional $F_k(t)$. Thus the time series $F_k(t)$ is presented on the lowest-binary-level of resolution.

An alarm is triggered at a time t_a, when a certain combination of patterns occurs; this combination is determined by application of the pattern recognition procedure.

Detailed description of pattern recognition methodology can be found in *Bongard* [1970], *Gelfand et al.* [1976], *Keilis-Borok and Lichtman* [1993], and *Keilis-Borok and Soloviev* [2003].

Four paradigms on premonitory patterns [*Keilis-Borok*, 2002] provide a guidance for the choice of functions $F_k(t)$.

Paradigm 1: Basic types of premonitory patterns. These are illustrated in Figure 22.4. As an extreme event draws near, the background activity becomes more intense and clustered in space-time, while the correlation range in space increases and size distribution (scaling relation) shifts in favor of relatively stronger events.

Paradigm 2: Large size of precursor manifestation area. Generation of an extreme event is not localized in its vicinity. For example, according to *Press and Allen* [1995], the Parkfield, California earthquake with the characteristic source dimension 10 km "*...is not likely to occur until activity picks up in the Great Basin or the Gulf of California,*" about 800 km away. Numerous evidences for that paradigm are described also by *Mogi* [1968], *Aki* [1996], *Press and*

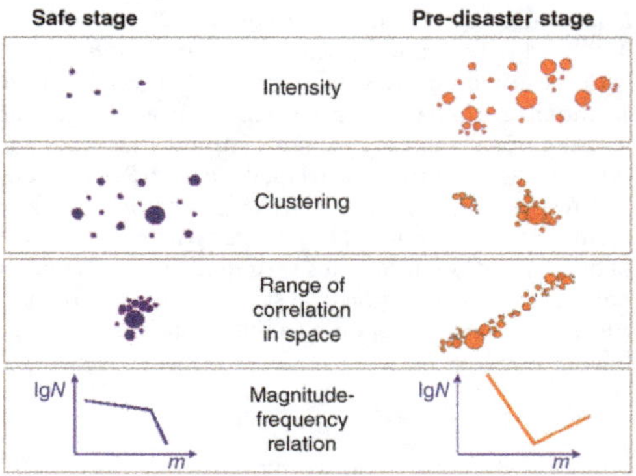

Figure 22.4 Four types of premonitory patterns.

Briggs [1975], *Keilis-Borok and Press* [1980], *Ma et al.* [1990], *Romanowicz* [1993], etc. In the case of seismicity these correlations may be explained by several mechanisms that range from microfluctuations of large-scale tectonic movements to impact of migrating fluids [*Barenblatt et al.*, 1983; *Barenblatt*, 1993; *Press and Allen*, 1995; *Sornette and Sammis*, 1995; *Aki*, 1996; *Bowman et al.*, 1998; *Pollitz et al.*, 1998; *Turcotte et al.*, 2000]. An example of the long-range correlation in a socioeconomic system is surge of ethnic violence in a French suburb, preceded by a rise in ethnic delinquency countrywide [*Bui Trong*, 2003].

Paradigm 3: Similarity. Quantitative definition of prediction algorithms is self-adjusting to regional conditions. For example, earthquake prediction algorithms developed for seismicity of California demonstrate the applicability to other active regions, with the magnitude of the target event varying from 3.0 to 6.6, for starquakes, and, at the other end of the spectrum, for fracturing in engineering constructions and laboratory samples [*Keilis-Borok et al.*, 1980; *Aki*, 1996; *Rotwain et al.*, 1997, 2006; *Keilis-Borok and Shebalin*, 1999; *Kossobokov et al.*, 2000; *Keilis-Borok and Soloviev*, 2003; *Kossobokov and Shebalin*, 2003]. The energy of a target event in these applications ranges from ergs (microfracture) to 10^{26} ergs (major earthquake), and even to 10^{41} ergs (starquake). Another example is an algorithm for predicting the surge of unemployment, applicable "as is" to the United States, France, Germany, and Italy [*Keilis-Borok et al.*, 2005].

Paradigm 4: Dual nature of premonitory patterns. The premonitory patterns shown in Figure 22.4 are "universal," common for hierarchical complex systems of different origin. They can be reproduced in the models of dynamic clustering [*Gabrielov et al.*, 2008], branching diffusion [*Gabrielov et al.*, 2010], percolation [*Zaliapin et al.*, 2005, 2006], direct, inverse, and colliding cascades [*Allègre et al.*, 1982; *Narkunskaya and Shnirman*, 1994; *Shnirman and Blanter*, 1999, 2003; *Gabrielov et al.*, 2000; *Zaliapin et al.*, 2003; *Yakovlev et al.*, 2005], as well as in certain system-specific models [*Soloviev and Ismail-Zadeh*, 2003; *Sornette*, 2004].

22.1.3. Coping with Risk of Data Fitting

Since our prediction algorithms cannot be defined unambiguously by an existing theory, they inevitably include some adjustable elements, from selecting the data used for prediction to the values of numerical parameters. These elements are adjusted retrospectively by "predicting" past extreme events.

Such data-fitting might be self-deceptive: As *J. von Neumann* put it "*with four exponents I can fit an elephant.*" Hence, the following tests are performed.

Sensitivity analysis: predictions should not be too sensitive to variations in adjustable parameters.

Out-of-sample analysis: application of an algorithm to past data that has not been used in the algorithm's development. The test is considered successful if the accuracy of prediction does not drop too much.

Prediction in advance is the ultimate test.

22.2. PREDICTING INDIVIDUAL EXTREME EVENTS

22.2.1. Earthquakes

Relatively better tested are the algorithms based on premonitory seismicity patterns [*Keilis-Borok*, 1990, 2002; *Keilis-Borok and Shebalin*, 1999 *Keilis-Borok and Soloviev*, 2003; *Kossobokov and Shebalin*, 2003; *Peresan et al.*, 2005]. The predictions are filed in advance on the following websites: http://www.mitp.ru/predictions.html; http://users.ictp.it/www_users/sand/index_files/DevelopmentofPrediction.html; and http://rtptest.org/.

Access to yet unexpired alarms on these websites is limited to about 200 scientists and professional experts worldwide. This is done in compliance with the UNESCO guidelines, since public release of prediction might trigger disruptive anxiety of population and profiteering. Predictions are made available for general public after a strong earthquake occurs or alarm expires, whichever comes first.

22.2.1.1. M8 and MSc Algorithms [Kossobokov and Shebalin, 2003]

Algorithm M8 provides intermediate-term predictions. Characteristic duration of an alarm is about 5 years. That algorithm was first developed for predicting the largest earthquakes ($M \geq 8$) worldwide. The algorithm MSc ("Mendocino Scenario") provides a second approximation to M8, considerably reducing the alarm area. Figure 22.5 shows an example of prediction by both algorithms.

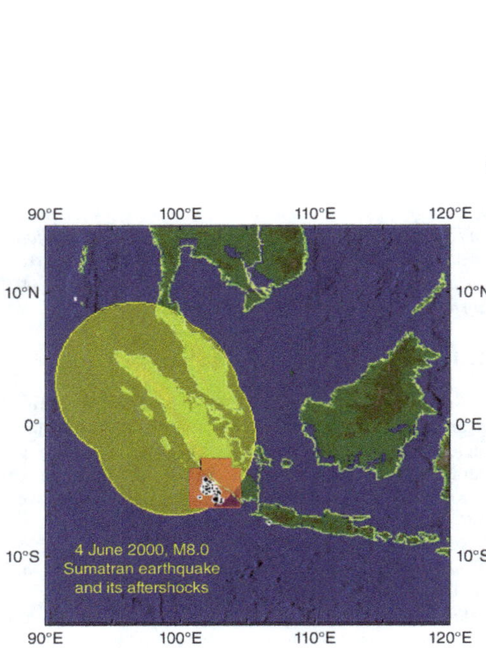

Figure 22.5 Alarms capturing the Sumatra earthquake, 4 June 2000. The yellow oval curve bounds the area of alarm by M8, put on record in July 1996, to expire on 1 July 2001. Red rectangle is its reducing made by MSc, put on record in January 1998. White circles indicate epicenters of Sumatra earthquake and its first-month aftershocks.

Table 22.1 Scoring of M8 and M8 and MSc Pedictions, 1992–2010

Algorithm	Total Number of Target Earthquakes	Number of Predicted Earthquakes	Space-Time Volume of Alarm (%)
M8	17	12	29
M8 and MSc	17	8	15

22.2.1.1.1. Scoring

Thus far, the algorithms have had most success in predicting future earthquakes in the magnitude range 8–8.5 (Table 22.1). Statistical significance of predictions exceeds 99%.

22.2.1.2. "Second Strong Earthquake" (SSE) Algorithm [Levshina and Vorobieva, 1992; Vorobieva, 1999, 2009]

That algorithm is applied when a strong earthquake of a certain magnitude M has occurred. The algorithm predicts whether a second strong earthquake with magnitude $(M-1)$ or more will occur within 18 months of the first one, within a distance $R=0.03\times10^{0.5M}$ km that is one and a half of the characteristic linear size of the aftershock area for an earthquake with magnitude M [*Tsuboi*, 1956]. Figure 22.6 shows an example of SSE prediction made after the Landers earthquake, 28 June 1992, $M=7.6$. Prediction was released in EOS, October 1992, with alarm ending on 28 December 1993. A second strong earthquake $M\geq6.6$ was predicted to occur within the yellow circle. On 17 January 1994, 20 days after the alarm expired, the Northridge earthquake ($M=6.8$) did occur and resulted in 57 deaths, more than 5000 injuries, and more than 20 billion USD in property damage. For the sake of rigorous scoring, this earthquake was counted as not predicted. On the practical side, escalation of preparedness measures in response to this prediction would be fully justified.

22.2.1.2.1. Scoring

Since 1989 the SSE algorithm is being tested in 10 regions worldwide [*Vorobieva*, 1999, 2009], with the magnitude of the first events ranging from 5 or above in the Dead Sea Rift region, to 7 or above in the Balkans. In total, 31 first events have been considered. Eight of them were followed by a second strong earthquake within 18 months; 6 have been predicted and 2 missed. For the remaining 23 single events, 19 correct predictions and 4 false alarms have been made. Statistical significance of these predictions is above 99%.

22.2.1.3. "Reverse Tracing of Precursors" (RTP) Algorithm [Shebalin et al., 2004, 2006; Keilis-Borok et al., 2004a]

The algorithm is aimed at predictions about 9 months in advance, much shorter than by the M8 and SSE algorithms. This algorithm, as its name suggests, traces precursors in the reverse order of their formation. First it identifies "candidates" for short-term precursors. These are long, quickly formed chains of earthquakes in the background seismicity. Such chains reflect an increase in the earthquake correlation range (Figure 22.7). Next, each chain is examined to determine whether there had been any preceding intermediate-term precursors in its vicinity within the previous five years. If so, the chain triggers an alarm.

22.2.1.3.1. Scoring

Since 2003, the RTP algorithm has been tested by predicting future earthquakes in five regions of the world (California and adjacent regions; Central and Northern Italy with adjacent regions; Eastern Mediterranean; Northern Pacific, Japan and adjacent regions). Till 2012, five out of seven target earthquakes have been predicted (captured by alarms) and two missed. Out of 19 alarms, 5 were correct and 14 false, 2 of the latter being near misses occurring close to alarm areas. The data are still insufficient for rigorous estimation of statistical significance.

Figure 22.6 Prediction of Northridge, California earthquake by SSE algorithm.

22.2.2. Presidential Elections in the United States

Tradition regards American elections as a trial by battle, where the goal of the competitors is to attract a maximum number of voting blocks with minimal alienation of other blocks. The outcome depends strongly on manipulation of public opinion and last-minute sensations. Accordingly, that tradition ascribes the victory of *George H. W. Bush* in 1988 to three brilliant campaigners: speechwriter Peggy Noonan, who improved his image overnight by the New Orleans speech; two hardball politicians staged mass-media exposure of failure of *M. Dukakis* as a Governor of Massachusetts. Furthermore *M. Dukakis* fired a good campaigner. As a result, he lost the elections, which were in his grasp (he led by 17% in opinion polls).

This notion tells us that a huge mass of voters can reverse its opinion through the influence of three campaigners and the loss of one, in a "for-want-of-a-nail-the-war-was-lost" fashion. In other words, it portrays Jane/Joe Voter as an excitable simpleton, manipulated by commercials, reversing the vote for transient reasons, irrelevant to the essence of the electoral dilemma. American elections deserve a more dignified explanation.

An alternative view, contrasting with that described above, was developed by holistic approach, understanding electorate as a hierarchical complex system [*Keilis-Borok and Lichtman*, 1981]. Here, election outcome depends on coarse-grained socioeconomic and political factors of the common-sense type. These factors were given a robust definition by means of the questionnaire (with *yes/no* responses) shown in Table 22.2. The data on the past 31 elections, 1860–1980 were considered. The actual electoral dilemma is found to be whether an incumbent party will win or lose rather than whether the

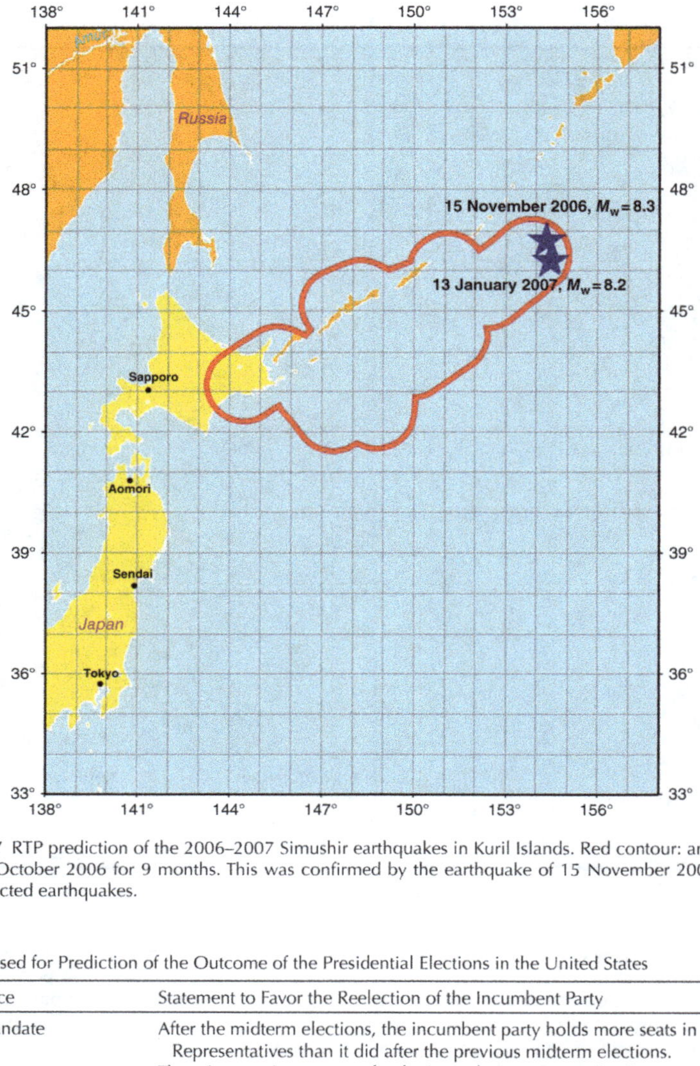

Figure 22.7 RTP prediction of the 2006–2007 Simushir earthquakes in Kuril Islands. Red contour: area of alarm issued on October 2006 for 9 months. This was confirmed by the earthquake of 15 November 2006, $M=8.3$. Stars: predicted earthquakes.

Table 22.2 Keys Used for Prediction of the Outcome of the Presidential Elections in the United States

Key #	Relevance	Statement to Favor the Reelection of the Incumbent Party
1	Party mandate	After the midterm elections, the incumbent party holds more seats in the U.S. House of Representatives than it did after the previous midterm elections.
2	Contest	There is no serious contest for the incumbent-party nomination.
3	Incumbency	The incumbent-party candidate is the sitting president.
4	Third party	There is no significant third party or independent campaign.
5	Short-term economy	The economy is not in recession during the election campaign.
6	Long-term economy	Real per-capita economic growth during the term equals or exceeds mean growth during the previous two terms.
7	Policy change	The incumbent administration effects major changes in national policy.
8	Social unrest	There is no sustained social unrest during the term.
9	Scandal	The administration is untainted by major scandal.
10	Foreign/military failure	The administration suffers no major failure in foreign or military affairs.
11	Foreign/military success	The administration achieves a major success in foreign or military affairs.
12	Incumbent charisma	The incumbent-party candidate is charismatic or a national hero.
13	Challenger charisma	The challenging-party candidate is not charismatic or a national hero.

Republicans or Democrats will win. A victory of a challenger party is considered as an extreme event. *The prediction algorithm* [*Lichtman and Keilis-Borok*, 1989; *Keilis-Borok and Lichtman*, 1993; *Lichtman*, 1996, 2000, 2005] was developed by pattern recognition analysis of the data. The extreme event (victory of the challenger) is predicted if the number of answers *no*, on the questions listed in Table 22.2, is 6 or more.

Results of retrospective analysis and advance predictions made by *A. Lichtman* for eight subsequent elections, 1984–2012, are shown in Table 22.3. All advance predictions happened to be correct. That includes *Al Gore's* victory by popular vote, which was reversed by electorate; this happened three times through the whole history. Timing and source of the advance predictions are listed in Table 22.4.

22.2.2.1. What have we Understood about the Elections?

Uniformity of prediction rules transcends the diversities of situations prevailing in individual elections. Accordingly, the same pattern of the choice of president has prevailed since 1860, that is, since election of *A. Lincoln*, throughout all the overwhelming changes of these 140 years. And in particular, note that the electorate of 1860 did not include the groups making up 75% of the present electorate: no women, blacks, or most of the citizens of Latin American, South European, East European, and Jewish descent (*Lichtman*, 2000). In a nutshell, we have found that the outcome of a presidential election is determined by collective estimation of performance of incumbent administration over the previous 4 years.

22.2.3. Surges in Unemployment

Here we describe uniform prediction of the sharp and long-term rise in unemployment in France, Germany, Italy, and the United States [*Keilis-Borok et al.*, 2005]—we term this FAU, which is an acronym for "Fast Acceleration of Unemployment." The data comprise macroeconomic indicators of national economy. In stability tests a variety of other indicators were also analyzed. Figure 22.8 shows retrospective alarms. Exactly the same self-adjusting algorithm was applied to all four countries.

Table 22.3 Prediction of US Presidential Election Years when the Incumbent Won Are Shown in Italic; Years when the Challenger Won, in Bold

Number of answers *yes* on questions listed in Table 22.2									
0	1	2	3	4	5	6	7	8	9
Predictions (published months in advance)									
					2000[a]				
		1984	*1988*	*2004*	*1996*	**1992**	**2012**	**2008**	
Retrospective analysis									
			1964					**1980**	
			1928					**1976**	
			1916					**1968**	
			1908					**1952**	
		1944	*1900*	*1972*				**1932**	
	1956	*1940*	*1872*	*1924*	*1948*	**1912**	**1884**	**1920**	**1960**
1904	*1936*	*1868*	*1864*	*1880*	*1888*[a]	**1892**	**1860**	**1896**	**1876**[a]

[a]Years when popular vote was reversed by electoral vote.

Table 22.4 Timing and Source of Predictions

Election	Date of Prediction	Source
1984	April 1982	"How to Bet in '84," *Washingtonian Magazine*, April 1982
1988	May 1988	"How to Bet in November," *Washingtonian Magazine*, May 1988
1992	September 1992	"The Keys to the White House," *Montgomery Journal*, 14 September 1992
1996	October 1996	"Who Will Be the Next President?" *Social Education*, October 1996
2000	November 1999	"The Keys to Election 2000," *Social Education*, November/December 1999.
2004	April 2003	"The Keys to the White House," *Montgomery Gazette*, April 25, 2003
2008	February 2006	"Forecast for 2008," *Foresight*, February 2006,
2012	March 2010	"Allan Lichtman's prediction: Obama wins re-election in 2012," *Gazette.net*, Friday, 26 March 2010.

Prediction-in-advance has been in progress since 1999, so far only for the United States. Two recent episodes of FAU have been predicted correctly, without failures to predict or false alarms (Figure 22.9).

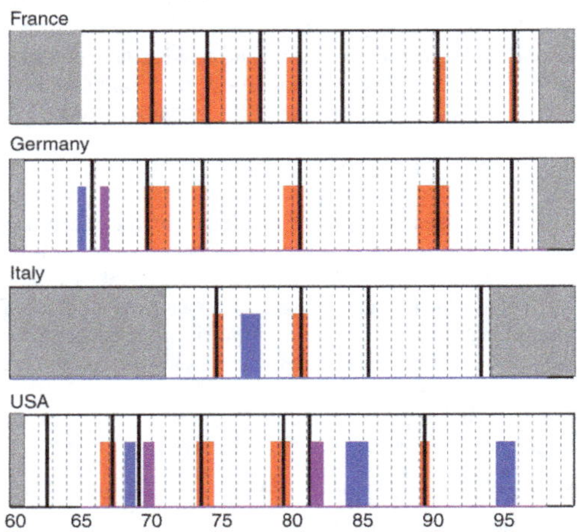

Figure 22.8 FAUs and alarms in the four countries. Vertical lines: prediction targets (FAUs). Red bars: correct alarms. Purple bars: alarms triggered within the periods of unemployment surge. Blue bars: false alarms. The 1968 alarm is scored as false since it expired 1 month before FAU.

22.2.4. US Economic Recessions

Prediction targets are the peaks and troughs of economic activity, that is, the first and last months of each recession as identified by the NBER. The data used in the prediction algorithm comprise six monthly leading economic indicators, reflecting interest rates, industrial production, inventories, and job market [*Keilis-Borok et al.*, 2000, 2008].

Retrospective alarms and recessions are shown together in Figure 22.10. We see that five recessions occurring between 1961 and 2000 were preceded by alarms. The sixth recession started in April 2001, one month before the corresponding alarm. In practice, this is not a failure-to-predict, since recessions are usually declared by NBER much later than they begin. The duration of each alarm was 1–14 months. Total duration of all alarms was 38 months, or 13.6% of the time interval considered. There was only one false alarm in 2003.

According to the NBER announcement issued in December 2008, the last recession in the United States began in January 2008. Our algorithm gave an alarm started in May 2008, that is, 4 months after the recession start but 7 months before the NBER announcement.

The same six macroeconomic indicators have been used to develop an algorithm for prediction of the recovery from a recession [*Keilis-Borok et al.*, 2008]. The algorithm declares alarms within 6 months before the end of each American recession since 1960 and at no other time during these recessions (Figure 22.10). This study is a natural

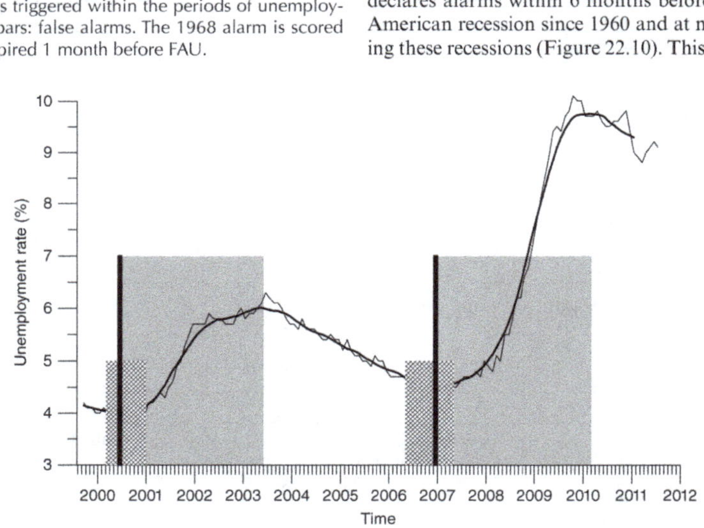

Figure 22.9 Advance predictions of *FAUs*. A thin curve shows the monthly unemployment rate in the United States, according to the data of the Bureau of Labor Statistics, U.S. Department of Labor (http://data.bls.gov). The thick curve shows this rate with seasonal variation smoothed out. Vertical lines: prediction targets (*FAUs*). Gray bars: periods of unemployment growth. Checkered bars: periods of alarms.

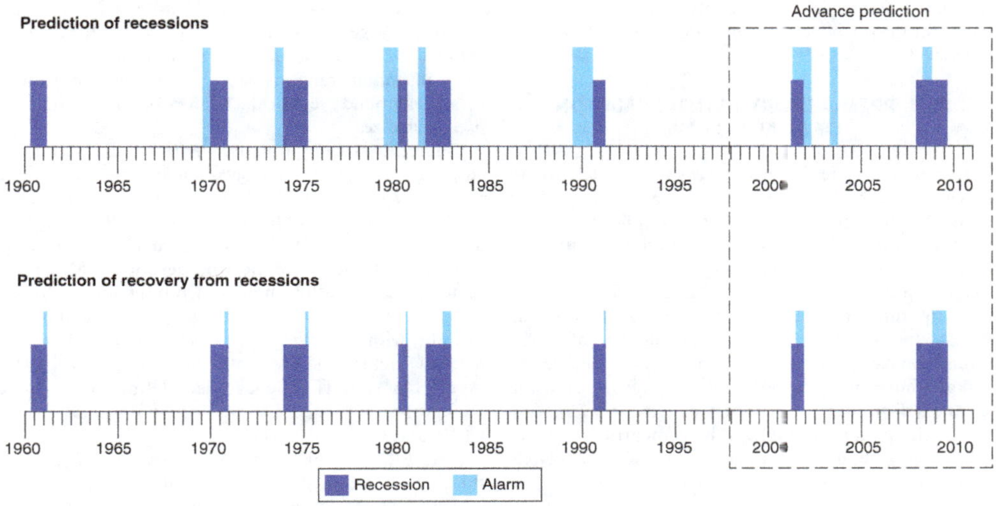

Figure 22.10 Prediction of economic recessions in the United States.

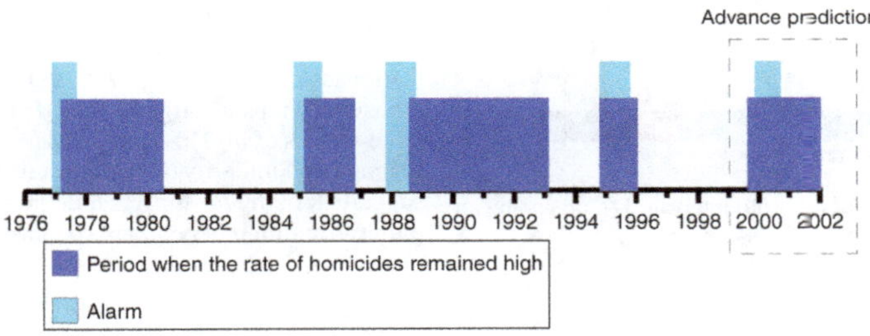

Figure 22.11 Prediction of homicide surges in Los Angeles.

continuation of the previous one, aimed at predicting the start of a recession. Comparing these cases we find that precursory trends of financial indicators are opposite during transition to a recession and recovery from it. To the contrary, precursory trends of economic indicators happen to have the same direction (upward or downward) but are steeper during recovery. The algorithm declares an alarm starting in November 2008 for the end of the last recession.

22.2.5. Homicide Surge

Prediction target is the start of a sharp and lasting rise ("a surge") of the homicide rate in an American megacity, Los Angeles.

Data comprise monthly rates of 11 types of low-level crimes: burglaries, assaults, and robberies. Statistics of these types of crime in Los Angeles over the period 1975–2002 has been analyzed to find an algorithm for predicting such a surge of the homicide rate [*Keilis-Borok et al.*, 2003].

Premonitory patterns include: first— an escalation of burglaries and assaults, but not of robberies; closer to a homicide surge, robberies also escalate.

It has been found in retrospective analysis that this algorithm is applicable through all the years considered despite substantial changes, both in socioeconomic conditions and in the procedure for the counting of crimes. Alarms and homicide surges are plotted together in Figure 22.11. In total, alarms occupy 15% of the time considered. Moreover, the algorithm gives satisfactory results for the prediction of homicide surges in New York city as well.

22.3. PREMONITORY TRANSFORMATION OF SCALING LAW

Here, we compare the scaling laws in the background activity of a complex system in the time periods of three kinds: **D**, preceding an extreme event; **X**, following it; and **N**, others (Figure 22.12). The background activity of a system is described by background events with sizes (magnitudes) that are defined for different complex systems by different ways. Scaling law is defined as $P(m) = N(m)/N^T$, where $N(m)$ is the number of background events of the size $\geq m$, N^T is the total number of background events (obviously, $P(m)$ is equivalent to the statistical distribution function). Changing in scaling law before strong events is analyzed by comparison of functions $P(m)$ obtained separately for D- and N-periods (functions $P^D(m)$ and $P^N(m)$, respectively).

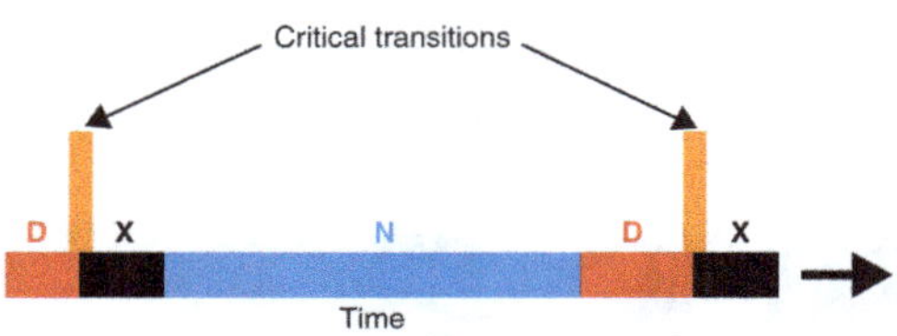

Figure 22.12 Division of time into intervals: **D**—preceding an extreme event; **X**—following it; **N**—others.

Figure 22.13 shows functions $P^D(m)$ and $P^N(m)$ in the case of the largest (with magnitude $M \geq 8.0$) of the world as extreme events and worldwide seismicity main shocks in lower magnitude range as background events. Magnitude of a background event or the number of its aftershocks is used as its size.

Figure 22.14 shows functions $P^D(m)$ and $P^N(m)$ in the case of the start of socioeconomic crises (economic recessions in the United States, unemployment surges in the United States, and homicide surges in Los Angeles) as extreme events. Background events are defined as changes in trends of indicators describing a relevant system: U.S. monthly industrial production for recessions and unemployment and the monthly number of assaults with firearms in Los Angeles for homicides. Sizes of background events are proportional to changes in relevant trends. The detailed definition of these background events is given by *Keilis-Borok and Soloviev* [2010] and *Ghil et al.* [2011].

Figure 22.15 shows functions $P^D(m)$ and $P^N(m)$ for three countries, in the case of months with extremely large score of casualties (killed and wounded) caused by terrorism as extreme events. Here **D**-period before only one extreme event (indicated in Figure 22.3) is considered for each country and background events are months when there are casualties caused by terrorism with the numbers of the casualties as event sizes.

In each case function $P^D(m)$ has distinctly higher ("heavier") tails at large m than $P^D(m)$, and extends to larger values of m. This demonstrates predictive power of scaling law. Similarity of this effect in so different systems suggests looking for universal definition of premonitory patterns in different complex systems.

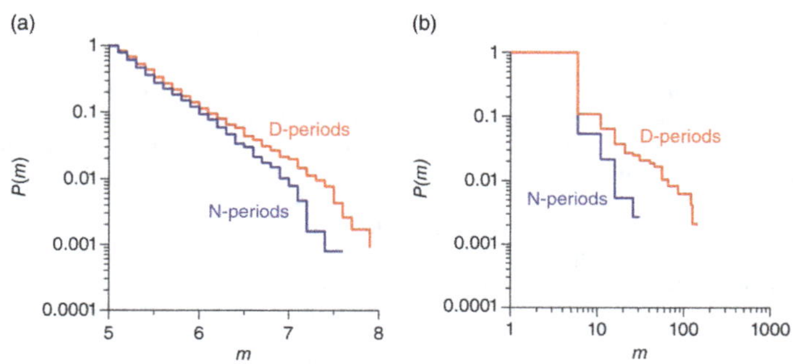

Figure 22.13 Premonitory transformation of a scaling law: earthquakes. Extreme events: main shocks with magnitude $M \geq 8$ worldwide, 1985–2002. Red curves correspond to D-periods and blue curves correspond to N-periods. Background activity: seismicity in lower magnitude range; m is magnitude of individual main shocks (a) or number of aftershocks (b). Courtesy of *L. Romashkova*.

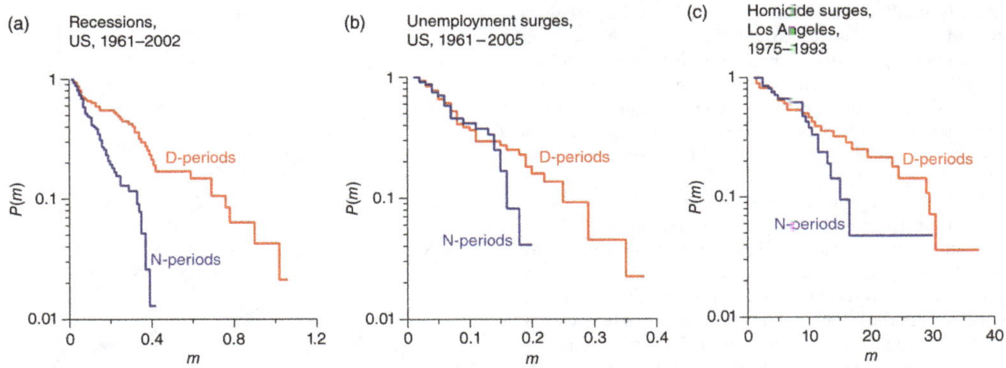

Figure 22.14 Premonitory transformation of scaling law: socioeconomic crises. Extreme events: starting points of respective crises. Red curves correspond to D-periods and blue curves correspond to N-periods. Background activity, change in trend of a monthly indicator: (a and b) denote industrial production and (c) denotes assaults with firearms.

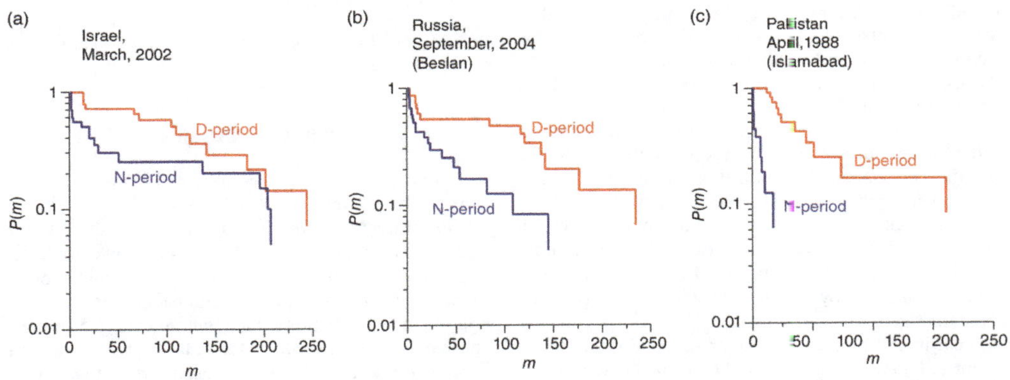

Figure 22.15 Premonitory transformation of scaling law: terrorism. Extreme events are the months with extremely large score of casualties (killed and wounded). Background activity is represented by the monthly number m of casualties.

22.4. PREDICTION AND DISASTER PREPAREDNESS

> Of course, things are complicated… But in the end every situation can be reduced to a simple question: Do we act or not? If yes, in what way?
>
> [*E. Burdick*, "480"]

What, if any, preparedness actions to undertake in response to a prediction, given its inherently limited accuracy? Methodology assisting decision maker in choosing optimal response to earthquake prediction is developed in *Kantorovich et al.* [1974], *Molchan* [1991, 1997, 2003], *Keilis-Borok et al.* [2004b], *Davis et al.* [2007, 2010], and *Molchan and Keilis-Borok* [2008].

Earthquakes might hurt population, economy, and environment in many different ways, from destruction of buildings, lifelines, and other constructions, to triggering other natural disasters, economic, and political crises. That diversity of damage requires a hierarchy of preparedness measures, from public safety legislation and insurance to simulation alarms, preparedness at home, and red alert. Different measures can be implemented on different timescales, from seconds to decades. They should be implemented in areas of different size, from selected sites to large regions; should be maintained for different time periods; and should belong to different levels of jurisdiction, from local to international. Such measures might complement, supersede, or mutually exclude each other. For this reason, optimizing preparedness involves comparison of a large number of combinations of possible measures [*Davis et al.*, 2007, 2010].

Disaster management has to take into account the cost/benefit ratio of possible preparedness measures. No single measure alone is sufficient. However, many efficient measures are inexpensive and do not require high accuracy of prediction. As is the case for all forms of disaster preparedness, including national defense, a prediction can be useful if its accuracy is known, even if it is not high.

Decision depends on specific circumstances in the area of alarm. At the same time, it depends on the prediction quality that can be described by two characteristics: n, rate of failures-to-predict (in percents) and τ, percent of total space-time considered occupied by alarms. Their values are determined as follows. Consider a prediction algorithm applied during the time period T. A certain number of alarms A are declared. N extreme events did occur, and N_f of them have been missed by alarms. Altogether, the alarms cover the time D. Then $\tau = (D \cdot 100\%)/T$ and $n = (N_f \cdot 100\%)/N$.

The choice of preparedness measures is not unique. Different measures may supersede or mutually exclude one another, leaving certain freedom of choice to a decision maker [*Keilis-Borok et al.*, 2004b; *Davis et al.*, 2007, 2010]. Designer of a prediction algorithm has certain freedom to choose the tradeoff between different characteristics of its accuracy (duration of alarms, and rate of failures to predict) by varying adjustable elements of the algorithm.

Accordingly, prediction and preparedness should be optimized jointly; *there is no "best" prediction per se* [*Molchan*, 1997, 2003; *Molchan and Keilis-Borok*, 2008]. A framework for such optimization is shown in Figure 22.16. Earthquake prediction strategies (composite strategies are meant) are characterized by points at (τ, n)—diagram and an envelope of these points (curve Γ) is constructed. If there is a loss function depending on τ and n, then the contours ("loss curves") with the constant value γ of this function can be constructed. Optimal strategy is the tangent point of contours Γ and γ [*Molchan*, 1997]. Thus disaster preparedness would be more flexible and efficient if prediction would be carried out in parallel with several versions of an algorithm. This has not yet been done.

Other software and know-how for assisting decision makers in choosing the best combination is described by *Kantorovich et al.* [1974], *Molchan* [1991, 2003], *Keilis-Borok* [2003], and [*Davis et al.*, 2010].

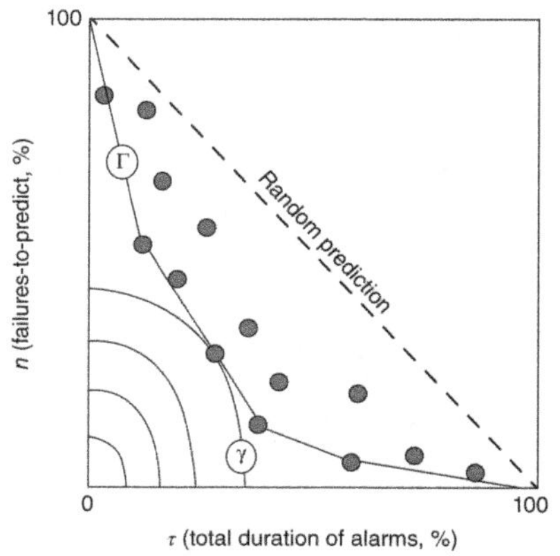

Figure 22.16 Joint optimization of prediction and preparedness based on theory of optimal control. Dots show points corresponding to results of application of different earthquake prediction strategies. Γ is their envelope. Thin contours (γ) show "loss curves" with constant value of a prevented "loss." Optimal strategy is the tangent point of contours Γ and γ. After *Molchan* [2003].

22.5. CONCLUSION

Major results of the studies described in the earlier sections could be summarized as follows:

1. *Prediction algorithms* are developed for the extreme events of each type considered. As required in complexity studies these algorithms are robust and self-adjusting to the scale of the system, level of its background activity, magnitude of prediction targets, etc. Accuracy of prediction is defined by the rate of false alarms, rate of failures to predict, and total time-space occupied by the alarms. The algorithms allow to choose the trade-off between these characteristics.
2. *New understanding of the origin of the extreme events* considered is also developed.
3. *Linking prediction with disaster preparedness.*

These results have large potential for further development. A wealth of available and highly relevant data, models, and practical experience for disaster prediction and preparedness remains as yet untapped. Further, less immediate results within reach [*Keilis-Borok and Soloviev*, 2006; *Keilis-Borok*, 2007] include, for example, considerable increase in prediction accuracy; prediction of other kinds of disasters and crises; and with luck, new tools for disaster control.

In the general scheme of things, the problems considered here belong to a much wider field of predictive understanding and control of crises and disasters: a key for the survival and sustainability of our civilization.

REFERENCES

Aki, K. (1996), Scale dependence in earthquake phenomena and its relevance to earthquake prediction, *Proc. Nat. Acad. Sci.*, *93*, 3740–3747.

Allègre, C. J., J.-L. Le Mouël, and V. Provost (1982), Scaling rules in rock fracture and possible implications for earthquake prediction, *Nature*, *297*, 47–49.

Barenblatt, G. I., V. I. Keilis-Borok, and A. S. Monin (1983), Filtration model of earthquake sequence, *Trans. (Doklady) Acad. Sci. SSSR*, *269*, 831–834.

Barenblatt, G. I. (1993), Micromechanics of fracture, in *Theoretical and Applied Mechanics*, edited by E. R. Bodner, J. Singer, A. Solan, and Z. Hashin, Elsevier, Amsterdam, pp. 25–52.

Bongard, M. M. (1970), *Pattern Recognition*, Spartan Books, New York.

Bowman, D. D., G. Ouillon, G. G. Sammis, A. Sornette, and D. Sornette (1998), An observational test of the critical earthquake concept, *J. Geophys. Res.*, *103*, 24359–24372.

Bui Trong, L. (2003), Risk of collective youth violence in French suburbs, in *Risk Science and Sustainability*, NATO Science Series. II. Mathematics, Physics and Chemistry, vol. *112*, edited by T. Beer and A. Ismail-Zadeh, pp. 199–221, Kluwer Academic Publishers, Boston, MA.

Crutchfield, J. P., J. D. Farmer, N. H. Packard, and N. H. Shaw (1986), Chaos, *Sci. Am.*, *255*, 46–57.

Davis, C., K. Goss, V. Keilis-Borok, G. Molchan, P. Lahr, and C. Plumb (2007), Earthquake prediction and tsunami preparedness, *Workshop on the Physics of Tsunami, Hazard Assessment Methods and Disaster Risk Management*, 14–18 May 2007, Trieste: ICTP.

Davis, C. A., V. Keilis-Borok, G. Molchan, P. Shebalin, P. Lahr, and C. Plumb (2010), Earthquake prediction and disaster preparedness: Interactive analysis, *Nat. Hazard. Rev.*, *11*(4), 173–184.

Gabrielov, A. M., I. V. Zaliapin, W. I. Newman, and V. I. Keilis-Borok (2000), Colliding cascade model for earthquake prediction, *Geophys. J. Int.*, *143*(2), 427–437.

Gabrielov, A., V. Keilis-Borok, Y. Sinai, and I. Zaliapin (2008), Statistical properties of the cluster dynamics of the systems of statistical mechanics, in *Boltzmann's Legacy*, ESI Lectures in Mathematics and Physics, edited by G. Gallavotti, W. L. Reiter, and J. Yngvason, pp. 203–215, European Mathematical Society, Zurich.

Gabrielov, A., V. Keilis-Borok, S. Olsen, and I. Zaliapin (2010): Predictability of extreme events in a branching diffusion model. arXiv:1003.0017, January 2010.

Gelfand, I. M., S. A. Guberman, V. I. Keilis-Borok, L. Knopoff, F. Press, E. Y. Ranzman, I. M. Rotwain, and A. M. Sadovsky (1976), Pattern recognition applied to earthquake epicenters in California, *Phys. Earth Planet. Inter.*, *11*, 227–283.

Gell-Mann, M. (1994), *The Quark and the Jaguar: Adventures in the Simple and the Complex*, Freeman and Company, New York.

Ghil, M., P. Yiou, S. Hallegatte, B. D. Malamud, P. Naveau, A. Soloviev, P. Friederichs, V. Keilis-Borok, D. Kondrashov, V. Kossobokov, O. Mestre, C. Nicolis, H. W. Rust, P. Shebalin, M. Vrac, A. Will, and I. Zaliapin (2011), Extreme events: dynamics, statistics and prediction, *Nonlinear Process. Geophys.*, *18*, 295–350.

Kantorovich, L. V., V. I. Keilis-Borok, and G. M. Molchan (1974), Seismic risk and principles of seismic zoning, In *Seismic Design Decision Analysis*, Internal Study Report 43, Department of Civil Engineering, MIT, Cambridge (Mass).

Keilis-Borok, V. I., L. Knopoff, and I. M. Rotwain (1980), Bursts of aftershocks, long-term precursors of strong earthquakes, *Nature*, *283*, 258–263.

Keilis-Borok, V. I. and F. Press (1980), On seismological applications of pattern recognition, in *Source Mechanism and Earthquake Prediction Applications*, edited by C. J. Allegre, pp. 51–60, Editions du Centre National de la Recherché Scientifique, Paris.

Keilis-Borok, V. I. and A. Lichtman (1981), Pattern recognition applied to presidential elections in the United States 1860-1980: Role of integral social, economic and political traits, *Proc. Nat. Acad. Sci.*, *78*(11), 7230–7234.

Keilis-Borok, V. I. (1990), The lithosphere of the Earth as a nonlinear system with implications for earthquake prediction, *Rev. Geophys.*, *28*, 19–34.

Keilis-Borok, V. I. and A. Lichtman (1993), The self-organization of American society in presidential and senatorial elections, in *Limits of Predictability*, edited by Yu. A. Kravtsov, pp. 223–237, Springer-Verlag, Berlin-Heidelberg.

Keilis-Borok, V. I. and P. N. Shebalin (Eds.) (1999), Dynamics of lithosphere and earthquake prediction, *Phys. Earth Planet. Inter.*, *111*(3–4special issue).

Keilis-Borok, V., J. H. Stock, A. Soloviev, and P. Mikhalev (2000), Pre-recession pattern of six economic indicators in the USA, *J Forecast.*, *19*, 65–80.

Keilis-Borok, V. I. (2002), Earthquake prediction: State-of-the-art and emerging possibilities, *Annu. Rev. Earth Planet. Sci.*, *30*, 1–33.

Keilis-Borok, V. I. and A. A. Soloviev (Eds.) (2003), *Nonlinear Dynamics of the Lithosphere and Earthquake Prediction*, pp. 337, Springer-Verlag, Berlin-Heidelberg.

Keilis-Borok, V. I. (2003), Basic science for prediction and reduction of geological disasters, in *Risk Science and Sustainability*, NATO Science Series. II. Mathematics, Physics and Chemistry, vol. *112*, edited by T. Beer and A. Ismail-Zadeh, pp. 29–38, Kluwer Academic Publishers, Boston, MA.

Keilis-Borok, V. I., D. J. Gascon, A. A. Soloviev, M. D. Intriligator, R. Pichardo, and F. E. Winberg (2003), On predictability of homicide surges in megacities, in *Risk Science and Sustainability*, NATO Science Series. II. Mathematics, Physics and Chemistry, vol. *112*, edited by T. Beer and A. Ismail-Zadeh, pp. 91–110, Kluwer Academic Publishers, Boston, MA.

Keilis-Borok, V., P. Shebalin, A. Gabrielov, and D. Turcotte (2004a), Reverse tracing of short-term earthquake precursors, *Phys. Earth Planet. Inter.*, *145*(1-4), 75–85.

Keilis-Borok, V., C. Davis, G. Molchan, P. Shebalin, P. Lahr, and C. Plumb (2004b), Earthquake prediction and disaster preparedness: Interactive algorithms, *Eos. Trans. AGU*, *85*(47Fall Meeting Suppl., Abstract S22B-02).

Keilis-Borok, V. I., A. A. Soloviev, C. B. Allègre, A. N. Sobolevskii, and M. D. Intriligator (2005), Patterns of macroeconomic indicators preceding the unemployment rise in Western Europe and the USA, *Pattern Recogn.*, *38*(3), 423–435.

Keilis-Borok, V. and A. Soloviev (2006) Earthquakes prediction: "The paradox of want amidst plenty", *26th IUGG Conference on Mathematical Geophysics*, 4–8 June 2006, Sea of Galilee, Israel. Book of Abstracts, p 28.

Keilis-Borok, V.I. (2007), Earthquake prediction: paradigms and opening possibilities, *Geophysical Research Abstracts*, Volume 9, 2007. Abstracts of the Contributions of the EGU General Assembly 2007, Vienna, Austria, 15–20 April 2007 (CD-ROM): EGU2007-A-06766.

Keilis-Borok, V. I., A. A. Soloviev, M. D. Intriligator, and F. E. Winberg (2008), Pattern of macroeconomic indicators preceding the end of an American economic recession, *J. Pattern Recogn. Res.*, *3*(1), 40–53.

Keilis-Borok, V. I. and A. A. Soloviev (2010), Variations of trends of indicators describing complex systems: Change of scaling precursory to extreme events, *Chaos*, *20*, 033104.

Kossobokov, V. G., V. I. Keilis-Borok, and B. Cheng (2000), Similarities of multiple fracturing on a neutron star and on the Earth, *Phys. Rev. E*, *61*(4), 3529–3533.

Kossobokov, V. and P. Shebalin (2003), Earthquake prediction, in *Nonlinear Dynamics of the Lithosphere and Earthquake Prediction*, edited by V. I. Keilis-Borok and A. A. Soloviev, pp. 141–207, Springer-Verlag, Berlin-Heidelberg.

Lichtman, A. J. and V. I. Keilis-Borok (1989), Aggregate-level analysis and prediction of midterm senatorial elections in the United States, 1974-1986, *Proc. Nat. Acad. Sci.*, *86*(24), 10176–10180.

Lichtman, A. J. (1996), *The Keys to the White House*, Madison Books, Lanham, MD.

Lichtman, A. J. (2000), *The Keys to the White House*, Lexington Books Edition, Lanham, MD.

Lichtman, A. J. (2005), The keys to the White House: Forecast for 2008, *Foresight*, *3*, 5–9.

Levshina, T. and I. Vorobieva, 1992: Application of algorithm for prediction of a strong repeated earthquake to the Joshua Tree and Landers, Fall Meeting AGU, Abstracts, p. 382.

Ma, Z., Z. Fu, Y. Zhang, C. Wang, G. Zhang, and D. Liu (1990), *Earthquake Prediction: Nine Major Earthquakes in China*, Springer-Verlag, New York.

Mogi, K. (1968), Migration of seismic activity, *Bull. Earthq Res. Inst. Univ. Tokyo*, *46*(1), 53–74.

Molchan, G. (1991), Structure of optimal strategies in earthquake prediction, *Tectonophysics*, *193*, 267–276.

Molchan, G. M. (1997), Earthquake prediction as a decision-making problem, *Pure Appl. Geophys.*, *149*, 233–237.

Molchan, G. M. (2003), Earthquake prediction strategies: A theoretical analysis, in *Nonlinear Dynamics of the Lithosphere and Earthquake Prediction*, edited by V. I. Keilis-Borok and A. A. Soloviev, Springer-Verlag, Berlin-Heidelberg, pp. 209–237.

Molchan, G. and V. Keilis-Borok (2008), Earthquake prediction: probabilistic aspect, *Geophys. J. Int.*, *173*(3), 1012–1017.

Narkunskaya, G. S. and M. G. Shnirman (1994), On an algorithm of earthquake prediction, in *Computational Seismology and Geodynamics*, vol. *1*, edited by D. K. Chowdhury, pp. 20–24, AGU, Washington, DC.

Peresan, A., V. Kossobokov, L. Romashkova, and G. F. Panza (2005), Intermediate-term middle-range earthquake predictions in Italy: a review, *Earth Sci. Rev.*, *69*(1-2), 97–132.

Pollitz, F. F., R. Burgmann, and B. Romanowicz (1998), Viscosity of oceanic asthenosphere inferred from remote triggering of earthquakes, *Science*, *280*, 1245–1249.

Press, F. and P. Briggs (1975), Chandler wobble, earthquakes, rotation and geomagnetic changes, *Nature*, *256*, 270–273.

Press, F. and C. Allen (1995), Patterns of seismic release in the southern California region, *J. Geophys. Res.*, *100*(B4), 6421–6430.

Romanowicz, B. (1993), Spatiotemporal patterns in the energy-release of great earthquakes, *Science*, *260*, 1923–1926.

Rotwain, I., V. Keilis-Borok, and L. Botvina (1997), Premonitory transformation of steel fracturing and seismicity, *Phys. Earth Planet. Inter.*, *101*, 61–71.

Rotwain, I., G. De Natale, I. Kuznetsov, A. Peresan, and G. F. Panza (2006), Diagnosis of time of increased probability (TIP) for volcanic earthquakes at Mt.Vesuvius, *Pure Appl. Geophys.*, *163*(1), 19–39.

Shebalin, P., V. Keilis-Borok, I. Zaliapin, S. Uyeda, T. Nagao, and N. Tsybin (2004), Advance short-term prediction of the large Tokachi-oki earthquake, September 25, 2003, M = 8.1. A case history, *Earth Planets Space*, *56*(8), 715–724.

Shebalin, P., V. Keilis-Borok, A. Gabrielov, I. Zaliapin, and D. Turcotte (2006), Short-term earthquake prediction by reverse analysis of lithosphere dynamics, *Tectonophysics*, *413*, 63–75.

Shnirman, M. G. and E. M. Blanter (1999), Mixed hierarchical model of seismicity: Scaling and prediction, *Phys. Earth Planet. Inter.*, *111*(3-4), 295–303.

Shnirman, M. G. and E. M. Blanter (2003), Hierarchical models of seismicity, in *Nonlinear Dynamics of the Lithosphere and Earthquake Prediction*, edited by V. I. Keilis-Borok and A. A. Soloviev, pp. 37–69, Springer-Verlag, Berlin-Heidelberg.

Soloviev, A. and A. Ismail-Zadeh (2003), Models of dynamics of block-and-fault systems, in *Nonlinear Dynamics of the Lithosphere and Earthquake Prediction*, edited by V. I. Keilis-Borok and A. A. Soloviev, pp. 71–139, Springer-Verlag, Berlin-Heidelberg.

Sornette, D. and C. G. Sammis (1995), Complex critical exponents from renormalization group theory of earthquakes: Implications for earthquake predictions, *J. Phys. I France*, *5*, 607–619.

Sornette, D. (2004), *Critical Phenomena in Natural Sciences: Chaos, Fractals, Selforganization, and Disorder. Concept and Tools*, 2nd ed., Springer-Verlag, Berlin-Heidelberg.

Tsuboi, C. (1956), Earthquake energy, earthquake volume, aftershock area and strength of the Earth's crust, *J. Phys. Earth*, *4*, 63–69.

Turcotte, D. L., W. I. Newman, and A. Gabrielov (2000), A statistical physics approach to earthquakes, in *Geocomplexity and the Physics of Earthquakes*, pp. 83–96, AGU, Washington, DC.

Vorobieva, I. A. (1999), Prediction of a subsequent large earthquake, *Phys. Earth Planet. Int.*, *111*, 197–206.

Vorobieva, I. (2009), Prediction of subsequent strong earthquake. Advanced School on Non-Linear Dynamics and Earthquake Prediction, 28 September to 10 October, 2009, Trieste: ICTP, 2060-49, 37 pp.

Yakovlev, G., W. I. Newman, D. L. Turcotte, and A. Gabrielov (2005), An inverse cascade model for self-organized complexity and natural hazards, *Geophys. J. Int.*, *163*, 433–442.

Zaliapin, I., V. Keilis-Borok, and M. Ghil (2003), A Boolean delay equation model of colliding cascades. Part II: Prediction of critical transitions, *J. Stat. Phys.*, *111*, 839–861.

Zaliapin, I., H. Wong, and A. Gabrielov (2005), Inverse cascade in percolation model: hierarchical description of time-dependent scaling, *Phys. Rev. E*, *71*, 066118.

Zaliapin, I., H. Wong, and A. Gabrielov (2006), Hierarchical aggregation in percolation model, *Tectonophysics*, *413*, 93–107.

www.ingramcontent.com/pod-product-compliance
Lightning Source LLC
LaVergne TN
LVHW062212200926
848484LV00006B/1375
9781940076447